Biomass Wastes for Sustainable Industrial Applications

In this edited volume, Verma and Dubey collate important discussions from international researchers to address major innovations in the sustainable industrial applications of biomass wastes, including processing fundamentals, extraction, purification, properties, and industrial applications.

The amount of biomass waste is rising quickly, and such waste offers numerous advantages for sustainable development, particularly for environmentally friendly industrial use. This book therefore addresses this situation by providing a comprehensive overview of the sustainable industrial uses of biomass wastes. To enable ease of use and to facilitate readers' ability to implement this information in real-world contexts, the book is divided into three sections. First, the introduction discusses biomass wastes and their classification, processing, sustainability, and more as well as the production of bioproducts. The second part addresses classification in more detail in contexts, including forestry, agriculture, animal, industrial, municipal, and food processing wastes. Last, the third section addresses applications in areas such as electricity generation; lubrication, adhesion, and anticorrosion; green energy storage; catalysis; and more. Through this approach, readers will gain a comprehensive understanding of the challenges and opportunities of biomass wastes and will be able to apply their knowledge in a range of contexts, whether in further research or in industrial and other real-world scenarios.

This book is a vital resource for a broad readership, including students, academics, research professionals, research enterprises, R&D, and defence research laboratories. Especially, those researching and working in fields such as chemical engineering, material science and engineering, nanotechnology, energy, and environmental engineering will benefit greatly from the discussions within.

Biomass Wastes for Sustainable Industrial Applications
A Waste-to-Wealth Approach

Edited by

Chandrabhan Verma and Shikha Dubey

CRC Press
Taylor & Francis Group
Boca Raton London New York

CRC Press is an imprint of the
Taylor & Francis Group, an **informa** business

Designed cover image: Tin container for storing biowaste from a gardening company; branches of leaves and cut grass open iron painted green strongly stinks; stock photo

First edition published 2025
by CRC Press
2385 NW Executive Center Drive, Suite 320, Boca Raton FL 33431

and by CRC Press
4 Park Square, Milton Park, Abingdon, Oxon, OX14 4RN

CRC Press is an imprint of Taylor & Francis Group, LLC

© 2025 selection and editorial matter, Chandrabhan Verma and Shikha Dubey; individual chapters, the contributors

Reasonable efforts have been made to publish reliable data and information, but the author and publisher cannot assume responsibility for the validity of all materials or the consequences of their use. The authors and publishers have attempted to trace the copyright holders of all material reproduced in this publication and apologize to copyright holders if permission to publish in this form has not been obtained. If any copyright material has not been acknowledged please write and let us know so we may rectify in any future reprint.

Except as permitted under U.S. Copyright Law, no part of this book may be reprinted, reproduced, transmitted, or utilized in any form by any electronic, mechanical, or other means, now known or hereafter invented, including photocopying, microfilming, and recording, or in any information storage or retrieval system, without written permission from the publishers.

For permission to photocopy or use material electronically from this work, access www.copyright.com or contact the Copyright Clearance Center, Inc. (CCC), 222 Rosewood Drive, Danvers, MA 01923, 978-750-8400. For works that are not available on CCC please contact mpkbookspermissions@tandf.co.uk

Trademark notice: Product or corporate names may be trademarks or registered trademarks and are used only for identification and explanation without intent to infringe.

ISBN: 978-1-032-73944-1 (hbk)
ISBN: 978-1-032-73948-9 (pbk)
ISBN: 978-1-003-46683-3 (ebk)

DOI: 10.1201/9781003466833

Typeset in Minion
by Apex CoVantage, LLC

Contents

Editors

Chandrabhan Verma is a researcher at the Department of Petroleum and Chemical Engineering, Khalifa University of Science and Technology, Abu Dhabi. He earned a BSc and MSc at Udai Pratap College, Varanasi, and a PhD at the Indian Institute of Technology (BHU) Varanasi in 2017. Dr. Verma is a member of the American Chemical Society (ACS) and a lifetime member of the World Research Council (WRC). He is a reviewer and editorial board member of various internationally recognized ACS, RSC, Elsevier, Wiley and Springer platforms. Dr. Verma has published numerous research and review articles in different areas of science and engineering at ACS, Elsevier, RSC, Wiley, Springer, etc. He has a total citation of more than 14,089 with an h-index of 65 and an i-10 index of 175. Dr. Verma's research focuses on designing and developing industrially applicable corrosion inhibitors. He has edited or authored over 40 books for ACS, Elsevier, RSC, CRC, Springer and Wiley. Dr. Verma has received several awards for his academic achievements, including a Gold Medal in materials science (Organic Chemistry; 2010) and Best Publication awards from the Global Alumni Association of IIT-BHU (Second Prize 2013).

Shikha Dubey is an Assistant Professor (Analytical Chemistry) at the Department of Chemistry, Hemvati Nandan Bahuguna Garhwal University Srinagar (Garhwal) Uttarakhand. She earned a BSc (Hons.) and MSc in analytical chemistry at the Institute of Science, Banaras Hindu University, Varanasi, and a PhD at the Indian Institute of Technology (BHU) Varanasi.

After qualifying for the CSIR-NET exam, she joined the Department of Chemistry, Indian Institute of Technology (BHU). Dr. Dubey was awarded a Junior Research Fellowship (JRF) and Senior Research Fellowship (SRF) by the Defence Research and Development Organization (DRDO), Ministry of Defence, Government of India, New Delhi, India. During her research, she synthesized various nanomaterials via simple precipitation methods and green routes to treat metal-laden water and wastewater. Her research interest involves nano-biomaterial synthesis and characterization, development of low-cost adsorbents, nano-adsorbents, magnetic nano-sorbents for water remediation and green-synthesis of materials, characterization and application in environmental remediation, photocatalytic degradation of organic and inorganic pollutants, etc. Dr. Dubey has attended several international and national conferences and delivered expert talks.

Contributors

Muhammad Muneeb Ahmad
Department of Industrial
  Engineering
University of Salerno
Fisciano, Italy

Md. Ahmaruzzaman
National Institute of Technology
Silchar, India

Nafseen Ahmed
Jamia Millia Islamia (Central
  University)
New Delhi, India

Kashif R. Ansari
Southwest Petroleum University
Xindu, Chengdu, China

Laziz Azimov
Institute of Pharmaceutical
  Education and Research
Yunusabad, Tashkent, Uzbekistan

Manoj K. Banjare
MATS University
Pandri, Raipur, Chhattisgarh, India

Kamalakanta Behera
University of Allahabad
Prayagraj, Uttar Pradesh, India

Khasan Berdimuradov
Tashkent Institute of Chemical
  Technology
Shahrisabz, Uzbekistan

Elyor Berdimurodov
National University of
  Uzbekistan
Tashkent, Uzbekistan

Dheeraj Singh Chauhan
King Fahd University of
  Minerals
Dhahran, Saudi Arabia

Divya Choudhary
Amity University
Gurugram, Haryana, India

Carolina Garín Correa
Colegio Luis Antonio Verney
Universidade de Evora
Évora, Portugal

Omar Dagdag
Gachon University
Seongnam, South Korea

Walid Daoudi
Laboratory of Molecular
    Chemistry
University Mohamed I
Nador, Morocco

Anirban Das
Amity University
Gurugram, Haryana, India

Saurav Dixit
Lovely Professional University
Phagwara, Punjab, India

Camila Dias dos Reis Barros
Colegio Luis Antonio Verney
Universidade de Evora
Évora, Portugal

Luana Barros Furtado
Federal University of Rio de
    Janeiro
Rio de Janeiro, Brazil

Luis Fernando Macias Gamboa
Facultad de Ciencias Químicas
Universidad Autónoma de Nuevo
    León
Nuevo León, México

Ritika Gera
Amity University
Gurugram, Haryana, India

Giuliana Gorrasi
Department of Industrial
    Engineering
University of Salerno
Fisciano, Italy

Rajesh Haldhar
Yeungnam University
Gyeongsan, Republic of Korea

Gavirangappa Hithamani
CSIR Central Food Technological
    Research Institute
Mysore, Karnataka, India

Víctor M. Jiménez Pérez
Universidad Autónoma
Nuevo León, Mexico

Kailash Juglan
Lovely Professional University
Phagwara, Punjab, India

Abbul Bashar Khan
Jamia Millia Islamia (Central
    University)
New Delhi, India

Hansang Kim
Gachon University
Seongnam, Republic
    of Korea

Ashish Kumar
Nalanda College of Engineering
Bihar Engineering University,
    Gokhulpur, Bihar, India

Saurabh Kumar
Amity University
Gurugram, Haryana, India

Lalit
Amity University
Gurugram, Haryana, India

Yuanhua Lin
Southwest Petroleum
    University
Xindu, Chengdu, China

Sheerin Masroor
A N College
Patliputra University
Patna, Bihar, India

Blanca M. Muñoz Flores
Facultad de Ciencias Químicas
Universidad Autónoma de Nuevo
    León
Nuevo León, México

**Anindita Bhuyan Rafaela C.
    Nascimento**
Colegio Luis Antonio Verney
Universidade de Evora
Évora, Portugal

Seema R. Pathak
Amity University Haryana
Pachgaon, Haryana, India

Anu Radha Pathania
Chandigarh University
Ajitgarh, Punjab, India

Apoorva Pathania
Chandigarh University
Gharuan, Mohali, Punjab,
    India

Raúl Segovia Pérez
Facultad de Ciencias Químicas
Universidad Autónoma de Nuevo
    León
Nuevo León, México

M. A. Quraishi
King Fahd University of Minerals
Dhahran, Saudi Arabia

Gyandshwar Kumar Rao
Amity University
Gurugram, Haryana, India

Akbarali Rasulov
National University of
    Uzbekistan
Tashkent, Uzbekistan

Anu Rathee
Amity University
Gurugram, Haryana, India

Matthew C. Risi
Te Aka Mātuatua - School of
    Science
University of Waikato
Hamilton, New Zealand

Shreya Sachdeva
Amity University
Gurugram, Haryana, India

Therola Sangtam
Nagaland University
Lumami, Zunheboto, Nagaland,
India

Graham C. Saunders
Te Aka Mātuatua - School of Science
University of Waikato
Hamilton, New Zealand

Kamal Nayan Sharma
Amity University
Gurugram, Haryana, India

Shveta Sharma
Government College Una
Una, Himachal Pradesh, India
and
Himachal Pradesh University
Shimla, Himachal Pradesh, India

Ambrish Singh
Nagaland University
Lumami, Zunhebotot, Nagaland,
India

Santosh Bahadur Singh
University of Allahabad
Prayagraj, Uttar Pradesh, India

Shivani Singh
Lovely Professional University
Phagwara, Punjab, India

Amit Somya
Amity University Bengaluru
Bengaluru, Karnataka, India

Akshima Soni
Amity University, Haryana
Pachgaon, Haryana, India

Abhinay Thakur
Lovely Professional University,
Phagwara,
Punjab, India

Yashika Verma
Amity University, Haryana
Pachgaon, Haryana, India

Gianluca Viscusi
University of Salerno,
Fisciano, Italy

Nisha Yadav
Amity University, Haryana
Pachgaon, Haryana, India

Husan Yaxshinorov
National University of Uzbekistan
Tashkent, Uzbekistan

Akhiu K. Yimchunger
Nagaland University
Lumami, Zünheboto, Nagaland,
India

Preface

THE AMOUNT OF BIOMASS waste generated is rising quickly, raising concerns about administration and disposal. As a result, value-added applications that recycle and reuse biomass wastes are becoming increasingly common. Biomass from various sources, such as animal, industrial, municipal solid, and waste from forestry and agriculture, has been used for multiple industrial purposes. Biomass wastes are considered nontoxic, biodegradable, and non-bioaccumulative because of their organic and biological composition. Biomass wastes benefit sustainable development in many ways, particularly for environmentally friendly industrial applications. The industrial use of biomass wastes is encouraged by strict environmental restrictions and growing demands for green chemistry. Biomass wastes are considered nontoxic, biodegradable, and non-bioaccumulative because of their organic and biological composition. Biomass wastes benefit sustainable development in many ways, particularly for environmentally friendly industrial applications. In the midst of growing environmental concerns and the search for sustainable solutions, biomass waste valuation appears to be a glimmer of light. *Biomass Wastes for Sustainable Industrial Applications: A Waste-to-Wealth Approach* summarizes using waste biomass to create a route toward sustainable development. This book takes the reader on a detailed exploration of the complex world of biomass wastes, revealing their qualities, classification, and wide range of industrial applications. Each chapter examines the origins of biomass wastes and their potential conversion into valuable resources, ranging from natural and biological beginnings to industrial residues and municipal solid wastes.

Overall, there are seventeen chapters in the book. The first three chapters provide readers with an overview of the fundamentals of biomass waste. The foundation for comprehending the wide range of applications for biomass wastes is laid forth in these chapters, which also address the basic concepts, classification systems, and characteristics of biomass wastes. This chapter examines their sources, both natural and biological, as well as the handling of biomass wastes. Chapter 3 discusses the creation and application of bioproducts from biomass wastes. The potential applications of biomass are highlighted as it examines the chemical and energy materials that can be extracted or synthesized from it. Their primary attributes, categorization, attributes, and wide range of uses in different industrial domains are examined in Chapter 4. Animal waste is covered in detail in Chapter 5, along with its basic characteristics, classification, qualities, and valuable applications in various industries, such as waste management, energy production, and agriculture. The industrial byproducts from biomass sources are explained in Chapter 6. It talks about their basic characteristics, classification, and uses in industrial processes, focusing on how they might be used to recover resources and reduce waste.

The methods for effective resource recovery and waste management in urban environments are covered in Chapter 7. Similarly, Chapter 8 explores residues from food processing. It examines their basic traits, categorization, attributes, and range of uses, emphasizing chances for resource optimization and waste valorization. The role of biomass wastes in environmental remediation is examined in Chapter 9, particularly highlighting how these wastes are used as adsorbents and in waste treatment procedures. Their function in chemical reactions and biodiesel synthesis is reviewed in Chapter 10. The use of catalysts obtained from biomass for sustainable chemical synthesis is covered, along with catalytic methods and reaction mechanisms. Chapter 11 examines their function in fuel cells, energy storage devices, and other electrochemical systems, focusing on environmentally friendly substitutes for traditional materials. Anaerobic digestion, pyrolysis, and fermentation are examples of bioenergy conversion methods covered in Chapter 12 and their potential for producing renewable energy. Biomass gasification, cogeneration, and renewable energy integration are examples of biomass-based power generation methods covered in Chapter 13.

The potential of biomass wastes in lubricating, adhesive, and anticorrosive compositions is examined in Chapter 14. With an emphasis on environmentally friendly substitutes for petroleum-based goods, it addresses their qualities, manufacturing processes, and performance attributes. Their applicability as structural elements, electrodes, and electrolytes in batteries, supercapacitors, and other energy storage devices is covered in Chapter 15. The development of catalysts, process optimization, and thermochemical and biochemical routes are covered in Chapter 16, which discusses sustainable hydrogen production. The modification of biopolymers made from biomass wastes for applications involving corrosion inhibition is examined in Chapter 17. It covers the synthesis, characterization, and performance assessment of modified biopolymer-based coatings and inhibitors to prevent corrosion in various industries.

This book is a significant resource for scholars, students, and practitioners and has been written for scholars in academia and industry. The editors and contributors are well-known researchers, scientists, and true professionals in academia and industry. On behalf of CRC Press, we are very thankful to the contributors to all chapters for their amazing and passionate efforts in making this book. Thanks to Dr. Gagandeep Singh (Senior Publisher) for his support and help during this project. In the end, all thanks to CRC Press for publishing the book.

Chandrabhan Verma
Shikha Dubey

I

Introduction

CHAPTER 1

Biomass Wastes

Fundamentals, Classification and Properties

Abhinay Thakur, Ashish Kumar
and Elyor Berdimurodov

1.1 INTRODUCTION

Biomass wastes, sourced from diverse origins such as agriculture, forestry, industry, and municipalities, represent a versatile and abundant resource with the potential to address multiple environmental challenges (1–3). These organic materials encompass a broad array of sources, such as agricultural residues like crop stalks and husks, forestry residues such as wood chips and sawdust, organic waste from industries, and municipal solid waste like food scraps and yard trimmings. The utilization of biomass wastes is crucial for environmental sustainability due to their multi faceted inclusion in lowering greenhouse gas emissions and slowing down global warming, managing waste effectively, and minimizing pollution (4, 5). First, biomass wastes offer a renewable and viable substitute for fossil fuels for energy production. Unlike finite fossil fuel reserves, biomass resources are replenishable, making them an attractive option for meeting energy demands while reducing reliance on environmentally harmful fuels. Biomass wastes can be effectively converted into energy, electricity, biofuels, and biogas through processes like fermentation, pyrolysis, and combustion. These energy conversion technologies offer ways to effectively use biomass wastes and lessen the carbon footprint related to the generation of energy. By displacing fossil fuels and capturing carbon dioxide through biomass growth and utilization, biomass waste utilization helps

DOI: 10.1201/9781003466833-2

to mitigate climate change and reduce greenhouse gas emissions, making it a critical component of global efforts to transition towards a low-carbon economy (6, 7). Furthermore, managing biomass wastes effectively contributes to waste reduction and resource conservation. Instead of allowing organic materials to decompose in landfills, where they emit methane, a potent greenhouse gas, biomass wastes can be repurposed for beneficial uses. For instance, composting turns organic waste into nutrient-rich soil additives that improve soil health and lessen the need for synthetic fertilizers. In a similar manner, the generation of biofuel from biomass wastes provides a sustainable substitute for fossil fuels, contributing to resource conservation and a decrease in greenhouse gas emissions (8). Utilizing biomass waste can also help with pollution and waste disposal problems. Organic waste streams can be diverted from dumpsters and incinerators to lessen the strain on waste management infrastructure, which can save money and benefit the environment. Additionally, by utilizing biomass wastes for energy production, air, soil, and water pollution associated with conventional energy sources can be mitigated. The combustion of biomass wastes produces lower emissions of sulfur dioxide, nitrogen oxides, and particulate matter compared to fossil fuels, thus improving air quality and reducing respiratory health risks. Alola et al. (9) delves into an analysis of waste management and industrial and agricultural greenhouse gas (GHG) emissions associated with the utilization of biomass, fossil fuels, and metallic ores in Iceland. The research aims to examine the drivers of sectoral GHG emissions in the country and assess the differential impacts of domestic material consumption (DMC) on aggregated GHG emissions, waste management GHG emissions (WGHG), industrial GHG emissions (IGHG), and agricultural GHG emissions (AGHG) over the period from 1990 to 2019. With Iceland's CAP 2020 initiative targeting significant improvements in environmental conditions by 2030, particularly in sectors including energy production, small industry, waste management, transportation, and agriculture, the study seeks to determine whether DMC, specifically of metallic ores, biomass, and fossil fuels, contributes differently to GHG emissions. Employing Fourier function approaches, the investigation reveals that metallic ores DMC lead to an increase in GHG emissions, while biomass and fossil fuel DMC mitigate GHG emissions in the long run. Furthermore, biomass DMC is found to mitigate AGHG and WGHG, with respective elasticities of 0.04 and 0.025 over the long term. Fossil fuel DMC significantly reduces IGHG with an elasticity of 0.18 in the long run, while AGHG and WGHG remain unaffected by fossil fuel consumption. Additionally, metallic ores DMC only spurs IGHG with an elasticity of approximately 0.24. Figure 1.1 depicts the potential utilization of rice straw for the generation of fuel and power. Rice straw, a byproduct of rice cultivation, shows promise as a renewable resource for producing energy through various methods such as combustion, gasification, or biofuel production. This illustration highlights the importance of exploring alternative sources of energy to meet growing energy demands while also addressing environmental concerns associated with traditional fossil fuel-based power generation.

Second, effective management of biomass wastes plays a pivotal role in waste reduction and resource conservation, offering significant environmental benefits. Instead of allowing organic materials to decompose in landfills, where they emit methane, a potent greenhouse gas, biomass wastes can be repurposed for beneficial uses (11). One effective method of using organic waste is composting, which turns garbage into nutrient-rich soil additions by allowing it to break down under regulated conditions. This procedure improves soil fertility and wellness by keeping organic waste out of landfills and lowering the demand for synthetic fertilizers. Composting improves soil structure, moisture retention, and nutrient cycling by reintroducing organic matter to the soil. This helps to support sustainable agriculture methods and lessens dependency on chemical inputs. In a similar vein, producing biofuel from biomass wastes offers a sustainable substitute for fossil fuels that lowers greenhouse gas emissions and conserves resources. Through processes such as anaerobic digestion, biomass wastes can be converted into biogas, biodiesel, or ethanol, offering renewable sources of energy for transportation, heating, and electricity generation. By utilizing biomass wastes as feedstock for biofuel production, finite fossil fuel resources are conserved, and dependency on non-renewable energy sources is reduced (12, 13). Additionally, biofuels produced from biomass wastes typically have lower carbon emissions compared to fossil fuels, contributing to climate change mitigation efforts and enhancing energy security. Moreover, problems with pollution and waste administration can be solved by using biomass wastes. Organic waste streams can be diverted from dumpsites to reduce the strain on waste management infrastructure, which can result in financial savings as well as environmental advantages. Through a variety of conversion technologies, biomass wastes can be turned from trash into useful goods like compost, biofuels, and biogas. This lowers the amount of garbage that needs to be disposed of, but it also produces income streams and job possibilities in the bioenergy

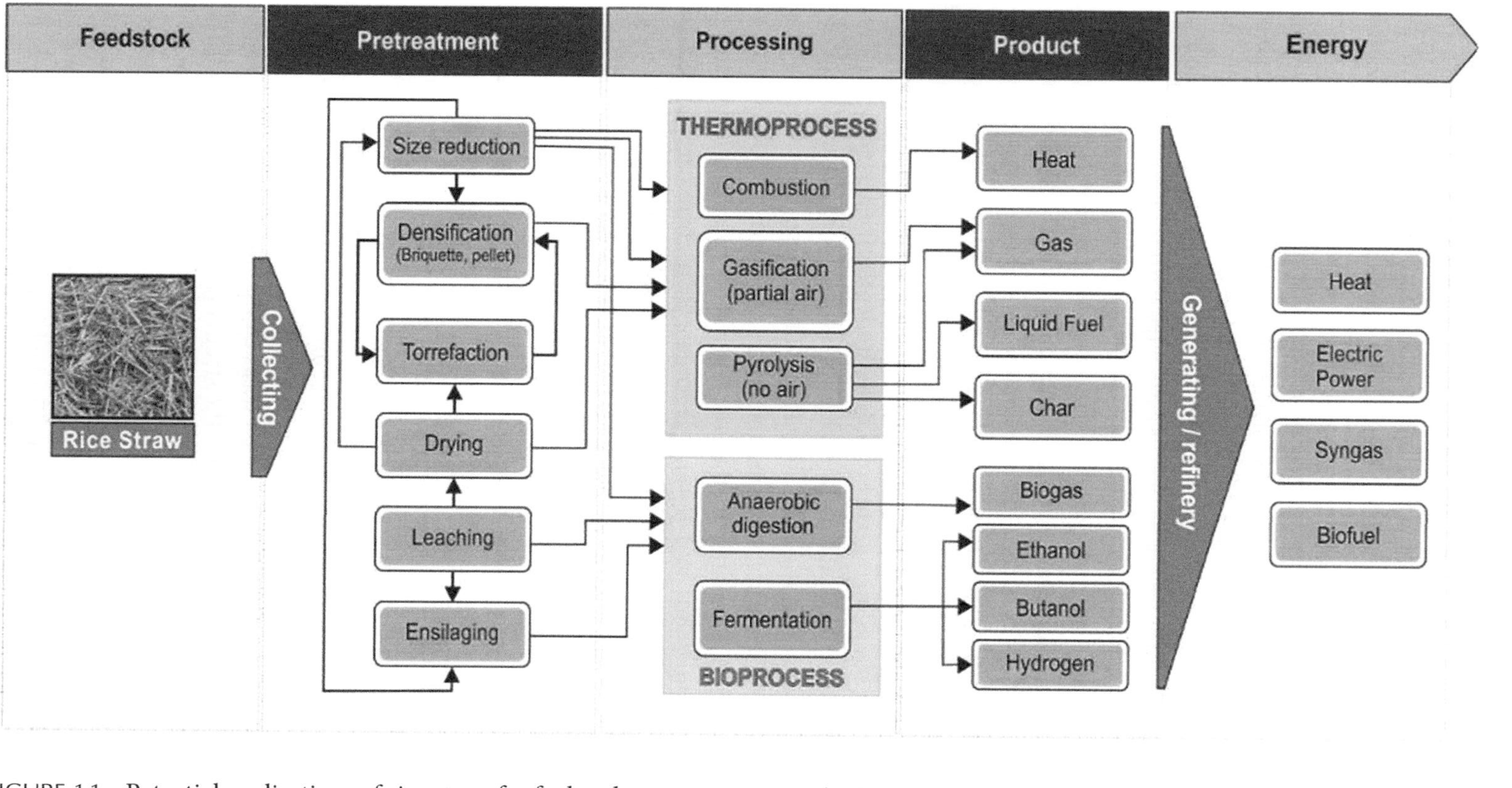

FIGURE 1.1   Potential applications of rice straw for fuel and power generation, highlighting its role as a renewable resource in meeting energy needs through various methods such as combustion, gasification, or biofuel production. Adapted from ref. (10) under CCBY 4.0.

and waste management industries. Moreover, utilizing biomass wastes for energy production can mitigate air, soil, and water pollution associated with conventional energy sources. For instance, the combustion of biomass wastes produces lower emissions of pollutants such as sulfur dioxide, nitrogen oxides, and particulate matter compared to fossil fuels, thereby improving air quality and reducing respiratory health risks. By replacing coal or oil-fired power plants with biomass-fired facilities, emissions of harmful pollutants are minimized, contributing to cleaner air and healthier environments for communities living near power generation facilities. Additionally, utilizing biomass residues for soil amendment and bioenergy production reduces the environmental footprint of waste management practices, leading to more sustainable and integrated approaches to resource utilization.

1.2 COMPOSITION AND GENERATION OF BIOMASS WASTES

1.2.1 Sources and Types of Biomass Wastes

Biomass wastes encompass a wide range of organic materials derived from various sources, including agriculture, forestry, industry, and municipalities. Understanding the sources and types of biomass wastes is crucial for identifying opportunities for their utilization and addressing environmental challenges (14). In this elaboration, we will delve into each of these sources in detail, exploring the types of biomass wastes they generate and their potential for sustainable use. Figure 1.2 shows the sources and types of biomass waste.

1.2.1.1 Agriculture

Agriculture stands as a primary source of biomass waste, generating a diverse range of organic materials throughout the production cycle. Crop residues, encompassing stalks, husks, leaves, and straw, emerge as common byproducts of harvesting operations across the agricultural landscape. These residues, generated in substantial quantities worldwide, constitute a significant resource with immense potential for bioenergy production and soil amendment (16, 17). Crop residues, abundant and widely available, offer various avenues for sustainable utilization. In terms of bioenergy production, these residues serve as valuable feedstock for processes like combustion, pyrolysis, and fermentation, enabling the generation of heat, electricity, biofuels, and biogas. Through advanced conversion

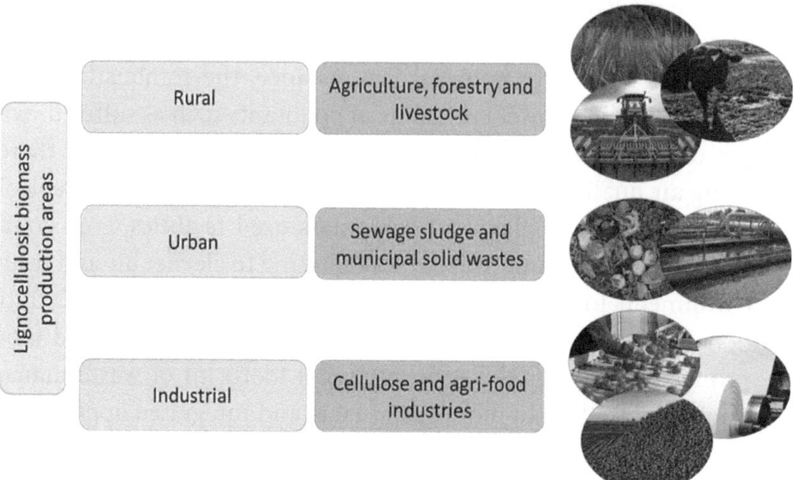

FIGURE 1.2 Depiction of the origins and classifications of biomass waste. Adapted from ref. (15) under CCBY 4.0.

technologies, such as gasification and anaerobic digestion, crop residues could be efficiently modified into renewable energy sources, offering alternatives to fossil fuels and leading to the diversification of energy portfolios. Moreover, crop residues play a vital role in soil management and fertility enhancement. When returned to the soil as organic amendments, these residues contribute to soil organic matter content, improving soil structure, moisture retention, and nutrient cycling. Incorporating crop residues into agricultural soils enhances soil health and fertility, reducing dependency on synthetic fertilizers and promoting sustainable agricultural practices. Additionally, crop residues act as a protective layer, mitigating soil erosion, and enhancing soil moisture conservation, particularly in arid and semi-arid regions prone to land degradation.

In addition to crop residues, agricultural activities generate various other organic wastes, further augmenting the biomass resource pool (18–20). Animal manure, a byproduct of livestock farming, represents a rich source of organic matter and nutrients, valuable for soil fertility enhancement and biogas production. Crop processing residues, arising from activities such as milling, pressing, and extraction, offer additional biomass resources for bioenergy production and value-added product manufacturing. Agro-industrial byproducts, exemplified by bagasse from sugarcane processing, present yet another avenue for biomass utilization, contributing to renewable energy production and waste valorization.

1.2.1.2 Forestry

Forestry operations constitute a major source of biomass waste, yielding substantial quantities of organic materials in various forms throughout the timber production and processing cycle (21, 22). These biomass wastes, including forest residues, logging residues, and wood processing residues, represent valuable resources with significant potential for utilization in diverse industrial applications. Forest residues, comprising branches, tops, and other woody debris remaining after timber harvesting, constitute a primary component of biomass wastes in forestry operations. These residues are often left on-site or scattered across the forest floor following logging activities. While historically considered as waste, forest residues are now recognized as valuable biomass resources for bioenergy production, soil enrichment, and forest ecosystem management. Through processes such as chipping, grinding, or pelletizing, forest residues can be converted into biomass fuels, including wood chips, pellets, and briquettes, suitable for combustion, gasification, or pyrolysis. These biomass fuels serve as renewable sources of heat and electricity, contributing to energy security and climate change mitigation.

Logging residues, another category of biomass wastes, encompass bark, branches, and stumps generated during timber harvesting operations (23, 24). Often left behind after logging activities, logging residues represent a significant biomass resource with potential for utilization in various industrial applications. Through chipping or grinding, logging residues can be processed into biomass feedstocks for pulp and paper manufacturing, engineered wood products, and bio-based chemicals. Additionally, logging residues can be utilized as feedstock for bioenergy production, contributing to renewable energy generation and reducing dependency on fossil fuels. Wood processing residues, generated during the manufacturing of lumber, plywood, and paper products, constitute another valuable source of biomass waste in forestry operations. These residues, including sawdust, wood chips, and bark, are produced in significant quantities and present opportunities for utilization in various industrial applications. Sawdust and wood chips, for example, can be utilized as raw materials for particleboard and fiberboard manufacturing, contributing to the production of sustainable building materials. Bark residues, rich in tannins and lignin, can be processed into value-added products such as mulch, compost, and bio-based chemicals.

1.2.1.3 Industry

Various industries generate biomass wastes as byproducts of their manufacturing processes and wastewater treatment operations, contributing to the overall biomass resource pool available for utilization. These organic waste streams originate from a diverse range of industrial activities, including food processing, agricultural processing, beverage production, pulp and paper manufacturing, textiles, pharmaceuticals, and chemicals (25). Organic waste streams from the food processing, agricultural processing, and beverage production industries encompass a wide variety of materials, including fruit and vegetable peels, food scraps, spent grains, and wastewater solids. These residues are generated during food preparation, packaging, and processing operations, as well as from cleaning and sanitation processes. While traditionally viewed as waste, these organic materials possess significant potential for utilization in various applications, including bioenergy production, animal feed, composting, and soil enrichment. Through procedures including anaerobic digestion, fermentation, or composting, organic waste streams from the food and beverage industries can be converted into biogas, biofuels, or nutrient-rich soil amendments, contributing to resource recovery and waste valorization efforts.

Similarly, the pulp and paper industry generates significant quantities of biomass wastes, including black liquor, a byproduct of the pulping process (26, 27). Black liquor, a mixture of lignin, cellulose, and other organic compounds, is produced during the chemical pulping of wood fibers to extract cellulose for paper production. While traditionally burned for energy recovery within pulp mills, black liquor can also be processed through gasification or pyrolysis to produce biofuels, such as syngas or bio-oil, suitable for transportation fuel or power generation. By valorizing black liquor as a renewable resource, the pulp and paper industry can enhance energy efficiency, reduce greenhouse gas emissions, and contribute to the development of bio-based products and fuels. Furthermore, other industrial sectors, such as textiles, pharmaceuticals, and chemicals, also generate organic waste streams suitable for biomass utilization. Textile manufacturing, for example, produces cotton waste, fabric scraps, and dye wastewater, which can be recycled or processed into bio-based materials, such as fibers or bio-based dyes. Pharmaceutical and chemical industries generate organic waste streams containing byproducts, intermediates, and unused compounds, which can be repurposed for bioenergy production, chemical synthesis, or pharmaceutical manufacturing. By exploring opportunities for biomass utilization, these industries can minimize waste generation, reduce environmental impacts, and contribute to the development of a circular economy.

1.2.1.4 Municipalities

Municipal solid waste (MSW) encompasses a broad spectrum of materials discarded by households, businesses, and institutions in urban areas. Among the diverse components of MSW, organic materials constitute a substantial portion, including food scraps, yard trimmings, paper products, and biodegradable plastics (28, 29). These organic waste streams contribute significantly to municipal waste volumes and present challenges for landfill management and waste disposal due to their potential for generating methane, a potent greenhouse gas, when decomposing anaerobically. However, despite the challenges they pose, organic materials in MSW also represent valuable resources that can be diverted from landfills and utilized in various sustainable waste management practices. For instance, composting helps organic waste break down into nutrient-rich manure, which may be applied as a soil additive to landscaping, urban greening, and agricultural endeavors. Composting is an environmentally responsible way to manage organic waste. Composting generates a beneficial soil conditioner that improves soil fertility, structure, and moisture retention while also reducing the amount of trash that needs to be disposed of. This is accomplished by controlled decomposition under aerobic circumstances.

Anaerobic digestion is another viable option for managing organic waste from MSW, particularly food scraps and other biodegradable materials. In anaerobic digestion facilities, organic waste undergoes microbial decomposition in the absence of oxygen, producing biogas (a mixture of methane and carbon dioxide) and digestate (30). Biogas can be utilized as a renewable energy source for electricity generation, heating, or vehicle fuel, while digestate can be used as a nutrient-rich fertilizer or soil conditioner. In addition to keeping organic waste out of dumpsters, anaerobic digestion produces renewable power and lowers greenhouse emissions from the breakdown of garbage. Additionally, organic material from MSW could be used in gasification or incineration procedures to produce bioenergy. Burning organic waste produces heat that could be utilized to produce steam or power. This process is known as incineration. Gasification, on the other hand, uses high temperatures in a controlled atmosphere to break down organic waste into synthesis gas, or syngas. Syngas could be processed into biofuels like ethanol or biodiesel or utilized as a fuel for electricity generation. Municipalities can lessen their dependency on fossil fuels, minimize greenhouse gas emissions, and encourage the growth of renewable energy sources by using organic waste for the generation of bioenergy (31, 32).

Each of these biomass waste sources generates a unique mix of organic materials with varying characteristics and potential applications. For example, crop residues and forestry residues are rich in cellulose, hemicellulose, and lignin, making them suitable feedstocks for biofuel production through processes like thermochemical conversion (e.g., pyrolysis, gasification) and biochemical conversion (e.g., fermentation). Animal manure and food processing wastes are rich in organic matter and nutrients, making them valuable inputs for composting and soil amendment in agriculture. Similarly, MSW components like food scraps and yard trimmings can be processed into compost or biogas through anaerobic digestion, reducing greenhouse gas emissions and promoting circular economy principles.

1.2.2 Factors Influencing Generation Rates

The generation rates of biomass wastes are influenced by various factors spanning agricultural, forestry, industrial, and municipal activities (33). Understanding these factors is crucial for assessing the magnitude of biomass waste generation and implementing effective waste management strategies. Agricultural practices significantly contribute to biomass waste generation, primarily through crop cultivation, harvesting, and post-harvest operations. Factors such as crop type, yield levels, cropping intensity, and farming practices influence the quantity and composition of agricultural residues produced. High-yielding crops like maize and rice generate larger quantities of crop residues compared to low-yielding crops, while mechanized farming practices, such as combine harvesting, can result in higher crop residue volumes due to increased plant biomass collection efficiency. Additionally, factors like crop rotation, tillage practices, and residue management techniques can affect the decomposition rates of crop residues and their subsequent conversion into biomass wastes. Forestry operations also generate significant volumes of biomass wastes, including forest residues, logging residues, and wood processing residues. The availability and composition of these residues are influenced by factors such as forest type, stand density, tree species, harvesting methods, and processing technologies. Clear-cutting practices result in larger quantities of logging residues, including branches, tops, and other woody debris, compared to selective harvesting methods. Similarly, the utilization of advanced wood processing technologies, such as sawmilling and pulp and paper manufacturing, can affect the types and volumes of wood processing residues generated. Figure 1.3 illustrates the fundamental cycle of biomass energy. This describes the entire process of obtaining energy from biomass sources. It includes steps like collecting or growing biomass, converting it into energy-dense forms (like heat, energy, or bioproducts), using it for different purposes, and eventually reusing or returning byproducts to the environment.

Various industrial sectors generate biomass wastes as byproducts of manufacturing processes, waste treatment, and product disposal. The

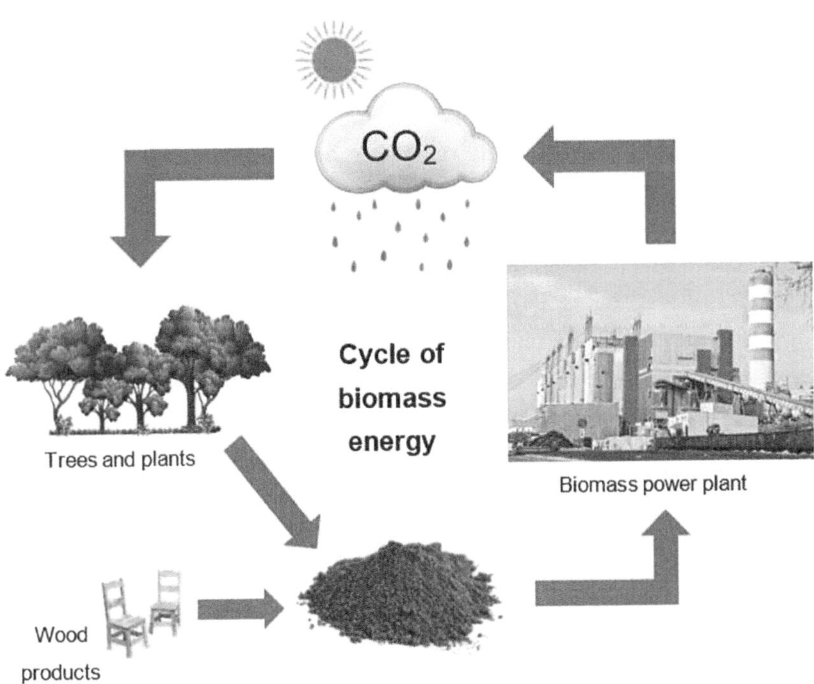

FIGURE 1.3　Schematic representation of the biomass energy cycle, illustrating the stages of biomass cultivation, conversion into usable energy, utilization, and environmental recycling. Adapted from ref. (34) under CCBY 4.0.

quantity and composition of industrial biomass wastes depend on factors such as production volumes, raw material inputs, production technologies, and waste management practices. For example, food processing industries produce organic waste streams like fruit and vegetable peels, food scraps, and wastewater solids during food preparation, packaging, and cleaning operations (35, 36). Similarly, the pulp and paper industry generates significant quantities of wood processing residues, such as sawdust, bark, and black liquor, as byproducts of pulp and paper manufacturing processes. Urbanization and population growth contribute to increased generation rates of municipal solid waste (MSW), including organic waste streams like food scraps, yard trimmings, and biodegradable plastics. Factors such as population density, consumption patterns, lifestyles, and waste management infrastructure influence the quantity and composition of MSW generated in urban areas. Densely populated cities with high levels of consumption and disposable incomes tend to generate larger volumes of organic waste compared to rural areas. Economic factors, including market demand, commodity prices, production costs, and waste management regulations, play a significant role in influencing biomass waste generation rates. Fluctuations in agricultural commodity prices can affect farmers' decisions regarding crop cultivation, harvesting, and residue management practices. Changes in energy prices and government incentives for renewable energy production can influence the utilization of biomass wastes for bioenergy generation. Waste management regulations, such as landfill bans on organic waste disposal and extended producer responsibility (EPR) schemes, can incentivize industries and municipalities to adopt sustainable waste management practices and reduce biomass waste generation. Technological advancements in agricultural, forestry, industrial, and waste management sectors can impact biomass waste generation rates by influencing production processes, resource utilization efficiency, and waste treatment technologies. Advancements in agricultural machinery, precision farming techniques, and crop breeding technologies can increase crop yields and optimize residue management practices, leading to changes in biomass waste generation patterns. Innovations in biomass conversion technologies, including anaerobic digestion, pyrolysis, and biorefining, could enhance the efficiency and viability of biomass waste utilization for energy production, biofuel manufacturing, and value-added product development.

1.3 ENVIRONMENTAL IMPLICATIONS

1.3.1 Impact on Air, Soil, and Water Quality

The impact of biomass waste on air, soil, and water quality is significant and multifaceted, with both positive and negative implications depending on management practices and environmental context. Understanding these impacts is crucial for developing effective waste management strategies that mitigate adverse effects and promote environmental sustainability.

1.3.1.1 Air Quality

Biomass waste management activities, including open burning, composting, and anaerobic digestion, can have significant impacts on air quality. Open burning of biomass wastes is a common practice in many regions, particularly in rural areas and agricultural settings. However, this practice releases pollutants such as particulate matter (PM), carbon monoxide (CO), volatile organic compounds (VOCs), and hazardous air pollutants (HAPs) into the atmosphere, contributing to air pollution and respiratory health problems (37, 38). The combustion of biomass in open fires or uncontrolled burning can result in the emission of high levels of these pollutants, posing hazards to human health and the environment. In addition to open burning, biomass combustion for energy production is another significant source of air emissions. When biomass wastes are burned to generate heat, electricity, or biofuels, greenhouse gases (GHGs) including methane (CH_4), carbon dioxide (CO_2), and nitrous oxide (N_2O) are released into the atmosphere. While biomass combustion emits carbon dioxide, a major contributor to climate change, the carbon released is part of the natural carbon cycle, as biomass absorbs carbon dioxide during its growth phase. Therefore, biomass combustion is often considered carbon-neutral over the long term, as long as the biomass is harvested sustainably and replanted to maintain a balance in carbon sequestration. However, it's important to note that biomass combustion can also emit methane and nitrous oxide, which are potent greenhouse gases with much higher global warming potentials than carbon dioxide. Additionally, biomass combustion can produce other air pollutants including nitrogen oxides (NO_x) and sulfur dioxide (SO_2), although typically at lower levels compared to fossil fuel combustion. These pollutants can contribute to acid rain, smog formation, and respiratory health problems, especially in areas with high levels of biomass-burning activity (39, 40).

Composting and anaerobic digestion are alternative biomass waste management techniques that can help mitigate air quality impacts compared to open burning or combustion for energy production. Composting involves the aerobic decomposition of organic waste materials, resulting in the production of compost, a nutrient-rich soil amendment (41). When compared to open burning or anaerobic digestion, the emissions from composting are usually lower, even if it can emit some greenhouse gases during the decomposition process, including carbon dioxide and methane. By breaking down organic waste in the lack of air, a biological mechanism known as anaerobic digestion creates digestate and biogas, a combination of carbon dioxide and methane. By capturing and using biogas as a sustainable energy source for the production of heat and power, greenhouse gas emissions can be reduced and dependency on fossil fuels is decreased. Additionally, anaerobic digestion can help reduce odor emissions associated with organic waste decomposition and provide opportunities for nutrient recovery and soil enrichment through the application of digestate as a fertilizer.

1.3.1.2 Soil Quality

Proper management of biomass wastes can indeed have significant positive effects on soil quality, ultimately enhancing soil fertility, structure, and microbial activity. One of the most common and beneficial practices is composting, which involves the decomposition of organic wastes into nutrient-rich compost. Compost is a valuable soil amendment that enriches the soil with essential nutrients, improves soil structure, and promotes microbial activity. When incorporated into soil, compost releases nutrients slowly over time, providing a steady and balanced supply of nitrogen, phosphorus, potassium, and other micronutrients necessary for plant growth. Additionally, compost helps to improve soil water retention and aeration, leading to better root development and overall plant health. Furthermore, the application of organic amendments such as crop residues and animal manure can significantly enhance soil fertility and structure (42–44). Crop residues, such as stalks, leaves, and husks left behind after harvest, contribute organic matter to the soil, which serves as a food source for soil microbes and promotes their activity. As these organic materials decompose, they release nutrients into the soil, improve soil structure by increasing aggregation, and enhance water infiltration and retention. Similarly, animal manure, when properly composted and applied to soil, adds valuable nutrients and organic matter, stimulating microbial activity and improving soil fertility over time.

However, it is crucial to note that improper management practices of biomass wastes can lead to adverse effects on soil quality. For instance, excessive application of organic amendments, such as compost or manure, can result in nutrient imbalances and soil pollution. Overapplication of certain nutrients, particularly nitrogen and phosphorus, can lead to nutrient runoff and leaching, contaminating groundwater and surface water bodies, and contributing to eutrophication and algal blooms. Additionally, the application of contaminated materials, such as compost containing heavy metals or pathogens, can pose risks to soil and environmental health. To mitigate these risks and ensure sustainable soil management practices, it is essential to adhere to guidelines for the application of organic amendments and monitor soil nutrient levels regularly. Proper composting techniques, such as maintaining appropriate carbon-to-nitrogen ratios, adequate moisture levels, and proper aeration, can help reduce the presence of pathogens and weed seeds in compost and ensure its safety for soil application. Furthermore, adding a variety of organic elements to soil management techniques like conservation tillage, intercropping, and cover crops can help strengthen soil structure, reduce erosion, and increase the sustainability of agriculture as a whole.

1.3.1.3 Water Quality

Biomass waste management practices have the potential to significantly impact water quality by introducing various pollutants into surface water and groundwater resources. One of the primary concerns is the leaching of nutrients, pathogens, and contaminants from biomass waste management activities into nearby water bodies (34, 45). Agricultural practices, such as the application of organic fertilizers or animal manure, can contribute to nutrient pollution in waterways. When rainfall or irrigation water runs off agricultural fields treated with these organic amendments, it can carry nutrients like nitrogen and phosphorus into streams, rivers, and lakes. Excessive nutrient inputs can lead to eutrophication, a process characterized by the overgrowth of algae and aquatic plants, which depletes oxygen levels in the water and harms aquatic ecosystems. Additionally, nutrient-rich runoff can fuel the growth of harmful algal blooms, some of which produce toxins harmful to aquatic life and human health. Composting and anaerobic digestion facilities also pose risks to water quality through the potential leaching of pollutants from decomposing organic materials (46, 47). Leachate generated during the composting or anaerobic digestion process can contain contaminants such as ammonia, heavy metals, and pathogens. If not properly managed, this leachate can infiltrate into the soil and reach groundwater or run off into surface water bodies, posing risks to aquatic ecosystems and human health. Ammonia, in particular, can be toxic to aquatic organisms and can lead to oxygen depletion in water bodies, further exacerbating water quality issues.

Improper disposal of biomass wastes in landfills represents another significant concern for water quality. Landfills serve as repositories for various types of waste, including organic materials, which can produce leachate as they decompose over time. If landfills lack adequate liner systems and monitoring measures, leachate containing a cocktail of pollutants can seep into the surrounding soil and groundwater, leading to groundwater contamination (48, 49). Heavy metals, organic pollutants, and pathogens present in landfill leachate can pose serious risks to human health and the environment if they reach drinking water sources or sensitive ecosystems. To mitigate the impacts of biomass waste management on water quality, it is essential to implement best management practices and regulatory measures. These may include proper nutrient management strategies in agriculture to minimize runoff, adequate siting and design of composting and anaerobic digestion facilities to prevent leachate contamination, and stringent landfill regulations to ensure proper containment and monitoring of landfill leachate. Additionally, promoting the use of sustainable waste management practices, such as composting, anaerobic digestion, and recycling, can help reduce the volume of biomass waste requiring disposal and minimize its potential impacts on water quality.

1.3.2 Contribution to Climate Change

Depending on how biomass is managed and used, the contribution of biomass waste to climate change is a complicated topic with both good and bad elements. Understanding these dynamics is crucial for evaluating the overall impact of biomass waste on climate change mitigation and adaptation efforts.

1.3.2.1 Carbon Sequestration and Emissions

Biomass waste, like other organic materials, contains carbon that is stored within its biomass. When biomass wastes decompose naturally or are burned, CO_2 is released into the atmosphere. But carbon from biomass is a part of the natural carbon cycle, contrasting carbon from fossil fuels, which releases carbon dioxide into the atmosphere after being stored underground for millions of years. Accordingly, the carbon absorbed during the growing phase of biomass balances off the carbon released during its breakdown or burning when it is cultivated and harvested sustainably (50, 51). Therefore, properly managed biomass waste can be considered carbon-neutral over the long term, as the carbon emitted is balanced by the carbon absorbed by new plant growth.

1.3.2.2 Greenhouse Gas Emissions

While biomass waste can be carbon-neutral when managed sustainably, certain biomass waste management practices can result in emissions of greenhouse gases (GHGs) such as methane (CH_4) and nitrous oxide (N_2O), which have higher global warming potentials than CO_2. For instance, methane, a strong greenhouse gas which leads to climate disruption, is produced during the anaerobic breakdown of organic waste in landfills. Similarly, incomplete combustion of biomass wastes, such as open burning or inefficient combustion in cookstoves, can result in emissions of particulate matter, black carbon, and other pollutants that contribute to regional and global warming.

1.3.2.3 Bioenergy and Renewable Energy

Biomass waste can also contribute to climate change mitigation by displacing fossil fuels and reducing greenhouse gas emissions from energy production. When biomass wastes are used as feedstock for bioenergy production, including gasification, combustion or anaerobic digestion, they can replace fossil fuels like coal, oil, and natural gas, which emit CO_2 when burned (52, 53). Bioenergy is a carbon-neutral or low-carbon energy source when compared to fossil fuels because, although biomass combustion generates CO_2, the carbon emitted is part of the natural carbon cycle and is compensated for by the carbon absorbed by new plant growth. Additionally, bioenergy systems can be designed to capture and sequester carbon dioxide emissions, further reducing their net impact on climate change.

1.3.2.4 Land Use Change and Sustainability

The sustainability of biomass waste utilization is influenced by land use change, particularly in the context of biomass cultivation for energy production. Converting natural ecosystems or agricultural lands into dedicated biomass plantations can have negative environmental consequences, including loss of biodiversity, soil degradation, and increased GHG emissions from land clearing and cultivation. Therefore, sustainable biomass utilization requires careful consideration of land use practices, conservation priorities, and ecosystem services to minimize negative impacts on climate change and other environmental indicators.

1.4 CLASSIFICATION OF BIOMASS WASTES

The classification of biomass wastes is essential for understanding their sources, characteristics, and potential applications, which can vary significantly depending on their origin and composition. Broadly, biomass wastes can be classified based on their origin into agricultural, forestry, industrial, and municipal categories. Additionally, biomass wastes can also be classified based on their characteristics and applications, which further delineates their suitability for various utilization pathways.

1.4.1 Classification Based on Origin
1.4.1.1 Agricultural Biomass Wastes
Agricultural biomass wastes originate from crop cultivation, harvesting, and post-harvest operations. These include crop residues (e.g., stalks, husks, straw), animal manure, crop processing residues (e.g., fruit and vegetable peels, pomace), and agro-industrial byproducts (e.g., bagasse, rice hulls). Agricultural biomass wastes are abundant and widely distributed, making them an important feedstock for bioenergy production, soil amendment, animal feed, and industrial applications (54, 55). Crop residues, in particular, are suitable for bioenergy production through combustion, pyrolysis, or anaerobic digestion, while animal manure could be utilized for biogas generation via anaerobic digestion.

1.4.1.2 Forestry Biomass Wastes
Forestry biomass wastes include residues generated from forest management, logging, and wood processing activities. These include forest residues (e.g., branches, tops, logging debris), wood processing residues (e.g., sawdust, bark, wood chips), and logging residues (e.g., stumps, slash). Forestry biomass wastes are abundant in forested regions and have significant potential for utilization in bioenergy production, pulp and paper manufacturing, and value-added products. Wood processing residues, in particular, are valuable feedstocks for bioenergy production through combustion, gasification, or pelletization, while forest residues can be used for soil stabilization, erosion control, and habitat restoration.

1.4.1.3 Industrial Biomass Wastes

Industrial biomass wastes are generated as byproducts of manufacturing processes, waste treatment, and product disposal in various industries. These include organic waste streams from food processing (e.g., food scraps, wastewater solids), agricultural processing (e.g., brewery spent grains, sugar beet pulp), pulp and paper manufacturing (e.g., black liquor, pulp sludge), and other industrial sectors (e.g., textile waste, pharmaceutical waste). Industrial biomass wastes have diverse compositions and applications, ranging from bioenergy production and composting to industrial symbiosis and resource recovery. For example, food processing wastes could be transformed into biogas via anaerobic digestion, while pulp and paper mill wastes can be utilized for energy generation, fiber recovery, and chemical production.

1.4.1.4 Municipal Biomass Wastes

Municipal biomass wastes, also known as municipal solid waste (MSW), comprise organic materials discarded by households, businesses, and institutions in urban areas. These include food scraps, yard trimmings, paper products, biodegradable plastics, and other organic waste components. Municipal biomass wastes are a significant fraction of the overall waste stream and pose challenges for landfill management and waste disposal (56, 57). However, they also represent valuable resources for composting, anaerobic digestion, and energy recovery. Municipal biomass wastes can be utilized for composting to produce nutrient-rich soil amendments, anaerobic digestion to generate biogas for energy production, or waste-to-energy technologies such as combustion or gasification to produce heat and electricity.

1.4.2 Classification Based on Characteristics and Applications

1.4.2.1 Feedstock for Bioenergy Production

Biomass wastes can serve as feedstock for bioenergy production through various conversion processes, including combustion, gasification, pyrolysis, and fermentation. Depending on their composition and properties, biomass wastes can be converted into biofuels such as biogas, bioethanol, biodiesel, and biochar, which can be used for heat, power, transportation, and industrial applications. Agricultural and forestry biomass residues are particularly suitable for bioenergy production due to their high energy content and availability (58).

1.4.2.2 Soil Amendment and Fertilizer

Organic biomass wastes, such as crop residues, animal manure, and composted organic matter, are valuable sources of nutrients and organic matter for soil amendment and fertilizer application in agriculture. These organic amendments improve soil structure, fertility, water retention, and microbial activity, leading to increased crop yields, reduced nutrient runoff, and improved environmental sustainability. Additionally, organic amendments can sequester carbon in soil, mitigating greenhouse gas emissions and enhancing soil carbon stocks.

1.4.2.3 Industrial Applications

Biomass wastes have diverse industrial applications, including pulp and paper manufacturing, bio-based chemicals production, and biorefining. Wood processing residues, for example, can be utilized for pulp and paper production, composite materials manufacturing, and biochemical conversion into platform chemicals and bioproducts. Similarly, agricultural and industrial biomass wastes can be processed into bio-based chemicals, polymers, and materials for use in various industries, helping to shift to a bio-based economy and lessen reliance on fossil fuels.

1.4.2.4 Waste-to-Energy and Resource Recovery

Biomass wastes can be utilized for waste-to-energy (WtE) technologies such as combustion, pyrolysis, and anaerobic digestion to recover energy and resources from organic waste streams. WtE technologies convert biomass wastes into heat, electricity, biogas, or biofuels, which can offset fossil fuel consumption, reduce greenhouse gas emissions, and contribute to renewable energy production. Additionally, WtE technologies can recover valuable materials and nutrients from biomass wastes, such as metals, phosphorus, and nitrogen, for recycling and reuse in various industrial processes.

1.5 PROPERTIES OF BIOMASS WASTES

The properties of biomass wastes are critical determinants of their suitability for various applications, including energy production, biofuel manufacturing, and agricultural utilization. Understanding these properties is essential for optimizing biomass waste utilization and maximizing resource recovery. Biomass wastes exhibit diverse physical, chemical, and thermal properties, each of which influences their behavior and performance in different processes and applications.

1.5.1 Physical Properties

The physical attributes of biomass wastes, including particle size, bulk density, moisture content, and porosity, play crucial roles in various aspects of biomass handling, storage, and conversion processes. Understanding these properties is essential for optimizing biomass utilization and maximizing efficiency in biomass-based energy production systems. Particle size is a key parameter that affects the handling and processing of biomass wastes (59, 60). Biomass materials can vary widely in particle size, ranging from fine powders to coarse chips or chunks. Particle size influences the flowability, handling, and feeding characteristics of biomass in conversion processes such as combustion, pyrolysis, and pelletization. Fine particles tend to flow more easily and may be more suitable for fluidized bed combustion or gasification systems, while larger particles may require crushing or grinding before processing. Bulk density, which refers to the mass of biomass per unit volume, is another important physical property of biomass wastes. Bulk density impacts storage capacity, transportation efficiency, and reactor design in biomass conversion systems. Biomass with higher bulk density requires less storage space and transportation volume, but it may have lower porosity and combustion efficiency. Conversely, biomass with lower bulk density may require more space and energy for handling and processing, but it may offer better combustion characteristics.

Moisture content is a critical factor that affects the heating value, combustion characteristics, and energy efficiency of biomass wastes. High moisture content biomass wastes require additional energy for drying before combustion or other conversion processes. Wet biomass may also have lower energy density and combustion efficiency compared to dry biomass (61, 62). Managing moisture content is essential for optimizing energy production and minimizing emissions in biomass-based systems. Porosity, which refers to the volume fraction of void spaces or pores within biomass structures, is another important physical property. Porosity influences gas diffusion, heat transfer, and reaction kinetics in biomass conversion processes. High porosity biomass wastes have greater surface area and reactivity, making them more suitable for thermochemical conversion processes like pyrolysis and gasification. Porosity also affects the permeability of biomass materials to gases and liquids, which can impact reaction rates and product yields in biomass conversion systems.

1.5.2 Chemical Properties

The chemical properties of biomass wastes play a critical role in determining their energy content, chemical reactivity, and environmental impact. Understanding these properties is essential for optimizing biomass utilization and minimizing the environmental footprint of biomass-based energy production systems. Composition is a fundamental chemical property of biomass wastes and refers to the types and proportions of organic compounds present in the biomass. Biomass wastes typically contain organic polymers such as cellulose, hemicellulose, and lignin, as well as proteins, lipids, and carbohydrates. The relative proportions of these components vary depending on the type and origin of the biomass. For instance, lignocellulosic biomass like wood and crop residues contains high levels of cellulose and lignin, while organic wastes from food processing may contain higher proportions of proteins and carbohydrates (63, 64). The composition of biomass influences its energy content, combustion characteristics, and suitability for different conversion processes. Elemental content is another important chemical property of biomass wastes and includes elements such as carbon, hydrogen, oxygen, nitrogen, sulfur, and trace minerals. These elements influence combustion efficiency, emission characteristics, and ash composition in biomass-based energy systems. High nitrogen and sulfur content biomass wastes, for example, may contribute to the formation of nitrogen oxides (NO_x) and sulfur dioxide (SO_2) emissions during combustion, which can have negative environmental impacts such as air pollution and acid rain.

Proximate analysis is a method used to characterize the moisture, volatile matter, fixed carbon, and ash content of biomass wastes. This analysis provides insights into the combustion behavior and energy yield of biomass fuels. Biomass with high volatile matter content tends to combust more readily, while biomass with high ash content may result in ash deposition and fouling issues in combustion systems. Understanding the proximate analysis of biomass wastes helps optimize combustion processes and maximize energy efficiency while minimizing emissions and ash-related problems. Ash composition is a critical chemical property of biomass wastes, particularly in combustion and thermal conversion processes. Biomass ash consists of inorganic minerals and trace elements derived from biomass feedstock, which can affect combustion efficiency, emission control, and ash management strategies. The composition of biomass ash varies depending on factors such as biomass type, processing conditions, and ash removal technologies. Certain trace elements in biomass ash, such as potassium, sodium, and chlorine, may lead to corrosion and erosion of boiler components, while others, such as calcium and magnesium, can enhance ash melting and reduce fouling tendencies. Managing biomass ash composition is essential for minimizing environmental impacts and optimizing the performance of biomass-based energy systems.

1.5.3 Thermal Properties

The thermal properties of biomass wastes play a crucial role in determining their behavior during combustion, thermal conversion processes, and energy recovery. These characteristics help to maximize the performance and sustainability of biomass-based energy systems by offering insightful information on the energy content, combustion kinetics, and effectiveness of these systems. Heating value, also known as calorific value or energy content, is a key thermal property of biomass wastes that measures the amount of heat released per unit mass or volume of biomass during combustion (65, 66). Higher heating value biomass wastes have greater energy density and potential for energy recovery, making them more desirable for use in combustion, gasification, and other thermal conversion processes. Ignition temperature is another important thermal property that defines the temperature at which biomass wastes undergo spontaneous combustion or ignition. Biomass wastes with lower ignition temperatures ignite more readily, posing potential risks of accidental fires or explosions. Understanding the ignition temperature of biomass wastes is essential for ensuring safe handling, storage, and processing in biomass-based energy systems.

Thermal conductivity is a measure of the ability of biomass wastes to conduct heat and influences heat transfer rates, combustion efficiency, and reactor design in biomass conversion systems. Biomass wastes with high thermal conductivity facilitate faster heat transfer and reaction kinetics, leading to improved combustion efficiency and energy conversion rates. This property is particularly important in applications such as fluidized bed combustion and gasification, where efficient heat transfer is critical for optimal performance. Specific heat capacity is a thermal property that measures the amount of heat required to raise the temperature of biomass wastes by one degree Celsius. It influences thermal storage, energy recovery, and process control in biomass conversion systems. Biomass wastes with high specific heat capacity absorb and release heat more slowly, contributing to thermal stability and energy efficiency. This property is relevant in applications where thermal buffering and temperature control are essential, such as biomass boilers, thermal storage systems, and heat exchangers.

1.6 CONVERSION PROCESSES

Conversion processes play a vital phase in transforming biomass wastes into valuable products, including heat, electricity, biofuels, and biochemicals. These processes utilize different mechanisms to break down biomass materials and extract energy or chemical compounds for various applications. The primary conversion processes for biomass wastes include combustion, pyrolysis, fermentation, and other emerging technologies.

1.6.1 Combustion

Combustion stands as the predominant conversion process for biomass wastes, notably for heat and power generation applications. In combustion, biomass wastes undergo controlled burning in the presence of oxygen, yielding heat energy that finds utility in heating, electricity generation, and industrial processes. This process unfolds through several distinct stages, each contributing to the overall energy release and conversion efficiency. Initially, biomass wastes undergo drying, where moisture content is reduced, ensuring efficient combustion. Following this, devolatilization occurs, involving the release of volatile compounds from the biomass matrix as gases. Char combustion ensues, where the remaining solid carbonaceous residue undergoes oxidation, liberating additional heat energy (67). Last, ash formation takes place, resulting from the incombustible mineral content of the biomass, which remains after combustion is completed. Biomass wastes are commonly combusted within specialized combustion systems such as boilers, furnaces, or stoves. These systems provide controlled environments conducive to efficient combustion reactions, ensuring maximum heat release and minimal emissions. Throughout the combustion process, biomass wastes react with oxygen to produce heat and combustion products, predominantly carbon dioxide (CO_2), water vapor (H_2O), and ash. The controlled nature of combustion systems enables the regulation of operating parameters to optimize energy production while minimizing environmental impacts.

Various combustion technologies exist, classified based on several factors including the combustion medium (e.g., air, oxygen), combustion temperature (e.g., grate combustion, fluidized bed combustion), and energy recovery efficiency (e.g., steam turbines, gas turbines). These technologies offer diverse approaches to biomass combustion, each with unique advantages and considerations regarding efficiency, emissions control, and operational flexibility (68–71). By selecting appropriate combustion technologies tailored to specific biomass feedstocks and end-use applications, stakeholders can maximize the value and sustainability of biomass waste utilization for energy generation (72).

1.6.2 Pyrolysis

Pyrolysis represents a thermochemical conversion pathway offering a promising route for the transformation of biomass wastes into valuable products such as syngas, biochar, and bio-oil, without the presence of oxygen (21). This process involves subjecting biomass wastes to elevated temperatures, typically ranging from 300 to 800°C, within an oxygen-limited environment. Under these conditions, thermal decomposition and vaporization occur, leading to the split of organic compounds present in the biomass. The pyrolysis procedure unfolds through several distinct stages, each contributing to the formation of specific end products. Initially, biomass wastes undergo drying, during which moisture content is reduced to facilitate subsequent thermal decomposition. Following this, devolatilization takes place, characterized by the release of volatile compounds from the biomass matrix in the form of gases and vapors. These volatile compounds comprise a range of organic molecules, including hydrocarbons, oxygenates, and other volatile organic compounds. As pyrolysis progresses, the remaining solid residue undergoes further decomposition, leading to the formation of biochar—a carbon-rich solid material. Biochar holds potential applications as a soil amendment to enhance soil fertility, improve water retention, and sequester carbon in agricultural soils (73, 74). Additionally, biochar can be activated to produce activated carbon, which finds utility in various environmental and industrial applications such as water purification and air filtration.

Simultaneously, liquid products known as bio-oil are generated during pyrolysis, comprising a complex mixture of organic compounds derived from the thermal breakdown of biomass. Bio-oil possesses energy density comparable to conventional fossil fuels and can be further processed or upgraded into biofuels, biochemicals, and specialty chemicals. Various upgrading techniques, such as hydrotreating, hydrodeoxygenation, and fractional distillation, are employed to refine bio-oil into higher-value products suitable for use in transportation, heating, and industrial applications. Furthermore, gaseous products known as syngas are formed during pyrolysis, primarily consisting of carbon monoxide (CO), hydrogen (H_2), and lesser amounts of methane (CH_4) and other gases. Potentially used as a flexible feedstock for a variety of processes including gasification, combustion, and Fischer-Tropsch synthesis, syngas can produce energy, heat, and synthetic fuels. The syngas can be utilized directly for heat and power generation or further processed to produce liquid fuels via Fischer-Tropsch synthesis, enabling the production of drop-in biofuels compatible with existing infrastructure.

1.6.3 Fermentation

Fermentation stands as a versatile biochemical conversion process that harnesses the metabolic activities of microorganisms, including bacteria, yeast, or fungi, to transform biomass wastes into valuable biofuels, biochemicals, and bioproducts. This process capitalizes on the ability of microorganisms to enzymatically degrade organic substrates present in biomass wastes, such as sugars, starches, or cellulose, and convert them into desired end products through metabolic pathways. In fermentation, biomass wastes are first prepared to make their organic constituents accessible to microbial enzymes. Substrates containing sugars, starches, or cellulose are typically used as feedstocks for fermentation (75, 76). These biomass wastes can originate from a range of resources, such as leftover agricultural materials, food processing byproducts, lignocellulosic biomass, and organic waste streams. Microorganisms such as bacteria, yeast, or fungi are then introduced into the fermentation process, where they metabolize the available organic compounds and produce desired end products. Depending on the specific microorganism and fermentation conditions employed, a wide range of biofuels, biochemicals, and bioproducts can be synthesized through fermentation.

A number of variables, such as the kind of microorganism employed, the make-up of the feedstock, and the fermentation environment, can be used to categorize fermentation processes. For example, yeast fermentation, which is widely used in the manufacturing of ethanol, converts glucose into carbon dioxide and ethanol by using yeast strains. In contrast, anaerobic digestion produces biogas that is mainly made of carbon dioxide and methane by using a group of microorganisms to break down complex organic components in an oxygen-limited environment. The composition of the biomass feedstock, the intended end products, the scalability of the process, and the process's viability economically all influence the choice of fermentation method. Ethanol fermentation, for example, is a well-established process for producing bioethanol from biomass wastes such as grains, sugarcane, or lignocellulosic biomass. This process offers significant potential for renewable energy production and carbon mitigation. Similarly, anaerobic digestion is widely utilized to treat organic wastes such as food scraps, animal manure, and sewage sludge, while simultaneously generating biogas as a renewable energy source. Anaerobic digestion produces biogas, which could be converted to biomethane and injected into natural gas pipelines for use as fuel in vehicles, a source of heat and power.

1.6.4 Other Technologies

In addition to the established conversion technologies like combustion, pyrolysis, and fermentation, the field of biomass waste utilization is witnessing the development of several innovative approaches, each offering unique advantages and applications. Gasification emerges as a promising thermochemical conversion process that transforms biomass wastes into synthesis gas, commonly known as syngas. This process involves subjecting biomass to high temperatures in a controlled environment with limited oxygen and steam. Through gasification, biomass undergoes a series of chemical reactions, resulting in the production of syngas composed of hydrogen, carbon monoxide, and other gases. Syngas produced from gasification hold versatile applications, including heat and power generation or conversion into liquid fuels through catalytic processes like Fischer-Tropsch synthesis (77, 78). Gasification offers several benefits, including high energy efficiency, flexibility in feedstock utilization, and reduced emissions compared to conventional combustion processes. Hydrothermal processing represents another promising avenue for biomass waste conversion, involving the treatment of biomass in high-pressure and high-temperature water-based reactions. This process facilitates the breakdown of biomass components into bio-oil, biogas, and biochar, depending on the reaction conditions. Hydrothermal processing offers several advantages, including fast reaction kinetics, high energy efficiency, and the ability to process wet biomass feedstocks without the need for drying. The resulting bio-oil can be further upgraded into biofuels or biochemicals, while biochar finds application as a soil amendment or carbon sequestration agent.

Algae cultivation stands as a sustainable approach to biomass waste utilization, leveraging photosynthetic microorganisms like microalgae or macroalgae to convert biomass wastes and carbon dioxide into biomass, lipids, proteins, and other valuable products. Through photosynthesis, algae assimilate carbon dioxide and nutrients from biomass wastes, converting them into biomass rich in lipids, proteins, and carbohydrates. Algal biomass could be harvested and processed through biorefining techniques to extract valuable products such as biofuels, feedstock for animal or aquaculture industries, and high-value chemicals. Algae cultivation offers several advantages, including high growth rates, carbon capture capabilities, and the ability to grow in diverse environments, including wastewater treatment facilities and marginal lands. These emerging conversion technologies hold significant promise for enhancing the sustainability and efficiency of biomass waste utilization, offering diverse pathways for energy production, waste management, and resource recovery. Continued research and development efforts in these areas are crucial to unlocking their full potential and accelerating the transition towards a bio-based circular economy.

1.7 APPLICATIONS AND UTILIZATION

Applications and utilization of biomass wastes encompass a wide range of sectors and industries, including energy production, biofuel manufacturing, and waste management strategies. Through the utilization of biomass wastes, stakeholders may effectively tackle environmental issues, decrease dependency on fossil fuels, and advance sustainable development by tapping into their energy and resource potential. Here's a discussion of the key applications and utilization pathways for biomass wastes:

1.7.1 Energy Production

In order to reduce greenhouse gas emissions and move toward renewable energy resources, energy generation using biomass wastes is essential. Various conversion technologies are employed to harness the energy potential of biomass wastes, including combustion, gasification, and anaerobic digestion (9, 79). Combustion stands as one of the most established methods for utilizing biomass wastes as a renewable energy source. In this process, biomass wastes such as agricultural residues, forestry residues, and municipal solid waste are burned in combustion systems such as boilers, furnaces, and power plants (80–86). The heat energy released during combustion is utilized for various applications, including space heating, district heating, and industrial processes. Combustion systems are versatile and can accommodate a wide range of biomass feedstocks, making them suitable for decentralized energy production in both rural and urban settings. Gasification represents another efficient method for converting biomass wastes into energy-rich syngas. During gasification, biomass wastes are subjected to high temperatures in the presence of a controlled amount of oxygen and steam, leading to the production of syngas composed of carbon monoxide, hydrogen, and other gases. Syngas can be utilized for heat and power generation in gas turbines or internal combustion engines. Moreover, syngas can undergo further processing to produce liquid fuels or chemicals, offering additional pathways for biomass waste utilization and valorization. Figure 1.4 illustrates a typical power plant fueled by rice husk. This depiction showcases the operational setup and components of a power generation facility that utilizes rice husk as its primary fuel source. Rice husk, an abundant byproduct of rice milling, is utilized in such power plants to generate electricity through combustion or gasification processes.

Anaerobic digestion offers a sustainable approach to biomass waste management while simultaneously producing renewable energy. In anaerobic digestion, microorganisms break down organic matter present in biomass wastes in the absence of oxygen, resulting in the production of biogas, primarily composed of methane and carbon dioxide (87, 88). Biogas can be used for heat and power generation in combined heat and power (CHP) systems, providing energy for residential, commercial, or industrial applications. Additionally, biogas can be upgraded into biomethane through purification processes and injected into natural gas pipelines or used as a transportation fuel, further diversifying its applications and contributing to decarbonization efforts. Waghmare et al. (89) isolated a cellulolytic bacterium, identified as Enterobacter sp. SUK-Bio, from plant litter soil and studied for its ability to utilize various cellulosic materials including carboxymethyl cellulose, sugarcane trash, grass powder, sorghum husk, wheat straw, and water hyacinth. Among these materials, sorghum husk demonstrated superior utilization, leading to higher production of cellulolytic and hemicellulolytic enzymes (including filter paperase, β-glucosidase, endoglucanase, exoglucanase, xylanase, and glucoamylase) on day 8 of incubation. Additionally, sorghum husk yielded the highest amount of reducing sugars (554 mg/L) at a rate of 3.84 mg/h/L. Fourier-transform infrared spectroscopy analysis of sorghum husk indicated changes in functional groups and a decrease in total crystallinity ratio following microbial degradation. This suggests effective enzymatic breakdown of cellulose and hemicellulose components present in sorghum husk by Enterobacter sp. SUK-Bio. The findings underscore the potential of Enterobacter sp. SUK-Bio for efficient bioconversion of agricultural residues like sorghum husk into valuable products such as

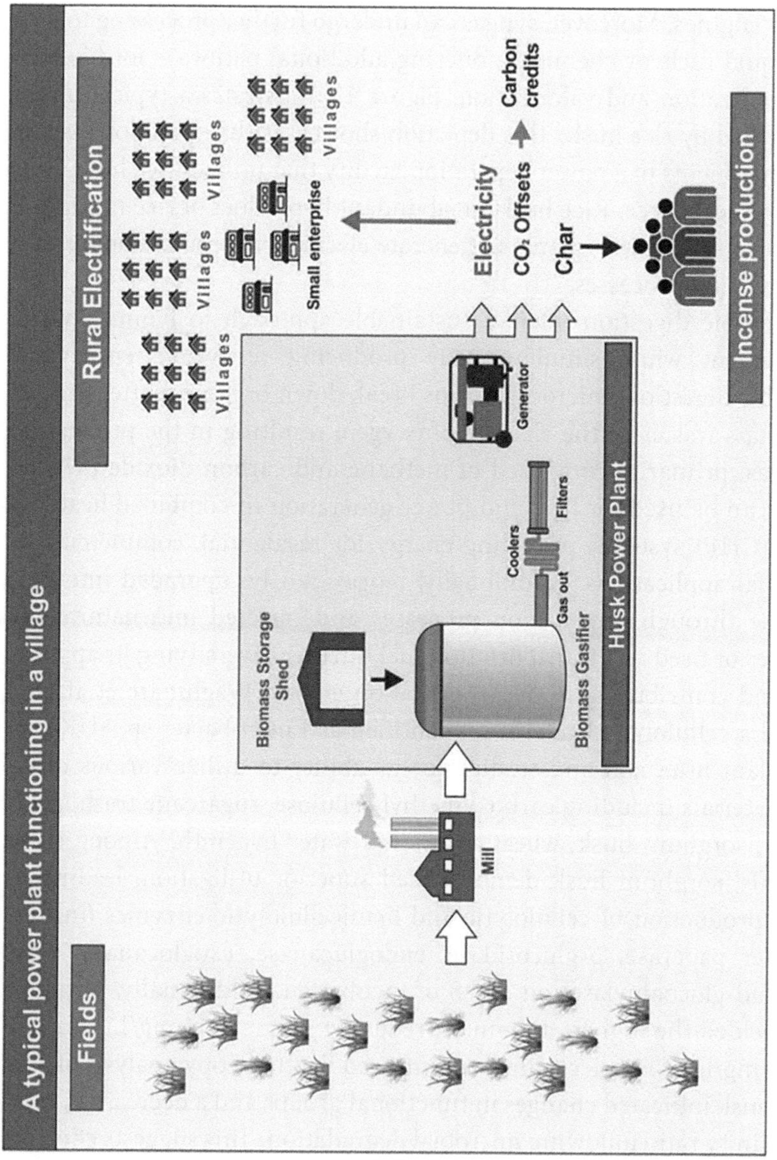

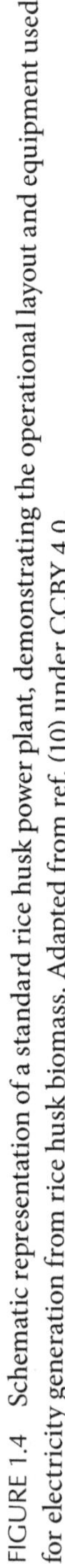

FIGURE 1.4 Schematic representation of a standard rice husk power plant, demonstrating the operational layout and equipment used for electricity generation from rice husk biomass. Adapted from ref. (10) under CCBY 4.0.

reducing sugars, highlighting its significance in waste valorization and biofuel production.

Suárez-Macías et al. (90) centered on investigating the possibility of using biomass bottom ash as a filler in bitumen emulsion-based intermittent grading bituminous mixes. This approach aims to address the increasing energy consumption driven by population growth, wherein renewable energy sources like biomass combustion generate waste, such as biomass bottom ash, posing an environmental challenge due to its disposal. The first step of the investigation is to determine whether biomass bottom ash is suitable as a filler material by analyzing its physical and chemical characteristics. Then, bituminous mixes containing biomass bottom ash were made, and Marshall tests and particle loss were used to examine the mixtures' mechanical and physical characteristics. A comparative analysis is conducted with mixtures containing conventional and ophite aggregates. Following the SEM analysis aimed at characterizing the surface of biomass bottom ashes in detail and providing qualitative insights, the resulting images (Figure 1.5) revealed a highly irregular surface texture of the ashes. The SEM images depict the presence of numerous micropores and cavities on the surface, contributing significantly to the higher specific surface area observed.

The presence of these micropores and cavities exhibits a vital role in increasing the specific surface area of the ashes. This, in turn, facilitates enhanced bitumen absorption when incorporated into bituminous mixtures. Consequently, the formation of a higher quality mastic occurs, which effectively coats the aggregates, thereby improving the tensile strength

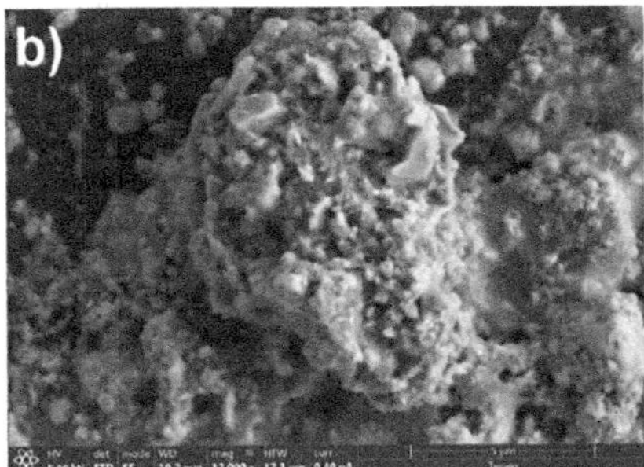

FIGURE 1.5 SEM images of biomass bottom ash at varying magnifications in the secondary setting. (a) depicts the ash at 3000X magnification, while (b) provides a closer view at 12000X magnification. Adapted from ref. (90) under CCBY 4.0.

of the bituminous mix. Moreover, the increased proportion of bitumen due to higher absorption by the biomass bottom ashes also contributes to extending the service life of the bituminous mix. This is attributed to the greater amount of mastic present, which enhances the durability and resistance of the mixture to wear and tear over time. The results show that using biomass bottom ash as a filler material might produce mixes with a greater bitumen percentage, improved mechanical behavior, and comparable physical properties to conventional mixtures. This suggests the potential for creating more sustainable road materials by repurposing waste that would otherwise be landfilled.

1.7.2 Biofuel Production

Biomass wastes serve as a versatile and sustainable feedstock for the production of biofuels, offering an alternative to fossil fuels and contributing to the reduction of greenhouse gas emissions in the transportation sector. Several biofuel options derived from biomass wastes include bioethanol, biodiesel, and biogasoline, each with unique production processes and applications. Bioethanol represents one of the most widely used biofuels worldwide, produced primarily through the fermentation of biomass sugars or starches (91–93). Feedstocks for bioethanol production include corn, sugarcane, and lignocellulosic biomass such as agricultural residues and energy crops. Microorganisms such as yeast or bacteria ferment the sugars present in these feedstocks to produce ethanol. For gasoline-powered vehicles, bioethanol could be mixed with gasoline at different ratios to lower pollutants and raise octane ratings. This makes bioethanol a sustainable and greener alternative fuel.

Biodiesel is another important biofuel derived from biomass wastes, typically produced from vegetable oils, animal fats, or recycled cooking oils. The production process involves a transesterification reaction, where triglycerides present in the feedstock are converted into fatty acid methyl esters (FAME), the primary component of biodiesel. Biodiesel can be used as a direct replacement for conventional diesel fuel in compression-ignition engines without requiring modifications to existing infrastructure or vehicle engines. When compared to petroleum diesel, it has benefits such as decreased emissions of sulfur, carbon monoxide, and particulate matter, which improves air quality and lessens environmental impact. Biogasoline represents an emerging biofuel option synthesized from biomass-derived syngas or bio-oil through advanced catalytic processes such as Fischer-Tropsch synthesis or hydroprocessing. These processes convert biomass-derived feedstocks into liquid hydrocarbons with properties similar to petroleum-based gasoline. Biogasoline can be used as a drop-in replacement for conventional gasoline in spark-ignition engines, offering a renewable alternative fuel option with potential reductions in greenhouse gas emissions and reliance on fossil fuels. In an experiment, Sudiyani et al. (94) investigated the utilization of biomass waste, specifically empty fruit bunch (EFB) fiber from the palm oil industry, for the production of bioethanol through a pilot-scale unit. EFBs are a significant byproduct of palm oil production, containing substantial amounts of cellulose (37.3–46.5%) and hemicelluloses (25.3–33.8%), making them promising raw materials for fermentation processes. Moreover, their abundance and lack of competition with the food supply render them cost-effective for bioethanol production. The bioconversion process involved several steps: pretreatment to reduce the crystallinity of cellulose and remove lignin, enzymatic hydrolysis (saccharification) to convert cellulose into fermentable sugars using a blend of enzymes, and fermentation of these sugars into bioethanol. In this study, a pilot-scale unit was established to develop and assess the ethanol production process based on enzymatic saccharification. The process involved pretreating EFBs with 10% NaOH, followed by saccharification using modified cellulase enzymes, and fermentation in a 350 L fermentor tank using local Saccharomyces cerevisiae Mk strains at 32°C for 48 hours.

1.7.3 Waste Management Strategies

Biomass wastes are managed and utilized through a variety of waste management strategies aimed at reducing environmental impacts, promoting resource recovery, and achieving sustainability objectives. One popular method of managing trash is composting, which is the process of biologically breaking down organic items including food scraps, yard waste, and agricultural residues (95–102). Through composting, these organic wastes are transformed into nutrient-rich compost, which serves as a valuable soil conditioner and fertilizer in agricultural and landscaping applications (103, 104). In addition to keeping organic waste out of landfills, composting enhances soil quality, increases moisture retention, and stimulates plant growth—all of which support ecological health and sustainable farming methods. Anaerobic digestion represents another effective waste management strategy, particularly for organic wastes with high moisture content, such as food scraps, animal manure, and sewage sludge. During anaerobic digestion, microorganisms break down organic matter in the absence of oxygen, producing biogas (primarily methane and carbon dioxide) and digestate. Biogas can be utilized as a renewable energy source for heat and power generation, while digestate serves as a nutrient-rich fertilizer or soil conditioner. Anaerobic digestion not only produces renewable energy but also mitigates greenhouse gas emissions from organic waste decomposition, offering environmental and economic benefits. Auteri et al. (55) explore the feasibility of utilizing cactus pear pruning waste (CPPW) as a cost-effective biomass adsorbent for removing phosphorus (P) from aqueous solutions in batch mode. Samples of biomass derived from cactus pear were collected and analyzed after enrichment with either calcium (Ca) or iron (Fe) to assess their P removal capabilities. The investigation revealed P removal capacities of 2.27 mg/g, 1.33 mg/g, and 1.87 mg/g for Ca^{2+}-enriched, Fe^{2+}-loaded, and Fe^{3+}-loaded biomass, respectively. Various models were examined to describe the P adsorption process, with the Langmuir isotherm model identified as the most suitable for accurately representing the adsorption behavior of the enriched biomass. Kinetic analysis of the adsorption process using pseudo-first-order, pseudo-second-order, and intraparticle diffusion models indicated that the pseudo-second-order model provided the best fit to the experimental data.

Additionally, the study investigated the desorption and regeneration process, revealing minimal P desorption (less than 8%) from Ca- or Fe-loaded biomass, indicating the strong stability of the biomass-cation-P system. Economic analysis estimated the cost of CPPW-based adsorption ranged from 8 to 161 euros per ton, with an additional 230 euros attributed to pruning costs inherent to the crop. These costs were found to be below the threshold (320 euros per ton) for economically viable P reuse at the farm level. Similarly, Sharma et al. (66) investigated the potential valorization of cellulosic fibers extracted from waste biomass of Canna indica for reinforcement in natural fiber polymeric composites, presenting a technological strategy toward achieving a circular economy. The Canna indica biomass, obtained from constructed wetlands treating municipal wastewater, was analyzed for physicochemical, mechanical, structural, crystallographic, and thermal properties to assess its suitability as a replacement for synthetic fibers in composites. The extracted Canna indica (CI) fibers demonstrated favorable characteristics, including a higher cellulose content (60 wt%) and a low wax fraction (0.5 wt%), making them advantageous for reinforcement purposes. With thermal stability up to 237°C and average dimensions of 4.3 mm in length, 842 μm in diameter, and a density of 0.75 g/cm3, the CI fibers exhibited promising mechanical properties, with a mean maximum tensile strength of 113 ± 6.82 MPa and Young's modulus of 0.8 ± 7.91 GPa. Nano-indentation tests further revealed a nano hardness of 0.3 ± 0.6 GPa and modulus of 1.62 ± 0.2 GPa. Crystallographic analysis indicated an 87.45% crystallinity index and a 3.2 nm crystallite size for CI fibers, while morphological examination revealed rough surfaces and shallow cavities, enhancing their suitability for reinforcement applications. The study suggests that this approach could contribute to a circular economy by valorizing Canna indica biomass sourced from natural wastewater treatment plants.

Similarly, Wojewódzki et al. (105) explored the utilization of biochar derived from various organic waste sources to enhance soil fertility, soil enzyme activity, and carbon sequestration. Conducted by researchers from the Department of Biogeochemistry and Soil Science at Bydgoszcz University of Science and Technology, Poland, along with collaborators from Minia University, Egypt, and the Slovak University of Agriculture in Nitra, Slovakia, the investigation aims to assess the impact of biochar produced via low-temperature pyrolysis from different organic waste materials. Upon mixing the biochars with the soil, there was a notable decrease in the TOC/TN ratio values, which varied from 10.04 (S + OMOC) to 53.17 (S + PB). Following a 60-day incubation period of soil mixed with biochars, a decrease in TOC content was observed, ranging from 4.48% (variant S + PB) to 10.77% (variant S + OMOC; Figure 1.6). The nitrogen losses

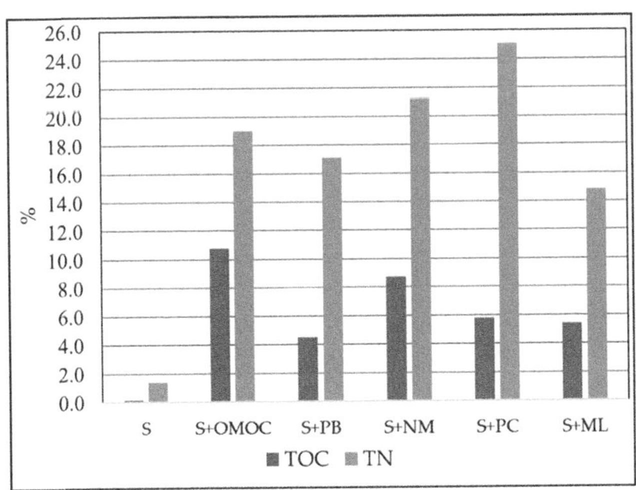

FIGURE 1.6 Depiction of the percentage change in TOC and TN content from the initial to the final incubation stages. Adapted from ref. (105) under CCBY 4.0.

ranged from 14.8% (variant S + ML) to 25.07% (variant S + PC), indicating higher intensity in nitrogen transformations compared to carbon transformations, thus widening the TOC/TN ratio. The greatest increase in the TOC/TN ratio was observed in the variant S + PC, while the lowest was in variant S + OMOC. Results indicate that the influence of biochar amendments varies depending on the type of feedstock material. This research provides valuable insights into the potential of biochar from waste biomass to improve soil health and promote sustainable agricultural practices, highlighting its role in enhancing nutrient availability and carbon sequestration in soils.

Waste-to-energy (WtE) technologies play a vital role in converting biomass wastes into usable energy forms, including heat, power, and electricity. These technologies encompass various processes such as combustion, gasification, and pyrolysis, each offering unique advantages for energy recovery from biomass wastes. Through the process of combustion, biomass wastes are burned to produce heat that may either be used directly for space heating or transformed into electricity using gas or steam turbines. Biomass wastes are gasified to create syngas, which can then be processed further to create chemicals or biofuels. Syngas is utilized for heat and power generation. Biochar, bio oil, and syngas are produced during the pyrolysis process of biomass wastes, which presents opportunities for energy production and the creation of value-added products.

1.8 CHALLENGES AND FUTURE DIRECTIONS

Addressing the myriad technological and economic challenges inherent in biomass waste utilization is crucial for realizing its full potential as a sustainable resource. Technological hurdles, such as suboptimal efficiency and the variability of feedstock properties, often impede the widespread adoption of biomass conversion technologies (104, 106). To overcome these challenges, significant investments in research and development are necessary to advance the state-of-the-art processes. This includes optimizing the design of conversion reactors, improving feedstock preprocessing techniques, and developing innovative catalysts to enhance conversion efficiency and product yields. Additionally, integrating circular economy principles into biomass utilization strategies can further enhance resource efficiency and minimize waste generation by maximizing the valorization of biomass feedstocks and byproducts. On the economic front, challenges such as market uncertainty and competition for resources pose significant barriers to investment and market growth in the biomass sector. Stable policy frameworks, coupled with financial incentives and regulatory support, are essential for providing the necessary confidence and certainty to attract investments in biomass projects. Policies such as renewable energy mandates, carbon pricing mechanisms, and feed-in tariffs can help create a favorable market environment for biomass utilization. Moreover, fostering collaboration among stakeholders, including government agencies, industry partners, research institutions, and local communities, is critical for promoting knowledge-sharing, capacity-building, and technology transfer in biomass waste utilization.

Looking ahead, future research directions should prioritize process optimization, feedstock characterization, sustainability assessment, and technology integration to address the remaining challenges and unlock the full potential of biomass waste utilization. Process optimization efforts should focus on refining conversion technologies to improve energy efficiency, reduce emissions, and enhance the overall economic viability of biomass projects (107, 108). Advanced characterization techniques and predictive modeling tools can provide valuable insights into feedstock properties, variability, and availability, enabling informed decision-making in biomass supply chain management. Sustainability assessments, including life cycle analyses and social impact assessments, are essential for evaluating the environmental, economic, and social implications of biomass utilization pathways and identifying opportunities for improvement. Last but not least, combining the use of biomass with other renewable energy sources like solar, breezes, and hydropower could improve the resilience, dependability, and flexibility of the energy system and open the door to a more varied and sustainable energy future.

1.9 CONCLUSION

Biomass wastes, derived from agricultural, forestry, industrial, and municipal sources, represent a valuable resource with significant potential for mitigating environmental pollution, mitigating reliance on fossil fuels, and elevating sustainable development. Throughout the chapter, several key themes have emerged, highlighting the importance and complexities of biomass waste utilization in the context of environmental sustainability, energy security, and waste management. First, biomass wastes play a critical role in environmental sustainability by offering an alternative to fossil fuels for energy production and reducing greenhouse gas emissions. Through processes such as combustion, pyrolysis, fermentation, and other emerging technologies, biomass wastes can be converted into heat, power, biofuels, and biochemicals, displacing conventional energy sources and contributing to climate change mitigation efforts. By harnessing the energy and resource potential of biomass wastes, stakeholders can achieve multiple environmental benefits, including reduced air pollution, improved soil health, and enhanced biodiversity conservation. Second, biomass waste utilization is essential for waste management and resource conservation, helping to divert organic waste streams from landfills and incinerators while recovering valuable materials and nutrients for recycling and reuse. Through composting, anaerobic digestion, and waste-to-energy technologies, biomass wastes can be converted into valuable products such as compost, biogas, and renewable electricity, reducing the burden on waste management infrastructure and minimizing pollution of air, soil, and water. Additionally, biomass waste utilization promotes circular economy principles by closing the loop on material flows and maximizing resource efficiency throughout the biomass value chain.

Despite the significant potential of biomass waste utilization, several challenges must be addressed to realize its full benefits and overcome barriers to implementation. Technological challenges, including low conversion efficiency, feedstock variability, and high capital costs, require continued research and development to improve process performance and economic viability. Economic challenges, such as market uncertainty, resource competition, and policy instability, necessitate supportive policies, financial incentives, and stakeholder engagement to create an enabling environment for biomass projects and investments. Moreover, addressing environmental and social concerns, such as land use conflicts, biodiversity impacts, and community engagement, is essential for ensuring the sustainability and acceptance of biomass waste utilization initiatives. Moving forward, future research directions in biomass waste utilization should focus on advancing conversion technologies, optimizing feedstock management practices, assessing sustainability impacts, and fostering collaboration among stakeholders. Process optimization efforts should aim to enhance energy efficiency, reduce emissions, and improve product yields across biomass conversion pathways, from feedstock preprocessing to product upgrading. Advanced characterization techniques and modeling tools can provide valuable insights into feedstock properties, variability, and supply chain dynamics, enabling informed decision-making and risk management in biomass projects. Sustainability assessments, including life cycle analyses and social impact assessments, are essential for evaluating the environmental, economic, and social implications of biomass utilization pathways and guiding policy and investment decisions. Additionally, fostering collaboration among governments, industries, research institutions, and local communities can promote knowledge-sharing, capacity-building, and technology transfer in biomass waste utilization, facilitating the transition to a more sustainable and inclusive economy.

REFERENCES

1. Angelini F, Bellini E, Marchetti A, Salvatori G, Villano M, Pontiggia D, et al. Efficient utilization of monosaccharides from agri-food byproducts supports Chlorella vulgaris biomass production under mixotrophic conditions. *Algal Res [Internet]*. 2024;77(November 2023):103358. Available from: https://doi.org/10.1016/j.algal.2023.103358

2. Ni W, Jing Y. Expanding the boundary of biorefinery: Long-chain heteroatom-containing chemicals from biomass. *Carbon Capture Sci Technol [Internet]*. 2024;10(November 2023):100158. Available from: https://doi.org/10.1016/j.ccst.2023.100158

3. Gonçalves MA, dos Santos HCL, Ribeiro TS, da Cas Viegas A, da Rocha Filho GN, Vieira da Conceição LR. Boosting biodiesel production of waste frying oil using solid magnetic acid catalyst from agro-industrial waste. *Arab J Chem*. 2024;17(2).

4. Jamil MAB, Hayano K, Kassa A, Sekine R, Mochizuki Y. Curing effects on geotechnical properties of clays treated with palm kernel shell ash and rice husk ash: Insights from water absorption characteristics of stabilizers. *Case Stud Constr Mater [Internet]*. 2024;20(October 2023):e02947. Available from: https://doi.org/10.1016/j.cscm.2024.e02947

5. Kostyniuk A, Likozar B. Catalytic wet torrefaction of biomass waste into bio-ethanol, levulinic acid. *Chem Eng J [Internet]*. 2024;149779. Available from: https://doi.org/10.1016/j.cej.2024.149779

6. Cabrera-Reyes P, Palomo J, García-Mateos FJ, Ruiz-Rosas R, Rosas JM, Rodríguez-Mirasol J, et al. Sustainable carbon-based nickel catalysts for the steam reforming of model compounds of pyrolysis liquids. *Fuel Process Technol*. 2024;253(October 2023).

7. Susanto B, Tosuli YT, Nami H, Surjosatyo A, Alandro D, Nugroho AD, et al. Characterization of sago tree parts from Sentani, Papua, Indonesia for biomass energy utilization. *Heliyon [Internet]*. 2024;10(1):e23993. Available from: https://doi.org/10.1016/j.heliyon.2024.e23993

8. Sharma S, Asolekar SR, Thakur VK, Asokan P. Unlocking the potential: A paradigm-shifting approach for valorizing lignocellulosic waste biomass of constructed wetland enabling environmental and societal sustainability. *Ind Crops Prod [Internet]*. 2024;207(P1):117709. Available from: https://doi.org/10.1016/j.indcrop.2023.117709

9. Alola AA, Adebayo TS. Analysing the waste management, industrial and agriculture greenhouse gas emissions of biomass, fossil fuel, and metallic ores utilization in Iceland. *Sci Total Environ [Internet]*. 2023;887(May):164115. Available from: https://doi.org/10.1016/j.scitotenv.2023.164115

10. Mofijur M, Mahlia TMI, Logeswaran J, Anwar M, Silitonga AS, Rahman SMA, et al. Potential of rice industry biomass as a renewable energy source. *Energies*. 2019;12(21).

11. Li C, Gillings MR, Zhang C, Chen Q, Zhu D, Wang J, et al. Ecology and risks of the global plastisphere as a newly expanding microbial habitat. *Innovation [Internet]*. 2024;5(1):100543. Available from: https://doi.org/10.1016/j.xinn.2023.100543

12. Yang H, Nurdiawati A, Gond R, Chen S, Wang S, Tang B, et al. Carbon-negative valorization of biomass waste into affordable green hydrogen and battery anodes. *Int J Hydrogen Energy*. 2024;49:459–471.

13. Koyunoğlu C. Biofuel production utilizing tenebrio molitor: A sustainable approach for organic waste management. *Int J Thermofluids [Internet]*. 2024;100603. Available from: https://doi.org/10.1016/j.ijft.2024.100603

14. González-Arias J, Torres-Sempere G, Villora-Picó JJ, Reina TR, Odriozola JA. Synthetic natural gas production using CO_2-rich waste stream from hydrothermal carbonization of biomass: Effect of impurities on the catalytic activity. *J CO_2 Util*. 2024;79(December 2023).

15. Ahmed B, Aboudi K, Tyagi VK, álvarez-Gallego CJ, Fernández-Güelfo LA, Romero-García LI, et al. Improvement of anaerobic digestion of lignocellulosic biomass by hydrothermal pretreatment. *Appl Sci*. 2019;9(18):1–17.

16. Sanglaow T, Prasert K, Chanthad C, Liangruksa M, Sutthibutpong T. A DFT study on the fundamental mechanisms of quantum capacitance enhancement within the carbon-based electrodes through different classes of doped configurations from biomass-derived elements. *Results Mater [Internet]*. 2024;21(December 2023):100529. Available from: https://doi.org/10.1016/j.rinma.2024.100529

17. Ding Z, Hamann KT, Grundmann P. Enhancing circular bioeconomy in Europe: Sustainable valorization of residual grassland biomass for emerging bio-based value chains. *Sustain Prod Consum [Internet]*. 2024;45(January):265–280. Available from: https://doi.org/10.1016/j.spc.2024.01.008

18. Bediako JK, Kudoahor E, Lim CR, Affrifah NS, Kim S, Song MH, et al. Exploring the insights and benefits of biomass-derived sulfuric acid activated carbon for selective recovery of gold from simulated waste streams. *Waste Manag [Internet]*. 2024;177(February):135–145. Available from: https://doi.org/10.1016/j.wasman.2024.02.002

19. Bisinella V, Schmidt S, Varling AS, Laner D, Christensen TH. Waste LCA and the future. *Waste Manag [Internet]*. 2024;174(November 2023):53–75. Available from: https://doi.org/10.1016/j.wasman.2023.11.021

20. Materazzi M, Chari S, Sebastiani A, Lettieri P, Paulillo A. Waste-to-energy and waste-to-hydrogen with CCS: Methodological assessment of pathways to carbon-negative waste treatment from an LCA perspective. *Waste Manag [Internet]*. 2024;173(November 2023):184–199. Available from: https://doi.org/10.1016/j.wasman.2023.11.020

21. Afraz M, Muhammad F, Nisar J, Shah A, Munir S, Ali G, et al. Production of value added products from biomass waste by pyrolysis: An updated review.

Waste Manag Bull [Internet]. 2024;1(4):30–40. Available from: https://doi.org/10.1016/j.wmb.2023.08.004

22. Phiri R, Mavinkere Rangappa S, Siengchin S, Oladijo OP, Dhakal HN. Development of sustainable biopolymer-based composites for lightweight applications from agricultural waste biomass: A review. *Adv Ind Eng Polym Res [Internet].* 2023;6(4):436–450. Available from: https://doi.org/10.1016/j.aiepr.2023.04.004

23. Pang YX, Sharmin N, Wu T, Pang CH. An investigation on plant cell walls during biomass pyrolysis: A histochemical perspective on engineering applications. *Appl Energy [Internet].* 2023;343(May):121055. Available from: https://doi.org/10.1016/j.apenergy.2023.121055

24. Salimbeni A, Lombardi G, Rizzo AM, Chiaramonti D. Techno-Economic feasibility of integrating biomass slow pyrolysis in an EAF steelmaking site: A case study. *Appl Energy [Internet].* 2023;339:120991. Available from: https://www.sciencedirect.com/science/article/pii/S0306261923003550

25. Nasran M, Khan N, Firdaus M, Yusop M, Faizal M, Mohamed P, et al. Utilization of landscape biomass waste as activated carbon to scavenge oxytetracycline: Attraction mechanism, batch and continuous studies. *Arab J Chem [Internet].* 2023;16(11):105256. Available from: https://doi.org/10.1016/j.arabjc.2023.105256

26. da Luz Corrêa AP, da Silva PMM, Gonçalves MA, Bastos RRC, da Rocha Filho GN, da Conceição LRV. Study of the activity and stability of sulfonated carbon catalyst from agroindustrial waste in biodiesel production: Influence of pyrolysis temperature on functionalization. *Arab J Chem.* 2023;16(8).

27. Ciurko D, Łaba W, Kancelista A, John Ł, Gudiña EJ, Lazar Z, et al. Efficient conversion of black cumin cake from industrial waste into lipopeptide biosurfactant by Pseudomonas fluorescens. *Biochem Eng J.* 2023;197(May).

28. Serafin J, Dziejarski B, Vendrell X, Kiełbasa K, Michalkiewicz B. Biomass waste fern leaves as a material for a sustainable method of activated carbon production for CO2 capture. *Biomass Bioenergy.* 2023;175(November 2022).

29. Silva JP, Teixeira S, Teixeira JC. Characterization of the physicochemical and thermal properties of different forest residues. *Biomass Bioenergy.* 2023;175(May).

30. Munisamy Sambasivam K, Kuppan P, Shashirekha V, Tamilarasan K, Abinandan S. Cascading utilization of residual microalgal biomass: Sustainable strategies for energy, environmental and value-added product applications. *Bioresour Technol Reports [Internet].* 2023;23(April):101588. Available from: https://doi.org/10.1016/j.biteb.2023.101588

31. Cobo-Ceacero CJ, Moreno-Maroto JM, Guerrero-Martínez M, Uceda-Rodríguez M, López AB, Martínez García C, et al. Effect of the addition of organic wastes (cork powder, nut shell, coffee grounds and paper sludge) in clays to obtain expanded lightweight aggregates. *Bol la Soc Esp Ceram y Vidr [Internet].* 2023;62(1):88–105. Available from: https://doi.org/10.1016/j.bsecv.2022.02.007

32. Cheng S, Ding X, Dong X, Zhang M, Tian X, Liu Y, et al. Immigration, transformation, and emission control of sulfur and nitrogen during gasification of MSW: Fundamental and engineering review. *Carbon Resour Convers [Internet]*. 2023;6(3):184–204. Available from: https://doi.org/10.1016/j.crcon.2023.03.003

33. Gani A, Adisalamun, Arkan DMR, Suhendrayatna, Reza M, Erdiwansyah, et al. Proximate and ultimate analysis of corncob biomass waste as raw material for biocoke fuel production. *Case Stud Chem Environ Eng [Internet]*. 2023;8(November):100525. Available from: https://doi.org/10.1016/j.cscee.2023.100525

34. Kalak T. Potential use of industrial biomass waste as a sustainable energy source in the future. *Energies*. 2023;16(4).

35. Bai H, Lin H, Singh Chauhan B, Amer AM, Fayed M, Ayed H, et al. Development of a novel power and freshwater cogeneration plant driven by hybrid geothermal and biomass energy. *Case Stud Therm Eng [Internet]*. 2023;52(November):103695. Available from: https://doi.org/10.1016/j.csite.2023.103695

36. Damayanti D, Wulandari YR, Marpaung DS, Supriyadi D, Yohana DT, Saputri DR, et al. Unlocking green potential: Enhancing biomass valorization via thermal depolymerization with spent catalysts. *Catal Commun [Internet]*. 2023;185(November):106813. Available from: https://doi.org/10.1016/j.catcom.2023.106813

37. Gómez JA, Matallana LG, Teixeira JA, Sánchez ÓJ. A framework for the design of sustainable multi-input second-generation biorefineries through process simulation: A case study for the valorization of lignocellulosic and starchy waste from the plantain agro-industry. *Chem Eng Res Des [Internet]*. 2023;195:551–571. Available from: https://doi.org/10.1016/j.cherd.2023.06.004

38. Azman NF, Katahira T, Nakanishi Y, Chisyaki N, Uemura S, Yamada M, et al. Sustainable oil palm biomass waste utilization in Southeast Asia: Cascade recycling for mushroom growing, animal feedstock production, and composting animal excrement as fertilizer. *Clean Circ Bioeconomy [Internet]*. 2023;6(June):100058. Available from: https://doi.org/10.1016/j.clcb.2023.100058

39. dos Reis GS, Bergna D, Grimm A, Lima EC, Hu T, Naushad M, et al. Preparation of highly porous nitrogen-doped biochar derived from birch tree wastes with superior dye removal performance. *Colloids Surfaces A Physicochem Eng Asp [Internet]*. 2023;669(February):131493. Available from: https://doi.org/10.1016/j.colsurfa.2023.131493

40. Tositti L, Masi G, Morozzi P, Zappi A, Chiara Bignozzi M. Cleaner, sustainable, and safer: Green potential of alkali-activated materials in current building industry, radiological good practice, and a few tips. *Constr Build Mater [Internet]*. 2023;409(October):133879. Available from: https://doi.org/10.1016/j.conbuildmat.2023.133879

41. Li Y, Qi B. Secondary utilization of jujube shell bio-waste into biomass carbon for supercapacitor electrode materials study. *Electrochem Commun*

[Internet]. 2023;152(May):107512. Available from: https://doi.org/10.1016/j.elecom.2023.107512

42. Shaku B, Mofokeng TP, Coville NJ, Ozoemena KI, Maubane-Nkadimeng MS. Biomass valorisation of marula nutshell waste into nitrogen-doped activated carbon for use in high performance supercapacitors. *Electrochim Acta [Internet]*. 2023;442(September 2022):141828. Available from: https://doi.org/10.1016/j.electacta.2023.141828

43. Sajdak M, Majewski A, Di Gruttola F, Gałko G, Misztal E, Rejdak M, et al. Evaluation of the feasibility of using TCR-derived chars from selected biomass wastes and MSW fractions in CO_2 sequestration on degraded and post-industrial areas. *Energies*. 2023;16(7).

44. Aworanti OA, Agbede OO, Agarry SE, Ajani AO, Ogunkunle O, Laseinde OT, et al. Decoding anaerobic digestion: A holistic analysis of biomass waste technology, process kinetics, and operational variables. *Energies*. 2023;16(8):1–36.

45. Goh QH, Zhang L, Ho YK, Chew IML. Modelling and multi-objective optimisation of sustainable solar-biomass-based hydrogen and electricity co-supply hub using metaheuristic-TOPSIS approach. *Energy Convers Manag [Internet]*. 2023;293(May):117484. Available from: https://doi.org/10.1016/j.enconman.2023.117484

46. Chua JY, Pen KM, Poi JV, Ooi KM, Yee KF. Upcycling of biomass waste from durian industry for green and sustainable applications: An analysis review in the Malaysia context. *Energy Nexus [Internet]*. 2023;10(May):100203. Available from: https://doi.org/10.1016/j.nexus.2023.100203

47. Gani A, Erdiwansyah E, Munawar E, Mahidin M, Mamat R, Rosdi SM. Investigation of the potential biomass waste source for biocoke production in Indonesia: A review. *Energy Reports [Internet]*. 2023;10:2417–38. Available from: https://doi.org/10.1016/j.egyr.2023.09.065

48. Morales-Paredes CA, Rodríguez-Linzán I, Saquete MD, Luque R, Osman SM, Boluda-Botella N, et al. Silica-derived materials from agro-industrial waste biomass: Characterization and comparative studies. *Environ Res*. 2023;231(March).

49. Guo Y, Wang Q. Exploring the adsorption potential of Na2SiO3-activated porous carbon materials from waste bamboo biomass for ciprofloxacin rapid removal in wastewater. *Environ Technol Innov [Internet]*. 2023;32:103318. Available from: https://doi.org/10.1016/j.eti.2023.103318

50. Jaiswal RK, Soren S, Jha G. A novel approach of utilizing the waste biomass in the magnetizing roasting for recovery of iron from goethitic iron ore. *Environ Technol Innov [Internet]*. 2023;31:103184. Available from: https://doi.org/10.1016/j.eti.2023.103184

51. Iannello S, Bond Z, Sebastiani A, Errigo M, Materazzi M. Axial segregation behaviour of a reacting biomass particle in fluidized bed reactors: Experimental results and model validation. *Fuel [Internet]*. 2023;338(October 2022):127234. Available from: https://doi.org/10.1016/j.fuel.2022.127234

52. Ikonen E, Liukkonen M, Hansen AH, Edelborg M, Kjos O, Selek I, et al. Fouling monitoring in a circulating fluidized bed boiler using direct and

indirect model-based analytics. *Fuel [Internet]*. 2023;346(March):128341. Available from: https://doi.org/10.1016/j.fuel.2023.128341

53. Adeniyi AG, Iwuozor KO, Emenike EC, Ajala OJ, Ogunniyi S, Muritala KB. Thermochemical co-conversion of biomass-plastic waste to biochar: A review. *Green Chem Eng [Internet]*. 2023;5(1):31–49. Available from: https://doi.org/10.1016/j.gce.2023.03.002

54. Wang K, Tester JW. Sustainable management of unavoidable biomass wastes. *Green Energy Resour [Internet]*. 2023;1(1):100005. Available from: https://doi.org/10.1016/j.gerr.2023.100005

55. Auteri N, Scalenghe R, Saiano F. Phosphorus recovery from agricultural waste via cactus pear biomass. *Heliyon*. 2023;9(9).

56. Zaini HM, Saallah S, Roslan J, Sulaiman NS, Munsu E, Wahab NA, et al. Banana biomass waste: A prospective nanocellulose source and its potential application in food industry—a review. *Heliyon [Internet]*. 2023;9(8):e18734. Available from: https://doi.org/10.1016/j.heliyon.2023.e18734

57. Calvo-Saad MJ, Solís-Chaves JS, Murillo-Arango W. Suitable municipalities for biomass energy use in Colombia based on a multicriteria analysis from a sustainable development perspective. *Heliyon*. 2023;9(10).

58. Dong B, Hu C, Skvaril J, Thorin E, Li H. Selecting the approach for dynamic modelling of CO_2 capture in biomass/waste fired CHP plants. *Int J Greenh Gas Control [Internet]*. 2023;130(May):104008. Available from: https://doi.org/10.1016/j.ijggc.2023.104008

59. Makwana J, Dhass AD, Ramana PV, Sapariya D, Patel D. An analysis of waste/biomass gasification producing hydrogen-rich syngas: A review. *Int J Thermofluids [Internet]*. 2023;20(October):100492. Available from: https://doi.org/10.1016/j.ijft.2023.100492

60. Cuenca-Moyano GM, Cabrera M, López-Alonso M, Martínez-Echevarría MJ, Agrela F, Rosales J. Design of lightweight concrete with olive biomass bottom ash for use in buildings. *J Build Eng [Internet]*. 2023;69(October 2022):106289. Available from: https://doi.org/10.1016/j.jobe.2023.106289

61. Mujtaba M, Fernandes Fraceto L, Fazeli M, Mukherjee S, Savassa SM, Araujo de Medeiros G, et al. Lignocellulosic biomass from agricultural waste to the circular economy: A review with focus on biofuels, biocomposites and bio-plastics. *J Clean Prod [Internet]*. 2023;402(March):136815. Available from: https://doi.org/10.1016/j.jclepro.2023.136815

62. Zeug W, Yupanqui KRG, Bezama A, Thrän D. Holistic and integrated life cycle sustainability assessment of prospective biomass to liquid production in Germany. *J Clean Prod*. 2023;418(March).

63. Khalid U, Rehman Z ur, Ullah I, Khan K, Kayani WI. Efficacy of geopolymerization for integrated bagasse ash and quarry dust in comparison to fly ash as an admixture: A comparative study. *J Eng Res [Internet]*. 2023;(May). Available from: https://doi.org/10.1016/j.jer.2023.08.010

64. Arias A, Feijoo G, Moreira MT. Biorefineries as a driver for sustainability: Key aspects, actual development and future prospects. *J Clean Prod*

[Internet]. 2023;418:137925. Available from: https://www.sciencedirect.com/science/article/pii/S0959652623020838

65. Jimenez-Paz J, Lozada-Castro JJ, Lester E, Williams O, Stevens L, Barraza-Burgos J. Solutions to hazardous wastes issues in the leather industry: Adsorption of Chromium iii and vi from leather industry wastewaters using activated carbons produced from leather industry solid wastes. *J Environ Chem Eng*. 2023;11(3).

66. Sharma S, Asolekar SR, Thakur VK, Asokan P. Valorization of cellulosic fiber derived from waste biomass of constructed wetland as a potential reinforcement in polymeric composites: A technological approach to achieve circular economy. *J Environ Manage [Internet]*. 2023;340(March):117850. Available from: https://doi.org/10.1016/j.jenvman.2023.117850

67. Arregi A, Santamaria L, Lopez G, Olazar M, Bilbao J, Artetxe M, et al. Appraisal of agroforestry biomass wastes for hydrogen production by an integrated process of fast pyrolysis and in line steam reforming. *J Environ Manage [Internet]*. 2023;347(October):119071. Available from: https://doi.org/10.1016/j.jenvman.2023.119071

68. Thakur A, Kumar A, Sharma S, Ganjoo R, Assad H. Computational and experimental studies on the efficiency of Sonchus arvensis as green corrosion inhibitor for mild steel in 0.5 M HCl solution. *Mater Today Proc [Internet]*. 2022;66:609–621. Available from: https://doi.org/10.1016/j.matpr.2022.06.479

69. Thakur A, Sharma S, Ganjoo R, Assad H, Kumar A. Anti-corrosive potential of the sustainable corrosion inhibitors based on biomass waste: A review on preceding and perspective research. *J Phys Conf Ser*. 2022;2267(1): 012079.

70. Thakur A, Kaya S, Kumar A. Recent innovations in nano container-based self-healing coatings in the construction industry. *Curr Nanosci*. 2021;18(2):203–216.

71. Kumar A, Thakur A. *Encapsulated Nanoparticles in Organic Polymers for Corrosion Inhibition [Internet]*. *Corrosion Protection at the Nanoscale*. Elsevier Inc.; 2020, 345–362. Available from: http://dx.doi.org/10.1016/B978-0-12-819359-4.00018-0

72. Ibarra-Esparza FE, González-López ME, Ibarra-Esparza J, Lara-Topete GO, Senés-Guerrero C, Cansdale A, et al. Implementation of anaerobic digestion for valorizing the organic fraction of municipal solid waste in developing countries: Technical insights from a systematic review. *J Environ Manage*. 2023;347(January).

73. Ibarra-Esparza FE, González-López ME, Ibarra-Esparza J, Lara-Topete GO, Senés-Guerrero C, Cansdale A, et al. Implementation of anaerobic digestion for valorizing the organic fraction of municipal solid waste in developing countries: Technical insights from a systematic review. *J Environ Manage*. 2023;347(September).

74. Babisk MP, Sales Barreto GN, Bornili Gadioli MC, Girondi Delaqua GC, Monteiro SN, Fontes Vieira CM. Incorporation of Eichhornia crassipes dry

biomass in red ceramics aiming at energy savings during the firing stage. *J Mater Res Technol*. 2023;25:522–531.

75. Mohamed AM, Tayeh BA, Abu Aisheh YI, Salih MNA. Utilising olive-stone biomass ash and examining its effect on green concrete: A review paper. *J Mater Res Technol [Internet]*. 2023;24:7091–7107. Available from: https://doi.org/10.1016/j.jmrt.2023.05.039

76. Hassan M, Mohanty AK, Misra M. 3D Printing in upcycling plastic and biomass waste to sustainable polymer blends and composites: A review. *Mater Des [Internet]*. 2023;237(November 2023):112558. Available from: https://doi.org/10.1016/j.matdes.2023.112558

77. Onfray C, Thiam A. Biomass-derived carbon-based electrodes for electrochemical sensing: A review. *Micromachines*. 2023;14(9).

78. Bennett JA, Ellett K, Middleton R, Winter S, Blumer E. Preliminary life cycle assessment of a net-zero power plant co-fired with waste coal and biomass. *Procedia CIRP [Internet]*. 2023;116:19–22. Available from: https://doi.org/10.1016/j.procir.2023.02.004

79. De la Torre-Bayo JJ, Zamorano M, Torres-Rojo JC, Rodríguez ML, Martín-Pascual J. Analyzing the production, quality, and potential uses of solid recovered fuel from screening waste of municipal wastewater treatment plants. *Process Saf Environ Prot [Internet]*. 2023;172(November 2022):950–970. Available from: https://doi.org/10.1016/j.psep.2023.02.083

80. Verma C, Thakur A, Ganjoo R, Sharma S, Assad H, Kumar A, et al. Coordination bonding and corrosion inhibition potential of nitrogen-rich heterocycles: Azoles and triazines as specific examples. *Coord Chem Rev [Internet]*. 2023;488(April):215177. Available from: https://doi.org/10.1016/j.ccr.2023.215177

81. Thakur A, SAVAŞ K, Kumar A. Recent trends in the characterization and application progress of nano-modified coatings in corrosion mitigation of metals and alloys. *Appl Sci*. 2023;13:730.

82. Thakur A, Kumar A, Kaya S, Benhiba F, Sharma S. Electrochemical and computational investigations of the Thysanolaena latifolia leaves extract: An eco-benign solution for the corrosion mitigation of mild steel. *Results Chem [Internet]*. 2023;6(September):101147. Available from: https://doi.org/10.1016/j.rechem.2023.101147

83. Kaya S, Lgaz H, Thakkur A, Kumar A, Özbakır D, Karakuş N, et al. Molecular insights into the corrosion inhibition mechanism of omeprazole and tinidazole: A theoretical investigation. *Mol Simul*. 2023;1–15.

84. Thakur A, Kumar A. Exploring the potential of ionic liquid-based electrochemical biosensors for real-time biomolecule monitoring in pharmaceutical applications: From lab to life. *Results Eng [Internet]*. 2023;20(July):101533. Available from: https://doi.org/10.1016/j.rineng.2023.101533

85. Dhonchak C, Agnihotri N. Computational Insights in the spectrophotometrically 4H-chromen-4-one complex using DFT method. *Biointerface Res Appl Chem*. 2023;13(4):357.

86. Thakur A, Sharma S, Ganjoo R, Assad H, Kumar A. Anti-corrosive potential of the sustainable corrosion inhibitors based on biomass waste: A review on preceding and perspective research. *J Phys Conf Ser.* 2022;2267(1).

87. Phang FJF, Soh M, Khaerudini DS, Timuda GE, Chew JJ, How BS, et al. Catalytic wet torrefaction of lignocellulosic biomass: An overview with emphasis on fuel application. *South African J Chem Eng [Internet].* 2023;43(October 2022):162–189. Available from: https://doi.org/10.1016/j.sajce.2022.10.008

88. Romero-Perdomo F, González-Curbelo MÁ. Integrating multi-criteria techniques in life-cycle tools for the circular bioeconomy transition of agri-food waste biomass: A systematic review. *Sustain.* 2023;15(6).

89. Waghmare PR, Patil SM, Jadhav SL, Jeon BH, Govindwar SP. Utilization of agricultural waste biomass by cellulolytic isolate Enterobacter sp. SUK-Bio. *Agric Nat Resour [Internet].* 2018;52(5):399–406. Available from: https://doi.org/10.1016/j.anres.2018.10.019

90. Suárez-Macías J, Terrones-Saeta JM, Corpas-Iglesias FA, Iglesias-Godino FJ. Study of the incorporation of biomass bottom ash as a filler for discontinuous grading bituminous mixtures with bitumen emulsion. *Appl Sci.* 2021;11(8).

91. Sun W, Sun Y, Hong X, Zhang Y, Liu C. Research on biomass waste utilization based on pollution reduction and carbon sequestration. *Sustain.* 2023;15(5).

92. Novotný M, Marković M, Raček J, Šipka M, Chorazy T, Tošić I, et al. The use of biochar made from biomass and biosolids as a substrate for green infrastructure: A review. *Sustain Chem Pharm [Internet].* 2023;32:100999. Available from: https://www.sciencedirect.com/science/article/pii/S2352554123000335

93. Shi Z, Ferrari G, Ai P, Marinello F, Pezzuolo A. Artificial intelligence for biomass detection, production and energy usage in rural areas: A review of technologies and applications. *Sustain Energy Technol Assessments [Internet].* 2023;60(November):103548. Available from: https://doi.org/10.1016/j.seta.2023.103548

94. Sudiyani Y, Styarini D, Triwahyuni E, Sudiyarmanto, Sembiring KC, Aristiawan Y, et al. Utilization of biomass waste empty fruit bunch fiber of palm oil for bioethanol production using pilot-scale unit. *Energy Procedia [Internet].* 2013;32:31–38. Available from: http://dx.doi.org/10.1016/j.egypro.2013.05.005

95. Thakur A, Kumar A. Unraveling the multifaceted mechanisms and untapped potential of activated carbon in remediation of emerging pollutants: A comprehensive review and critical appraisal of advanced techniques. *Chemosphere [Internet].* 2024;346(November 2023):140608. Available from: https://doi.org/10.1016/j.chemosphere.2023.140608

96. Sharma D, Thakur A, Sharma MK, Jakhar K, Kumar A, Sharma AK, Hari OM. Synthesis, electrochemical, morphological, computational and

corrosion inhibition studies of 3-(5-Naphthalen-2-yl-[1,3,4]oxadiazol-2-yl)-pyridine against mild steel in 1 M HCl. *Asian J Chem.* 2023;35(5):1079–1088.

97. Sharma D, Thakur A, Kumar M, Sharma R, Kumar S, Om H. Effective corrosion inhibition of mild steel using novel 1, 3, 4-oxadiazole-pyridine hybrids: Synthesis, electrochemical, morphological, and computational insights. *Environ Res [Internet].* 2023;234(July):116555. Available from: https://doi.org/10.1016/j.envres.2023.116555

98. Thakur A, Kumar A. Ecotoxicity analysis and risk assessment of nanomaterials for the environmental remediation. *Macromol Symp.* 2023;410(1):1–23.

99. Arifa Farzana B, Mujafarkani N, Thakur A, Kumar A, Mushira Banu A, Shifana M. Evaluating (p-Semidine-Guanidine-Formaldehyde) terpolymer resin efficiency as anti-corrosive agent for mild steel in 1 M H2SO4: An experimental and computational approach. *Inorg Chem Commun [Internet].* 2023;158(P1):111572. Available from: https://doi.org/10.1016/j.inoche.2023.111572

100. Thakur A, Kumar A. Recent trends in nanostructured carbon-based electrochemical sensors for the detection and remediation of persistent toxic substances in real-time analysis. *Mater Res Express.* 2023;10(3).

101. Thakur A, Kumar A, Singh A. Adsorptive removal of heavy metals, dyes, and pharmaceuticals: Carbon-based nanomaterials in focus. *Carbon N Y [Internet].* 2023;217(November 2023):118621. Available from: https://doi.org/10.1016/j.carbon.2023.118621

102. Kaya S, Thakur A, Kumar A. The role of in Silico/DFT investigations in analyzing dye molecules for enhanced solar cell efficiency and reduced toxicity. *J Mol Graph Model.* 2023;124(May).

103. Pazzaglia A, Gelosia M, Giannoni T, Fabbrizi G, Nicolini A, Castellani B. Wood waste valorization: Ethanol based organosolv as a promising recycling process. *Waste Manag [Internet].* 2023;170(February):75–81. Available from: https://doi.org/10.1016/j.wasman.2023.08.003

104. Kabir Ahmad R, Anwar Sulaiman S, Yusup S, Sham Dol S, Inayat M, Aminu Umar H. Exploring the potential of coconut shell biomass for charcoal production. *Ain Shams Eng J [Internet].* 2022;13(1):101499. Available from: https://doi.org/10.1016/j.asej.2021.05.013

105. Wojewódzki P, Lemanowicz J, Debska B, Haddad SA, Tobiasova E. The application of biochar from waste biomass to improve soil fertility and soil enzyme activity and increase carbon sequestration. *Energies.* 2023;16(1).

106. Liu S, Zhu Y, Liao Y, Wang H, Liu Q, Ma L, et al. Advances in understanding the humins: Formation, prevention and application. *Appl Energy Combust Sci [Internet].* 2022;10(February):100062. Available from: https://doi.org/10.1016/j.jaecs.2022.100062

107. Ochieng R, Gebremedhin A, Sarker S. Integration of waste to bioenergy conversion systems: A critical review. *Energies.* 2022;15(7).

108. Gutiérrez Ortiz FJ. Biofuel production from supercritical water gasification of sustainable biomass. *Energy Convers Manag X.* 2022;14(December 2021).

CHAPTER 2

Biomass Wastes

Natural and Biological Origin, Processing, Sustainability, and Recycling

Khasan Berdimuradov, Elyor Berdimurodov, Abhinay Thakur, Ashish Kumar, Husan Yaxshinorov, Akbarali Rasulov and Laziz Azimov

2.1 INTRODUCTION

Biomass wastes are organic materials from plants and animals produced as a byproduct of forestry, farming, and industrial processes and are not used for food or feed. When utilized properly, these materials can be considered a renewable energy source since they contain solar energy saved during photosynthesis. The term "biomass wastes" refers to a broad range of commodities, including sawdust and wood from forestry operations, agricultural residues like straw and husk, food scraps and yard trimmings from municipal solid waste, animal slurry, and manure. The idea of biomass wastes includes solid and liquid organic materials that can be transformed via various procedures into energy or other beneficial products [1, 2]. These procedures can be as basic as burning fuel to generate heat or as complex as thermochemical and biochemical conversions to create a variety of biofuels, biogases, and biochemicals. In a circular economy, where the goal is to minimize waste and maximize the usefulness of all commodities, biomass wastes are an essential resource. By managing them and turning them into useful resources, they can help promote sustainable development by lowering dependency on fossil fuels, lowering greenhouse

DOI: 10.1201/9781003466833-3

gas emissions, and boosting the economy in rural regions. The importance of biomass wastes comes from their ability to support a robust and sustainable bioeconomy in addition to their capacity to generate energy and lessen environmental effects. Biomass wastes present a viable route in the global effort to shift toward more environmentally friendly energy sources and lessen the impact of human activity on the environment [3, 4]. They serve as a link between current waste management techniques and the sustainable and resource-efficient manufacturing of materials, chemicals, and energy of the future.

For a number of reasons, the study of biomass wastes is critical to both economic growth and environmental sustainability. Resource management, energy production, and environmental protection may all significantly progress with an understanding of and utilization of biomass wastes. Biofuels, biogas, and biomass pellets are just a few of the renewable energy sources that may be created from biomass waste. These energy sources are essential for lowering dependency on fossil fuels, which are limited and a significant source of greenhouse gas emissions worldwide. Through the study of biomass wastes, we may enhance conversion technologies and increase the economic and technological efficiency of producing renewable energy. Effective management solutions are essential as the amount of waste generated worldwide keeps growing. Research on biomass waste provides information on ways to keep organic waste streams out of landfills, where they would otherwise break down anaerobically and release the potent greenhouse gas methane [5, 6]. Biomass wastes can be converted into useful resources by procedures like anaerobic digestion and composting, which lowers the overall amount of waste produced and its negative effects on the environment. Researching biomass wastes helps to create mitigation methods for the harmful environmental effects of waste accumulation and improper handling. We can stop soil, water, and air pollution, maintain biodiversity, and shield ecosystems from the damage brought on by conventional waste disposal techniques by identifying sustainable uses for biomass wastes. Opportunities for economic growth are presented by the biomass waste industry, especially in rural areas where the majority of the trash is produced. By generating new businesses and technological advancements, the processing of biomass wastes can boost regional economies and generate employment at facilities that handle collection, processing, and conversion. A circular economy minimizes the demand for new raw materials by reusing, recycling, or converting waste resources into various kinds of value. Researching biomass wastes is essential to developing these ideas. The aforementioned approach fosters sustainability and resource efficiency by guaranteeing that organic resources are fully utilized prior to disposal. The production of energy from biomass wastes and other uses can make a substantial contribution to the fight against climate change. Biomass waste management and conversion technologies can assist in lowering the overall carbon footprint of energy consumption and waste management procedures by substituting fossil fuels and lowering methane emissions from waste [7, 8]. Advanced biofuels, bioplastics, and biochemicals are just a few of the valuable goods that may be produced from trash thanks to new technologies and procedures developed as a result of research into biomass wastes. This encourages the development of new technology and the uptake of environmentally friendly ones. In conclusion, researching biomass wastes is critical to accomplishing environmental issues, advancing sustainable development, and creating new avenues for economic expansion. It embodies a comprehensive strategy for resource management that satisfies human needs and ambitions for a sustainable future while respecting the ecological constraints of the earth.

2.2 SOURCES OF BIOMASS WASTES

Table 2.1 lists the natural origins of biomass wastes, with an emphasis on forestry residues, aquatic biomass, and agricultural residues. Each category is unique in terms of where it comes from, the kinds of waste it produces, and the possible uses for these resources. Following the harvest of the main crop, agricultural leftovers are the leftover byproducts. These include plant parts such stems, leaves, husks, and shells that are not used for food or feed. They are produced by processing operations as well as in the field.

Common examples include bagasse, the fibrous residue left over from sugarcane after the juice is extracted, corn stover (leaves and stalks of maize plants), straw from cereals like wheat and rice, and shells from nuts like peanuts and coconuts. These residues have a number of potential uses, such as gasification or combustion to produce bioenergy, feedstock for anaerobic digestion to produce biofuels (such as ethanol and biodiesel), composting processes, animal feed (especially for residues with nutritional value), and the production of natural fiber-based materials (like biodegradable pots and packaging) [9, 10].

TABLE 2.1   Sources of Biomass Wastes of Natural Origin, Including Agricultural Residues, Forestry Residues, and Aquatic Biomass

Source	Description	Examples	Potential Uses
Agricultural Residues	These are by-products generated from farming activities after the primary crop has been harvested. They include materials left in the field or processing residues.	Straw, husk, corn stover, bagasse, and shells	Bioenergy production, biofuels, biogas, composting, animal feed, and natural fiber-based materials.
Forestry Residues	Residues generated from forestry operations include materials left over from logging activities and the processing of wood. This category also covers tree thinning and underbrush from forest management activities.	Sawdust, bark, branches, and wood chips	Biomass pellets, bioenergy, paper production, construction materials, and mulch.
Aquatic Biomass	This refers to organic material derived from aquatic environments, both freshwater and marine. It includes algae, seaweed, and aquatic plants, as well as residues from the aquaculture industry.	Algae, seaweed, aquatic plants	Biofuels (especially biodiesel from algae), bioplastics, pharmaceuticals, dietary supplements, and animal feed.

Forestry residues are the organic components that remain after logging and wood processing, as well as after forest management tasks like underbrush removal and thinning. Sawdust, bark, branches, wood chips, and trimmings fall under this category. Forestry leftovers are utilized as mulch for landscaping and soil conservation, as well as for the manufacturing of biomass pellets, a renewable energy source, bioenergy, paper, and construction materials.

Organic elements from water settings, including freshwater and marine sources, comprise aquatic biomass. It includes bigger plants like seaweed and microorganisms like algae. It can also contain leftovers from aquaculture, which is the rearing of aquatic creatures. Important examples include seaweed, algae, and different aquatic plants. Particularly noteworthy are algae's fast growth rates and high oil content, which make them ideal for the manufacture of biofuel. Due to the oil content of algae, aquatic biomass has a variety of uses, most notably in the production of biofuels, particularly biodiesel. Additionally, it's utilized in the production of bioplastics, medications (algae and seaweed include chemicals with medical characteristics), dietary supplements (algae contain omega-3 fatty acids), and aquaculture feed [11, 12].

In conclusion, investigating and making use of these biomass waste sources is critical to the development of waste reduction, renewable and sustainable energy solutions, and the advancement of a circular economy. Every source is different and has advantages of its own, ranging from lowering greenhouse gas emissions to offering sustainable substitutes for conventional materials and fossil fuels.

The process of turning organic materials into heat, electricity, or different types of biofuels (solid, liquid, or gaseous) through thermochemical or biochemical processes is known as bioenergy generation from biomass. The simplest and least effective technique of this process is direct burning. This renewable energy cycle provides a response to energy demands by balancing the absorption and re-emission of CO_2 through biomass burning and plant growth. Biofuels, which are especially important for the transportation industry, are a green substitute that can be generated anywhere in the world by taking advantage of the forestry and agriculture potential of places like Latin America. The division of biofuels into first, second, and third generations according to the source of their biomass emphasizes the significance of aquatic organisms such as algae and non-edible lignocellulosic biomass for the production of biofuels in a sustainable manner. Particular attention is paid to second-generation biofuels since they offer an economically feasible and scalable alternative due to their reduced resource requirements, little environmental effect, and lack of competition with food sources. In rising economies, where there is a growing need for chemical products and biofuels, lignocellulosic biomass has the potential to be an abundant and low-cost feedstock. This potential is reinforced by environmental regulations that aim to reduce reliance on petroleum and promote rural development. Technical-economic evaluations and instruments such as life cycle assessments are used to analyze economic competitiveness, environmental consequences, and social sustainability aspects like food and energy security, job possibilities, and health concerns in order to determine how sustainable the production of biofuels is [13]. This all-encompassing strategy highlights the world's transition to sustainable energy sources and points to a bright future for the biofuel industry and its ability to contribute to the attainment of social, economic, and environmental sustainability objectives.

Animal byproducts, municipal organic waste, and industrial organic waste are just a few sources of the diverse range of organic elements that

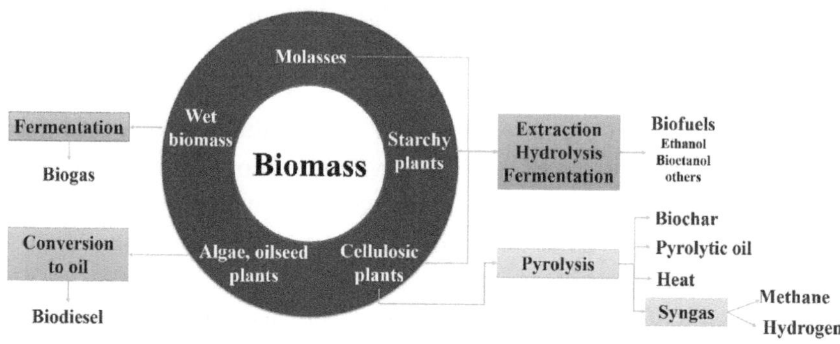

FIGURE 2.1   Different types of biofuel products from lignocellulosic biomass [13].

make up biomass wastes of biological origin, each having its own special qualities and possible applications (Figure 2.1 and Table 2.2). Animal by-products are parts of animals that are not edible, such as bones, blood, feathers, and manure from farming and meat processing. These by-products can be processed into organic fertilizers, bioenergy (such as biogas), and animal feed supplements and can even be used to make some chemicals and medicines. Municipal organic waste provides opportunities for composting, anaerobic digestion to produce biogas and biofertilizers, and energy recovery through sophisticated incineration techniques. Biodegradable materials such as food scraps, yard waste, paper, and cardboard are collected from residential, commercial, and public services. Industrial organic waste can be used to produce bioenergy, including biofuels and biogas, industrial composting, soil amendments, and as raw materials for bio-based products [14, 15]. This trash comes from the food processing, brewing, paper, and other industries. When taken as a whole, these biological origin biomass waste sources demonstrate the possibility for environmentally friendly, energy-producing, and circular economy-promoting sustainable waste management techniques.

2.3 PROCESSING OF BIOMASS WASTES

2.3.1 Bioprocessing of Waste Biomass

Key elements of bioprocessing waste biomass are outlined in Table 2.3 and Figure 2.2, along with methodologies, financial concerns, environmental effects, and future directions. The two main methods, direct microbial conversion and enzymatic bioconversion, use cellulases and other particular enzymes along with bacterial and fungal strains to break down

TABLE 2.2 Sources of Biomass Wastes of Biological Origin, Including Animal By-Products, Municipal Organic Waste, and Industrial Organic Waste

Source	Description	Examples	Potential Uses
Animal By-Products	These are the non-edible parts of animals generated from agricultural activities, slaughterhouses, and meat processing plants. They include materials that are not used for human consumption but can still be processed for other valuable uses.	Bones, blood, feathers, manure, and meat processing waste.	Bioenergy production (biogas), organic fertilizers, animal feed supplements, and in the production of certain chemicals and pharmaceuticals.
Municipal Organic Waste	Organic waste collected from residential, commercial, and public services. It primarily consists of biodegradable waste, including food waste, yard trimmings, and paper products. This type of waste is a significant component of municipal solid waste (MSW).	Food scraps, yard waste (leaves, grass clippings), paper, and cardboard.	Composting, anaerobic digestion to produce biogas and biofertilizers, and energy recovery through incineration with energy capture.
Industrial Organic Waste	Organic waste generated from industrial processes, excluding those from the agricultural and forestry sectors. This includes waste from food processing industries, breweries, paper manufacturing, and other sectors where organic materials are a byproduct.	Food processing residues, pulp and paper mill sludge, effluents from breweries and distilleries.	Bioenergy (biofuels, biogas), industrial composting, soil amendments, and as raw materials for the production of bio-based products.

biomass into useful components. The techno-economic elements highlight the necessity of cost control and profit maximization while concentrating on the bioprocessing industry's financial sustainability [16]. Environmental impact assessments emphasize how important sustainable waste management is to lowering greenhouse gas emissions, especially when done through life-cycle assessment (LCA). To improve sustainability and efficiency, future perspectives emphasize the necessity for developing bioprocessing technologies, such as genetic alterations and affordable enzyme synthesis. Using both direct microbial conversion and enzymatic bioconversion procedures, bioprocessing approaches for biomass waste valorization use naturally occurring microorganisms to convert resistant biomass structures into value-added products. These techniques make use

TABLE 2.3 The Key Aspects of Bioprocessing of Waste Biomass

Bioprocessing Aspect	Details
Direct Microbial Conversion	Utilizes various bacterial and fungal strains for biomass degradation. Examples include Streptomyces sp., Bacillus sp., Clostridium sp. for bacteria, and Trichoderma reesei, Phanerochaete chrysosporium for fungi.
Enzymatic Bioconversion	Employs specific enzymes for the breakdown of biomass into fermentable sugars. Common enzymes include cellulases, ligninolytic enzymes, and hemicellulases.
Techno-Economic Aspects	Addresses capital and operating costs, including fermenter costs, raw material costs, labor, utilities, and waste treatment. Profitability analysis is crucial for assessing economic viability.
Environmental Impact	Life-cycle assessment (LCA) evaluates the environmental impacts of biomass waste management, including GHG emissions from mismanaged biomass waste and the potential reduction through proper bioprocessing.
Future Perspectives	Highlights the need for optimized bioprocessing pathways, cost-effective enzyme production, and advanced microbial strains through genetic modifications and omics approaches for improved efficiency and sustainability.

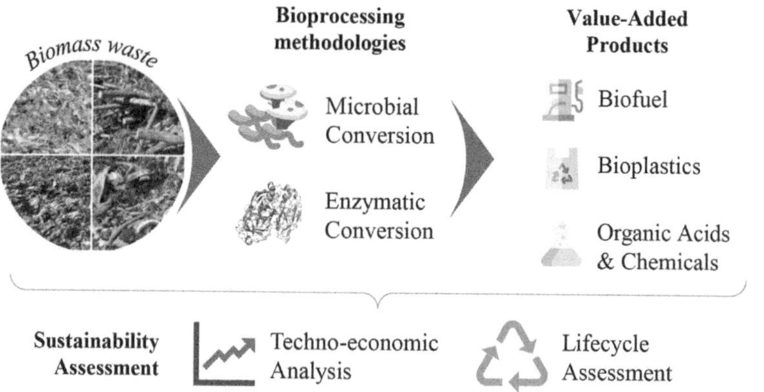

FIGURE 2.2 General comparisons in bioprocessing of waste biomass [16].

of a range of bacterial and fungal strains that are able to produce particular enzymes, like hydrolytic and ligninolytic enzymes, which decompose cellulose and hemicellulose into sugars that can be fermented. It is essential to perform this pretreatment step because it increases the biomass's susceptibility to subsequent processes like fermentation and hydrolysis, which

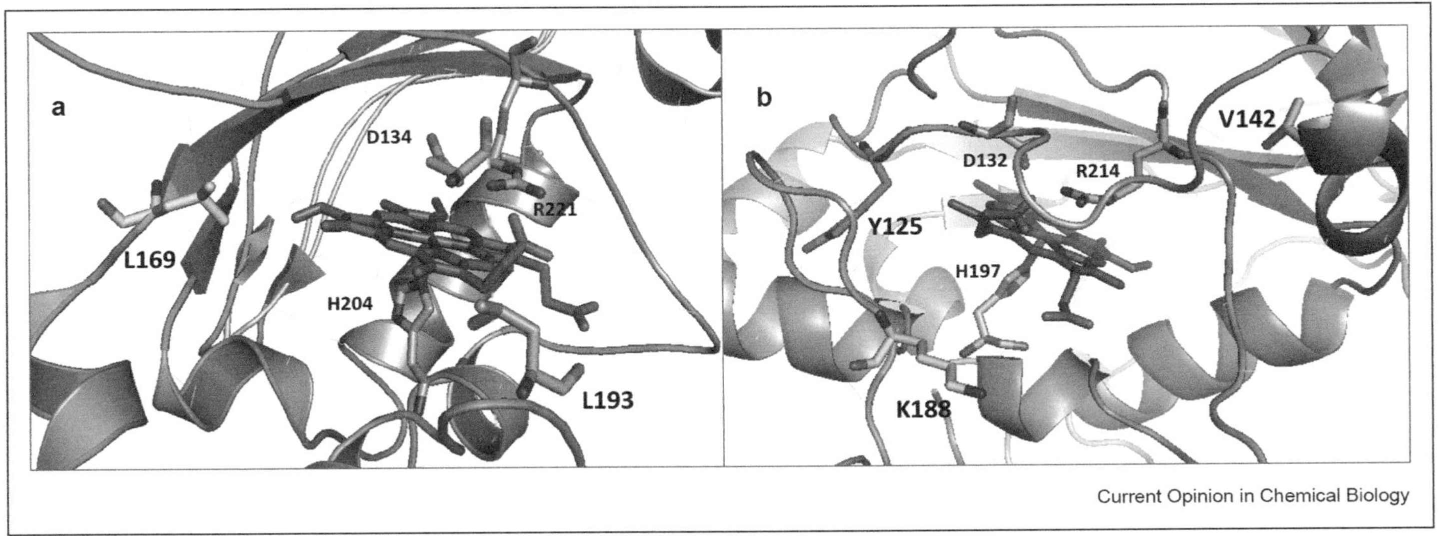

FIGURE 2.3   The locations of residues that increase the activity of designed DyP-type peroxidases and their structures. a. Focused libraries were utilized to identify residues (L169 and L193) in Pseudomonas fluorescens Dyp1B. b. Y125, V142, and K188 residues identified by error-prone polymerase chain reaction in Pseudomonas putida DyP [17].

produce organic acids, biofuels, and other bioproducts. By concentrating on turning waste into valuable resources, the method is environmentally beneficial and supports a circular economy. The biotechnological developments in this subject are highlighted by the application of particular microbial strains and enzymes, which enable the effective breakdown of biomass waste. By addressing waste management concerns and providing a route to the production of renewable energy and chemicals, this strategy highlights the significance of improving these bioprocessing methods for long-term sustainability and financial feasibility.

It draws attention to the identification of auxiliary enzymes that capture phenoxy radicals to prevent the unwanted dimerization of lignin, such as dihydrolipoamide dehydrogenase from T. fusca. It also discusses the identification of oxidase enzymes that are essential for producing hydrogen peroxide, which is required for peroxidase enzymes that oxidize lignin. It also explores protein engineering, presenting mutations in bacterial DyP-type peroxidase enzymes that greatly increase their specificity and catalytic effectiveness for molecules generated from lignin, opening the door to more effective lignin conversion into useful compounds (Figure 2.3) [17].

2.3.2 Mechanical and Thermal Processing of Biomass Wastes

The mechanical processing of waste biomass includes techniques to get the biomass ready for more effective energy conversion (Table 2.4). Size reduction—which includes milling, grinding, and chipping—is essential to boosting the surface area of the biomass and, consequently, its reactivity in later processes. This first stage is energy-intensive, which presents a problem in terms of equipment wear and operating expenses, but it also makes the material easier to handle and carry. Pelletizing and briquetting, which follow size reduction, densify the biomass into pellets and briquettes, respectively, enhancing storage, transit effectiveness, and energy density [18–20]. However, these procedures could call for the use of binders, which could introduce undesirable pollutants into the biomass, and they also demand large initial investments in technology.

In terms of thermal processing, pyrolysis, gasification, and combustion convert biomass waste into energy and useful products (Table 2.4). Pyrolysis yields bio-oil, syngas, and biochar at high temperatures without the presence of oxygen, providing a broad spectrum of products from a single process. Pyrolysis has advantages, but it also requires careful control over operating conditions, and the bio-oil that is produced may require additional refining. Gasification, on the other hand, effectively creates a

TABLE 2.4 Clearly Distinguish between Mechanical and Thermal Processing Methods of Biomass Wastes, Their Descriptions, Advantages, and Challenges

Processing Type	Method	Description	Advantages	Challenges
Mechanical	Size reduction	Decreasing the physical size of biomass through chipping, grinding, and milling.	Increases surface area for further processing. Improves homogeneity for handling and transportation.	High energy consumption. Equipment wear and tear.
	Pelletizing and briquetting	Densifying biomass into compact shapes through extrusion (pellets) or compression (briquettes).	Enhanced storage and transport efficiency. Higher energy density for efficient combustion.	Initial investment in equipment. Potential need for binders, which may introduce contaminants.
Thermal	Pyrolysis	Decomposing organic material at high temperatures in the absence of oxygen to produce bio-oil, syngas, and biochar.	Produces a variety of products. Can process diverse biomass types.	Requires careful control of process conditions. Bio-oil may require further refining.
	Gasification	Converting biomass into a combustible gas mixture through partial oxidation at high temperatures.	Converts biomass to usable gas efficiently. Syngas can be used for electricity or synthetic fuel production.	High temperatures and control systems needed. Tar production necessitates gas cleaning.
	Combustion	Burning biomass in the presence of oxygen to produce heat and power.	Direct conversion into heat and power. Can handle a variety of feedstocks.	Lower efficiency compared to other thermal processes. Emission control for pollutants is necessary.

flexible syngas fit to produce synthetic fuels and power by converting biomass into a flammable gas combination. To reduce the amount of tar produced, this process requires high temperatures and complex management systems. Finally, combustion is the process of burning biomass directly to produce heat and electricity. It is a simple method, but it has drawbacks, including poorer efficiency and the requirement for strict emission controls to reduce environmental damage [21, 22]. Together, these thermal and mechanical procedures provide a thorough method for managing biomass waste; each has unique benefits and difficulties and helps to ensure the production of energy in a sustainable manner.

The usage of renewable biomass resources for the production of bioenergy and biomaterials is based on the critical role that mechanical size reduction plays in the conversion of biomass feedstock into energy and polymer biomaterials (Figure 2.4). The procedure, which includes several techniques such as cutting, crushing, ultra-fine grinding, and nanogrinding, considerably improves the productivity and potency of biomass conversion technologies. Size reduction improves the volume calorific value, which not only makes densification procedures easier but also optimizes the raw material supply chain and storage conditions for biomass. Moreover, it increases the biomass's overall surface area that is accessible, which enhances the components' bioaccessibility. This improvement in accessibility is essential for minimizing the mass and heat transfer limits encountered during these reactions as well as for the effective conversion of saccharides during hydrolysis. As a result, the energy inputs needed for biomass processing are reduced, highlighting the many advantages of size reduction in the bio-based product industry. The application of various mechanical stressors, such as impact, compression, friction, and shear, determines how effective mechanical size reduction is. These stresses play a crucial role in attaining the intended reduction in biomass size and can coexist in commercial equipment such as centrifugal mills, extruders, hammer mills, pin mills, and jet mills. All types of mills and grinders are designed to produce particular stresses, such as shear forces in extruders and impact and friction in jet mills, in order to meet different needs related to biomass processing. The physical and chemical characteristics of the biomass, its moisture content, the intended final particle size, and the particular application targets all have an impact on the equipment selection. Notably, some mills work well with wet materials that have a greater moisture content, while others work well with dry biomass [23, 24]. Without sacrificing the materials' inherent qualities, technologies such as fluidized bed superfine grinders have shown to be extremely effective at increasing the surface area and bioavailability of materials. Mechanical size reduction techniques play a crucial role in the production of bio-based goods due to their variety and adaptability, which enable the conversion of renewable biomass resources into valuable materials and energy.

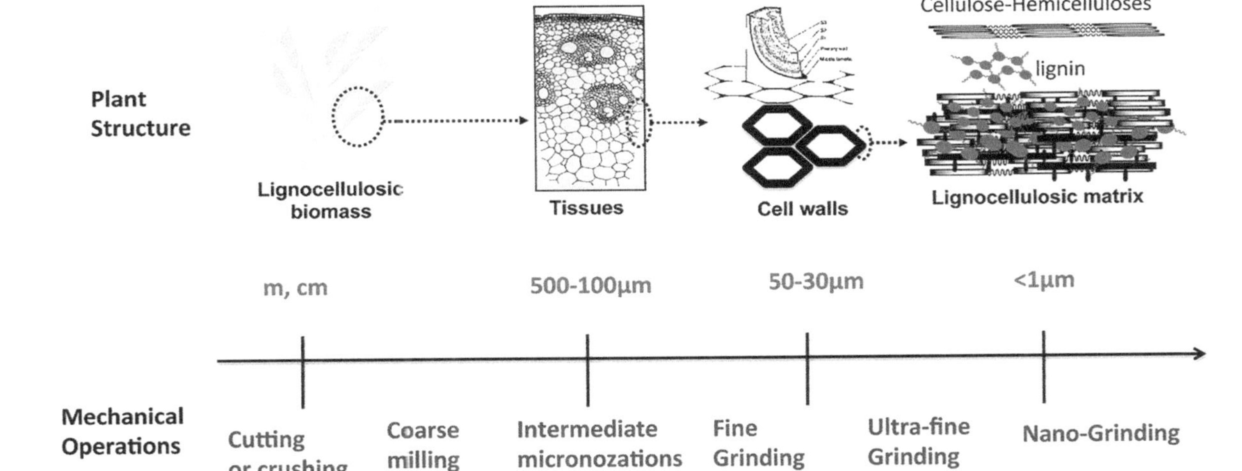

FIGURE 2.4   (a) The many mechanical processes used to reduce the size of components associated with plant construction; (b) a schematic illustration of a few pieces of commercial milling machinery showing the various mechanical stresses that are produced [25].
(*Continued*)

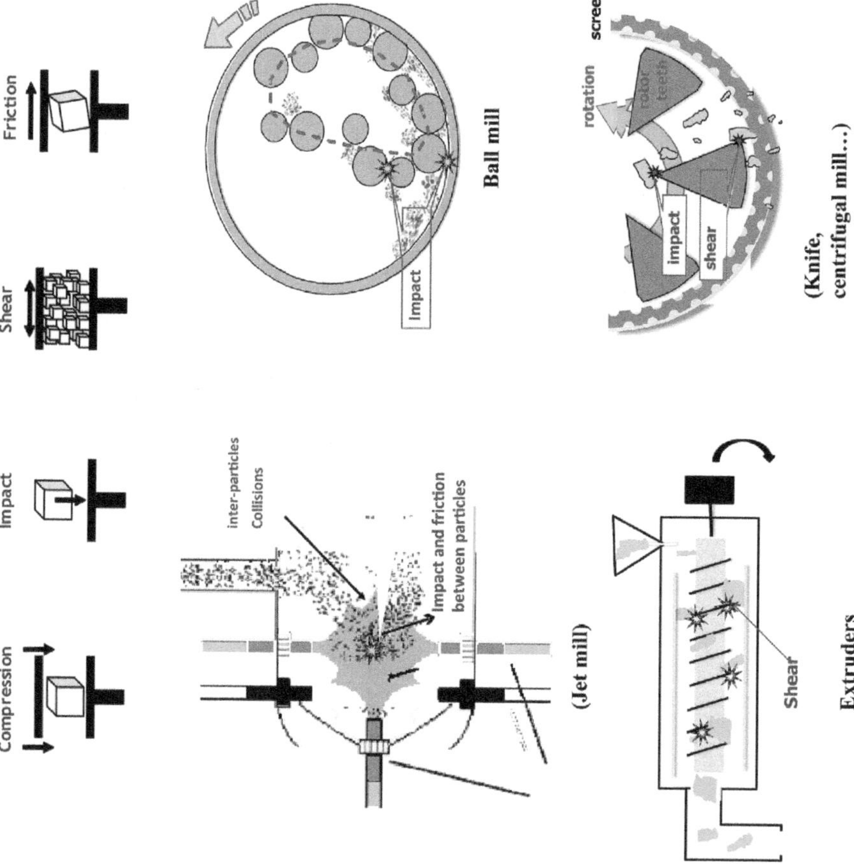

FIGURE 2.4 (Continued)

2.4 SUSTAINABILITY CONSIDERATIONS, RECYCLING, AND UTILIZATION OF BIOMASS WASTES

Biorefineries, which strategically use renewable biomass to produce biofuels, chemicals, and electricity, have a multidimensional influence that is highlighted by the sustainable development framework within the bioeconomy and circular economy (Table 2.5). This strategy shows a promising development in the mitigation of climate change as it considerably lowers greenhouse gas emissions. In addition, it guarantees biodiversity protection and encourages sustainable land-use practices, upholding ecological equilibrium [26, 27]. From an economic perspective, biorefineries help make bio-based enterprises more viable and competitive by making use of cheap and readily accessible biomass. They also help to capitalize on the expanding market for biofuels, especially bioethanol, which is expected to develop significantly by 2050. Socially, the move away from fossil fuels and toward renewable biomass resources strengthens energy security

TABLE 2.5 Environmental Impact, Economic Aspects, and Social Implications Associated with the Implementation of Biorefineries to Produce Biofuels and Energy from Renewable Biomass Resources

Category	Subcategory	Details	Implications
Environmental Impact	**Greenhouse gas emissions**	Reducing GHG through renewable biomass utilization.	Positive impact on climate change mitigation.
	Biodiversity and land use	Sustainable biomass sourcing to preserve biodiversity and land.	Minimizes ecological footprint and supports ecosystem health.
Economic Aspects	**Cost-effectiveness**	Biorefineries' strategic role in the bioeconomy, leveraging low-cost, highly available biomass.	Enhances economic viability and competitiveness of bio-based industries.
	Market potential	Growth potential in biofuel markets, with bioethanol expected to significantly increase by 2050.	Opportunities for expansion and innovation in renewable energy sectors.
Social Implications	**Energy security**	Renewable biomass as an alternative to fossil fuels, enhancing energy independence.	Contributes to national and global energy security.
	Rural development and employment	Creating manufacturing jobs and fostering regional development through biomass utilization.	Stimulates socioeconomic growth and stability in rural areas.

and lessens reliance on non-renewable energy sources [13]. Additionally, it creates jobs and stimulates rural development, which promotes socio-economic stability and progress in rural areas. When taken as a whole, these components show the many advantages of incorporating sustainable practices into the exploitation of biomass and emphasize the critical role that biorefineries play in promoting social welfare, economic progress, and environmental sustainability.

The sustainability of using biomass to generate power is a complex topic that takes into account a number of important factors, including cost, effectiveness, greenhouse gas emissions, availability, constraints, land and water use, and social effects. Although power generated from biomass generally has advantages in terms of cost, effectiveness, emissions, availability, and constraints, it also has serious drawbacks due to excessive water and land use and negative social effects [28]. The type of biomass and the location of its production have a significant impact on how sustainable biomass is as a source of electricity. Growing resilient crops on marginal or underutilized land and making use of waste products are seen to be more sustainable methods than planting specific energy crops on arable land that would otherwise be good for food production, particularly if high fertilizer rates are used. Because biomass is renewable and can replace fossil fuels, it is gaining attention globally for use in the production of power. This will improve energy security and help reduce carbon pollution. However, a thorough examination of the environmental, social, technical, and financial elements of biomass energy generation must be done to determine its sustainability. The processes of pyrolysis, gasification, and direct combustion—all of which have differing levels of efficiency and development—are essential for turning biomass into electricity. The type of biomass chosen—residues or dedicated energy crops—determines how sustainable it is; residues from forestry and agricultural processes have a more promising future because they require less money and transportation than dedicated energy crops, which require a lot of land, water, and maintenance resources. Additionally, the sustainability factors consider how biomass is grown and processed in terms of land use, water use, and social ramifications [29, 30]. This emphasizes how crucial it is to choose biomass sources that minimize competition with food production and prevent negative effects on biodiversity and nearby communities. The study emphasizes the necessity of a balanced strategy for producing power from biomass, one that puts energy security first while reducing negative social and environmental effects and eventually contributes to a sustainable energy future.

The cutting edge of alternative energy solutions is bioenergy, which is produced from a variety of biomass sources and provides a sustainable route for energy generation and waste management. This energy can be expressed as heat, electricity, or biofuels, depending on whether biomass is converted into a useful form by thermochemical or biochemical processes (Figure 2.5) [31]. Pretreatment of biomass is the first step in the production of bioenergy and is necessary to get it ready for subsequent conversion processes. Electricity, gaseous fuels like hydrogen and synthesis gas, and liquid biofuels like methanol and ethanol can all be produced using biomass conversion technology. While waste lipids and vegetable oil are used in the creation of biodiesel, specific crops like corn and sugar cane are usually used to produce bioethanol. Direct combustion is a simple process that turns biomass into thermal energy, which can then be used immediately for heating or transformed into electrical power by steam turbines. In order to fulfill the high energy market needs, this procedure

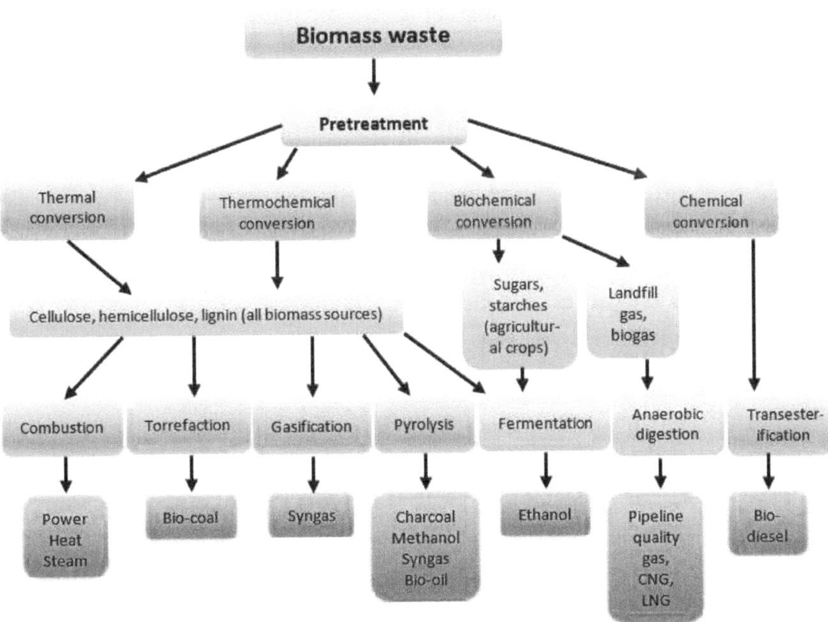

FIGURE 2.5 Biomass conversion into bioproducts [31].

occasionally calls for conversion techniques to increase the energy density of the biomass fuel. The main technique of conversion for biomass is mechanical processing, which includes briquetting, pressing, pelleting, and grinding. By increasing the bulk density of biomass and lowering its cost of transportation, these methods hope to improve its suitability for energy production. Biomass is further converted into energy through thermal conversion processes like pyrolysis, gasification, carbonization, and combustion. Two effective techniques for recovering energy from biomass are co-combustion with coal and incineration in fluidized bed boilers. Biofuels with improved calorific values and physical characteristics similar to coal are produced through thermal carbonization and pyrolysis, which change the physicochemical characteristics of biomass. Anaerobic fermentation of biomass to produce alcohols or biogas and transesterification of fats and oils to produce biodiesel are examples of biochemical conversion processes. Biomass constituents including cellulose, hemicellulose, and lignin are transformed into useful liquid fuels and gasses through these processes. In the search for sustainable energy solutions, each approach offers distinct opportunities and problems. This emphasizes the significance of choosing the right conversion technology based on the type of biomass that is available and the required energy form. With the help of these many conversion techniques, biomass is shown as a flexible and sustainable energy source that can effectively manage waste and make a substantial contribution to the world's energy mix.

The primary focus of the growing global energy crisis is the harmful effects of traditional fossil fuels on the environment. Although historically essential for promoting industrial expansion and energy security, these sources are now widely acknowledged for their contribution to greenhouse gas emissions, which play a major role in contributing to global climate change [32]. The search for alternative energy sources, especially those derived from organic or plant wastes, has become necessary at this pivotal point in the pursuit of sustainability and environmental stewardship. These substitutes present a viable way to fulfill the growing energy demand while reducing the negative effects on the environment and the shortage of fossil fuels. The use of agricultural residue as a strong substitute for non-renewable resources signals a move toward more ecologically friendly and sustainable energy sources. By offering a sustainable source for the manufacture of biofuel and value-added chemicals, using agricultural waste as feedstock for biorefinery operations not only supports energy security but also adheres to environmentally favorable standards. This methodology highlights the shift towards a circular bioeconomy, wherein the production of biochemicals and green fuels from renewable biomass results in sustainable and economically viable consequences. By intelligently utilizing a range of conversion strategies to harness different agricultural biomass, the biorefinery idea produces bioenergy. By guaranteeing the sustainable use of resources, this integrated method blends in perfectly with the circular bioeconomy framework. In order to prepare the path for value addition using cutting-edge technology, the article discusses in detail the origins and production of agricultural waste. The bioeconomy is further enhanced by biorefinery solutions that show the way to a wide range of value-added products and are supplemented by a life cycle assessment of biomass from agricultural waste. A good feedstock for the manufacturing of sustainable fuel and chemicals is agricultural waste biomass, which is high in structural polymers including cellulose, hemicellulose, and lignin. Plant secondary metabolites are transformed into readily available energy sources when these wastes are broken down by different microbes, which releases carbon. A good example of this resource's potential is the biochemical or thermochemical processes used to produce biofuels from agricultural waste, such as bioethanol, butanol, biohydrogen, biomethanol, and biogas. Nonetheless, feedstock choices and process variables have an impact on the amount and quality of biofuels generated [33, 34].

Biomass "waste-to-energy" is a concept that addresses all aspects of biomass waste management from production to distribution and is proving to be a persuasive way to turn low-value biomass into high-value, economically viable goods (Figure 2.6). Although the future looks bright, obstacles stand in the way of commercialization, such as the need for the necessary infrastructure, the availability of feedstock, technological developments, efficient management techniques, and gaining public acceptability. The efficiency of energy conversion is acknowledged for both thermochemical processes like incineration, pyrolysis, gasification, plasma, and torrefaction, and biochemical processes like composting, bioethanol fermentation, anaerobic digestion, and landfill methane capture. When compared to conventional fossil fuels, these strategies have the potential to drastically reduce greenhouse gas emissions by more than 50%. They also work well with the infrastructure that is already in place. This chapter explores the most recent developments in these conversion technologies, their global growth, and the scientific challenges that must be solved in order to properly utilize these green technologies in the promotion of sustainable energy solutions.

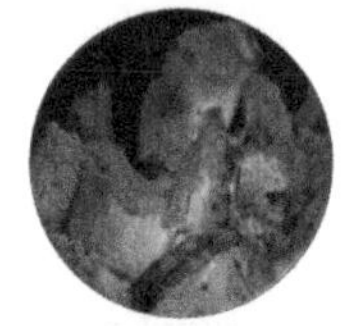
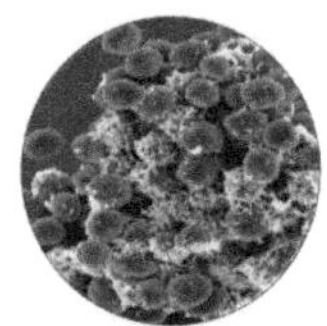

FIGURE 2.6   Pre-treatment methods of raw biomass [35].

The 'waste-to-energy' concept for biomass, which converts low-value biomass into lucrative goods, is becoming more and more popular due to its potential economic benefits. This method addresses waste management from biomass throughout its whole lifecycle, from production to delivery [36]. But infrastructure, feedstock availability, technology, management, and public acceptance are some of the obstacles that face commercialization. Promising techniques for efficient energy conversion include biochemical processes like composting, bioethanol fermentation, anaerobic digestion, and landfill methane collection, as well as thermochemical conversion methods like incineration, pyrolysis, gasification, plasma, and torrefaction. When compared to fossil fuels, these approaches have the potential to reduce greenhouse gas emissions by over 50% due to their high efficiency and compatibility with current infrastructure. This research examines the global growth of these conversion technologies, their improvements, and the scientific obstacles to their adoption for sustainable energy solutions.

2.5 CONCLUSION

The significance of biomass wastes from many sources, their processing methods, and their function in recycling and sustainability initiatives were all emphasized in this chapter. We talked about how biomass waste can be used to produce biofuels, bioenergy, and bioproducts, and we emphasized the need for creative and effective processing techniques. Biomass wastes offer a significant chance to progress recycling and sustainability. Their use promotes economic growth, energy security, and environmental preservation in addition to economic development. Sustainable Development Goals are aided by the strategic recycling and valuation of biomass wastes, which in turn facilitate the shift to a circular economy. To solve obstacles pertaining to biomass waste management and valorization, researchers, policymakers, and practitioners must work together urgently. To fully utilize biomass wastes, regulations that are supportive, cutting-edge research, and useful applications are necessary. In order to encourage the effective and sustainable use of biomass wastes for a more environmentally friendly future, stakeholders are urged to fund research, create sustainable technologies, and put policies into place.

REFERENCES

1. Amalina, F., et al., Biochar production techniques utilizing biomass waste-derived materials and environmental applications–A review. *Journal of Hazardous Materials Advances*, 2022: p. 100134.
2. Arpia, A.A., et al., Sustainable biofuel and bioenergy production from biomass waste residues using microwave-assisted heating: A comprehensive review. *Chemical Engineering Journal*, 2021. **403**: p. 126233.
3. Chen, R., et al., Sustainable utilization of biomass waste-rice husk ash as a new solidified material of soil in geotechnical engineering: A review. *Construction and Building Materials*, 2021. **292**: p. 123219.
4. Chew, K.W., et al., Abatement of hazardous materials and biomass waste via pyrolysis and co-pyrolysis for environmental sustainability and circular economy. *Environmental Pollution*, 2021. **278**: p. 116836.
5. Cho, E.J., et al., Bioconversion of biomass waste into high value chemicals. *Bioresource Technology*, 2020. **298**: p. 122386.
6. Chua, S.Y., et al., Biodiesel synthesis using natural solid catalyst derived from biomass waste—A review. *Journal of Industrial and Engineering Chemistry*, 2020. **81**: p. 41–60.
7. Escalante, J., et al., Pyrolysis of lignocellulosic, algal, plastic, and other biomass wastes for biofuel production and circular bioeconomy: A review of thermogravimetric analysis (TGA) approach. *Renewable and Sustainable Energy Reviews*, 2022. **169**: p. 112914.
8. Ethaib, S., et al., Microwave-assisted pyrolysis of biomass waste: A mini review. *Processes*, 2020. **8**(9): p. 1190.
9. Gayathiri, M., et al., Activated carbon from biomass waste precursors: Factors affecting production and adsorption mechanism. *Chemosphere*, 2022. **294**: p. 133764.
10. Jjagwe, J., et al., Synthesis and application of Granular activated carbon from biomass waste materials for water treatment: A review. *Journal of Bioresources and Bioproducts*, 2021. **6**(4): p. 292–322.
11. Kang, C., et al., A review of carbon dots produced from biomass wastes. *Nanomaterials*, 2020. **10**(11): p. 2316.

12. Khairol Anuar, N.K., et al., A review on multifunctional carbon-dots synthesized from biomass waste: Design/fabrication, characterization and applications. *Frontiers in Energy Research*, 2021. **9**: p. 67.

13. Clauser, N.M., et al. Biomass waste as sustainable raw material for energy and fuels. *Sustainability*, 2021. **13**, DOI: 10.3390/su13020794.

14. Lim, B.A., et al., Critical review on the development of biomass waste as precursor for carbon material as electrocatalysts for metal-air batteries. *Renewable and Sustainable Energy Reviews*, 2023. **184**: p. 113451.

15. Ma, M., et al., Co-pyrolysis re-use of sludge and biomass waste: Development, kinetics, synergistic mechanism and industrialization. *Journal of Analytical and Applied Pyrolysis*, 2022: p. 105746.

16. Usmani, Z., et al., Bioprocessing of waste biomass for sustainable product development and minimizing environmental impact. *Bioresource Technology*, 2021. **322**: p. 124548.

17. Bugg, T.D.H., J.J. Williamson, and G.M.M. Rashid, Bacterial enzymes for lignin depolymerisation: New biocatalysts for generation of renewable chemicals from biomass. *Current Opinion in Chemical Biology*, 2020. **55**: p. 26–33.

18. Manasa, P., S. Sambasivam, and F. Ran, Recent progress on biomass waste derived activated carbon electrode materials for supercapacitors applications—A review. *Journal of Energy Storage*, 2022. **54**: p. 105290.

19. Romani, A., R. Suriano, and M. Levi, Biomass waste materials through extrusion-based additive manufacturing: A systematic literature review. *Journal of Cleaner Production*, 2023. **386**: p. 135779.

20. Safian, M.T.-U., U.S. Haron, and M.N.M. Ibrahim, A review on bio-based graphene derived from biomass wastes. *BioResources*, 2020. **15**(4): p. 9756.

21. Seow, Y.X., et al., A review on biochar production from different biomass wastes by recent carbonization technologies and its sustainable applications. *Journal of Environmental Chemical Engineering*, 2022. **10**(1): p. 107017.

22. Soffian, M.S., et al., Carbon-based material derived from biomass waste for wastewater treatment. *Environmental Advances*, 2022. **9**: p. 100259.

23. Sri Shalini, S., et al., Biochar from biomass waste as a renewable carbon material for climate change mitigation in reducing greenhouse gas emissions—A review. *Biomass Conversion and Biorefinery*, 2021. **11**: p. 2247–2267.

24. Su, G., et al., Co-pyrolysis of microalgae and other biomass wastes for the production of high-quality bio-oil: Progress and prospective. *Bioresource Technology*, 2022. **344**: p. 126096.

25. Barakat, A., et al., Mechanical pretreatments of lignocellulosic biomass: Towards facile and environmentally sound technologies for biofuels production. *RSC Advances*, 2014. **4**(89): p. 48109–48127.

26. Sun, J., L. Zhang, and K.-C. Loh, Review and perspectives of enhanced volatile fatty acids production from acidogenic fermentation of lignocellulosic biomass wastes. *Bioresources and Bioprocessing*, 2021. **8**(1): p. 68.

27. Tiwari, S.K., et al., Methods for the conversion of biomass waste into value-added carbon nanomaterials: Recent progress and applications. *Progress in Energy and Combustion Science*, 2022. **92**: p. 101023.

28. Evans, A., V. Strezov, and T.J. Evans, Sustainability considerations for electricity generation from biomass. *Renewable and Sustainable Energy Reviews*, 2010. **14**(5): p. 1419–1427.

29. Yaqoob, H., et al., The potential of sustainable biogas production from biomass waste for power generation in Pakistan. *Journal of Cleaner Production*, 2021. **307**: p. 127250.

30. Yu, S., et al., Nanocellulose from various biomass wastes: Its preparation and potential usages towards the high value-added products. *Environmental Science and Ecotechnology*, 2021. **5**: p. 100077.

31. Kalak, T., Potential use of industrial biomass waste as a sustainable energy source in the future. *Energies*, 2023. **16**, DOI: 10.3390/en16041783.

32. Kumar Sarangi, P., et al., Utilization of agricultural waste biomass and recycling toward circular bioeconomy. *Environmental Science and Pollution Research*, 2023. **30**(4): p. 8526–8539.

33. Yusuf, A.A. and F.L. Inambao, Characterization of Ugandan biomass wastes as the potential candidates towards bioenergy production. *Renewable and Sustainable Energy Reviews*, 2020. **117**: p. 109477.

34. Zhou, C. and Y. Wang, Recent progress in the conversion of biomass wastes into functional materials for value-added applications. *Science and Technology of Advanced Materials*, 2020. **21**(1): p. 787–804.

35. Dyjakon, A. and T. Noszczyk, Alternative fuels from forestry biomass residue: Torrefaction process of horse chestnuts, oak acorns, and spruce cones. *Energies*, 2020. **13**, DOI: 10.3390/en13102468.

36. Siwal, S.S., et al., Recovery processes of sustainable energy using different biomass and wastes. *Renewable and Sustainable Energy Reviews*, 2021. **150**: p. 111483.

CHAPTER 3

Bioproducts from Biomass Wastes

Chemicals and Energy Materials Production and Utilization

Omar Dagdag, Sheerin Masroor, Walid Daoudi, Rajesh Haldhar, Elyor Berdimurodov and Hansang Kim

3.1 INTRODUCTION

The depletion of fossil fuels and growing concern about environmental sustainability have prompted the quest for renewable and sustainable alternatives [1]. Biomass, or organic matter obtained from living or recently living organisms, has emerged as a promising feedstock for manufacturing bioproducts such as chemicals and energy resources [2]. Biomass wastes, including agricultural residues, forestry byproducts, and industrial waste, are a largely untapped resource. These wastes frequently pose environmental issues, including landfilling and incineration. However, various biological, chemical, and thermochemical processes can transform them into valuable bioproducts. Biomass waste can be used to make a variety of chemicals and energy resources, including biofuels like ethanol, biodiesel, and biogas, which can replace fossil fuels in transportation and heating [3]. Biochemicals such as lactic acid, succinic acid, and bioplastics are employed in the food, pharmaceutical, and packaging sectors. Biomaterials include activated carbon, biocomposites, and bio-based adhesives used in environmental protection, construction, and manufacturing [4]. The manufacture

DOI: 10.1201/9781003466833-4

and use of bioproducts derived from biomass waste provides numerous environmental and economic benefits, including the ability to substitute fossil fuel-based products, reducing greenhouse gas emissions. Biomass waste can be diverted from landfills and incineration plants, lowering environmental pollution and conserving natural resources. The bioproducts business has the potential to produce new jobs while also stimulating economic growth in both rural and urban areas [5–8].

This introduction provides an overview of the use of biomass wastes to produce bioproducts, focusing on chemical and energy applications. Various conversion techniques, including biochemical, thermochemical, and hybrid approaches, are examined, along with their efficiency and obstacles. Additionally, the use of bioproducts in various industries and sectors is investigated, focusing on their role in sustainable development and the transition to a bio-based economy. Biomass waste can be converted into valuable resources through innovative methods and technical breakthroughs, benefiting the environment and the economy (see Figure 3.1).

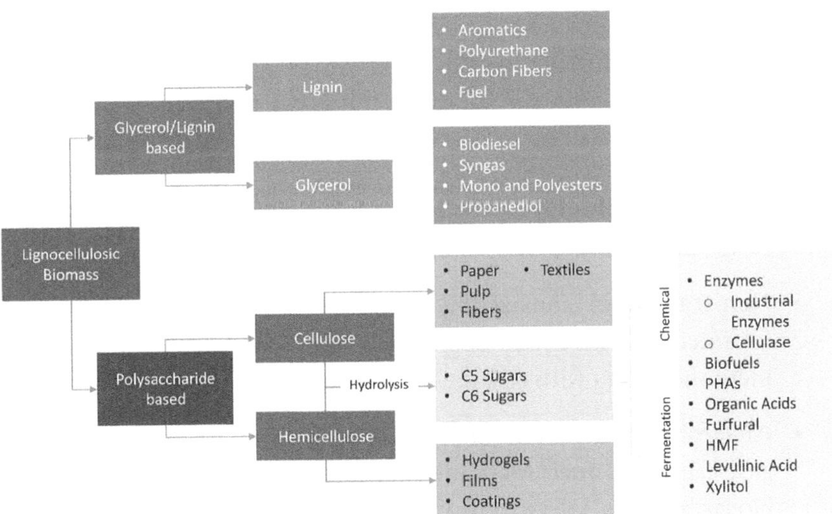

FIGURE 3.1 Valorization of lignocellulosic biomass as bioproducts and chemicals. Reproduced with permission from reference [9]. Copyright 2022 Elsevier.

3.2 BIOMASS WASTES AS A SOURCE OF BIOPRODUCTS

Natural resources classified as renewable are those that, mostly via natural reproduction, can replenish themselves to make up for the portion eaten or used up [10]. Their cost, accessibility, and sustainability set them apart. Since these substances are frequently present in the environment, they aid in keeping the ecosystem livable. After application, it's imperative to prevent careless usage. Thus, they should be well maintained worldwide. Renewable energy sources are those materials that satisfy the specified requirements, such as water, wind, and other naturally occurring elements in the biomass categories. Non-renewable resources, on the other hand, are those materials that do not fit into the categories of renewable groups, such as coal, gasoline, petroleum, and fuel. As mentioned, photosynthesis, facilitated by solar energy conversion, produces biomass naturally [11]. Equation (1) may be used to express the photosynthesis procedure in the presence of solar radiation acting as a catalyst.

$$H_2O + CO_2 \rightarrow CH_2O + O_2$$

The reaction mechanism in Eq. (1) highlights the significance of solar energy in developing advantageous elements and streamlines the process of creating essential components in biomass materials [12]. However, there are other ways to create fuel and energy, along with oil production from biomass, such as pyrolysis, digestion, solid fuel combustion, and catalyzed fermentation [12]. Biomass-based materials typically fall into three main categories:

- The traditional solid mass form of biomass [13] (wood and agricultural waste, either raw or altered).

- Non-traditional biomass [14] (translated as "biomass in liquid form for internal combustion engine use, after which it is converted into methanol along with ethanol).

- Anaerobic fermentation of biomass results in the production of biogas, a gaseous fuel. Methane makes up between 55% and 65% of biogas biomass, followed by CO_2 (30% to 40%) and various contaminants including H2, N2, and H_2S [10, 15].

Four primary sources of biomass have been certified for production and consumption. These are chosen based on the clarity and effectiveness of their offerings. The sources are shown in Figure 3.2.

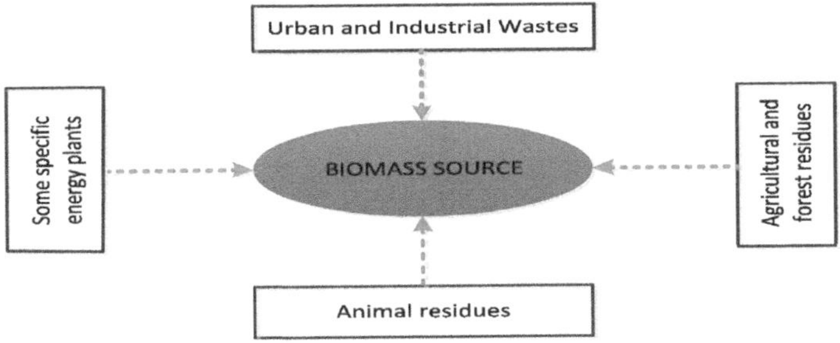

FIGURE 3.2 Genuine biomass sources. Reproduced with permission from reference [10]. Copyright 2021 Elsevier.

3.3 BIOMASS CONVERSION

It is acknowledged that the initial stage in creating essential functional products from various biomass components is biomass conversion. The main goal of biomass is to satisfy industrial needs by producing fuel, energy, and oil lubricants. This is a big chance to make the most of and benefit from biomass's worth in our ecosystem. The many conversion strategies being used in the industry provide possibility and success. It offers a way to turn trash into commodities that may be used, such as chemicals, energy, fuels, and oils, including biomaterials, in addition to disposing of materials [16]. It can operate straight out of the container before conversion. Still, it can also operate after being changed [17] to produce liquid or gaseous fuel along with additional products that can be considered lubricant (raw with no alteration or transformation). Figure 3.3 illustrates these conversion methods.

3.3.1 Thermal-Chemical

In an oxygen-poor environment, lignocellulosic derivatives undergo thermal-chemical breakdown in inert circumstances [18, 19]. This method of converting biomass is widely utilized to produce heat, power, liquid and gaseous chemicals, and liquid fuel for various industrial uses [20, 21]. It comprises extra sub-techniques that vary according to the outcome (Figure 3.4).

A restricted oxygen supply is needed for gasification, which is accomplished by burning the volatile material, devitalizing the biomass, and resulting in a fuel gas with high hydrogen and carbon monoxide content. After certain particles and tars are removed, the resulting gas has a decreased calorific value, making it suitable for turbine combustion.

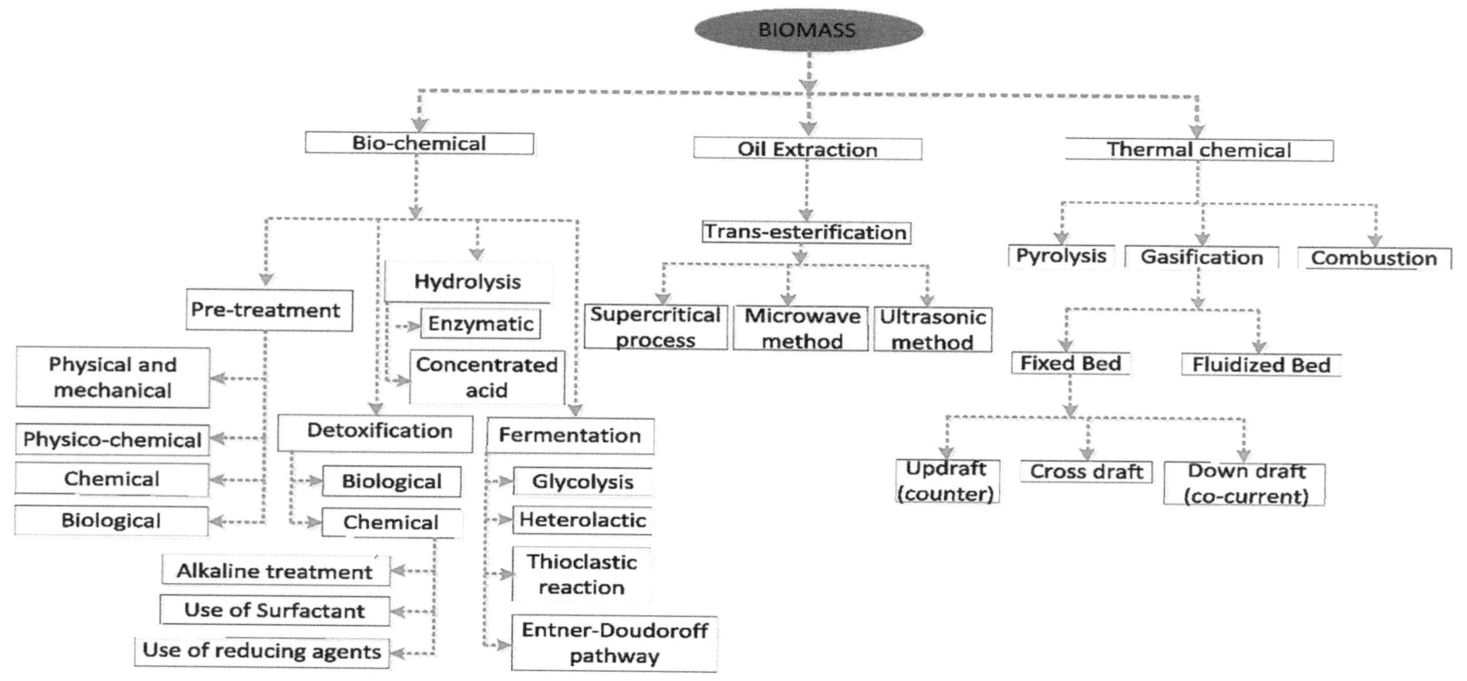

FIGURE 3.3 Categorization of biomass conversion. Reproduced with permission from reference [10]. Copyright 2021 Elsevier.

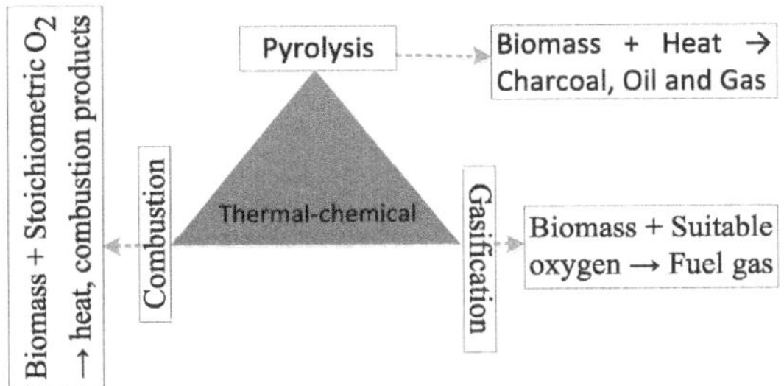

FIGURE 3.4 Sub-methods for thermochemical transformation, including the resultant products. Reproduced with permission from reference [10]. Copyright 2021 Elsevier.

Pyrolysis is the primary method to convert biomass into solid char, gas, and liquid bio-oil. It may be categorized as slow or quick based on how long it takes to change the feedstock into the different outputs [22].

3.3.2 Bio-Chemical

Using enzymes and bacteria, alongside additional microorganisms, heat, and other chemicals, biochemical conversion processes break down cellulose, including hemicellulose, into sugar and other intermediate products [23]. When it comes to using biomass to produce chemicals, oils, fuels, and other commodities, this is the most sought-after and efficient technological method [20]. As shown in Figure 3.5, this is further divided into many branches due to the formulation products. The structure of lignin is generally altered, cellulose is recrystallized, and the accessible surface area is increased by lignocellulosic biomass conversion procedures such as pretreatment, hydrolysis, and fermentation [23]. It also partly depolymerizes cellulose, solubilizes hemicellulose and lignin, lowers formulation operating costs, and minimizes end-product losses [10, 23, 24].

3.3.3 Oil Extraction

Most bio-lube/biodiesel used in commercial settings comes from feedstocks containing vegetable or animal fats used in transesterification [25]. This is the catalytic reaction between an aliphatic short-chain alcohol (ethanol or methanol) and an oil product [25, 26]. This process yields phospholipids, triglycerides, and glycerol [14], which are beneficial in tribology applications [10]. By using this method, "enol" derivatives that are hard to get through other channels can be produced [27]. High-quality vinyl

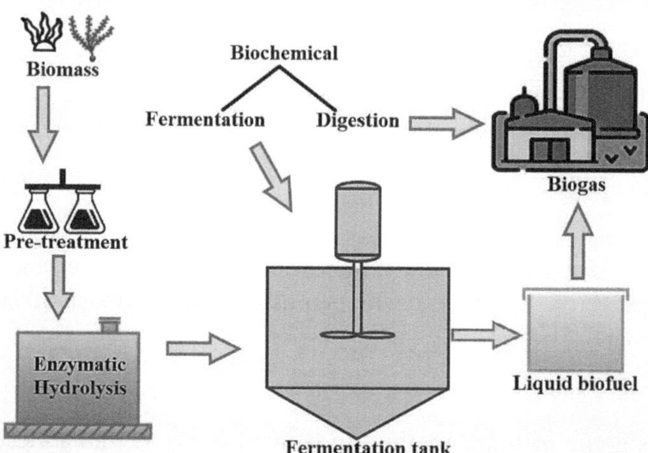

FIGURE 3.5 Biochemical conversion of biomass into value-added products. Reproduced with permission from reference [29]. Copyright 2024 Elsevier.

ethers are produced in large numbers using the transesterification procedure, which depends on the cost and accessibility of certain ingredients like vinyl acetate [27, 28].

3.4 POTENTIAL COMPONENTS OF BIOMASS

The following discussion of the three main elements that enable and sustain biomass explains how they support sustainability initiatives, as seen in Figure 3.6.

3.4.1 Cellulose

A polysaccharide made up of β (1-4) linked D-glucose units is called cellulose [30]. Industry usage of ethanol fuel has increased due to cellulose from energy crops being converted into biofuels like ethanol (Figure 3.7) [31]. Its amphipathic nature makes it helpful in lubricating additives and oils alike. It is advised because of the hydrophilic head's capacity to adhere to polar materials and the hydrophobic tail found in lubricants in the nonpolar zone that dissolve during operation [32]. The main biomass components that show potential as industrial materials are shown in Figure 3.6. Additionally, it strongly resists the effects of alkalis and acids. Most

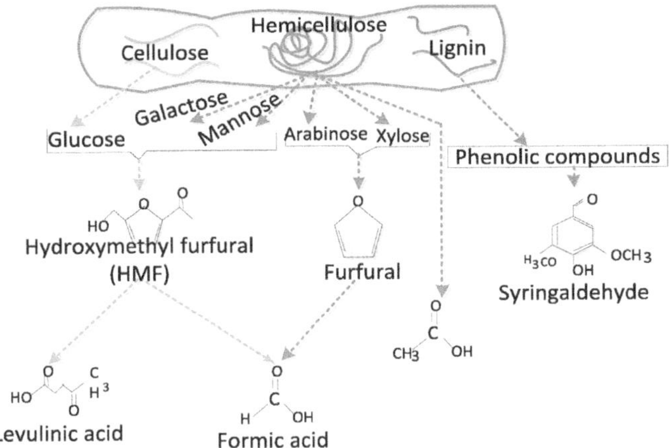

FIGURE 3.6 Schematic representation of significant components of biomass. Reproduced with permission from reference [10]. Copyright 2021 Elsevier.

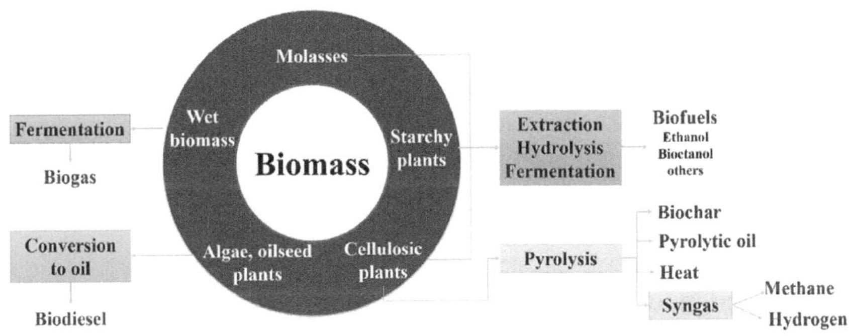

FIGURE 3.7 Different types of biofuels are derived from lignocellulosic biomass. Reproduced with permission from reference [35]. Copyright 2021 MDPI.

biomass contains 6.17% hydrogen, 61.44% carbon, and 32.39% oxygen as cellulose components [10]. On the other hand, pectin, lignin, hemicellulose, and other minor substances make up plant matter or biomass [33]. Because of its long chain length, cellulose has a high tensile strength, making it appropriate for mechanical and thermal applications [34]. Moreover, because of its mechanical and thermal qualities, monocrystalline cellulose is employed as an additive filler phase in bio-based polymer matrices, mainly for constructing nanocomposites [34].

3.4.2 Hemicellulose

Hemicellulose, a polysaccharide found in biomass, is naturally connected to cellulose but has a distinct structure, content, and function [10]. Hemicellulose comprises many different sugar monomers, whereas cellulose only contains or has anhydrous glucose. Because of its composition, hemicellulose can only characterize the difference between acid detergent fiber (ADF) and neutral detergent fiber (NDF) [10].

3.4.3 Lignin

Lignin is an essential structural material that provides vascular plants with support tissue. It resembles a complicated organic polymer. Lignin's constituents render it appropriate for use as a detergent, dispersion, and emulsion stabilizer, and it is now employed in the water treatment industry. Strong mechanical strength is provided, and its durability makes it desirable [10].

3.5 UTILIZATION OF BIOPRODUCTS IN VARIOUS SECTORS

Scientists are especially interested in using "biomass" for renewable energy sources. Biomass is the term for plant or animal waste, not for food or fuel but as building blocks for industrial processes like creating and storing energy or as a chemical processing input. These energies can be converted into useful energy either directly or indirectly. The production of different value-added items from biomass is shown in Figure 3.8.

3.5.1 Food Industries

Biomass may be used to create biofuels like bioethanol along with biodiesel. These biofuels can support the food supply chain's long-term sustainability and help lower greenhouse gas emissions. They apply to both transportation and industrial settings. Biogas, mainly consisting of methane and carbon dioxide, may be produced via anaerobic digestion using biomass. Biogas may be used in food processing facilities to produce electricity, heat buildings, or cook. It can reduce dependency on fossil fuels by being burnt directly or transformed into thermal energy. Some biomass resources may be used to make animal feed, giving chickens and cattle vital nutrients. Food scraps and agricultural waste may be used to make compost, a crucial organic fertilizer. Composting biomass waste completes the nutrient cycle, lessening the demand for artificial fertilizers. Packaging that is compostable and biodegradable and constructed of

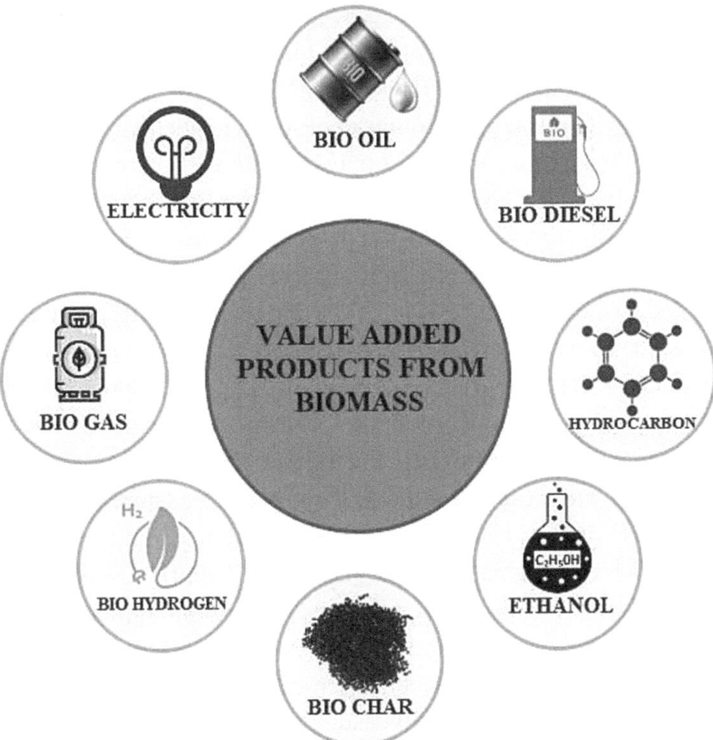

FIGURE 3.8 Development of different biomass-based products with additional value. Reproduced with permission from reference [29]. Copyright 2024 Elsevier.

materials produced from biomass may assist in reducing the adverse environmental effects of food packaging waste. Specific biomass components, such plant extracts, can be employed as natural flavorings and additives in the food business. This enhances the nutritious content of food goods and gives them distinct tastes. Food items can benefit from adding starch and sugars generated from biomass to improve texture and provide useful qualities. In the food industry, microorganisms may be cultivated to create biomass high in proteins and additional nutrients, possibly as a protein substitute. Biomass contains bioactive substances like phytochemicals and antioxidants that may be utilized to make functional food items and may even improve health. Nutrient-dense foods like carotenoids, omega-3 fatty acids, and even proteins obtained from algae may be made from algae, a biomass [36].

3.5.2 Bio-Oil

Bio-oil may be used as a substitute for fossil fuels in power stations and industrial operations, or it can be used as a sustainable energy source to produce heat with electricity. Boilers or gasifiers explicitly made for this use may be burnt to generate energy. It may be further enhanced by producing advanced biofuels like biodiesel or renewable jet fuel through hydrodeoxygenation or esterification. Bio-oil contains various organic components that may be used as feedstock to make chemicals and additional goods with added value. Plastics, resins, and a range of industrial products can be produced from these materials. Because bio-oil has a more significant energy density than raw biomass, it is simpler to transport and store to central processing facilities. As biomass stores carbon in the bio-oil and biochar created during pyrolysis, which happens when it grows, the technique can help with carbon sequestration if the feedstock for the bio-oil is sourced sustainably. Compared to gaseous (syngas) and solid (biochar) fuels, bio-oil has a higher power density and is more straightforward to transport and store. Bio-oil is a dark, viscous organic liquid with a pH of 3.5 to 4.2. The quality and amount of biomass types change depending on the operation conditions, as do biochemical oil components and physical properties. High acids and methyl esters can inhibit the growth of bacteria, beetles, and other organisms. Because bio-oil contains carbon, it can be used as a long-term supply of nutrients and, hence, a fertilizer [37].

3.5.3 Biodiesel

Collecting unaffected alcohol, neutralization glycerine, extraction of oil, feedstock manufacturing, (trans) the esterification process responses, product segregation, biodiesel washing, and biodiesel refining are the processes involved in a typical biodiesel manufacturing process. Bio-waste-produced catalysts have been employed in the commercial production of biodiesel due to their availability, cost, and—above all—environmental friendliness. Biomass is the primary feedstock used to make biodiesel. The most often utilized materials include biomass sources, vegetable oils, animal fats, algae, and numerous other lipid-rich materials. Through a procedure called transesterification, these feedstocks mix with alcohol in addition to the presence of a catalyst to create glycerin and biodiesel. Algae may be utilized as a biomass source to produce biodiesel. Ponds, tanks, or bioreactors can grow algae with a higher lipid content than other plant-based feedstocks. It is possible to produce algae on non-arable terrain, although they do not compete with food crops for resources. Like lipid-rich biomass, lignocellulosic biomass may be processed into biofuels, such as biodiesel [29].

3.5.4 Biohydrogen

Various biomass resources may be converted into biohydrogen by thermochemical processes such as simple gasification, steam gasification, bio-based oils, supercritical water gasification, and pyrolysis steam reforming [38]. Due to the limited availability of fossil fuels, the continuous use of these sources of greenhouse gases—especially CO_2, is unsustainable. Since hydrogen gas ignites with a large quantity of energy per unit weight and offers a limitless supply of energy, it can be considered a possible energy carrier for future demands with the benefits of a carbon-negative approach. There are several benefits to producing biohydrogen as a substitute for many alternative hydrogen sources [39].

3.5.5 Biogas

Around the world, biogas is recognized as a traditional kind of off-grid energy. Biogas may also be used to create power. Biogas may be converted into electricity in energy facilities using internal combustion engines (ICEs) or gas turbines (GTs), the two most popular power production methods. Another strategy that seems promising is the use of microgas turbines because of their minimal NOx emissions along with adaptability to shifting load needs. Various microturbines can meet low- to medium-power demands, whose capabilities range from 70 kW upwards to greater than 250 kW. The surrounding businesses and industries might be able to get the electricity they need [40].

3.5.6 Biomedical

It is highly poisonous in its traditional form and has a poor absorption rate. Thanks to nanotechnology, this issue has been resolved; selenium at the nanoscale has better biological impacts, including biocompatibility. Due to its low toxicity, stability, and ability to deliver nanomedicine to specific sites, the biological method of producing selenium nanoparticles offers many opportunities in the biomedical area [41]. Excellent antimicrobial, cancer prevention, anti-diabetic, and anti-inflammatory qualities make selenium a necessary vitamin. Its conventional form has a broad spectrum of toxicity and poor absorption rate. An even polysaccharide derived from Ulva lactuca was used to create a conjugate of selenium (Se). The production, in addition to the stability of the nanoparticles, was assessed using

UV-vis spectroscopy, and TEM analysis showed that the average size of the particles was 85 nm. Se NPs have shown outstanding antioxidant qualities alongside negligible harm to cell lines. The anti-microbial effectiveness of the mouthwash showed that it was similarly effective against Lactobacillus along with Candida albicans and more effective than antibiotics towards S. mutans and S. aureus [42].

3.6 CONCLUSION

Developing and using bioproducts derived from biomass wastes is a possible answer to the worldwide waste management challenges and the growing need for renewable resources. This chapter reviews the most recent research and advances in this sector, covering topics ranging from biomass pretreatment and conversion methods to bioproduct characterization and applications. The chapter's conclusions are as follows:

- Biomass waste is a significant and untapped resource for developing value-added bioproducts.

- Pretreatment procedures are critical to improving biomass reactivity and conversion efficiency.

- Thermochemical, biochemical, and electrochemical conversion methods provide many paths for bioproduct generation, each having advantages and disadvantages.

- Bioproducts from biomass waste have many potential applications as chemicals and energy sources.

- Using bioproducts helps create a more circular and sustainable economy by minimizing waste and encouraging the use of renewable resources.

3.7 CHALLENGES AND PROSPECTS

The manufacture and use of bioproducts from biomass waste have a bright future because of the growing need for clean and renewable alternatives to fossil fuels and traditional materials. Future essential areas of development may include:

- Ongoing R&D efforts aim to increase the efficiency and cost-effectiveness of bioproduct production methods. This includes creating new and more efficient enzymes for biomass conversion, enhancing fermentation and thermochemical processes, and investigating new feedstocks and methods for bioproduct manufacturing.

- The bioproducts market is predicted to expand rapidly in the future years, driven by rising demand for biofuels, biochemicals, and biomaterials; government policies and incentives to support the economy; and consumer desire for sustainable and ecologically friendly products.

- The production and use of bioproducts derived from biomass waste can provide several environmental and social benefits, such as lower greenhouse gas emissions, better waste management, job creation and economic development, and increased energy security.

Major obstacles in accomplishing resource recovery using biomass are:

- Ensuring that biomass is produced and collected sustainably without depleting natural resources or hurting ecosystems.

- Scaling up bioproduct production and lowering prices to compete with fossil fuel-based products.

- Improving the efficiency and scalability of bioproduct manufacturing methods to satisfy rising demand.

- Overcoming technical and logistical barriers to integrating bioproduct manufacturing into current agriculture, forestry, and energy industries.

- Educating customers on the benefits of bioproducts and increasing market demand for these products.

- Creating supportive policies and regulatory frameworks to stimulate investment in bioproduct production and consumption.

- Minimizing the environmental effects of bioproduct production, such as greenhouse gas emissions, water consumption, and land use modification.

- Balancing biomass demand for bioproduct production against other land uses, such as food cultivation and forestry.

- Overcoming technical hurdles in biomass conversion to bioproducts, such as the recalcitrance of lignocellulosic biomass.

- Building the infrastructure for biomass collection, transportation, and processing.

REFERENCES

1. M. Galbe and G. Zacchi, "Pretreatment of lignocellulosic materials for efficient bioethanol production," *Biofuels,* pp. 41–65, 2007.
2. F. Cherubini, "The biorefinery concept: Using biomass instead of oil for producing energy and chemicals," *Energy Conversion and Management,* vol. 51, pp. 1412–1421, 2010.
3. M. E. Himmel, S.-Y. Ding, D. K. Johnson, W. S. Adney, M. R. Nimlos, J. W. Brady, *et al.,* "Biomass recalcitrance: Engineering plants and enzymes for biofuels production," *Science,* vol. 315, pp. 804–807, 2007.
4. A. J. Ragauskas, C. K. Williams, B. H. Davison, G. Britovsek, J. Cairney, C. A. Eckert, *et al.,* "The path forward for biofuels and biomaterials," *Science,* vol. 311, pp. 484–489, 2006.
5. H. M. Junginger, T. Mai-Moulin, V. Daioglou, U. Fritsche, R. Guisson, C. Hennig, *et al.,* "The future of biomass and bioenergy deployment and trade: A synthesis of 15 years IEA bioenergy task 40 on sustainable bioenergy trade," *Biofuels, Bioproducts and Biorefining,* vol. 13, pp. 247–266, 2019.
6. U. B. Strategy, "A sustainable bioeconomy for Europe: Strengthening the connection between economy, society and the environment," *European Commission,* 2018, https://www.qualenergia.it/wp-content/uploads/2018/10/ec_bioeconomy_strategy_2018.pdf
7. F. Cherubini and A. H. Strømman, "Principles of biorefining," in *Biofuels,* ed: Elsevier, 2011, pp. 3–24.
8. A. J. Ragauskas, G. T. Beckham, M. J. Biddy, R. Chandra, F. Chen, M. F. Davis, *et al.,* "Lignin valorization: Improving lignin processing in the biorefinery," *Science,* vol. 344, p. 1246843, 2014.
9. V. K. Gupta, A. Pandey, M. Koffas, S. I. Mussatto, and S. Khare, "Biobased biorefineries: Sustainable bioprocesses and bioproducts from biomass/bioresources special issue," *Renewable and Sustainable Energy Reviews,* vol. 167, ed: Elsevier, 2022, p. 112683.
10. A. C. Opia, M. K. B. A. Hamid, S. Syahrullail, A. B. Abd Rahim, and C. A. Johnson, "Biomass as a potential source of sustainable fuel, chemical and tribological materials–overview," *Materials Today: Proceedings,* vol. 39, pp. 922–928, 2021.
11. F. H. Isikgor and C. R. Becer, "Lignocellulosic biomass: A sustainable platform for the production of bio-based chemicals and polymers," *Polymer Chemistry,* vol. 6, pp. 4497–4559, 2015.
12. M. Bertero, G. de la Puente, and U. Sedran, "Fuels from bio-oils: Bio-oil production from different residual sources, characterization and thermal conditioning," *Fuel,* vol. 95, pp. 263–271, 2012.

13. S. Karekezi, K. Lata, and S. T. Coelho, "Traditional biomass energy: Improving its use and moving to modern energy use," in *Renewable Energy*, ed: Routledge, 2012, pp. 258–289.

14. N. E.-A. El-Naggar, S. Deraz, and A. Khalil, "Bioethanol production from lignocellulosic feedstocks based on enzymatic hydrolysis: Current status and recent developments," *Biotechnology*, vol. 13, pp. 1–21, 2014.

15. E. Kovács, R. Wirth, G. Maróti, Z. Bagi, G. Rákhely, and K. L. Kovács, "Biogas production from protein-rich biomass: Fed-batch anaerobic fermentation of casein and of pig blood and associated changes in microbial community composition," *PLoS One*, vol. 8, p. e77265, 2013.

16. Q. Zhang, J. Chang, T. Wang, and Y. Xu, "Review of biomass pyrolysis oil properties and upgrading research," *Energy Conversion and Management*, vol. 48, pp. 87–92, 2007.

17. V. P. Soudham, "Biochemical conversion of biomass to biofuels: Pretreatment–detoxification–hydrolysis–fermentation," Umeå universitet, 2015, https://www.diva-portal.org/smash/record.jsf?pid=diva2%3A809239&dswid=-5850

18. C. Z. Zaman, K. Pal, W. A. Yehye, S. Sagadevan, S. T. Shah, G. A. Adebisi, *et al.*, "Pyrolysis: A sustainable way to generate energy from waste," *Pyrolysis*, vol. 1, pp. 3–36, 2017.

19. L. Ma, T. Wang, Q. Liu, X. Zhang, W. Ma, and Q. Zhang, "A review of thermal–chemical conversion of lignocellulosic biomass in China," *Biotechnology Advances*, vol. 30, pp. 859–873, 2012.

20. B. Joffres, D. Laurenti, N. Charon, A. Daudin, A. Quignard, and C. Geantet, "Thermochemical conversion of lignin for fuels and chemicals: A review," *Oil & Gas Science and Technology–Revue d'IFP Energies Nouvelles*, vol. 68, pp. 753–763, 2013.

21. N. Canabarro, J. F. Soares, C. G. Anchieta, C. S. Kelling, and M. A. Mazutti, "Thermochemical processes for biofuels production from biomass," *Sustainable Chemical Processes*, vol. 1, pp. 1–10, 2013.

22. W. M. Lewandowski, M. Ryms, and W. Kosakowski, "Thermal biomass conversion: A review," *Processes*, vol. 8, p. 516, 2020.

23. L. Canilha, A. K. Chandel, T. Suzane dos Santos Milessi, F. A. F. Antunes, W. Luiz da Costa Freitas, M. das Graças Almeida Felipe, *et al.*, "Bioconversion of sugarcane biomass into ethanol: An overview about composition, pretreatment methods, detoxification of hydrolysates, enzymatic saccharification, and ethanol fermentation," *BioMed Research International*, vol. 2012, 2012.

24. L. Lemée, D. Kpogbemabou, L. Pinard, R. Beauchet, and J. Laduranty, "Biological pretreatment for production of lignocellulosic biofuel," *Bioresource Technology*, vol. 117, pp. 234–241, 2012.

25. G. Griffin, D. Batten, and T. Beer, "A review of physical properties of biomass pyrolysis oil," *International Journal of Renewable Energy Research—HOME*, vol. 5, pp. 2004–20018, 2015.

26. J. Hu, F. Yu, and Y. Lu, "Application of Fischer–Tropsch synthesis in biomass to liquid conversion," *Catalysts,* vol. 2, pp. 303–326, 2012.
27. B. Bongfa, S. Syahrullail, M. Abdul Hamid, and P. Samin, "Suitable additives for vegetable oil-based automotive shock absorber fluids: An overview," *Lubrication Science,* vol. 28, pp. 381–404, 2016.
28. Y. Okimoto, S. Sakaguchi, and Y. Ishii, "Development of a highly efficient catalytic method for synthesis of vinyl ethers," *Journal of the American Chemical Society,* vol. 124, pp. 1590–1591, 2002.
29. P. Swaminaathan, A. Saravanan, and P. Thamarai, "Utilization of bioresources for high-value bioproducts production: Sustainability and perspectives in circular bioeconomy," *Sustainable Energy Technologies and Assessments,* vol. 63, p. 103672, 2024.
30. V. S. Sikarwar, M. Zhao, P. S. Fennell, N. Shah, and E. J. Anthony, "Progress in biofuel production from gasification," *Progress in Energy and Combustion Science,* vol. 61, pp. 189–248, 2017.
31. M. M. Zainol, N. A. S. Amin, and M. Asmadi, "Optimization studies of oil palm empty fruit bunch liquefaction for carbon cryogel production as catalyst in levulinic acid esterification," *Jurnal Teknologi,* vol. 80, pp. 137–145, 2018.
32. H. Luo and M. M. Abu-Omar, "Chemicals from lignin," *Encyclopedia of Sustainable Technologies,* vol. 3, pp. 573–585, 2017.
33. M. M. Rodríguez, "Lignin biomass conversion into chemicals and fuels," 2016, https://orbit.dtu.dk/en/publications/lignin-biomass-conversion-into-chemicals-and-fuels
34. S. Iijima, C. Brabec, A. Maiti, and J. Bernholc, "Structural flexibility of carbon nanotubes," *The Journal of Chemical Physics,* vol. 104, pp. 2089–2092, 1996.
35. N. M. Clauser, G. González, C. M. Mendieta, J. Kruyeniski, M. C. Area, and M. E. Vallejos, "Biomass waste as sustainable raw material for energy and fuels," *Sustainability,* vol. 13, p. 794, 2021.
36. X. Hu and M. Gholizadeh, "Progress of the applications of bio-oil," *Renewable and Sustainable Energy Reviews,* vol. 134, p. 110124, 2020.
37. T. S. Ahamed, S. Anto, T. Mathimani, K. Brindhadevi, and A. Pugazhendhi, "Upgrading of bio-oil from thermochemical conversion of various biomass–mechanism, challenges and opportunities," *Fuel,* vol. 287, p. 119329, 2021.
38. J.-H. Park, K. Chandrasekhar, B.-H. Jeon, M. Jang, Y. Liu, and S.-H. Kim, "State-of-the-art technologies for continuous high-rate biohydrogen production," *Bioresource Technology,* vol. 320, p. 124304, 2021.
39. S. Mona, S. S. Kumar, V. Kumar, K. Parveen, N. Saini, B. Deepak, *et al.,* "Green technology for sustainable biohydrogen production (waste to energy): A review," *Science of the Total Environment,* vol. 728, p. 138481, 2020.
40. S. Abanades, H. Abbaspour, A. Ahmadi, B. Das, M. Ehyaei, F. Esmaeilion, *et al.,* "A critical review of biogas production and usage with legislations framework across the globe," *International Journal of Environmental Science and Technology,* pp. 1–24, 2022.

41. S. S. Salem, M. S. E. Badawy, A. A. Al-Askar, A. A. Arishi, F. M. Elkady, and A. H. Hashem, "Green biosynthesis of selenium nanoparticles using orange peel waste: Characterization, antibacterial and antibiofilm activities against multidrug-resistant bacteria," *Life*, vol. 12, p. 893, 2022.

42. M. Vikneshan, R. Saravanakumar, R. Mangaiyarkarasi, S. Rajeshkumar, S. Samuel, M. Suganya, *et al.*, "Algal biomass as a source for novel oral nano-antimicrobial agent," *Saudi Journal of Biological Sciences*, vol. 27, pp. 3753–3758, 2020.

II

Classification

CHAPTER 4

Forestry and Agricultural Residues-Based Wastes

Fundamentals, Classification, Properties, and Applications

Abhinay Thakur, Ashish Kumar and Amit Somya

4.1 INTRODUCTION

Forestry and agricultural residues serve as valuable biomass resources that arise as byproducts of land management practices (1–3). These residues from forestry and agricultural activities offer immense potential for various applications, ranging from energy production to farm use and industrial processes. This elaboration focuses on forestry residues, highlighting their characteristics, variability, and suitability for applications. Forestry residues encompass diverse organic materials, including branches, tree tops, and stumps, which accumulate in forested areas following timber harvesting operations. These residues result from thinning, logging, and clear-cutting, where trees are harvested for timber production (4–6). The composition and properties of forestry residues are influenced by several factors, including tree species, harvesting methods, and site conditions. One key factor affecting the characteristics of forestry residues is the tree species from which they originate. Different tree species have distinct chemical compositions and structural characteristics, impacting the properties of their residues. For example, softwood residues derived from

DOI: 10.1201/9781003466833-6

coniferous trees such as pine, spruce, and fir typically have higher lignin content than hardwood residues from oak, maple, and beech. Lignin, a complex polymer in plant cell walls, provides structural support and rigidity to woody biomass. The higher lignin content in softwood residues makes them more suitable for structural integrity and durability applications, such as construction materials and engineered wood products. Moreover, the harvesting methods employed in forestry operations influence the size and composition of residues generated (7, 8). Mechanized harvesting techniques, such as whole-tree harvesting and mechanized felling, produce residues with varying sizes and shapes, including branches, tops, and logging residues. In contrast, selective harvesting methods, such as partial cutting and thinning, may result in residues consisting mainly of branches and small-diameter trees. The choice of harvesting method also affects the moisture content of forestry residues, as mechanized harvesting processes may result in higher moisture levels due to increased exposure to rainfall and environmental conditions. Additionally, site conditions, such as soil type, climate, and topography, play a significant role in determining the characteristics of forestry residues (9, 10). Forest ecosystems exhibit spatial and temporal variability in biomass production, influenced by soil fertility, water availability, and climatic conditions. Consequently, forestry residues may show differences in chemical composition, moisture content, and decomposition rates across forest types and regions. Forestry residues present several challenges, primarily related to their accumulation in forested areas. One significant challenge is the increased risk of wildfires, as dry residues fuel forest fires, posing threats to ecosystems and human communities. Additionally, the accumulation of forestry residues can disrupt natural habitats and nutrient cycling processes, affecting soil health and biodiversity. Moreover, logistical challenges such as transportation and storage hinder the efficient utilization of forestry residues, particularly in remote or inaccessible forested areas.

Agricultural residues constitute a significant portion of biomass resources globally, arising as remnants of crop cultivation and processing activities. These residues encompass diverse organic materials, each with unique properties and potential applications. This elaboration delves into common agricultural residues' characteristics, variability, and suitability for various uses, highlighting examples such as cereal crop residues and sugarcane bagasse (11, 12). Common agricultural residues include crop stalks, husks, straw, and bagasse, which are left behind after harvesting and processing crops such as grains, fruits, and sugarcane. These residues represent a renewable and abundant source of biomass with considerable potential for valorization in bioenergy, biochemical, and biorefinery applications, as shown in Figure 4.1.

Alatzas et al. (1) assessed the potential of biomass resources for renewable energy generation in Greece, specifically emphasizing the regions of Crete, Thessaly, and Peloponnese. By aligning Greek legislation with EU directives, the country has shown an increased interest in renewable energy projects, particularly those involving combined heat and power (CHP) systems. The research estimates that Greece possesses substantial lignocellulosic biomass resources, totaling approximately 2,132,286 tons annually, comparable to figures reported in other Mediterranean countries like Italy and Portugal (13). Moreover, the study highlights the significant contribution of agricultural residues to the overall biomass supply, with Crete and Thessaly alone producing millions of tons of residues

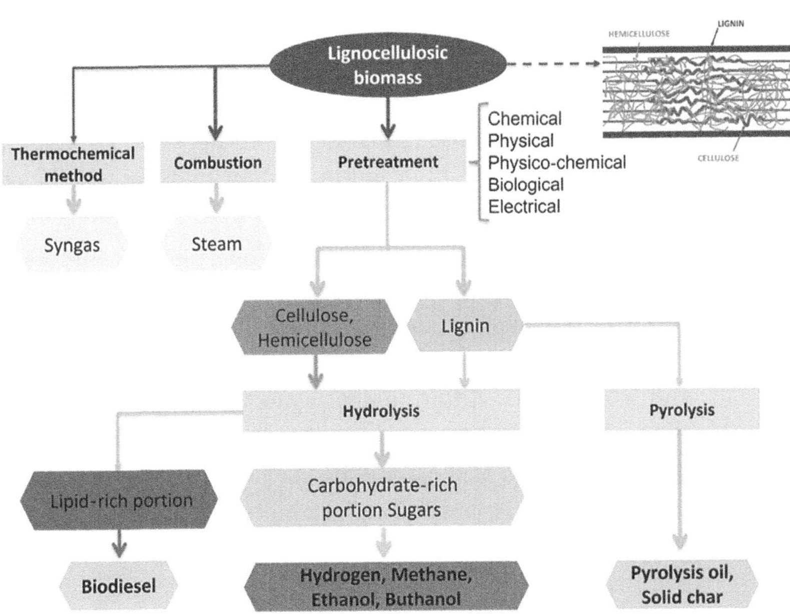

FIGURE 4.1　The primary methodologies for an energetic valorization of lignocellulosic biomass for the production of biofuels. Adapted from ref. (5) under CCBY 4.0.

annually. Noteworthy agricultural residues include olive pits, olive pruning, and cotton ginning remnants, each accounting for over 100,000 tons per year. The research also presented a case study for a proposed CHP gasification facility in Messenia, illustrating the region's promising biomass energy potential of around 3,800,000 GJ/year. This facility, employing small-scale gasification technology, is anticipated to utilize approximately 7956 tons of biomass annually, producing 6,630 MWh of electricity and 8,580 MWh of thermal energy (14, 15). The composition and properties of agricultural residues vary depending on factors such as crop type, cultivation practices, and processing methods. For instance, cereal crop residues like wheat straw and rice husks are rich in cellulose and hemicellulose, two polysaccharides that serve as valuable feedstocks for bioenergy production. Cellulose, the most abundant organic compound on Earth, provides structural support to plant cell walls and can be enzymatically hydrolyzed into glucose for bioethanol production or converted into other biobased products through biochemical processes. Similarly, a heteropolymer composed of various sugars, hemicellulose, can be depolymerized into fermentable sugars for biofuel and biochemical production. In contrast, sugarcane bagasse, the fibrous residue left behind after extracting juice from sugarcane stalks, contains significant lignin content, influencing its properties and potential applications. Lignin, a complex polymer that binds cellulose and hemicellulose in plant cell walls, provides rigidity and resistance to degradation. The high lignin content in sugarcane bagasse makes it less accessible to enzymatic hydrolysis than cellulose-rich residues, necessitating pretreatment processes to enhance its digestibility and conversion efficiency (16, 17). However, lignin-rich residues like bagasse offer advantages in biochemical and biorefinery processes, where lignin can be valorized as a feedstock for producing value-added chemicals, materials, and bioproducts. Additionally, sugarcane bagasse can be used as a bioenergy production feedstock through combustion, gasification, or pyrolysis, providing renewable heat and power for industrial processes or electricity generation. Furthermore, the variability in agricultural residues extends beyond their chemical composition to include physical properties such as particle size, moisture, and ash, which influence their handling, storage, and conversion processes. Effective utilization of agricultural residues requires a comprehensive understanding of their characteristics and properties, coupled with appropriate technologies and strategies for biomass conversion and valorization. Despite their abundance, agricultural residues pose challenges related to handling, storage, and seasonality (18,

19). Harvesting and collecting residues from farm fields require efficient machinery and logistical infrastructure, adding to operational costs for farmers and bioenergy producers. Moreover, the high moisture content of freshly harvested residues and their susceptibility to degradation necessitates proper storage and handling practices to prevent spoilage and maintain biomass quality. Additionally, the seasonal nature of agricultural activities results in intermittent availability of residues, requiring strategic planning for year-round utilization. Figure 4.2 illustrates a schematic depiction of an integrated biorefinery catering to the first, second, third, and fourth generations of biomass.

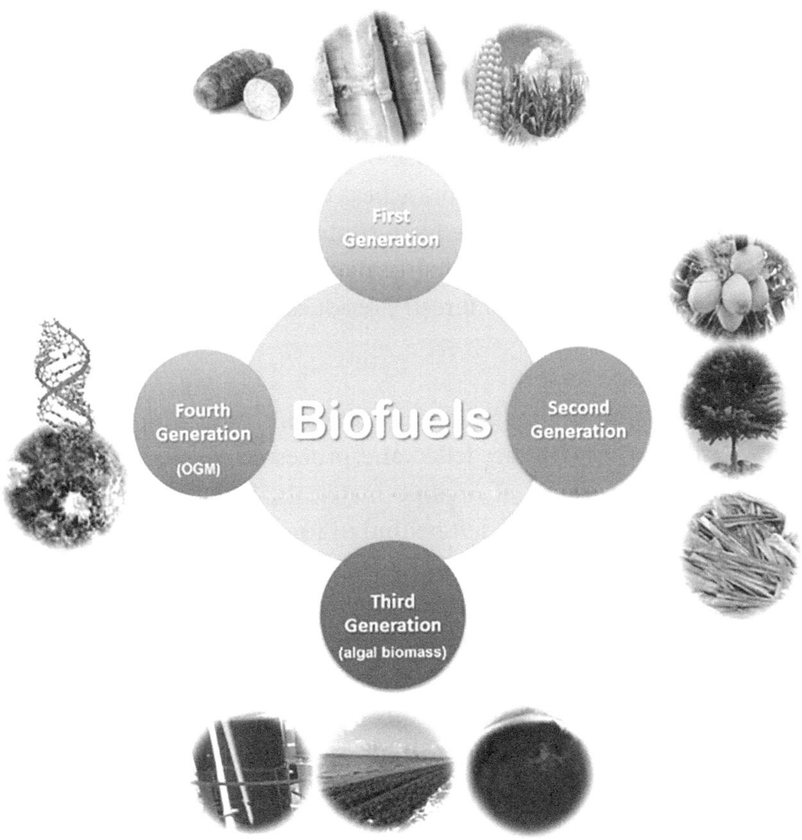

FIGURE 4.2 Schematic depiction of an integrated biorefinery catering to the first, second, third, and fourth generations of biomass. Adapted from ref. (2) under CCBY 4.0.

Forestry and agricultural residues represent abundant biomass resources with significant untapped potential for various applications. These residues can be used as bioenergy production feedstocks, including generating heat, electricity, and biofuels. Advanced conversion technologies such as gasification, pyrolysis, and anaerobic digestion enable the efficient conversion of residues into renewable energy carriers, contributing to energy security and greenhouse gas mitigation. Furthermore, forestry and agricultural residues serve as valuable feedstocks for biochemical and biorefinery processes, enabling the production of biobased materials, chemicals, and bioproducts (20). Additionally, residues can be utilized in agricultural practices as soil amendments, mulch, and livestock bedding, enhancing soil fertility and water retention while reducing the need for synthetic fertilizers and pesticides.

4.2 FUNDAMENTAL CONCEPTS

4.2.1 Origins and Sources of Forestry and Agricultural Residues

Forestry residues encompass diverse organic materials generated during forestry operations, including branches, tree tops, stumps, and logging debris. These residues originate from timber harvesting, forest thinning, and logging operations conducted in natural or managed forest ecosystems. The primary sources of forestry residues include:

4.2.1.1 Timber Harvesting

Timber harvesting, a fundamental practice in forestry operations, generates residues after trees are felled and processed for timber. These residues encompass a range of organic materials, including branches, tree tops, and stumps, which are left behind in forested areas post-harvesting. Branches and tree tops are typically removed during tree processing to obtain merchantable timber, while stumps remain in the ground after tree removal. Traditionally, these residues were left to decompose naturally in the forest, returning nutrients to the soil and providing habitat for various organisms. However, with the advancement of biomass-based industries, there has been growing interest in collecting and utilizing these residues for multiple applications (21–23). In contemporary forestry practices, the management of timber harvesting residues varies depending on site conditions, market demand, and available infrastructure. In some cases, residues are left behind in forested areas to decompose naturally, contributing to nutrient cycling and ecosystem processes. Alternatively, residues may be collected and transported to biomass utilization facilities, where they can be processed into various products such as wood chips, wood pellets, or bioenergy. Using timber harvesting residues in biomass-based industries offers several benefits, including reducing waste, diversifying revenue streams for forest owners, and providing renewable sources of energy and materials.

However, the collection and utilization of timber harvesting residues also pose logistical and economic challenges. Transporting bulky and heterogeneous residues from remote forested areas to processing facilities can be costly and energy-intensive, requiring efficient collection, handling, and transportation systems (24). Moreover, market demand and pricing for residues may fluctuate, affecting the economic viability of collection and utilization activities. Despite these challenges, using timber harvesting residues represents a promising opportunity to enhance the sustainability and value of forestry operations while contributing to the transition towards a more circular and resource-efficient economy.

4.2.1.2 Forest Thinning

Forest thinning is a silvicultural practice essential for maintaining healthy and productive forests. Thinning operations entail the selective removal of trees, typically smaller or less desirable, to optimize stand density, improve tree growth, and enhance overall forest health. By reducing competition for resources such as sunlight, water, and nutrients, thinning promotes the development of more resilient and vigorous trees, leading to improved forest structure and biodiversity. Thinning residues comprise various organic materials generated during forest management, including small-diameter trees, branches, and woody debris (25, 26). These residues result from removing trees and vegetation during thinning operations and may include logging slash and understory vegetation. Thinning residues play important roles in ecosystem functioning and management. While traditionally considered waste or byproducts of forest management, these residues are increasingly recognized for their potential value in biomass utilization and ecosystem services. Thinning residues contain organic matter, nutrients, and biomass that can be utilized for various purposes, including bioenergy production, soil improvement, and carbon sequestration. Additionally, leaving residues on the forest floor can provide habitat for wildlife, enhance soil fertility, and promote nutrient cycling, contributing to overall ecosystem health and resilience.

However, effectively managing thinning residues requires careful planning and considering ecological, economic, and social factors. Removing residues from forest stands can impact soil erosion, nutrient cycling, and biodiversity, highlighting the importance of sustainable harvesting practices and residue management strategies. Integrating thinning operations with biomass utilization initiatives can optimize the value of residues while minimizing environmental impacts and maximizing ecosystem benefits. Collaborative approaches involving forest managers, researchers, policymakers, and local communities are essential for developing and implementing sustainable thinning practices that balance ecological conservation with economic viability and social well-being.

4.2.1.3 Logging Operations

Logging operations encompass a series of activities to harvest timber from forests, each generating its own residues. Skidding involves dragging felled trees from the harvest site to loading areas using specialized equipment, which can result in the accumulation of logging slash, branches, and small-diameter trees along skid trails (27–29). Delimbing is removing branches from fallen trees, producing additional residues in the form of branches and tops. Bucking involves cutting felled trees into smaller sections, resulting in residues such as logging slash, bark, and sawdust. These residues accumulate at logging sites or along logging roads and trails, posing challenges related to waste management and environmental impacts. The accumulation of logging residues presents several environmental and logistical challenges. Environmental concerns include the risk of soil erosion, habitat fragmentation, and water pollution associated with residue accumulation. Additionally, logging residues can create barriers to natural regeneration, hinder access to logging sites, and increase wildfire risk. Logistical challenges arise from the need to manage and dispose of residues effectively, which terrain, weather conditions, and regulatory requirements can complicate. Furthermore, the transportation of residues to processing facilities or disposal sites can incur additional costs and logistical constraints. Despite these challenges, logging residues also present opportunities for utilization in biomass-based industries. Residues such as logging slash, bark, and sawdust can be collected and processed into various products, including wood chips, mulch, bioenergy, and engineered wood products. By valorizing logging residues, forest managers can minimize waste, generate additional revenue streams, and promote sustainable forest management practices (30, 31). However, effectively utilizing logging residues requires infrastructure, technology, and market development investment to overcome logistical, economic, and regulatory barriers. Moreover, sustainable residue management strategies, such as leaving residues on-site to enhance soil fertility and biodiversity, should be considered to minimize environmental impacts and promote ecosystem resilience.

Similarly, agricultural residues comprise organic materials after crop cultivation, harvesting, and processing activities of farm landscapes. Familiar sources of agricultural residues include:

4.2.1.4 Crop Harvesting

Crop harvesting marks the culmination of the agricultural cycle, leaving behind residues such as stalks, stems, leaves, and husks in the field. These residues, collectively known as crop residues, are crucial in soil conservation, moisture retention, and nutrient cycling. The composition and quantity of crop residues vary depending on crop type, cultivation practices, and harvesting methods. For example, cereal crop residues like wheat and rice straw are rich in cellulose and hemicellulose, while legume crop residues such as soybean stalks contain higher protein content. Furthermore, post-harvest residues left on the field can serve as valuable organic matter, enhancing soil fertility and structure while reducing erosion and runoff.

4.2.1.5 Crop Processing

Crop processing involves a series of activities to convert raw agricultural commodities into processed products such as grains, oils, and fibers. Residues such as straw, husks, chaff, and bran are generated as byproducts during processing. These residues vary in composition and properties depending on the crop type and processing methods employed. For instance, rice milling produces rice husks and bran, while wheat milling generates wheat straw and bran. These residues can be utilized for various purposes, including animal bedding, composting, and bioenergy production (32, 33). Additionally, residues from oilseed processing, such as soybean husks and meal, are rich in protein and can be used as feed ingredients for livestock and poultry.

4.2.1.6 Agroindustrial Operations

Agroindustrial operations encompass value-added processing industries that transform agricultural commodities into higher-value products. Examples include sugar milling, oil extraction, and bioethanol production. These processes generate residues such as bagasse, oil cake, and distillers' grains, which are byproducts of the extraction and refining processes. Bagasse, the fibrous residue remaining after sugarcane crushing, is used as a biofuel in sugar mills or for paper pulp production. Oil cake, a byproduct of oilseed extraction, is rich in protein and used as animal feed. Similarly, distillers' grains, a byproduct of bioethanol production, are rich in protein and fiber and are used as livestock feed. These residues offer opportunities for valorization in various industries, contributing to waste reduction, resource efficiency, and circular economy principles.

4.2.2 Composition and Characteristics of Residues

Forestry and agricultural residues exhibit diverse compositions and characteristics influenced by feedstock type, plant species, harvesting methods, and environmental conditions. These residues consist primarily of organic compounds such as cellulose, hemicellulose, lignin, proteins, lipids, sugars, minerals, and varying amounts of water and volatile matter. The composition and properties of residues determine their suitability for different applications, including bioenergy production, biochemical processing, and agricultural uses.

- Cellulose: Cellulose is the most abundant organic compound in forestry and agricultural residues, accounting for the structural integrity of plant cell walls. It is a linear polymer composed of glucose units linked by β-1,4-glycosidic bonds, providing strength, rigidity, and durability to biomass materials. Cellulose-rich residues like wood, straw, and stalks are valuable feedstocks for bioenergy production and biochemical processing, as cellulose can be enzymatically hydrolyzed into fermentable sugars for ethanol production or converted into platform chemicals and biopolymers (34, 35).

- Hemicellulose: Hemicellulose is a heterogeneous polysaccharide found in plant cell walls, comprising a diverse array of sugar monomers such as xylose, arabinose, mannose, and galactose. It acts as a cementing material between cellulose fibrils, contributing to the flexibility and cohesion of biomass materials. Hemicellulose-rich residues like straw, husks, and bran are suitable feedstocks for bioenergy production, as hemicellulose can be hydrolyzed into fermentable sugars for bioethanol or biogas production.

- Lignin: Lignin is a complex phenolic polymer that fills the spaces between cellulose and hemicellulose in plant cell walls, providing structural support and impermeability to water. It is highly resistant to enzymatic degradation and is a barrier to biomass utilization in biochemical and biorefinery processes. Lignin-rich residues like bark, bagasse, and straw pose challenges for biomass conversion due to their recalcitrance and high energy requirements for depolymerization and valorization.

- Proteins, lipids, and sugars: Forestry and agricultural residues contain proteins, lipids, and sugars as minor components, contributing to their nutritive value and potential applications in animal feed, human nutrition, and bioprocessing (36). Proteins and lipids in residues like bran, oil cake, and distillers' grains can be recovered and utilized as value-added co-products in agroindustrial processes. At the same time, sugars derived from enzymatic hydrolysis of lignocellulosic biomass can be fermented into biofuels and biochemicals.

- Minerals and ash: Residues may contain varying amounts of minerals and ash derived from plant tissues, soil contamination, and processing residues. The ash content of biomass materials affects their combustion properties, deposition, and emissions during thermal conversion. High ash content in residues like straw and bagasse can pose challenges for combustion and gasification technologies, necessitating ash management strategies to mitigate operational issues and environmental impacts.

4.2.3 Abundance and Distribution of Residue Resources

Forestry and agricultural residues represent abundant renewable biomass resources with significant untapped potential for sustainable utilization. The abundance and distribution of residue resources vary geographically and temporally, influenced by land use patterns, crop production systems, forest management practices, and climatic conditions.

- Forest residues: Forested areas globally produce substantial forestry residues, including logging slash, tree tops, branches, and stumps, as byproducts of timber harvesting and forest management activities. The abundance of forest residues depends on factors such as forest type, stand age, tree density, and harvesting intensity. Regions with extensive forest cover, such as boreal and temperate ecosystems, tend to have higher forestry residues available for utilization. Additionally, forestry residues may be more abundant in regions with active forest industries and logging operations, where timber harvesting activities generate significant quantities of logging residues and processing residues.

- Agricultural residues: Agricultural landscapes produce vast quantities of agricultural residues, comprising crop residues, processing residues, and agroindustrial byproducts (37–39). The abundance of agricultural residues varies seasonally and spatially, influenced by crop yields, cropping intensity, cropping systems, and agricultural practices. Regions with intensive crop production, such as the Midwestern United States, the European Union, and parts of Asia and South America, generate significant agricultural residues from cereal crops, oilseeds, pulses, and sugar crops. Additionally, agroindustrial regions with processing industries such as sugar mills, oilseed crushers, and bioethanol plants produce significant quantities of processing residues like bagasse, oil cake, and distillers' grains.

- Distributional patterns: The distribution of forestry and agricultural residues exhibits spatial variability, reflecting regional differences in land use, natural resource availability, and socio-economic factors. In forestry-dominated regions, forestry residues are concentrated in forested areas and logging sites, where timber harvesting activities occur. In agricultural landscapes, agricultural residues are dispersed across croplands, orchards, and agroindustrial facilities, reflecting the spatial distribution of crop production and processing activities. The distribution of residue resources is influenced by proximity to biomass conversion facilities, transportation infrastructure, and markets for biomass products and bioenergy.

- Temporal dynamics: The availability of forestry and agricultural residues varies temporally, following seasonal patterns of crop growth, harvesting, and processing cycles. Agricultural residues are typically available following crop harvesting seasons, with peak volumes for cereals, oilseeds, and pulses in the autumn months. Forestry residues may accumulate throughout the year due to continuous forest management activities, with fluctuations in availability depending on timber harvesting schedules and forest management practices (40, 41). Understanding the temporal dynamics of residue availability is essential for biomass supply chain management, logistics, and utilization planning.

4.3 CLASSIFICATION OF RESIDUES

4.3.1 Categorization Based on Botanical Origin

Residues derived from forestry and agriculture can be classified based on their botanical origin, with distinctions made between wood residues and crop residues, each with unique properties and potential applications. Wood residues originate from trees and woody plants, encompassing branches, tree tops, stumps, bark, and sawdust. These residues are typically generated during timber harvesting and processing operations in forestry practices. Wood residues are characterized by their high cellulose, hemicellulose, and lignin content, making them valuable feedstocks for various industries. Due to their composition, wood residues are particularly suitable for bioenergy production, biochemical processing, and manufacturing value-added products such as pulp and paper. In contrast, crop residues are derived from crops and include stalks, husks, straw, and bagasse, which are left behind after crop harvesting and processing activities. The composition and properties of crop residues vary depending on crop type, cultivation practices, and processing methods. For example, cereal crop residues like wheat straw and rice husks are rich in cellulose and hemicellulose, making them suitable for bioenergy production and biochemical processing. Sugarcane bagasse, a residue from sugar milling operations, contains significant lignin content, influencing its potential applications in bioenergy and biorefinery processes (42, 43). Crop residues can be utilized for various purposes, including animal feed, soil amendment, biofuel production, and biochemical processing, reflecting their versatility and importance in sustainable agriculture and biomass utilization initiatives.

Effective classification of residues based on their botanical origin allows for targeted utilization strategies and resource management practices. By understanding the composition and properties of wood and crop residues, stakeholders can identify optimal pathways for their utilization, maximize resource efficiency, and minimize environmental impacts. Moreover, integrating wood and crop residues into circular bioeconomy frameworks contributes to developing sustainable, resilient, and resource-efficient systems that reduce reliance on fossil-based resources and promote the transition towards a low-carbon, circular economy. Through innovative technologies, policies, and partnerships, wood and crop residues can significantly address global challenges such as climate change, energy security, and resource scarcity while fostering economic growth and social development in rural communities.

4.3.2 Classification by Physical and Chemical Properties

Residues can also be classified based on their physical and chemical properties, such as moisture content, particle size, chemical composition, and calorific value.

- Moisture content: Residues with high moisture content may require drying or preprocessing before utilization to improve combustion efficiency, reduce transportation costs, and prevent microbial degradation. Moisture content affects the energy content, storage stability, and handling characteristics of residues, influencing their suitability for different applications (44, 45). High moisture content can decrease the energy density of biomass materials, leading to lower combustion efficiency and increased emissions. Additionally, moisture content influences residues' storage and handling requirements, with wetter materials often requiring specialized storage facilities and handling equipment to prevent spoilage and degradation during storage and transportation.

- Particle size: Residues exhibit a wide range of particle sizes, from fine powders to coarse chips or chunks, influencing their processing requirements and suitability for various biomass conversion technologies. Particle size affects combustion behavior, gasification efficiency, and feedstock compatibility with biomass conversion systems. Fine particles may be prone to dust formation, combustion instabilities, and poor flow characteristics, while larger particles may require size reduction or densification for efficient handling and utilization. Optimizing particle size distribution is essential for maximizing biomass conversion efficiency and minimizing operational challenges in biomass processing facilities.

- Chemical composition: The chemical composition of residues, including cellulose, hemicellulose, lignin, proteins, lipids, sugars, and minerals, is critical in determining their suitability for specific applications. Residues with high cellulose and hemicellulose content are well-suited for bioenergy production and biochemical processing, as these polymers can be readily hydrolyzed into fermentable sugars. Conversely, residues rich in lignin may be more challenging to convert due to their recalcitrance and resistance to degradation (46). Understanding the chemical composition of residues allows for targeted utilization strategies and process optimization to maximize the value and efficiency of biomass conversion processes.

- Calorific value: The calorific value of residues, measured in terms of energy content per unit mass or volume, indicates their potential for heat generation and energy recovery. Residues with higher calorific values have more energy potential and may be preferred for combustion or thermal conversion processes. Calorific value is influenced by moisture content, chemical composition, and density, with dry, high-density residues typically exhibiting higher calorific values. Assessing the calorific value of residues is essential for determining their energy potential and optimizing the selection and operation of biomass conversion technologies for heat, power, or biofuel production.

4.3.3 Importance of Accurate Classification for Effective Utilization

Accurate classification of residues is crucial for selecting appropriate utilization pathways, optimizing process efficiency, and maximizing value-added opportunities.

- Resource assessment: Classification provides a structured framework for quantifying and characterizing residue resources, enabling comprehensive resource assessments. By categorizing residues based on their botanical origin, physical properties, and chemical composition, stakeholders can accurately estimate the availability and potential yields of different residues. This information is invaluable for resource planning, investment decisions, and policy development to promote sustainable biomass utilization. Additionally, resource assessment allows for identifying regional variations in residue availability, enabling targeted interventions and infrastructure development to maximize resource utilization and minimize logistical challenges.

- Technology selection: Classification guides the selection of appropriate technologies and processing methods tailored to residues' specific properties and characteristics (47, 48). Stakeholders can optimize process efficiency, product quality, and economic viability by categorizing residues based on their suitability for different utilization pathways, such as combustion, gasification, pyrolysis, biochemical conversion, or biorefinery processes. Moreover, matching residues to compatible technologies facilitates innovation and the development of customized solutions to address specific biomass resources and end-use applications. This approach fosters the integration of advanced technologies and the development of sustainable biomass utilization practices.

- Product development: Classification catalyzes product development and innovation by providing insights into different residues' unique properties and characteristics. Researchers and engineers can explore novel applications and value-added opportunities in bioenergy, bioproducts, and biomaterials by identifying residues with desirable attributes, such as high cellulose content, low ash content, or unique chemical functionalities. Leveraging residue classification enables the optimization of product formulations, process parameters, and supply chain logistics to maximize value capture and market competitiveness (49, 50). Additionally, it fosters collaboration among industry stakeholders, research institutions, and government agencies to accelerate the commercialization of innovative products and technologies derived from residues.

- Environmental impact: Accurate classification is essential for assessing the environmental impact of residue utilization activities and implementing effective mitigation measures. By quantifying environmental indicators such as greenhouse gas emissions, air pollution, water usage, and land use change associated with different utilization pathways, stakeholders can identify potential environmental hotspots and prioritize strategies for minimizing adverse effects. Furthermore, classification enables the development of sustainability metrics, lifecycle assessments, and certification standards to ensure responsible resource management and compliance with regulatory requirements. This holistic approach fosters transparency, accountability, and continuous improvement in biomass utilization practices, ultimately contributing to the preservation of ecosystems and the well-being of communities.

4.4 PROPERTIES OF RESIDUES

4.4.1 Physical Properties

4.4.1.1 Density

The density of residues is a critical parameter that influences various aspects of biomass utilization, including handling, transportation, and storage considerations. Density refers to the mass per unit volume of the biomass material and plays a significant role in determining its physical properties and behavior during processing. High-density residues like wood chips or pellets have a greater mass per unit volume than low-density residues like straw or husks. As a result, high-density residues typically exhibit higher energy density, meaning they contain more energy per unit volume, which can be advantageous for transportation and storage efficiency. For example, wood pellets, which are densely packed and have a high energy content, can be transported and stored more efficiently than bulkier low-density residues like straw or husks. Additionally, high-density residues may require less storage space, making them more suitable for large-scale biomass handling facilities and transportation logistics. Furthermore, density influences combustion characteristics and reactor design in biomass conversion technologies (51, 52). High-density residues tend to have better combustion properties, including improved particle flow and combustion stability, compared to low-density residues. High-density residues can form a more uniform and compact fuel bed, facilitating better air distribution and heat transfer during combustion. Moreover, density plays a crucial role in designing and operating biomass conversion reactors, such as gasifiers or pyrolysis units. The density of the feedstock affects particle flow dynamics, residence time, and heat transfer within the reactor, ultimately influencing process efficiency and product quality. Therefore, understanding the density of residues is essential for optimizing biomass conversion processes and ensuring the successful operation of biomass-based energy systems.

4.4.1.2 Moisture Content

Moisture content is a critical parameter in biomass residues that directly influences combustion efficiency, energy content, and storage stability. It refers to the amount of water in the biomass material and is significant in determining its overall quality and suitability for various applications. High moisture content in residues can have several detrimental effects on their utilization. First, high moisture content reduces the calorific value of residues, meaning that a more significant proportion of the energy content is tied up in water rather than being available for combustion. This lowers the efficiency of energy production processes, such as biomass combustion or gasification, as less energy is released per unit mass of biomass burned (53, 54). Moreover, high moisture content increases transportation costs and logistical challenges associated with handling and storage. Wet biomass materials are heavier and bulkier, requiring more energy and resources for transportation. Additionally, excessive moisture can lead to issues such as leaching during transport, increasing the risk of water pollution and environmental contamination.

Furthermore, high moisture content in residues can promote microbial degradation and spontaneous combustion. Microorganisms thrive in moist environments and can degrade biomass materials, leading to quality deterioration, foul odors, and reduced shelf life. Additionally, the heat generated by microbial activity in moist biomass piles can cause spontaneous combustion, posing safety hazards and increasing the risk of fires. Proper drying or preprocessing of biomass residues is often required to mitigate these challenges to reduce moisture content to acceptable levels. This can involve various techniques such as air, mechanical, or thermal drying. By reducing moisture content, biomass feedstocks' quality and energy value can be improved, enhancing their suitability for combustion, gasification, biofuel production, and other applications.

4.4.2 Chemical Properties

4.4.2.1 Lignin Content

Lignin is a complex polymer found in plant cell walls, providing structural support and rigidity to plant tissues. It is one of the main components of lignocellulosic biomass, alongside cellulose and hemicellulose. Lignin's presence significantly influences the properties and characteristics of residues, impacting their suitability for various biomass conversion processes. Residues with higher lignin content, such as wood chips or bark, tend to be more recalcitrant and resistant to degradation than those with lower lignin content (55, 56). This is because lignin forms a dense and complex network within the cell wall matrix, creating physical barriers that hinder the accessibility of cellulose and hemicellulose to enzymes and other degrading agents. As a result, lignin-rich residues may be more challenging to convert into biofuels or biochemicals through enzymatic hydrolysis or fermentation processes, which rely on breaking down cellulose and hemicellulose into fermentable sugars. However, despite its recalcitrance, lignin-rich residues still hold significant potential for valorization through thermochemical conversion processes such as pyrolysis or gasification. These processes involve the thermal decomposition of biomass in the absence of oxygen to produce a range of valuable products, including syngas (a mixture of carbon monoxide, hydrogen, and other gases), bio-oil (a liquid mixture of organic compounds), and biochar (a carbon-rich solid residue). With its high energy content and complex molecular structure, lignin can be efficiently converted into these valuable products through thermochemical processes. Pyrolysis, for example, involves heating biomass at high temperatures without oxygen to break down organic compounds into volatile gases, liquids, and char (57, 58). Lignin-rich residues undergo thermal decomposition, yielding bio-oil, syngas, and biochar, which can be further processed into biofuels, chemicals, or soil amendments. Similarly, gasification converts lignin-rich residues into syngas through partial oxidation at elevated temperatures, offering a versatile feedstock for producing heat, power, fuels, and chemicals.

4.4.2.2 Cellulose and Hemicellulose Content

Cellulose and hemicellulose are two major polysaccharides in plant cell walls, contributing to plant tissue's structural integrity and rigidity. Cellulose is a linear polymer composed of glucose units linked by β-1,4-glycosidic bonds, while hemicellulose is a heterogeneous polymer consisting of various sugar monomers such as xylose, arabinose, mannose, and galactose. Together, these polysaccharides represent valuable carbohydrate resources that can be enzymatically hydrolyzed into fermentable sugars for biofuel production. Residues with high cellulose and hemicellulose content, such as straw, bagasse, or wood chips, are preferred feedstocks for biochemical conversion processes like enzymatic hydrolysis or fermentation (59, 60). During enzymatic hydrolysis, cellulose and hemicellulose are broken down into constituent sugar molecules by cellulase and hemicellulase enzymes. These fermentable sugars, primarily glucose and xylose, serve as the primary carbon and energy source for microbial fermentation into biofuels such as bioethanol, biobutanol, or other bioproducts. The enzymatic hydrolysis of cellulose and hemicellulose requires the presence of cellulase and hemicellulase enzymes, which can efficiently break down the complex polysaccharide structures into soluble sugars. Residues with high cellulose and hemicellulose content provide abundant substrates for enzymatic hydrolysis, resulting in higher yields of fermentable sugars than lignin-rich residues. Additionally, the composition and structure of cellulose and hemicellulose in different biomass feedstocks can influence the efficiency of enzymatic hydrolysis and fermentation processes.

Biochemical conversion processes offer several advantages for biofuel production, including high specificity, mild reaction conditions, and compatibility with existing infrastructure. By utilizing cellulose and hemicellulose as feedstocks, biochemical conversion technologies enable the efficient production of biofuels from renewable biomass resources, contributing to developing sustainable and environmentally friendly energy systems. Moreover, the versatility of cellulose and hemicellulose allows for producing a wide range of bioproducts beyond biofuels, including platform chemicals, bioplastics, and pharmaceuticals, further expanding the potential applications of biomass-derived carbohydrates.

4.4.2.3 Ash Content

Ash content in biomass refers to the inorganic mineral residue that remains after biomass materials' combustion or thermal conversion. This ash comprises minerals and other inorganic compounds in the original biomass feedstock. The ash content of biomass significantly influences various aspects of thermal conversion processes, including combustion efficiency, ash deposition, slag formation, and emissions of particulate matter and pollutants. Residues with high ash content, such as agricultural residues (e.g., straw, husks) or tree bark, may pose challenges for combustion technologies for several reasons. First, during combustion, the minerals in the ash undergo chemical reactions and form ash particles that can accumulate on heat transfer surfaces, leading to ash deposition. This ash accumulation can reduce heat transfer efficiency, increase maintenance requirements, and decrease overall combustion efficiency. Moreover, high-ash biomass residues can contribute to slag formation in combustion systems. Slag is a molten or partially fused deposit that forms on combustion chamber walls or heat exchange surfaces due to the melting of ash particles at high temperatures. Slag formation can obstruct airflow and heat transfer, reducing combustion efficiency and increasing maintenance costs.

In addition to operational challenges, high-ash biomass residues can also result in emissions of particulate matter and pollutants during combustion. The combustion of ash-rich biomass can release fine particles into the atmosphere, contributing to air pollution and adverse health effects. Furthermore, certain minerals present in the ash, such as sulfur and chlorine compounds, can lead to the formation of harmful emissions such as sulfur dioxide (SO2) and hydrogen chloride (HCl). Ash management strategies may be implemented to mitigate the challenges associated with high ash content in biomass (61, 62). These strategies may include ash recycling, where the ash is reused as a soil amendment or fertilizer, or ash removal, where the ash is collected and disposed of in a controlled manner. Additionally, developing advanced combustion technologies and emission control systems can help minimize the impact of high-ash biomass residues on combustion efficiency and environmental emissions.

4.4.3 Thermal Properties and Calorific Value

4.4.3.1 Thermal Properties

The thermal properties of residues play a crucial role in determining their behavior and performance during thermal conversion processes such as combustion, gasification, or pyrolysis. These properties, which include ignition temperature, combustion kinetics, and ash fusion characteristics, provide valuable insights into how residues will react under different operating conditions and help optimize process performance. Ignition temperature refers to when a material begins to undergo combustion. Residues with low ignition temperatures ignite more readily and require less external energy input to initiate combustion. This property is significant for efficient energy recovery and heat generation in combustion or gasification systems, where rapid and consistent ignition is desired. For example, dry wood chips or sawdust, which typically have low ignition temperatures, are preferred feedstocks for biomass boilers or gasifiers due to their ease of ignition and high combustion efficiency. Combustion kinetics describe the rate at which a material undergoes combustion and release of heat energy. Residues with fast combustion kinetics combust rapidly, releasing energy quickly and efficiently (63, 64). This property is beneficial for achieving high thermal efficiency and minimizing residence time in combustion or gasification systems. Conversely, residues with slower combustion kinetics may require longer residence times and higher temperatures to complete combustion, leading to lower overall process efficiency.

Ash fusion characteristics refer to the temperature at which the ash residue from combustion softens and fuses. This property is critical for preventing slag formation and ash deposition on heat transfer surfaces within combustion or gasification systems. Residues with high ash fusion temperatures are preferred, as they are less likely to form molten deposits that can obstruct airflow and reduce heat transfer efficiency. Understanding the thermal behavior and combustion kinetics of residues is essential for optimizing process parameters, improving energy efficiency, and reducing emissions in thermal conversion systems. By selecting residues with favorable thermal properties and adjusting operating conditions accordingly, stakeholders can maximize energy recovery, minimize environmental impacts, and ensure the sustainable utilization of biomass resources for heat and power generation (65, 66). Additionally, advancements in thermal analysis techniques and computational modeling allow for more accurate prediction and optimization of thermal conversion processes, further enhancing the efficiency and sustainability of biomass utilization technologies.

4.4.3.2 Calorific Value

Calorific value—or heating value or energy content—is a fundamental parameter that quantifies the amount of energy released per unit mass or volume of biomass residues during combustion, gasification, or pyrolysis. It is a critical indicator of biomass fuel's energy potential and efficiency, providing valuable information for process optimization, energy planning, and economic assessment. Residues with high calorific values, such as wood chips or pellets derived from dense and energy-rich biomass sources, possess more significant energy potential and can produce more heat or power per unit mass compared to residues with lower calorific values, such as straw or husks. This makes high calorific value residues preferred feedstocks for applications where energy density and efficiency are critical, such as biomass boilers, gasifiers, or cogeneration systems. Various factors, including moisture content, chemical composition, and ash content, influence the calorific value of biomass residues. Moisture content is particularly significant, as water has a high heat capacity and must be evaporated before combustion, reducing the effective energy content of the biomass. Therefore, residues with lower moisture content generally exhibit higher calorific values and are more desirable for energy production (67, 68).

Additionally, the chemical composition of biomass residues, including the relative proportions of cellulose, hemicellulose, lignin, proteins, lipids, and other organic compounds, can impact their calorific values. Biomass materials rich in carbon and hydrogen, such as lignocellulosic materials, typically have higher calorific values due to their higher energy content per unit mass. Conversely, residues with higher ash content may have lower calorific values, as the inorganic minerals in the ash do not contribute to energy generation during combustion. Furthermore, contaminants or impurities in biomass residues, such as soil, sand, or non-combustible materials, can also affect their calorific values by diluting the energy content of the biomass. Therefore, proper preprocessing and quality control measures are essential to ensure consistent and reliable calorific values for biomass feedstocks.

4.5 ENVIRONMENTAL SUSTAINABILITY

4.5.1 Role of Forestry and Agricultural Residues in Carbon Sequestration

4.5.1.1 Biomass Carbon Sequestration

Forestry and agricultural residues serve as significant repositories of organic carbon, originating from the assimilation of atmospheric carbon dioxide (CO_2) during photosynthesis. This organic carbon, stored within the biomass of trees, crops, and other plant materials, represents a vital carbon sink crucial in mitigating climate change. When these residues are utilized for bioenergy production or biochemical processing, the carbon stored in biomass is effectively diverted from the atmosphere temporarily. Through processes such as combustion, gasification, or biochemical conversion, the carbon in biomass is converted into energy, biofuels, or bioproducts (69, 70). This utilization of residues for energy purposes serves to offset the use of fossil fuels, which would otherwise release additional carbon dioxide into the atmosphere. By utilizing forestry and agricultural residues as renewable energy sources, we can effectively reduce net greenhouse gas emissions by avoiding the release of carbon dioxide that would occur from the combustion of fossil fuels. This carbon diversion process helps to mitigate climate change by reducing the overall carbon footprint associated with energy production and consumption. Furthermore, by promoting biomass-derived energy, we can contribute to the transition to a low-carbon economy and reduce our reliance on finite and environmentally harmful fossil fuel resources.

However, it's important to note that it's not a permanent solution while using forestry and agricultural residues for bioenergy production can help mitigate climate change in the short to medium term. The carbon stored in biomass is only temporarily sequestered from the atmosphere and will eventually be released into the environment through decomposition or combustion. Therefore, to achieve long-term climate goals, it's essential to complement biomass utilization with efforts to reduce overall greenhouse gas emissions, increase energy efficiency, and promote the adoption of sustainable land management practices.

4.5.1.2 Soil Carbon Storage

Incorporating forestry and agricultural residues into the soil as organic amendments or mulch represents a valuable strategy for enhancing soil health and promoting carbon sequestration. These residues serve as a source of organic matter, enriching the soil with essential nutrients and fostering microbial activity. As organic matter decomposes slowly, it releases carbon into the soil, contributing to the buildup of soil organic carbon (SOC) stocks (71–73). One of the primary benefits of incorporating residues into soil is the improvement of soil fertility. The organic matter contained in residues provides a source of nutrients, such as nitrogen, phosphorus, and potassium, that are essential for plant growth. By replenishing these nutrients, residues support plant growth and enhance crop productivity, ultimately contributing to sustainable agricultural practices.

Furthermore, residues play a crucial role in enhancing soil structure and stability. As organic matter decomposes, it forms stable aggregates that help to create pore spaces within the soil. These pore spaces improve soil aeration, water infiltration, and root penetration, enhancing soil structure and reducing erosion risk. Additionally, the presence of organic matter in soil promotes the development of soil aggregates, which helps to prevent soil compaction and improve overall soil resilience (74, 75). Moreover, residues contribute to moisture retention in soil by increasing its water-holding capacity. The organic matter in residues acts like a sponge, absorbing and holding water within the soil profile. This helps to mitigate the effects of drought and improve water availability for plant uptake, especially in arid or drought-prone regions. By improving soil moisture retention, residues enhance crop resilience and reduce irrigation requirements, leading to more sustainable agricultural practices.

4.5.1.3 Long-Term Carbon Storage

Residues utilized for producing durable products such as biochar or engineered wood products play a significant role in long-term carbon storage and climate change mitigation. These products offer innovative solutions for capturing and storing carbon from forestry and agricultural residues, thereby reducing the concentration of greenhouse gases in the atmosphere. Biochar, produced through pyrolysis, involves heating biomass residues in the absence of oxygen, converting organic matter into a stable form of carbon-rich material. This biochar can then be incorporated into the soil as a long-term carbon sink. Biochar has a high surface area and porous structure, which provide habitat for beneficial microorganisms and enhance soil fertility and moisture retention (76, 77). Biochar can sequester carbon in the soil for extended periods, ranging from centuries to millennia, thereby contributing to long-term carbon storage and climate change mitigation. In addition to biochar, residues can be utilized to produce engineered wood products, such as cross-laminated timber (CLT) or glued-laminated timber (glulam). These products involve the compression and bonding of forestry residues into durable structural materials used in construction applications. Engineered wood products sequester carbon within the material itself and displace carbon-intensive materials such as concrete or steel, further reducing carbon emissions associated with construction activities (78–81).

By converting forestry and agricultural residues into durable products such as biochar or engineered wood, we can effectively lock up carbon for extended periods, mitigating the impacts of climate change. These products offer sustainable alternatives to conventional practices and provide opportunities for enhancing soil fertility, promoting biodiversity, and supporting resilient ecosystems. Using residues for long-term carbon storage represents a promising pathway for achieving climate change mitigation goals and advancing toward a more sustainable future.

4.5.2 Impact on Soil Health and Erosion Prevention

- Soil organic matter: Incorporating forestry and agricultural residues into soil represents a sustainable practice that enhances soil organic matter content, improving soil health and productivity. These residues are a valuable source of organic carbon, energy, and nutrients for soil microorganisms, stimulating microbial activity and diversity.

As microorganisms decompose the organic matter in residues, they release essential nutrients into the soil, such as nitrogen, phosphorus, and potassium, vital for plant growth and development (82, 83). This process promotes nutrient cycling within the soil, facilitating the breakdown of organic matter and the formation of stable soil aggregates. By enhancing soil structure, residues improve soil porosity, water infiltration, and root penetration, increasing water retention and nutrient availability for plants. Moreover, organic matter in soil promotes microbial resilience against environmental stresses, such as drought, disease, and extreme temperatures, thereby contributing to the overall resilience of agroecosystems.

- Erosion control: Applying forestry and agricultural residues as mulch or ground cover is crucial in erosion control and soil conservation. By covering the soil surface, residues reduce the impact of erosive forces such as water runoff, surface crusting, and wind erosion. Residues act as a protective barrier, shielding the soil from the effects of rainfall and preventing soil particles from being dislodged and transported by flowing water or wind. Additionally, residues help maintain soil moisture levels by reducing evaporation and minimizing soil temperature fluctuations, thereby improving soil water retention and reducing erosion risk. Furthermore, residues stabilize soil aggregates, preventing them from breaking apart and being washed away by runoff. By minimizing soil erosion, residues contribute to preserving soil fertility, preventing sedimentation in water bodies, and reducing nonpoint source pollution, ultimately supporting environmental conservation efforts and sustainable land management practices.

- Nutrient cycling: Forestry and agricultural residues play a vital role in nutrient cycling within agroecosystems, facilitating the recycling and redistribution of essential nutrients. As residues decompose and mineralize, nutrients stored within their organic matter are released back into the soil, becoming available for plant uptake and utilization. This nutrient recycling process enhances nutrient availability in the soil, reducing the reliance on synthetic fertilizers and minimizing nutrient losses to the environment through leaching or runoff (84, 85). By recycling nutrients from residues, agroecosystems can maintain soil fertility, support plant growth, and sustain agricultural productivity over time. Moreover, incorporating residues into soil promotes the development of nutrient-rich organic matter, which serves as a nutrient reservoir and enhances soil fertility in the long term.

4.5.3 Contribution to Greenhouse Gas Emissions Reduction

- Substitution for fossil fuels: Using forestry and agricultural residues for bioenergy production offers a sustainable alternative to fossil fuels, mitigating climate change by reducing greenhouse gas emissions. Biomass-derived fuels, such as wood pellets, bioethanol, or biogas, emit lower levels of greenhouse gases per unit of energy than traditional fossil fuels like coal, oil, or natural gas (86, 87). By substituting fossil fuels with biomass-derived fuels, emissions of CO_2 and other pollutants associated with fossil fuel combustion are minimized, contributing to global efforts to combat climate change and transition towards cleaner, renewable energy sources.

- Carbon neutral energy: Residues utilized for bioenergy production are often considered carbon-neutral when accounting for carbon emissions and sequestration throughout their lifecycle. While the combustion or conversion of residues releases CO_2 into the atmosphere, this carbon emission is offset by the carbon uptake during the growth of biomass feedstocks. As trees and crops used for biomass production absorb CO_2 from the atmosphere through photosynthesis, the net carbon emissions over the biomass lifecycle are nearly zero, resulting in a carbon-neutral energy source. This carbon-neutral characteristic makes biomass-derived energy an attractive option for reducing net greenhouse gas emissions and addressing climate change concerns.

- Avoided emissions: Utilizing forestry and agricultural residues for bioenergy or bioproducts also helps avoid methane (CH4) and nitrous oxide (N2O) emissions that would otherwise occur from biomass decomposition or open burning. When residues are left to decompose naturally or are burned in open fields, they release methane and nitrous oxide, which are potent greenhouse gases with significantly higher global warming potentials than CO_2. By capturing and utilizing these residues for energy production or conversion into value-added products, methane and nitrous oxide emissions are avoided, thereby reducing the overall greenhouse gas emissions associated with biomass utilization. This enhances the climate change mitigation potential of biomass utilization and contributes to improved air quality and environmental health.

4.6 APPLICATIONS

4.6.1 Energy Production

Forestry and agricultural residues are pivotal in meeting energy demands through various conversion processes. Biomass combustion is a traditional and widely employed method where residues are directly burned to generate heat and power. Whether in biomass boilers, stoves, or combined heat and power (CHP) plants, controlled burning of residues offers renewable solutions for space heating, industrial operations, and electricity generation, reducing reliance on fossil fuels and mitigating greenhouse gas emissions. Gasification presents another avenue where residues undergo thermal conversion in a low-oxygen environment to produce synthesis gas (syngas) comprising hydrogen, carbon monoxide, and methane. This versatile energy carrier finds applications in engines, turbines, or fuel cells for electricity production, biofuel, and biochemical synthesis (88–91). Pyrolysis offers a distinct approach, heating residues without oxygen to yield biochar, bio-oil, and syngas. Biochar enriches soil fertility and carbon sequestration as a soil amendment, while bio-oil contributes to renewable fuel and chemical production (92–98). Syngas serves multiple purposes, from heat and power generation to industrial feedstock. Sweden is known for extensively using biomass combustion for heat and power generation. In 2020, the country's biomass-fired power plants and district heating systems utilized over 30 million cubic meters of forestry residues, such as wood chips and bark, along with agricultural residues like straw and energy crops (99, 100). This biomass combustion accounted for approximately 15% of Sweden's total energy consumption, providing heat to homes, industries, and municipalities while significantly reducing CO_2 emissions. Finland has been investing in gasification technologies to convert forestry and agricultural residues into syngas for energy production. In 2019, a large-scale gasification plant in Finland's Pori region processed around 300,000 metric tons of forest residues annually to produce syngas. This syngas was then used in combined heat and power (CHP) plants to generate electricity and district heating, contributing to Finland's renewable energy targets and reducing reliance on imported fossil fuels. The United States has been exploring pyrolysis as a promising technology for converting forestry and agricultural residues into valuable products. In 2020, several pilot-scale pyrolysis projects were initiated nationwide to produce biochar, bio-oil, and syngas from residues such as forest slash, crop residues, and animal waste (101, 102). These pyrolysis initiatives contribute to renewable energy production and offer opportunities for soil improvement through biochar application and biofuel synthesis. In an experiment, Cahyanti et al. (7) studied a thorough comparison of how various operational parameters affect the resulting fuel properties. Torrefaction, recognized as a thermal pretreatment method, was explored for its potential to enhance the suitability of biomass for energy utilization. Through meticulous analysis, the research assesses a spectrum of physiochemical characteristics, changes in composition, interactions between moisture and biomass, and the behavior of ash during melting. The outcomes of the study underscored that elevating torrefaction temperatures and extending the duration of treatment correlate with heightened lignin levels and diminished concentrations of hemicellulose and cellulose. Furthermore, it was observed that torrefied biomass exhibits decreased moisture absorption compared to its raw counterpart, indicative of enhanced resistance to moisture. The study also revealed insights from moisture adsorption isotherm assessments, which aligned with a type II isotherm classification, predominantly described by the Oswin model. Moreover, torrefaction's impact on biomass ash's melting characteristics was discernible, with agricultural waste demonstrating a notable reduction in fouling tendencies post-treatment. Similarly, Enes et al. (103) explored the thermal properties of residual agroforestry biomass in Northern Portugal. The study aimed to assess the potential of biomass from both forestry and agricultural sectors in contributing to the goals set by the government for increasing bioenergy production. To achieve this, samples of agricultural and forest wastes and shrubs were collected from two specific sites in Northern Portugal, namely the Ave and Sabor basins. These samples were then analyzed to determine their higher heating value (HHV) and chemical composition. The evaluation of HHV followed established protocols outlined in Standard DD CEN/TS14918:2005, while lignin content was determined using the Klason method, and extractive content was assessed through the Soxhlet method. The study's findings revealed crucial insights into the thermal characteristics of the biomass samples. Notably, the HHV values for agricultural and forest wastes exhibited a consistent range of 17 to 21 MJ/kg, indicating relatively uniform energy potential across these materials. However, shrub biomass demonstrated slightly higher HHV values, ranging from 19 to 21 MJ/kg, statistically distinct from those of agricultural and forest wastes. Moreover, the analysis revealed that forest wastes contained higher holocellulose levels than agricultural wastes, while the opposite trend was observed for extractive contents.

4.6.2 Biofuels and Bioproducts

Agricultural and forestry residues are crucial in producing biofuels and bioproducts, offering sustainable alternatives to fossil fuels and petrochemical-based products (104). Bioethanol, derived from agricultural residues rich in cellulose and hemicellulose, serves as a renewable transportation fuel. Enzymatic hydrolysis and fermentation processes convert these residues into bioethanol, which can be blended with gasoline or used independently, thereby reducing greenhouse gas emissions and promoting energy security. Anaerobic digestion of organic residues, such as crop residues and food waste, produces biogas—a renewable energy source comprising methane and carbon dioxide. Biogas finds applications in heat and power generation, transportation fuel, and as biomethane for injection into natural gas pipelines, contributing to methane emission mitigation and nutrient-rich soil fertilization. Additionally, residues serve as valuable feedstocks for biochemical production, including organic acids, enzymes, and platform chemicals, through microbial fermentation or enzymatic conversion. In the United States, the Department of Energy's Bioenergy Technologies Office (BETO) reported that in 2020 approximately 6.8 billion gallons of bioethanol were produced from agricultural residues, such as corn stover and wheat straw (105–108). This contributed significantly to the country's renewable fuel targets and helped reduce greenhouse gas emissions by an estimated 35 million metric tons of CO_2 equivalent. In Germany, a study by the Federal Ministry for Economic Affairs and Energy found that 2019 the country's biogas sector utilized around 10.5 million metric tons of organic residues, including crop residues and food waste, to produce biogas. This biogas production accounted for approximately 10% of Germany's total renewable energy generation and reduced methane emissions by an estimated 20 million metric tons of CO_2 equivalent. A study analyzed global data on biochemical production from agricultural and forestry residues. The study found that in 2018 over 15 million metric tons of organic acids, enzymes, and platform chemicals were produced globally through microbial fermentation or enzymatic conversion of residues (109, 110). This biochemical production supported various industries, including bioplastics, pharmaceuticals, and cosmetics and contributed to reducing reliance on petrochemical-based products. Brazil is a global leader in bioethanol production, primarily from sugarcane residues. The country's ethanol industry utilizes bagasse, the fibrous residue left after sugarcane processing, as a feedstock for bioethanol production. In 2020, Brazil produced over 30 billion liters of bioethanol, contributing significantly to the country's transportation fuel market and reducing reliance on fossil fuels. This large-scale bioethanol production has helped Brazil achieve substantial greenhouse gas emission reductions and promote sustainable development in the bioenergy sector. Denmark has made significant strides in biogas utilization, mainly from agricultural and organic residues. The country's biogas sector utilizes livestock manure, food waste, and organic residues from agro-industrial activities to produce biogas for heat, power, and transportation fuel. In 2019, Denmark's biogas plants produced over 1.5 billion cubic meters of biogas, equivalent to around 15% of the country's natural gas consumption. This widespread adoption of biogas technology has helped Denmark reduce methane emissions from organic waste and transition towards a more sustainable energy system. China has emerged as a critical player in biochemical production from agricultural residues, driven by the country's growing demand for sustainable alternatives to petrochemical-based products (111–115). Chinese researchers and industries have developed innovative bioprocesses for converting crop residues, such as rice straw and corn stalks, into biochemicals, including organic acids, enzymes, and biopolymers. In 2018, China's biochemical industry produced over 5 million metric tons of biochemicals from agricultural residues, supporting various sectors, such as biodegradable plastics, pharmaceuticals, and industrial enzymes (116, 117). This shift towards bio-based chemicals has helped China reduce its reliance on fossil resources and mitigate environmental pollution associated with traditional chemical manufacturing processes.

Zambon et al. (118) presented an innovative approach in agro-forestry, which involves repurposing agricultural waste to produce biochar fuel. Traditionally, agricultural residues, such as crop leftovers, are burnt in fields to create nutrient-rich ashes that can fertilize the soil. However, this method often lacks sufficient nitrogen and phosphorus and may even introduce heavy metal contaminants, adversely affecting soil quality and crop yield. To address this challenge, the research explored the viability of establishing a new supply chain that utilizes olive and hazelnut pruning residues, abundant in the Viterbo region of Italy, for biochar fuel production, as shown in Figure 4.3. The study comprehensively analyzed the physicochemical properties of the biochar derived from these residues, adhering to European Biochar Certificate (EBC) standards. The results demonstrated that high-quality biochar can be produced cost-effectively

FIGURE 4.3   Illustration of the pruning remnants from (a) olive and (b) hazelnut following crop activities, alongside (c) bio-shredding. It is adapted from ref. (118) under CCBY 4.0.

from olive and hazelnut biomass residues, offering promising opportunities for sustainable waste management and agricultural soil enhancement practices.

4.7 CHALLENGES AND FUTURE OUTLOOKS

Using forestry and agricultural residues confronts many technical, economic, environmental, and social challenges. Yet, it also presents promising prospects for sustainable development and climate mitigation. Technical hurdles encompass optimizing biomass processing technologies to efficiently convert residues into valuable products while addressing the variability in residue composition arising from species diversity and geographic location (119, 120). Economic challenges revolve around achieving cost competitiveness in fluctuating market conditions and investment risks, requiring innovative financing mechanisms and supportive policies to stimulate market growth and investment. Environmental concerns emphasize the imperative of mitigating the impacts of biomass utilization on ecosystems, biodiversity, and soil health, necessitating sustainable land management practices and ecosystem-based approaches to balance biomass production with conservation objectives. Social challenges encompass fostering stakeholder engagement, ensuring inclusive decision-making processes, and addressing socio-economic disparities to promote equitable access to benefits and opportunities from biomass utilization. However, amidst these challenges, there are opportunities for progress driven by technological innovation, supportive policy frameworks, market development initiatives, and integrated landscape approaches. Advances in biomass conversion technologies, such as gasification, pyrolysis, and biochemical processing, offer new pathways for converting residues into bioenergy, biofuels, biochemicals, and bioproducts, contributing to resource efficiency and circularity in the bioeconomy. Policy support, including renewable energy targets, carbon pricing mechanisms, and sustainability standards, provides regulatory certainty and market incentives encouraging investment and innovation in biomass utilization. Moreover, the growing demand for renewable products and sustainable alternatives to fossil fuels underscores the potential for biomass to play a significant role in the transition to a low-carbon and circular bioeconomy (121, 122). By addressing barriers, seizing opportunities, and embracing collaborative approaches, stakeholders can unlock the full potential of forestry and agricultural residues to drive sustainable development, mitigate climate change, and foster a more resilient and equitable future for society and the environment.

4.8 CONCLUSION

Forestry and agricultural residues represent abundant and renewable organic materials generated as byproducts of land management practices and agricultural activities. These residues encompass various organic materials, including branches, tree tops, crop stalks, husks, and straw, each possessing unique characteristics that influence their suitability for various applications. The chapter highlights the importance of sustainable resource management in maximizing the potential of residues while minimizing environmental impacts and promoting economic viability. Throughout the chapter, the significance of forestry and agricultural residues in contributing to environmental sustainability, carbon sequestration, soil health improvement, and reduction of greenhouse gas emissions is underscored. Residues are crucial in mitigating climate change by sequestering carbon in biomass and soils, reducing greenhouse gas emissions through bioenergy production, and enhancing soil fertility and resilience. Additionally, residues offer diverse applications across energy production, biofuels and bioproducts, and agricultural practices, contributing to renewable energy generation, bio-based industries, and sustainable agriculture.

However, using forestry and agricultural residues faces various challenges that must be addressed to realize their full potential and overcome barriers to adoption. Technical challenges include optimizing biomass processing technologies, addressing residue variability, and improving efficiency and cost-effectiveness. Economic challenges center on achieving cost competitiveness, navigating market uncertainties, and securing investment and financing for biomass projects. Environmental concerns emphasize the importance of mitigating environmental impacts, promoting sustainable land management, and ensuring biodiversity conservation. Social challenges include fostering stakeholder engagement, addressing socio-economic disparities, and promoting equitable distribution of benefits from biomass utilization. Despite these challenges, the future outlook for the utilization of forestry and agricultural residues is promising, driven by technological innovation, supportive policies, market development, and integrated landscape approaches. Advances in biomass conversion technologies, policy incentives, and growing demand for renewable products offer opportunities for overcoming barriers and promoting sustainable development. By addressing technical, economic, environmental, and social challenges through collaborative efforts and holistic approaches, stakeholders can harness the potential of residues to drive the transition towards a low-carbon and circular bioeconomy.

REFERENCES

1. Alatzas S, Moustakas K, Malamis D, Vakalis S. Biomass potential from agricultural waste for energetic utilization in Greece. *Energies*. 2019;12(6).
2. Olguin-Maciel E, Singh A, Chable-Villacis R, Tapia-Tussell R, Ruiz HA. Consolidated bioprocessing, an innovative strategy towards sustainability for biofuels production from crop residues: An overview. *Agronomy*. 2020;10(11).
3. Duque-Acevedo M, Belmonte-Ureña LJ, Yakovleva N, Camacho-Ferre F. Analysis of the circular economic production models and their approach in agriculture and agricultural waste biomass management. *Int J Environ Res Public Health*. 2020;17(24):1–34.
4. Bharti A, Paritosh K, Mandla VR, Chawade A, Vivekanand V. Gis application for the estimation of bioenergy potential from agriculture residues: An overview. *Energies*. 2021;14(4):1–15.
5. Blasi A, Verardi A, Lopresto CG, Siciliano S, Sangiorgio P. Lignocellulosic agricultural waste valorization to obtain valuable products: An overview. *Recycling*. 2023;8(4):1–46.
6. Casau M, Dias MF, Matias JCO, Nunes LJR. Residual biomass: A comprehensive review on the importance, uses and potential in a circular bioeconomy approach. *Resources*. 2022;11(4):1–16.
7. Cahyanti MN, Doddapaneni TRKC, Madissoo M, Pärn L, Virro I, Kikas T. Torrefaction of agricultural and wood waste: Comparative analysis of selected fuel characteristics. *Energies*. 2021;14(10):1–19.
8. Moosavi S, Manta O, El-Badry YA, Hussein EE, El-Bahy ZM, Fawzi NFBM, et al. A study on machine learning methods' application for dye adsorption

prediction onto agricultural waste activated carbon. *Nanomaterials.* 2021;11(10).

9. Liuzzi S, Rubino C, Stefanizzi P, Martellotta F. The agro-waste production in selected EUSAIR regions and its potential use for building applications: A review. *Sustain.* 2022;14(2).

10. Aliaño-González MJ, Gabaston J, Ortiz-Somovilla V, Cantos-Villar E. Wood waste from fruit trees: Biomolecules and their applications in agrifood industry. *Biomolecules.* 2022;12(2):1–46.

11. Xu P, Shu L, Li Y, Zhou S, Zhang G, Wu Y, et al. Pretreatment and composting technology of agricultural organic waste for sustainable agricultural development. *Heliyon [Internet].* 2023;9(5):e16311. Available from: https://www.sciencedirect.com/science/article/pii/S2405844023035181

12. Gani A, Erdiwansyah E, Munawar E, Mahidin M, Mamat R, Rosdi SM. Investigation of the potential biomass waste source for biocoke production in Indonesia: A review. *Energy Reports [Internet].* 2023;10:2417–2438. Available from: https://www.sciencedirect.com/science/article/pii/S2352484723013033

13. Fernanda Rojas Michaga M, Michailos S, Akram M, Cardozo E, Hughes KJ, Ingham D, et al. Bioenergy with carbon capture and storage (BECCS) potential in jet fuel production from forestry residues: A combined techno-economic and life cycle assessment approach. *Energy Convers Manag [Internet].* 2022;255:115346. Available from: https://www.sciencedirect.com/science/article/pii/S019689042200142X

14. Lima AR, Cristofoli NL, Rosa da Costa AM, Saraiva JA, Vieira MC. Comparative study of the production of cellulose nanofibers from agro-industrial waste streams of Salicornia ramosissima by acid and enzymatic treatment. *Food Bioprod Process [Internet].* 2023;137:214–225. Available from: https://www.sciencedirect.com/science/article/pii/S0960308522001481

15. Mir KA, Park C, Purohit P, Kim S. Comparative analysis of greenhouse gas emission inventory for Pakistan: Part II agriculture, forestry and other land use and waste. *Adv Clim Chang Res [Internet].* 2021;12(1):132–144. Available from: https://www.sciencedirect.com/science/article/pii/S1674927821000198

16. Duque-Acevedo M, Belmonte-Ureña LJ, Plaza-Úbeda JA, Camacho-Ferre F. The management of agricultural waste biomass in the framework of circular economy and bioeconomy: An opportunity for greenhouse agriculture in Southeast Spain. *Agronomy.* 2020;10(4).

17. Moura P, Henriques J, Alexandre J, Oliveira AC, Abreu M, Gírio F, et al. Sustainable value methodology to compare the performance of conversion technologies for the production of electricity and heat, energy vectors and biofuels from waste biomass. *Clean Waste Syst [Internet].* 2022;3:100029. Available from: https://www.sciencedirect.com/science/article/pii/S277291252200029X

18. La Picirelli de Souza L, Rajabi Hamedani S, Silva Lora EE, Escobar Palacio JC, Comodi G, Villarini M, et al. Theoretical and technical assessment of

agroforestry residue potential for electricity generation in Brazil towards 2050. *Energy Reports [Internet]*. 2021;7:2574–2587. Available from: https://www.sciencedirect.com/science/article/pii/S2352484721002420

19. Ro JW, Zhang Y, Kendall A. Developing guidelines for waste designation of biofuel feedstocks in carbon footprints and life cycle assessment. *Sustain Prod Consum [Internet]*. 2023;37:320–330. Available from: https://www.sciencedirect.com/science/article/pii/S2352550923000532

20. Tran TH, Nguyen TBT, Le HST, Phung DC. Formulation and solution technique for agricultural waste collection and transport network design. *Eur J Oper Res [Internet]*. 2024;313(3):1152–1169. Available from: https://www.sciencedirect.com/science/article/pii/S0377221723006811

21. Lizundia E, Luzi F, Puglia D. Organic waste valorisation towards circular and sustainable biocomposites. *Green Chem [Internet]*. 2022;24(14):5429–5459. Available from: https://www.sciencedirect.com/science/article/pii/S1463926222004873

22. Siol C, Thrän D, Majer S. Utilizing residual biomasses from agriculture and forestry: Different approaches to set system boundaries in environmental and economic life-cycle assessments. *Biomass Bioenergy [Internet]*. 2023;174:106839. Available from: https://www.sciencedirect.com/science/article/pii/S096195342300137X

23. Sani MNH, Amin M, Siddique AB, Nasif SO, Ghaley BB, Ge L, et al. Waste-derived nanobiochar: A new avenue towards sustainable agriculture, environment, and circular bioeconomy. *Sci Total Environ [Internet]*. 2023;905:166881. Available from: https://www.sciencedirect.com/science/article/pii/S0048969723055067

24. Sabaruddin FA, Megashah LN, Shazleen SS, Ariffin H. Emerging trends in the appliance of ultrasonic technology for valorization of agricultural residue into versatile products. *Ultrason Sonochem [Internet]*. 2023;99:106572. Available from: https://www.sciencedirect.com/science/article/pii/S1350417723002845

25. Wang J, Azam W. Natural resource scarcity, fossil fuel energy consumption, and total greenhouse gas emissions in top emitting countries. *Geosci Front [Internet]*. 2024;15(2):101757. Available from: https://www.sciencedirect.com/science/article/pii/S1674987123002244

26. Tolisano C, Del Buono D. Biobased: Biostimulants and biogenic nanoparticles enter the scene. *Sci Total Environ [Internet]*. 2023;885:163912. Available from: https://www.sciencedirect.com/science/article/pii/S0048969723025330

27. Brosowski A, Thrän D, Mantau U, Mahro B, Erdmann G, Adler P, et al. A review of biomass potential and current utilisation—Status quo for 93 biogenic wastes and residues in Germany. *Biomass Bioenergy [Internet]*. 2016;95:257–272. Available from: https://www.sciencedirect.com/science/article/pii/S0961953416303415

28. Gil A. Challenges on waste-to-energy for the valorization of industrial wastes: Electricity, heat and cold, bioliquids and biofuels. *Environ*

Nanotechnology, Monit Manag [Internet]. 2022;17:100615. Available from: https://www.sciencedirect.com/science/article/pii/S2215153221001902

29. Zhang Z, Ma Z, Song L, Farag MA. Maximizing crustaceans (shrimp, crab, and lobster) by-products value for optimum valorization practices: A comparative review of their active ingredients, extraction, bioprocesses and applications. *J Adv Res [Internet].* 2024;57:59–76. Available from: https://www.sciencedirect.com/science/article/pii/S2090123223003259

30. Siegrist A, Bowman G, Burg V. Energy generation potentials from agricultural residues: The influence of techno-spatial restrictions on biomethane, electricity, and heat production. *Appl Energy [Internet].* 2022;327:120075. Available from: https://www.sciencedirect.com/science/article/pii/S0306261922013320

31. Miettinen J, Ollikainen M, Nieminen M, Valsta L. Cost function approach to water protection in forestry. *Water Resour Econ [Internet].* 2020;31:100150. Available from: https://www.sciencedirect.com/science/article/pii/S2212428418300252

32. Hameed Ologunde O, Kehinde Bello S, Abolanle Busari M. Integrated agricultural system: A dynamic concept for improving soil quality. *J Saudi Soc Agric Sci [Internet].* 2024; Available from: https://www.sciencedirect.com/science/article/pii/S1658077X24000146

33. Brosowski A, Krause T, Mantau U, Mahro B, Noke A, Richter F, et al. How to measure the impact of biogenic residues, wastes and by-products: Development of a national resource monitoring based on the example of Germany. *Biomass Bioenergy [Internet].* 2019;127:105275. Available from: https://www.sciencedirect.com/science/article/pii/S0961953419302247

34. Yuan ZL, Gerbens-Leenes PW. Biogas feedstock potentials and related water footprints from residues in China and the European Union. *Sci Total Environ [Internet].* 2021;793:148340. Available from: https://www.sciencedirect.com/science/article/pii/S0048969721034112

35. Awogbemi O, Kallon DV Von. Application of biochar derived from crops residues for biofuel production. *Fuel Commun [Internet].* 2023;15:100088. Available from: https://www.sciencedirect.com/science/article/pii/S2666052023000043

36. Taquetti VB, Silva VV, Chaves ILS, Oliveira RGE, Maffioletti FD, Ferreira G, et al. Performance of eucalypt particleboard with the addition of farm waste. *Heliyon [Internet].* 2023;9(12):e22760. Available from: https://www.sciencedirect.com/science/article/pii/S2405844023099681

37. Kainth S, Sharma P, Pandey OP. Green sorbents from agricultural wastes: A review of sustainable adsorption materials. *Appl Surf Sci Adv [Internet].* 2024;19:100562. Available from: https://www.sciencedirect.com/science/article/pii/S2666523923001964

38. Auteri N, Scalenghe R, Saiano F. Phosphorus recovery from agricultural waste via cactus pear biomass. *Heliyon [Internet].* 2023;9(9):e19996. Available from: https://www.sciencedirect.com/science/article/pii/S2405844023072043

39. Lisbona P, Pascual S, Pérez V. Waste to energy: Trends and perspectives. *Chem Eng J Adv [Internet]*. 2023;14:100494. Available from: https://www.sciencedirect.com/science/article/pii/S2666821123000510

40. Muizniecea I, Dace E, Blumberga D. Dynamic modeling of the environmental and economic aspects of bio-resources from agricultural and forestry wastes. *Procedia Earth Planet Sci [Internet]*. 2015;15:806–812. Available from: https://www.sciencedirect.com/science/article/pii/S1878522015003926

41. Seah CC, Tan CH, Arifin NA, Hafriz RSRM, Salmiaton A, Nomanbhay S, et al. Co-pyrolysis of biomass and plastic: Circularity of wastes and comprehensive review of synergistic mechanism. *Results Eng [Internet]*. 2023;17:100989. Available from: https://www.sciencedirect.com/science/article/pii/S2590123023001160

42. Röder M, Chong K, Thornley P. The future of residue-based bioenergy for industrial use in Sub-Saharan Africa. *Biomass Bioenergy [Internet]*. 2022;159:106385. Available from: https://www.sciencedirect.com/science/article/pii/S0961953422000460

43. Tippayawong KY, Santiteerakul S, Ramingwong S, Tippayawong N. Cost analysis of community scale smokeless charcoal briquette production from agricultural and forest residues. *Energy Procedia [Internet]*. 2019;160:310–316. Available from: https://www.sciencedirect.com/science/article/pii/S1876610219312524

44. Parthasarathy P, Alherbawi M, Shahbaz M, Al-Ansari T, McKay G. Developing biochar from potential wastes in Qatar and its revenue potential. *Energy Convers Manag X [Internet]*. 2023;20:100467. Available from: https://www.sciencedirect.com/science/article/pii/S259017452300123X

45. Nabi BG, Mukhtar K, Ansar S, Hassan SA, Hafeez MA, Bhat ZF, et al. Application of ultrasound technology for the effective management of waste from fruit and vegetable. *Ultrason Sonochem [Internet]*. 2024;102:106744. Available from: https://www.sciencedirect.com/science/article/pii/S13504 1772300456X

46. Singh Yadav SP, Bhandari S, Bhatta D, Poudel A, Bhattarai S, Yadav P, et al. Biochar application: A sustainable approach to improve soil health. *J Agric Food Res [Internet]*. 2023;11:100498. Available from: https://www.sciencedirect.com/science/article/pii/S2666154323000054

47. Martinho VJPD, Rodrigues RN. Bioenergy relations with agriculture, forestry and other land uses: Highlighting the specific contributions of artificial intelligence and co-citation networks. *Heliyon [Internet]*. 2024;10(4):e26267. Available from: https://www.sciencedirect.com/science/article/pii/S2405844024022989

48. Rahimi Z, Anand A, Gautam S. An overview on thermochemical conversion and potential evaluation of biofuels derived from agricultural wastes. *Energy Nexus [Internet]*. 2022;7:100125. Available from: https://www.sciencedirect.com/science/article/pii/S2772427122000808

49. Prasanna Kumar DJ, Mishra RK, Chinnam S, Binnal P, Dwivedi N. A comprehensive study on anaerobic digestion of organic solid waste: A review on configurations, operating parameters, techno-economic analysis and

current trends. *Biotechnol Notes [Internet]*. 2024;5:33–49. Available from: https://www.sciencedirect.com/science/article/pii/S2665906924000047

50. Xie J, Zhong W, Jin B, Shao Y, Liu H. Simulation on gasification of forestry residues in fluidized beds by Eulerian–Lagrangian approach. *Bioresour Technol [Internet]*. 2012;121:36–46. Available from: https://www.science direct.com/science/article/pii/S0960852412009947

51. Bao K, Bieber LM, Kürpick S, Radanielina MH, Padsala R, Thrän D, et al. Bottom-up assessment of local agriculture, forestry and urban waste potentials towards energy autonomy of isolated regions: Example of Réunion. *Energy Sustain Dev [Internet]*. 2022;66:125–139. Available from: https:// www.sciencedirect.com/science/article/pii/S0973082621001460

52. Rowan M, Umenweke GC, Epelle EI, Afolabi IC, Okoye PU, Gunes B, et al. Anaerobic co-digestion of food waste and agricultural residues: An overview of feedstock properties and the impact of biochar addition. *Digit Chem Eng [Internet]*. 2022;4:100046. Available from: https://www.sciencedirect. com/science/article/pii/S2772508122000369

53. Pinheiro WBS, Pinheiro Neto JR, Botelho AS, Dos Santos KIP, Da Silva GA, Muribeca AJB, et al. The use of Bagassa guianensis aubl. forestry waste as an alternative for obtaining bioproducts and bioactive compounds. *Arab J Chem [Internet]*. 2022;15(6):103813. Available from: https://www.science direct.com/science/article/pii/S1878535222001290

54. De Pinto A, Li M, Haruna A, Hyman GG, Martinez MAL, Creamer B, et al. Low emission development strategies in agriculture. An agriculture, forestry, and other land uses (AFOLU) perspective. *World Dev [Internet]*. 2016;87:180–203. Available from: https://www.sciencedirect.com/science/ article/pii/S0305750X16304041

55. Yang L, Guan Z, Chen S, He Z. Re-measurement and influencing factors of agricultural eco-efficiency under the 'dual carbon' target in China. *Heliyon [Internet]*. 2024;10(3):e24944. Available from: https://www.sciencedirect. com/science/article/pii/S2405844024009757

56. Khatri-Chhetri A, Costa Junior C, Wollenberg E. Greenhouse gas mitigation co-benefits across the global agricultural development programs. *Glob Environ Chang [Internet]*. 2022;76:102586. Available from: https://www. sciencedirect.com/science/article/pii/S0959378022001248

57. Li S, Skelly S. Physicochemical properties and applications of biochars derived from municipal solid waste: A review. *Environ Adv [Internet]*. 2023;13:100395. Available from: https://www.sciencedirect.com/science/arti cle/pii/S2666765723000558

58. Maria Coelho Vianna L, de Oliveira L, Durante Mühl D. Waste valorization in agribusiness value chains. *Waste Manag Bull [Internet]*. 2024;1(4):195–204. Available from: https://www.sciencedirect.com/science/article/pii/S2949 750723000408

59. Zhou H, Shen Y, Zhang N, Liu Z, Bao L, Xia Y. Wood fiber biomass pyrolysis solution as a potential tool for plant disease management: A review. *Heliyon [Internet]*. 2024;10(3):e25509. Available from: https://www.sciencedirect. com/science/article/pii/S2405844024015408

60. Wen Y, Wang S, Shi Z, Nuran Zaini I, Niedzwiecki L, Aragon-Briceno C, et al. H2-rich syngas production from pyrolysis of agricultural waste digestate coupled with the hydrothermal carbonization process. *Energy Convers Manag [Internet]*. 2022;269:116101. Available from: https://www.sciencedirect.com/science/article/pii/S0196890422008871

61. Kircher M, Aranda E, Athanasios P, Radojcic-Rednovnikov I, Romantschuk M, Ryberg M, et al. Treatment and valorization of bio-waste in the EU. *EFB Bioeconomy J [Internet]*. 2023;3:100051. Available from: https://www.sciencedirect.com/science/article/pii/S266704102300006X

62. Pisanó I, Gottumukkala L, Hayes DJ, Leahy JJ. Characterisation of Italian and Dutch forestry and agricultural residues for the applicability in the bio-based sector. *Ind Crops Prod [Internet]*. 2021;171:113857. Available from: https://www.sciencedirect.com/science/article/pii/S092666902100621X

63. Gao Z, Alshehri K, Li Y, Qian H, Sapsford D, Cleall P, et al. Advances in biological techniques for sustainable lignocellulosic waste utilization in biogas production. *Renew Sustain Energy Rev [Internet]*. 2022;170:112995. Available from: https://www.sciencedirect.com/science/article/pii/S1364032122008760

64. Kumar P, Subbarao PM V, Kala LD, Vijay VK. Experimental assessment of producer gas generation using agricultural and forestry residues in a fixed bed downdraft gasifier. *Chem Eng J Adv [Internet]*. 2023;13:100431. Available from: https://www.sciencedirect.com/science/article/pii/S2666821122001910

65. Dandabathula G, Chintala SR, Ghosh S, Balakrishnan P, Jha CS. Exploring the nexus between Indian forestry and the sustainable development goals. *Reg Sustain [Internet]*. 2021;2(4):308–323. Available from: https://www.sciencedirect.com/science/article/pii/S2666660X22000020

66. Rico JJ, Pérez-Orozco R, Patiño Vilas D, Porteiro J. TG/DSC and kinetic parametrization of the combustion of agricultural and forestry residues. *Biomass Bioenergy [Internet]*. 2022;162:106485. Available from: https://www.sciencedirect.com/science/article/pii/S0961953422001465

67. Cruz IA, Santos Andrade LR, Bharagava RN, Nadda AK, Bilal M, Figueiredo RT, et al. Valorization of cassava residues for biogas production in Brazil based on the circular economy: An updated and comprehensive review. *Clean Eng Technol [Internet]*. 2021;4:100196. Available from: https://www.sciencedirect.com/science/article/pii/S2666790821001567

68. Barbero-López A, López-Gómez YM, Carrasco J, Jokinen N, Lappalainen R, Akkanen J, et al. Characterization and antifungal properties against wood decaying fungi of hydrothermal liquefaction liquids from spent mushroom substrate and tomato residues. *Biomass Bioenergy [Internet]*. 2024;181:107035. Available from: https://www.sciencedirect.com/science/article/pii/S0961953423003343

69. Arias A, Costa CE, Feijoo G, Moreira MT, Domingues L. Process modeling, environmental and economic sustainability of the valorization of whey and eucalyptus residues for resveratrol biosynthesis. *Waste Manag [Internet]*.

2023;172:226–234. Available from: https://www.sciencedirect.com/science/article/pii/S0956053X23006414

70. Hassan M, Mohanty AK, Misra M. 3D printing in upcycling plastic and biomass waste to sustainable polymer blends and composites: A review. *Mater Des [Internet]*. 2024;237:112558. Available from: https://www.sciencedirect.com/science/article/pii/S0264127523009747

71. Thakur A, Kaya S, Kumar A. Recent innovations in nano container-based self-healing coatings in the construction industry. *Curr Nanosci*. 2021;18(2):203–216.

72. Thakur A, Kumar A. Ecotoxicity analysis and risk assessment of nanomaterials for the environmental remediation. *Macromol Symp*. 2023;410(1):1–23.

73. Thakur A, Kumar A. Exploring the potential of ionic liquid-based electrochemical biosensors for real-time biomolecule monitoring in pharmaceutical applications: From lab to life. *Results Eng [Internet]*. 2023;20(July):101533. Available from: https://doi.org/10.1016/j.rineng.2023.101533

74. Saharudin DM, Jeswani HK, Azapagic A. Biochar from agricultural wastes: Environmental sustainability, economic viability and the potential as a negative emissions technology in Malaysia. *Sci Total Environ [Internet]*. 2024;919:170266. Available from: https://www.sciencedirect.com/science/article/pii/S0048969724004017

75. Ali N, Rashid MI, Rehan M, Shah Eqani SAMA, Summan ASA, Ismail IMI, et al. Environmental evaluation of polyhydroxyalkanoates from animal slaughtering waste using material input per service unit. *N Biotechnol [Internet]*. 2023;75:40–51. Available from: https://www.sciencedirect.com/science/article/pii/S1871678423000080

76. Narayana Sarma R, Vinu R. An assessment of sustainability metrics for waste-to-liquid fuel pathways for a low carbon circular economy. *Energy Nexus [Internet]*. 2023;12:100254. Available from: https://www.sciencedirect.com/science/article/pii/S2772427123000840

77. Corona B, Shen L, Sommersacher P, Junginger M. Consequential life cycle assessment of energy generation from waste wood and forest residues: The effect of resource-efficient additives. *J Clean Prod [Internet]*. 2020;259:120948. Available from: https://www.sciencedirect.com/science/article/pii/S0959652620309951

78. Kaya S, Lgaz H, Thakkur A, Kumar A, Özbakır D, Karakuş N, et al. Molecular insights into the corrosion inhibition mechanism of omeprazole and tinidazole: A theoretical investigation. *Mol Simul*. 2023:1–15.

79. Sharma D, Thakur A, Sharma MK, Jakhar K, Kumar A, Sharma AK, Hari OM. Synthesis, electrochemical, morphological, computational and corrosion inhibition studies of 3-(5-Naphthalen-2-yl-[1,3,4]oxadiazol-2-yl)-pyridine against mild steel in 1 M HCl. *Asian J Chem*. 2023;35(5):1079–1088.

80. Thakur A, Savaş K, Kumar A. Recent trends in the characterization and application progress of nano-modified coatings in corrosion mitigation of metals and alloys. *Appl Sci*. 2023;13:730.

81. Thakur A, Kumar A, Kaya S, Benhiba F, Sharma S. Electrochemical and computational investigations of the Thysanolaena latifolia leaves extract:

An eco-benign solution for the corrosion mitigation of mild steel. *Results Chem [Internet]*. 2023;6(September):101147. Available from: https://doi.org/10.1016/j.rechem.2023.101147

82. Duque-Acevedo M, Belmonte-Ureña LJ, Cortés-García FJ, Camacho-Ferre F. Agricultural waste: Review of the evolution, approaches and perspectives on alternative uses. *Glob Ecol Conserv [Internet]*. 2020;22:e00902. Available from: https://www.sciencedirect.com/science/article/pii/S2351989419307516

83. Mishra RK, Naik SU, Chistie SM, Kumar V, Narula A. Pyrolysis of agricultural waste in a thermogravimetric analyzer: Studies of physicochemical properties, kinetics behaviour, and gas compositions. *Mater Sci Energy Technol [Internet]*. 2022;5:399–410. Available from: https://www.sciencedirect.com/science/article/pii/S2589299122000210

84. Shalma S, Shabbirahmed AM, Haldar D, Patel AK, Singhania RR. Influence of reaction conditions on synthesis and applications of lignin nanoparticles derived from agricultural wastes. *Environ Technol Innov [Internet]*. 2023;31:103163. Available from: https://www.sciencedirect.com/science/article/pii/S2352186423001591

85. Cucina M. Integrating anaerobic digestion and composting to boost energy and material recovery from organic wastes in the circular economy framework in Europe: A review. *Bioresour Technol Reports [Internet]*. 2023;24:101642. Available from: https://www.sciencedirect.com/science/article/pii/S2589014X23003134

86. Islam MK, Khatun MS, Arefin MA, Islam MR, Hassan M. Waste to energy: An experimental study of utilizing the agricultural residue, MSW, and e-waste available in Bangladesh for pyrolysis conversion. *Heliyon [Internet]*. 2021;7(12):e08530. Available from: https://www.sciencedirect.com/science/article/pii/S2405844021026335

87. Sharma P, Sharma N. RSM approach to pre-treatment of lignocellulosic waste and a statistical methodology for optimizing bioethanol production. *Waste Manag Bull [Internet]*. 2024;2(1):49–66. Available from: https://www.sciencedirect.com/science/article/pii/S2949750723000500

88. da Costa TP, Murphy F, Roldan R, Mediboyina MK, Chen W, Sweeney J, et al. Technical and environmental assessment of forestry residues valorisation via fast pyrolysis in Ireland. *Biomass Bioenergy [Internet]*. 2023;173:106766. Available from: https://www.sciencedirect.com/science/article/pii/S0961953423000648

89. Oliveira M, Ramos A, Monteiro E, Rouboa A. Modeling and simulation of a fixed bed gasification process for thermal treatment of municipal solid waste and agricultural residues. *Energy Reports [Internet]*. 2021;7:256–269. Available from: https://www.sciencedirect.com/science/article/pii/S2352484721005886

90. Cox R, Narisetty V, Castro E, Agrawal D, Jacob S, Kumar G, et al. Fermentative valorisation of xylose-rich hemicellulosic hydrolysates from agricultural waste residues for lactic acid production under non-sterile conditions. *Waste Manag [Internet]*. 2023;166:336–345. Available from: https://www.sciencedirect.com/science/article/pii/S0956053X23003562

91. Bedoić R, Ćosić B, Duić N. Technical potential and geographic distribution of agricultural residues, co-products and by-products in the European Union. *Sci Total Environ [Internet]*. 2019;686:568–579. Available from: https://www.sciencedirect.com/science/article/pii/S0048969719322582

92. Kumar A, Thakur A. *Encapsulated Nanoparticles in Organic Polymers for Corrosion Inhibition [Internet]. Corrosion Protection at the Nanoscale.* Elsevier Inc., 2020, 345–362. Available from: http://dx.doi.org/10.1016/B978-0-12-819359-4.00018-0

93. Thakur A, Kumar A. Unraveling the multifaceted mechanisms and untapped potential of activated carbon in remediation of emerging pollutants: A comprehensive review and critical appraisal of advanced techniques. *Chemosphere [Internet]*. 2024;346(November 2023):140608. Available from: https://doi.org/10.1016/j.chemosphere.2023.140608

94. Thakur A, Sharma S, Ganjoo R, Assad H, Kumar A. Anti-corrosive potential of the sustainable corrosion inhibitors based on biomass waste: A review on preceding and perspective research. *J Phys Conf Ser*. 2022;2267(1).

95. Kaya S, Thakur A, Kumar A. The role of in Silico/DFT investigations in analyzing dye molecules for enhanced solar cell efficiency and reduced toxicity. *J Mol Graph Model*. 2023;124(May).

96. Thakur A, Sharma S, Ganjoo R, Assad H, Kumar A. Anti-corrosive potential of the sustainable corrosion inhibitors based on biomass waste: A review on preceding and perspective research. *J Phys Conf Ser*. 2022;2267(1):012079.

97. Dhonchak C, Agnihotri N. Computational insights in the spectrophotometrically 4H-chromen-4-one complex using DFT method. *Biointerface Res Appl Chem*. 2023;13(4):357.

98. Thakur A, Kumar A, Singh A. Adsorptive removal of heavy metals, dyes, and pharmaceuticals: Carbon-based nanomaterials in focus. *Carbon N Y [Internet]*. 2023;217(November 2023):118621. Available from: https://doi.org/10.1016/j.carbon.2023.118621

99. Yafetto L. Application of solid-state fermentation by microbial biotechnology for bioprocessing of agro-industrial wastes from 1970 to 2020: A review and bibliometric analysis. *Heliyon [Internet]*. 2022;8(3):e09173. Available from: https://www.sciencedirect.com/science/article/pii/S2405844022004613

100. Corré WJ, Conijn JG. Biogas from agricultural residues as energy source in hybrid concentrated solar power. *Procedia Comput Sci [Internet]*. 2016;83:1126–1133. Available from: https://www.sciencedirect.com/science/article/pii/S1877050916302666

101. Karić N, Maia AS, Teodorović A, Atanasova N, Langergraber G, Crini G, et al. Bio-waste valorisation: Agricultural wastes as biosorbents for removal of (in)organic pollutants in wastewater treatment. *Chem Eng J Adv [Internet]*. 2022;9:100239. Available from: https://www.sciencedirect.com/science/article/pii/S266682112100154X

102. Khedulkar AP, Dang VD, Thamilselvan A, Doong R an, Pandit B. Sustainable high-energy supercapacitors: Metal oxide-agricultural waste biochar composites paving the way for a greener future. *J Energy Storage*

[*Internet*]. 2024;77:109723. Available from: https://www.sciencedirect.com/science/article/pii/S2352152X23031213

103. Enes T, Aranha J, Fonseca T, Lopes D, Alves A, Lousada J. Thermal properties of residual agroforestry biomass of northern Portugal. *Energies*. 2019;12(8).

104. Lee SH, Lum WC, Boon JG, Kristak L, Antov P, Pędzik M, et al. Particleboard from agricultural biomass and recycled wood waste: A review. *J Mater Res Technol [Internet]*. 2022;20:4630–4658. Available from: https://www.sciencedirect.com/science/article/pii/S2238785422013990

105. Ameh VI, Ayeleru OO, Nomngongo PN, Ramatsa IM. Bio-oil production from waste plant seeds biomass as pyrolytic lignocellulosic feedstock and its improvement for energy potential: A review. *Waste Manag Bull [Internet]*. 2024;2(2):32–48. Available from: https://www.sciencedirect.com/science/article/pii/S2949750724000221

106. Khoshnodifar Z, Ataei P, Karimi H. Recycling date palm waste for compost production: A study of sustainability behavior of date palm growers. *Environ Sustain Indic [Internet]*. 2023;20:100300. Available from: https://www.sciencedirect.com/science/article/pii/S2665972723000776

107. Reyes Molina EA, Park S, Park S, Kelley SS. Effective toluene removal from aqueous solutions using fast pyrolysis-derived activated carbon from agricultural and forest residues: Isotherms and kinetics study. *Heliyon [Internet]*. 2023;9(5):e15765. Available from: https://www.sciencedirect.com/science/article/pii/S2405844023029729

108. Qi H, Dong Z, You X, Li Y, Zhao Y, Sun X. Extended exergy accounting for assessing the sustainability of agriculture: A case study of Hebei Province, China. *Ecol Indic [Internet]*. 2023;150:110240. Available from: https://www.sciencedirect.com/science/article/pii/S1470160X23003825

109. Arias A, Feijoo G, Moreira MT. Process modelling and environmental assessment on the valorization of lignocellulosic waste to antimicrobials. *Food Bioprod Process [Internet]*. 2023;137:113–123. Available from: https://www.sciencedirect.com/science/article/pii/S0960308522001444

110. Aslam N, Hassan SA, Mehak F, Zia S, Bhat ZF, Yıkmış S, et al. Exploring the potential of cashew waste for food and health applications: A review. *Futur Foods [Internet]*. 2024;9:100319. Available from: https://www.sciencedirect.com/science/article/pii/S266683352400025X

111. Arifa Farzana B, Mujafarkani N, Thakur A, Kumar A, Mushira Banu A, Shifana M. Evaluating (p-Semidine-Guanidine-Formaldehyde) terpolymer resin efficiency as anti-corrosive agent for mild steel in 1 M H2SO4: An experimental and computational approach. *Inorg Chem Commun [Internet]*. 2023;158(P1):111572. Available from: https://doi.org/10.1016/j.inoche.2023.111572

112. Sharma D, Thakur A, Kumar M, Sharma R, Kumar S, Om H. Effective corrosion inhibition of mild steel using novel 1, 3, 4-oxadiazole-pyridine hybrids: Synthesis, electrochemical, morphological, and computational insights. *Environ Res [Internet]*. 2023;234(July):116555. Available from: https://doi.org/10.1016/j.envres.2023.116555

113. Thakur A, Kumar A, Sharma S, Ganjoo R, Assad H. Computational and experimental studies on the efficiency of Sonchus arvensis as green corrosion inhibitor for mild steel in 0.5 M HCl solution. *Mater Today Proc [Internet]*. 2022;66:609–621. Available from: https://doi.org/10.1016/j.matpr.2022.06.479

114. Thakur A, Kumar A. Recent trends in nanostructured carbon-based electrochemical sensors for the detection and remediation of persistent toxic substances in real-time analysis. *Mater Res Express*. 2023;10(3).

115. Verma C, Thakur A, Ganjoo R, Sharma S, Assad H, Kumar A, et al. Coordination bonding and corrosion inhibition potential of nitrogen-rich heterocycles: Azoles and triazines as specific examples. *Coord Chem Rev [Internet]*. 2023;488(April):215177. Available from: https://doi.org/10.1016/j.ccr.2023.215177

116. Calvin KV, Beach R, Gurgel A, Labriet M, Loboguerrero Rodriguez AM. Agriculture, forestry, and other land-use emissions in Latin America. *Energy Econ [Internet]*. 2016;56:615–624. Available from: https://www.sciencedirect.com/science/article/pii/S0140988315001127

117. Gallego-García M, Moreno AD, Manzanares P, Negro MJ, Duque A. Recent advances on physical technologies for the pretreatment of food waste and lignocellulosic residues. *Bioresour Technol [Internet]*. 2023;369:128397. Available from: https://www.sciencedirect.com/science/article/pii/S0960852422017308

118. Zambon I, Colosimo F, Monarca D, Cecchini M, Gallucci F, Proto AR, et al. An innovative agro-forestry supply chain for residual biomass: Physicochemical characterisation of biochar from olive and hazelnut pellets. *Energies*. 2016;9(7):1–11.

119. Pröll T, Afif R Al, Schaffer S, Pfeifer C. Reduced local emissions and long-term carbon storage through pyrolysis of agricultural waste and application of pyrolysis char for soil improvement. *Energy Procedia [Internet]*. 2017;114:6057–6066. Available from: https://www.sciencedirect.com/science/article/pii/S1876610217319446

120. Perveen F, Farooq M, Naeem A, Humayun M, Saeed T, Khan IW, et al. Catalytic conversion of agricultural waste biomass into valued chemical using bifunctional heterogeneous catalyst: A sustainable approach. *Catal Commun [Internet]*. 2022;171:106516. Available from: https://www.sciencedirect.com/science/article/pii/S1566736722001212

121. Wang X, Leng W, Nayanathara RMO, Milsted D, Eberhardt TL, Zhang Z, et al. Recent advances in transforming agricultural biorefinery lignins into value-added products. *J Agric Food Res [Internet]*. 2023;12:100545. Available from: https://www.sciencedirect.com/science/article/pii/S2666154323000522

122. Singh Karam D, Nagabovanalli P, Sundara Rajoo K, Fauziah Ishak C, Abdu A, Rosli Z, et al. An overview on the preparation of rice husk biochar, factors affecting its properties, and its agriculture application. *J Saudi Soc Agric Sci [Internet]*. 2022;21(3):149–159. Available from: https://www.sciencedirect.com/science/article/pii/S1658077X21001041

CHAPTER 5

Animal Wastes

Fundamentals, Classification, Properties, and Applications

Apoorva Pathania and Anu Radha Pathania

5.1 INTRODUCTION

The escalating environmental repercussions, depletion of energy resources, and the emission of carbon dioxide resulting from the combustion of fossil fuels have spurred an increasing inclination towards research into sustainable fuels [1, 2]. As an integral component of defensible incendiary, green fuels derived from atrophy biomass offer inexhaustible vitality solutions with environmentally friendly benefits [3, 4]. "Green fuel" refers to congealed, liquid, or gaseous propellant primarily derived from transforming bio-inexhaustible staples [5, 6]. The term "biomass" is commonly employed to denote biological resources derived from animals and plants [7, 8]. These are considered fundamental and pertinent inexhaustible vitality sources applicable for competitive generation, heat exertion, and bio-fuel production, among other uses [9]. A significant portion of biomass is discarded as residual biomass through human activities, such as sewage mire [10], wastage of food, manure, agriculture and forestry waste or residue, biodiesel waste, marine residue, and fermentation process waste contribute significantly to the pool of available biomass and bio-atrophy.

Biomass and bio-atrophy offer various scrutinization areas for vitality recovery using multiple processes, techniques, and technologies, potentially addressing inevitable vitality demands. Natural reserves are crucial for vitality production, but sustainability is critical. Technology can help maintain sustainability by addressing vitality generation, initial and

DOI: 10.1201/9781003466833-7

end-use, and vitality's features. Primary vitality sources include fossil fuels, thermal, hydroelectric, and nuclear reservoirs [11, 12]. Inexhaustible energy sources offer sustainable energy while addressing side effects like resource consumption, atrophy control, and atmospheric damage [13, 14].

Sustainability promotes global and intergenerational equality, focusing on the long-term durability of societies and encompassing cultural, administrative, commercial, and environmental aspects [15]. Technological advancements enhance labor productivity and miniaturization, causing community anxiety. Engineers should engage in discussions to ensure technical understanding. Fossil fuels' unlimited accessibility may be addressed by estimating their cost when economically viable resources are consumed, leading to new energy management methods. Efficient utilization of finite resources is crucial to mitigate CO_2 production and manage waste temperature, but their intertwined, non-renewable nature and potential depletion of the global energy reservoir pose threats. Atomic power stores could potentially supply fissionable substances, but the scientific and economic dilemma raises environmental concerns. Primary energy reservoirs need transformation into alternative energy forms, such as thermal energy, for industrial or professional output [16]. The original condition method involves heat dissipation, requiring a capacity description to evaluate losses in exchange processes. Vitality represents valuable work, and entropy is related to vitality destroyed, indicating irreversibility. Lower-cost outlining, lifestyle adjustments, automobile efficiency, and minimizing heating and cooling in residential buildings can help meet vitality demands, reduce electricity dissipation, and mitigate oil dissipation. Exhaustible vitality sources include biomass, hydroelectric, geothermal, solar, and wind, with hydropower being the primary source, using water energy to rotate turbines and harvest electricity [17–20]. This chapter presents potential approaches for evaluating the technological feasibility of pretreatment systems for biomass energy generation, considering active, financial, and environmental aspects. It aims to provide policies to discover and implement economically and environmentally viable methods for biomass treatment. It explores the conversion of biomass and bio-waste into renewable energy sources, addressing the types of biomasses, the need for biomass vitality, recovery processes, and methods for harnessing energy from these resources. It also explains biomass energy pretreatment options, analyzing system mechanisms, progress, potentials, financial developments, and profits. It explores new pretreatments and innovative strategies, focusing on economic-environmental advantages. It compiles various trash administration techniques, offers fundamental postulates, and correlates them. It is based on environmental salvation and guarding, focusing on adulteration sway, atrophy management, biomass vitality harvesting, renewable energy production, and bioenergy utilization.

5.2 RENEWABLE ENERGY INTEGRATION FOR SUSTAINABLE DEVELOPMENT

Actions for defensible headway are constantly needed to address environmental challenges, with renewable energy sources being a dynamic and efficient solution, highlighting a close relationship between inexhaustible atrophy and defensible refinement. Problems with energy supplies and jobs go beyond climate change; they also include environmental issues, including acid rain, air pollution, ozone depletion, deforestation, and chemical spills. These problems highlight the necessity of a quick energy transition with the least adverse environmental effects. The energy industry and society as a whole are becoming more environmentally concerned. There is growing acceptance of the idea that consumers are accountable for pollution and its expenses. The costs linked to different energy sources have increased recently, indicating a recognition of the importance of the environment. With sustained economic growth predicted and a considerable increase in the world's population by the middle of the twenty-first century [21], it is estimated that the demand for energy services will rise globally and may perhaps reach an order of magnitude by the year 2050. The growing energy-related worries also include environmental concerns, like acid rain [22], ozone depletion in the stratosphere [23], and global climate change [24]. Globally, installed renewable energy capacity increased steadily to 820 GW between 2006 and 2017, hitting 2079 GW in 2016 at an annual growth rate of 8%, and more growth is predicted. Many countries actively support national and local policy measures to raise the share of renewable energy. Various forms of renewable energy sources are depicted in Figure 5.1. The sun is the source of all Earth's energy resources. A constant flow of energy from the sun heats the earth, promotes photosynthesis, heats the land and oceans differently, causes winds and waves, and eventually results in rainfall, which helps with hydropower [25]. While geothermal heat originates from radioactive decay deep inside the Earth, tidal movements are caused by the moon's and sun's gravitational attraction. Although these are feasible energy sources, science underpins our understanding of them. The difficulty is not in comprehending the procedures technically but rather in putting these complex systems into

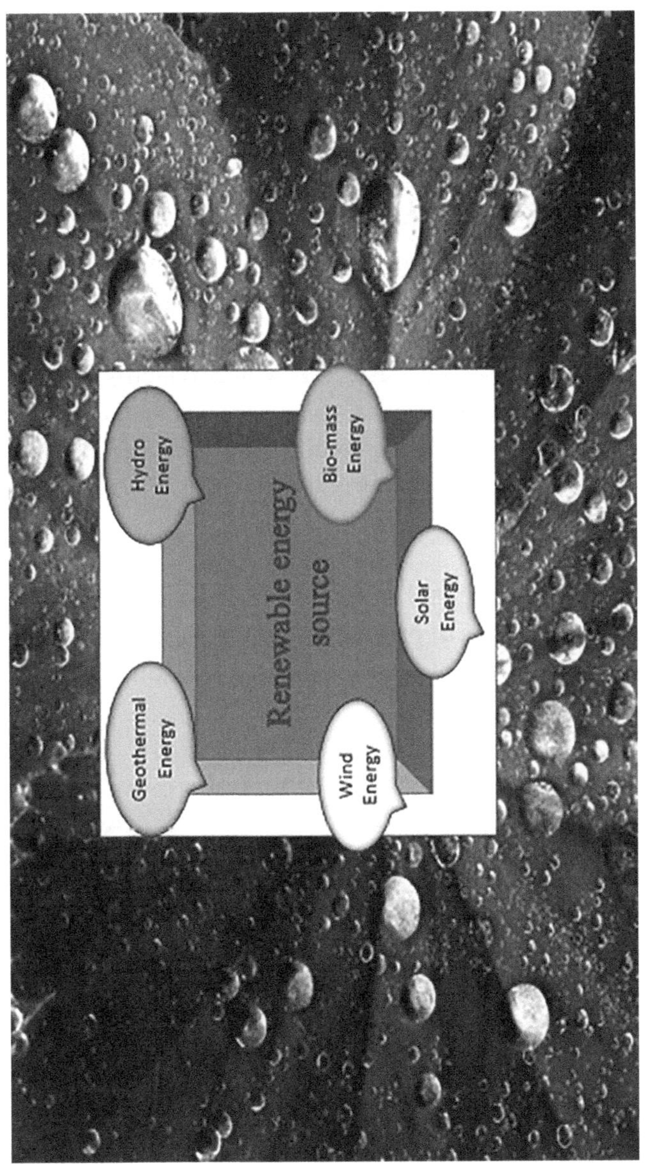

FIGURE 5.1 Reusable energy sources.

practice. Sustainable progress requires a defensible stock of vitality capital and efficient utilization of resources. Renewable energy sources are closely linked to sustainable growth, and a secure stock is essential for societal advancement [26]. Countries must plan energy transition strategies towards defensible zero-carbon development by replacing fossil fuels with renewable energy. This chapter suggests biomass energy harvesting as a solution, addressing environmental and social issues and promoting socio-economic development for future generations.

5.3 TYPES OF BIOMASS WASTE

Biomass, an inexhaustible energy source, includes plant materials, crops, and animal compost, with the potential for sustainable applications in various sectors like agriculture, wood waste, and marine plants. A comprehensive classification of bio-waste is provided in Figure 5.2.

Biomass, a reusable source with a chemical makeup of carbon, hydrogen, oxygen, and nitrogen, is a potential alternative to non-renewable vitality sources like livestock manure, sewage mire, and agriculture atrophy [27–29].

5.3.1 Municipal Solid Waste

Biomass is a feasible source of plant minerals, crops, and animal compost. It can be used for sustainable energy production, including herbaceous and

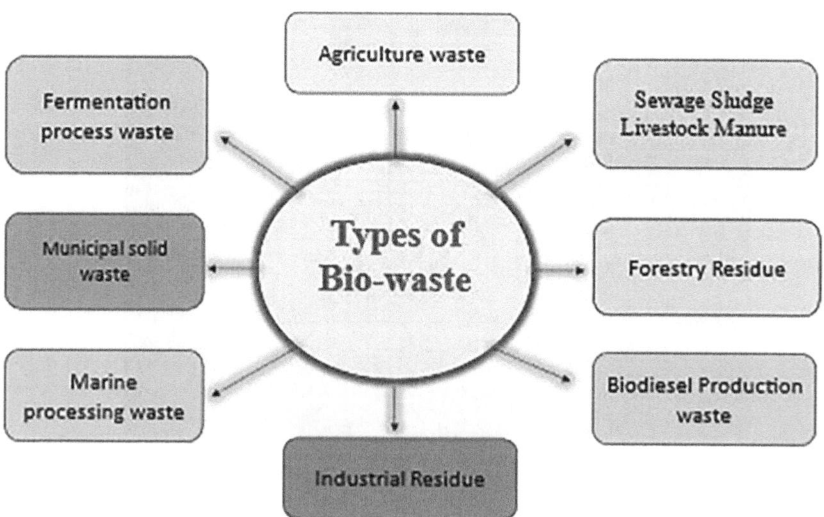

FIGURE 5.2   Types of bio-mass waste.

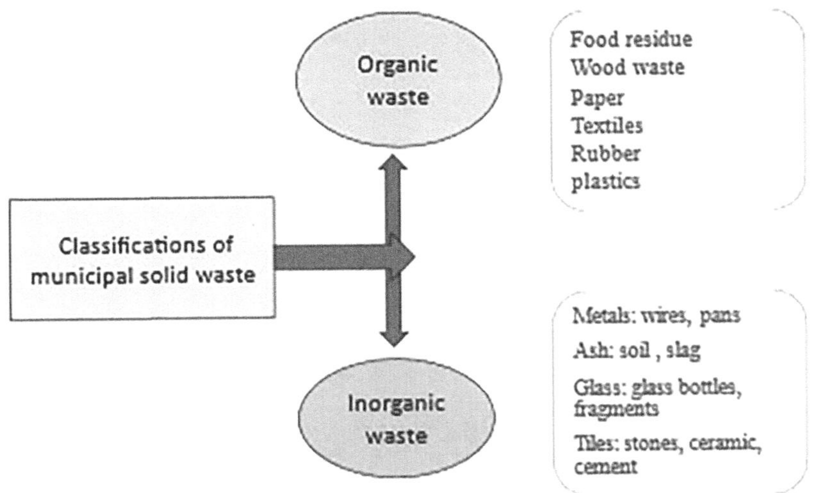

FIGURE 5.3 Classification of MSW.

wooden vitality products, farm food, and atrophy from various sources. Biomass is a heterogeneous and chemically combined livable radix, aiding in recovery processes. Municipal solid waste management (MSW) is a complex process involving atrophy generation, organization, collection, recycling, pre-treatment, and disposal. It requires collaboration from various professionals, including legal, commercial, regulatory, executive, environmental, and rights professionals. The authority and role in waste management are specific to each site and influenced by socio-economic, discernible, gregarious, academic, and directorate formation. Moguls are crucial in supporting the administration's continuity [30]. MSW consists of inorganic and organic debris in houses and buildings, such as food waste, cellulose commodities, rubber, plastic, glass, metals, and other materials.

MSW characteristics vary across nations and regions, with physical classification shown in Figure 5.3. Some fabrications action limit stubs sections to on-site events and vitality refinement, while some mites are fetched to municipal dross dispensation equipment [31, 32].

5.3.2 Agricultural Waste

Agricultural waste (AW) is residue from farming crops, including cornstalks, grass fodder, and sugarcane bagasse, containing lignocellulose, a carbon-rich biomass. AW has higher biomethane potential than food waste. Carbohydrates have a lower biochemical methane potential (BMP)

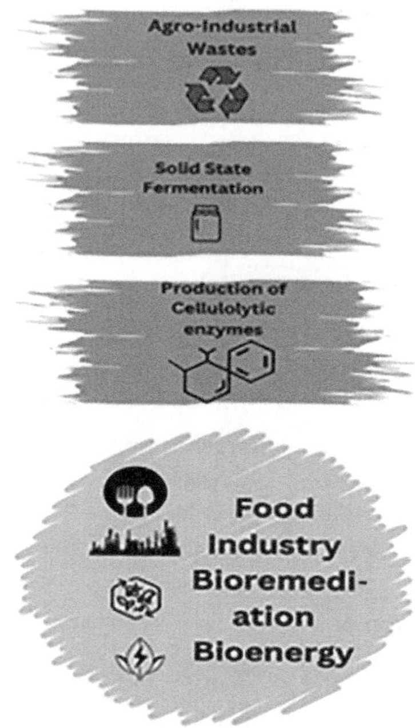

FIGURE 5.4   Effective management of AW.

than proteins or fats and may not decompose effectively under anaerobic digestion (AD). The efficiency of bioenergy recovery from agricultural wastes (AW) is lower than that of direct incineration. Biogas produced through AD is a valuable biofuel, and energy yields are prioritized over food production. The carbon-neutral approach has significantly improved energy production, reducing external fertilizers. Carbon-rich biomass, including AW, is co-digested with nitrogen-rich resources, concentrating additional nutrients like potassium in the residual filtrate. However, effective management of AWs has limited environmental and planning opportunities as shown in Figure 5.4. Long-term benefits include converting timber and pulp plant residues into usable resources [33, 34].

5.3.3 ■ Sewage Sludge

The wastewater execution method generates considerable sludge, with 60–90 grams of dehydrated mire created per capita daily. Sewage mire (SM) has the maximum biogas generation achievable in AD. However, 50–65% of combustible particles in new residue are demolished during mesophilic AD, even after a one-month hydraulic preservation interval [35].

Fresh sludge contains a high concentration of extracellular polymeric substances (EPS) and stiff cells, which limits the destruction efficacy of anaerobic digestion (AD) processes. Systems for experimentation have been created to provide AD microorganisms with nutrition. Numerous experiments have been carried out to increase CH_4 output and decrease residue, but costs and high energy requirements are involved. When applied to forests, sewage sludge (SS) under AD can accumulate heavy metals and non-biodegradable organic compounds, which might cause phytotoxic problems [36, 37].

Several factors, such as sewage sludge and livestock manure, cause variations in the recoverable inorganic resources in biomass waste. Research has concentrated on creating efficient methods for recovering phosphorus from manure for use in treatment procedures. These procedures use the residue from burned trash incineration, including thermochemical approaches, adsorption, and precipitation. Several protocols have been developed to improve the recovery of phosphorus. According to the study, there are three phases of weight loss in SS char (shown in Figure 5.5) during pyrolysis, starting with a 5% mass loss due to moisture extraction. With an overall mass loss of 9%, the derivative thermogravimetry curve

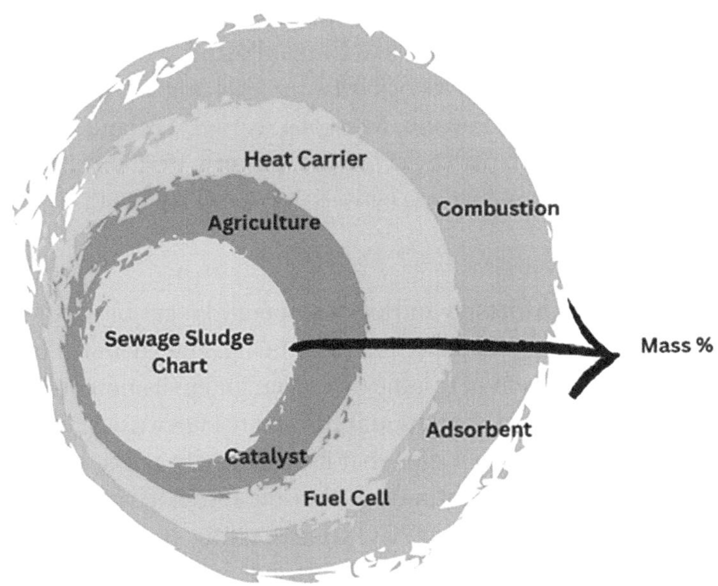

FIGURE 5.5   Chart of sewage mire (sludge).

displays two different stages. Significant mass loss occurs between 200 and 600°C due to an endothermic process in the differential scanning calorimetry curve linked to the poly-dankness of C–C knots and C–O chains [38–40]. The unique urban, underused water handling facility generates 90% of phosphorus drives through enhanced organic phosphorus abolition and sleet, creating the most abundant phosphorus retrieval reserve.

5.3.4 Animal Waste

5.3.4.1 Manure

One anticipated byproduct of cattle rearing is livestock manure (LM). Apart from releasing a foul odor, stored LM contributes to the emission of greenhouse gases, mainly methane (CH_4), into the environment if not frequently controlled. In addition, LM contains heavy metals, bacteria, illnesses, and pests, which increases the possibility of secondary contamination if misapplied to the ground [41, 42]. In Japan, after municipal solid waste (MSW) consumption (1.7 billion m^3), there is expected to be a significant potential for yearly CH_4 recovery (1.6 billion m^3) [43]. Numerous factors, including the type of cow, their diet, their capacity to consume particular foods, the amount of protein and fiber in their diet, the animal's life stage, its habitat, the environment, and its life cycle phase, all have an impact on the characteristics of LM (shown in Figure 5.6). The Food and Agriculture Organization shows that changes in manure output are linked to variations in cattle numbers (shown in Figure 5.7). Countries like China, Poland, Norway, and Canada experienced an increase in livestock between 1994 and 2004, while Austria, Belarus, Canada, China, Finland, Mongolia, Norway, Poland, Switzerland, and the United States experienced growth between 1994 and 2014. Poland experienced a significant increase between 1994 and 2014 [44].

5.3.4.2 Forestry Residue

Forestry residue from forestry and logging operations can be used for bioenergy production, pulp and paper manufacturing, and chemical processes. Sustainable management of residues is crucial for environmental and economic reasons, minimizing waste and efficient resource use. It is a vital raw material for bioenergy generation, but it poses fire hazards. Pyrolysis can convert FRs into valuable outputs like bio-oil and biochar, with catalytic pyrolysis enhancing bio-oil for VGO co-processing. Biorefineries are crucial in the bio-economy. Biochemical methods include fermentation and anaerobic digestion, while thermochemical methods include pyrolysis,

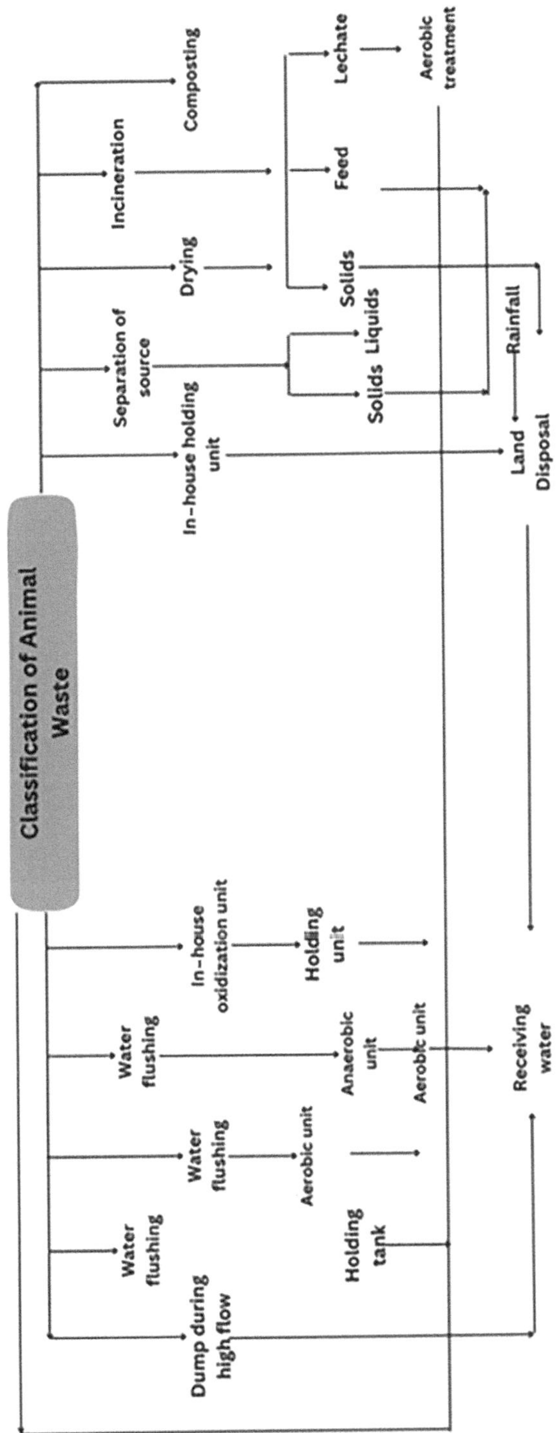

FIGURE 5.6 Classification of animal waste through manure disposal.

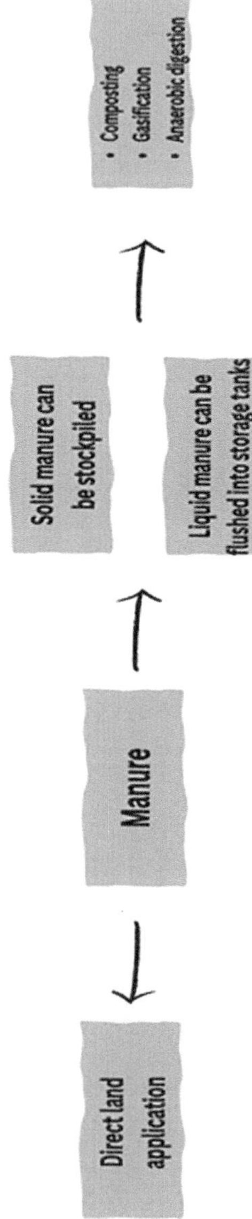

FIGURE 5.7 Manure decomposition strategies.

gasification, and combustion. Studies have explored pyrolysis of forestry residues to produce bio-oil, but it has high oxygen and water content, making it less suitable for fossil fuel replacement. Enhancing bio-oil through catalytic pyrolysis and hydro treatment is preferable. Common catalysts like H-ZSM-5 zeolites are effective but expensive and prone to deactivation in high water content bio-oil. Indispensable metal oxide catalysts like calcium oxide are less costly and resistant to deactivation [45–51]. Dyjakon and Noszczyk conducted a study on forestry biomass residues using the torrefaction process at 200–320°C. They found that the torrefaction process improves fuel qualities and minimizes unwanted features, making it essential to use various pre-treatment techniques to use biomass feedstock effectively. The study highlights the importance of a non-oxidative atmosphere and effective biomass feedstock utilization.

Forestry Residue Processing and Energy Generation

- Mechanical, thermal, or biological processing enhances feedstock biomass's physical properties.

- Optimal thermal treatment for forestry residue (FR) is crucial.

- FRs are gaining attention due to rising global energy consumption and GHG emissions reduction.

- Forestry biomass is a renewable energy source that cycles carbon.

- Portugal, with forests covering 35% of its area, is ideal for forest growth [52, 53].

5.3.5 Industrial Residue

The global society is grappling with resource issues and environmental catastrophes, particularly in China and India [54]. The petrochemical, pharmaceutical, agricultural, and food industries generate industrial leftovers, leading to significant greenhouse gas emissions. As a result, efforts are being made to provide environmentally friendly energy sources and substitute chemical synthesis techniques, often under the guidance of academia. The forestry, fishery, and agriculture sectors in the European Union generated around 21 million tons of waste in 2018, and more waste is anticipated as a result of rising food production and world population growth. According to a study by Paulina and her colleagues, the agro-food sectors produce biodegradable residues, including organic remnants from processed feedstock materials. Despite substantial studies into changing

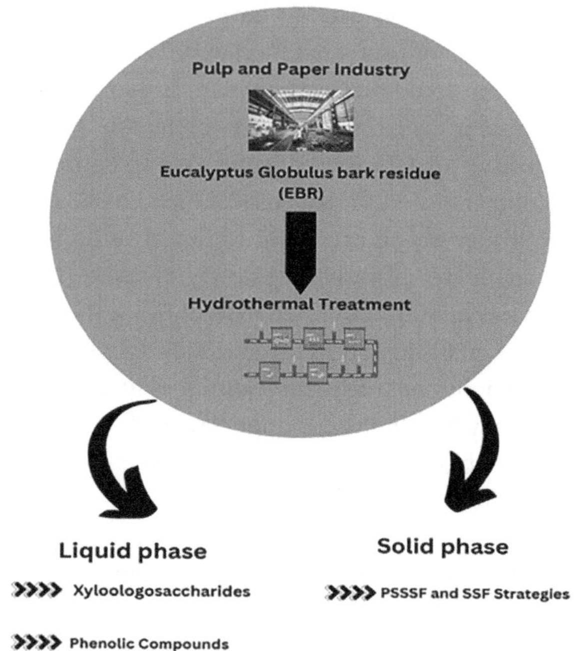

FIGURE 5.8 Classification of industry waste.

lignocellulosic materials, they still struggle to compete with petrochemical manufacturing methods. Alternative sources, such as industrial, agricultural, or forestry leftovers and municipal solid waste (shown in Figure 5.8), have been investigated to minimize prices. In the Portuguese market, the pulp and paper sector mainly relies on eucalyptus globulus lumber, producing enormous amounts of eucalyptus bark yearly [55, 56]. Gomes et al. scrutinized hydrothermal treatment (HT) for fractionating industrial Eucalyptus globulus bark material into biofuels and valuable chemicals. The study investigated the use of HT to co-produce biofuels and chemicals from industrial EBR wastes [57].

5.3.6 Biodiesel Production Waste (Animal Atrophy)

Vegetable lube, animal cellulite, or microalgae-derived lipids are transesterified to make biodiesel, which yields crude glycerol (CG). Every 100 kg of bio-diesel fabrication generates biodiesel waste (BW), which has applications in industries such as toiletries, personal care, pharmaceuticals, and food goods. Cost-effective and efficient BW treatment techniques are required, with CG potentially acting as ethanol and biohydrogen. Current advancements in BW stewardship technology are being utilized for the production of biochemicals like eicosapentaenoic acid, docosahexaenoic acid, glycolipid, biosurfactant, 1,3-propanediol, and cephalosporin C. BW can be converted into glycerin, which is a bedrock for fermentative propanediol yield and can also be utilized as a solvent in organic reactions or antifreeze. Traditional wastewater treatment procedures, however, may not adequately eradicate organic molecules such as esters [58–61].

5.3.7 Marine Processing Waste (Animal Waste)

Marine fisheries account for over 50% of global fish proffering, with over 70% used for processing. Most of the catch is categorized as filtering dregs, counting trimmings, fins, frames, heads, skin, and viscera [62, 63]. Additionally, byproducts like crustacean and shellfish shells accumulate. As they are typically used in fish production, MPWs, leftover fish elements, are often used in low-value products like fish oil, compost, pet food, and fish silage. Fish waste can yield high-value bioactive compounds like fish muscle-derived peptides, collagen, gelatin, and fish oil, which are rich in omega-3 fatty acids. These lipid-based compounds can be used as compost or feedstuff in pig foods, providing an alternative to conventional protein sources. Monsivais-Alonso et al. conferred about life cycle appraisals of enhanced recipes for a circular thrift in omega-3 production from discarded fish oil. According to estimates, fisheries waste reaches 20 million tons, or 25% of total marine harvest. Marine bioprocessing offers tremendous potential for converting and using these byproducts as augmented

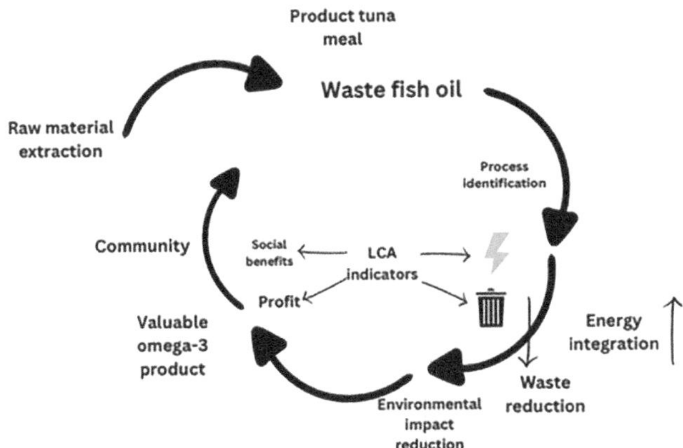

FIGURE 5.9 Marine waste management.

goods, although the word "outgrowth" is not expressly defined to separate it from garbage (shown in Figure 5.9) [64, 65].

5.3.8 Fermentation Process Waste (Organic Waste)

Byproducts obtained from the fermentation processes (shown in Figure 5.10) of lipids, proteins and carbohydrates can be recast into valuable materials. For instance, fatty acids and alcohol are manufactured from acetic acid, propionic acid, ethanol and butanol. This method turns atrophied materials' lignocellulosic biomass into valuable chemicals and fuels for biorefinery, providing a plentiful, feasible, and cost-effective vitality radix. Lignocellulosic biomass, derived from forestry, agriculture, and agro-industrial atrophy, consists of cellulose, hemicellulose, and lignin in see-saw proportions, along with a small dose of proteins and lipids, and ash is chemically composed of various materials.

Lignocellulose biomass is transformed into valuable biochemicals through pre-handling, alchemical hydrolysis, and turmoil. Pretreatment breaks the 3D tracery structure of lignocellulose biomass, allowing enzyme access and high fermentable sugar yield. This effortless, cost-effective, and efficient sugar convalescence method addresses environmental concerns and offers a viable alternative for valorizing residues from lignocellulosic waste [66, 67].

5.3.9 Food Processing Waste (Part of Animal Waste)

Efficient food processing waste management involves minimizing landfill disposal by adhering to the three Rs: Reduce, Reuse, and Recycle. This approach conserves vitality and water resources, and the EPA's food recovery hierarchy provides valuable guidance for food processors and beverage manufacturers. Baiano's study reveals that beverage manufacturers account for 26% of garbage, followed by dairy (21%), fruit/vegetable generation and treatment (14.8%), grain production (12.9%), meat goods refining and nurture (8%), vegetable and beast fats (3.9%), and fish commodity refining and storage (0.4%) [68–70]. Figure 5.11 summarizes current trends in stim valorization methodology for agro-food debris. Sustainable methods like AD, fermentation, and composting are used to recover food scraps in biorefinery commodities like bio-fuels, biomass, and bio-fertilizers. Agro-food waste can also be transformed into bio-dependent adsorbents for pollutant removal, and valuable ingredients can be repurposed as food flavors and therapeutics [71–73].

Food waste is a serious problem that affects natural resources and the supply chain at every stage. By using food recovery practices, we can

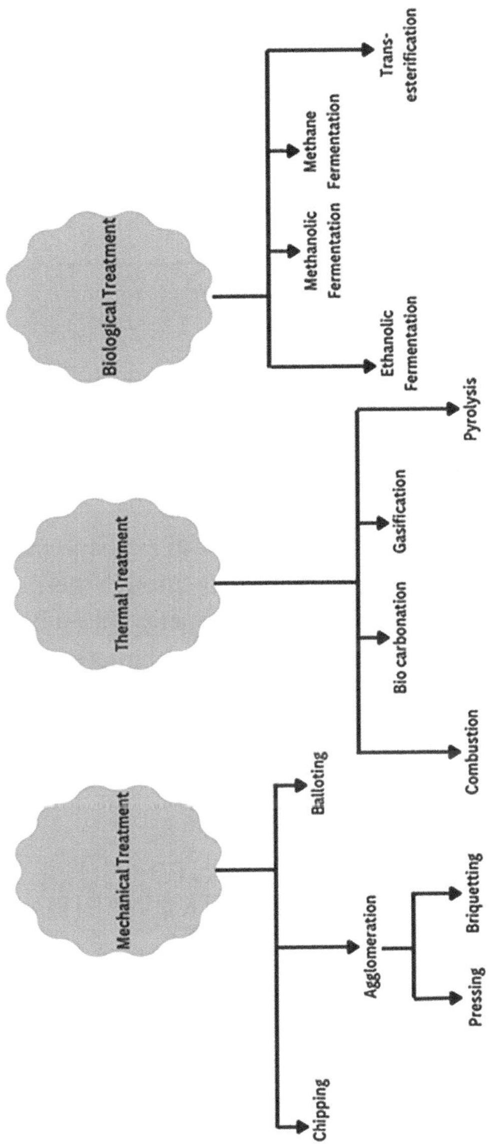

FIGURE 5.10   Process of fermentation waste.

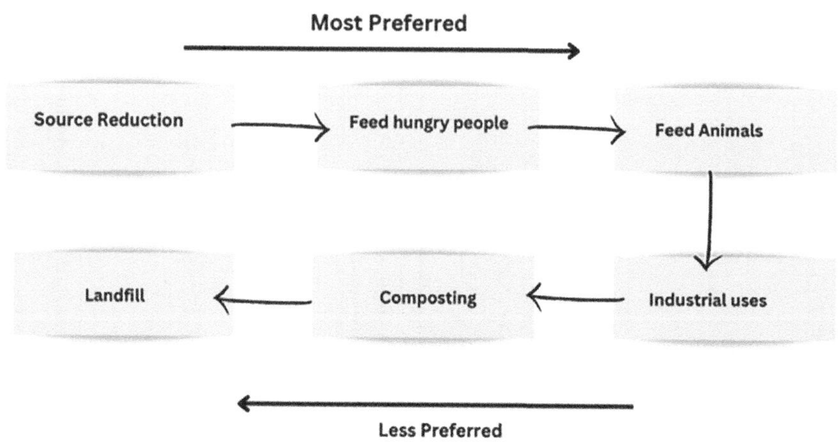

FIGURE 5.11   Methodology of FPW.

reduce waste while meeting the food demands of nine billion people by 2050. This method not only saves resources, but it also creates new avenues for food instability [74].

Ethanol fermentation with AD and landfill methane capture are methods for converting atrophy into heat vitality through thermochemical processes. Biochemical processes convert idle energy into subordinate vitality transporters like syngas, torrefied bits, bio-gas, bio-ethanol, and argan-oil, which can be used in ovens, vapor, gas turbines, or locomotives for heat and power generation. Advanced thermochemical processes like incineration, pyrolysis, gasification, and plasma-based technology are also being explored [75–77].

5.4 WAYS TO EXCHANGE WASTE INTO USEFUL ENERGY (ANIMAL AND FORESTRY WASTE)

Plasma processing, gasification, thermo-chemical incineration, torrefaction, biochemical composting, ethanol fermentation, and landfilling methane are few technologies or methods to exchange biochemical waste into the useful product that we can use in daily life. Thermal machinery converts waste directly into heat vitality during thermochemical processes, while biochemical ones convert unused waste into subordinate vitality transporters like syngas, torrefied bits, bio-gas, bio-ethanol, and bio-oil. These transporters can then be burned to harvest energy. This clean and effective energy yoking process allows for dynamic valorization of garbage.

5.4.1 Thermo-Chemical Conversion Technologies in Biomass Utilization

The method involves oxidative combustion of biomass atrophy to produce high-temperature energy, usually without direct ignition or incineration. This method converts plant chemical energy into heat and power. Direct burning of dry biomass has been used for heating and lighting for decades. Biomass is burned in furnaces to generate thermal power. Incineration procedures, such as open ignition and thermochemical modification, utilize chemical reactions at varying temperatures and may need incomplete oxidation or the absence of oxygen. These procedures include aeration, gas liberating, pyrolysis, gasification, and full oxidative ignition to break down organic surplus in residues (shown in Figure 5.12). To avoid thermochemical reactions, these approaches require accurate temperature regulation and appropriate air control. Pyrolysis and gasification, unlike incineration, may recover both the chemical and energy value of scrap. Chemical harvests can serve as fuel or raw materials for fuel generation [77, 78].

Pyrolysis is an ancillary gasification technique that employs indolent vapors as a gasification agent, such as carbon-based compounds as shown in Figure 5.13. It emits CO_2 and is analogous to combustion. The link between pyrolysis, gasification, and ignition is determined by the crops. The following sections discuss the aim of thermochemical transformation technologies that provide energy from biomass [79, 80].

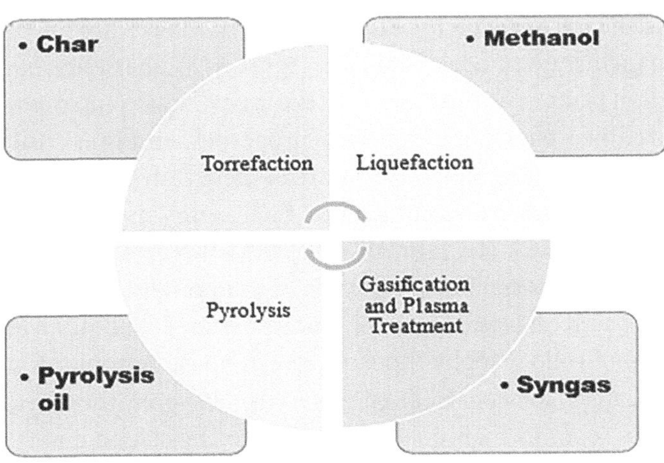

FIGURE 5.12 Thermochemical conversion techniques.

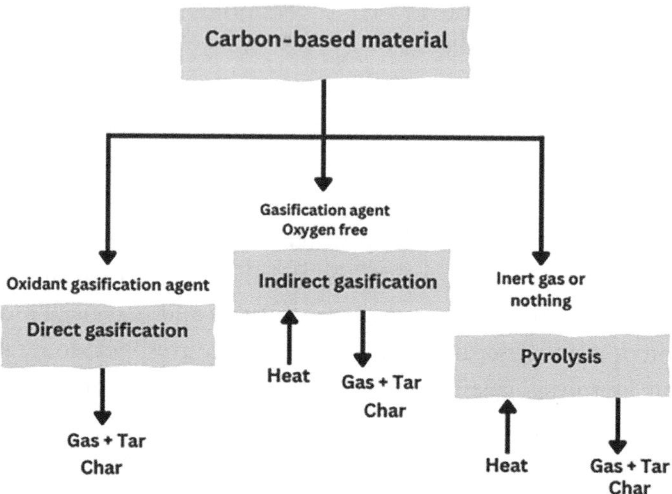

FIGURE 5.13   Steps for utilization of carbon-based materials.

5.4.1.1 Incineration

Incineration is a method for oxidizing combustible substances in waste, which is a heterogeneous substance made up mostly of organic components, ores, elements, and water. The process consists of evaporation and degassing, pyrolysis, and gasification. However, these processes frequently overlap, making it difficult to separate them. To limit contaminated emissions, models, furnace design, air circulation, and instrument engineering can be used. Grates, revolving kilns, and fluidized beds are the three most common incinerator models (shown in Figure 5.13). The design of a garbage incineration plant may differ based on the type of trash employed, with the chemical edifice, physical, and thermal properties serving as critical operators because of the unpredictable nature of these restrictions. Incineration generates flue gases, containing most fuel energy as heat, while fewer amounts of CO, HCl, HF, HBr, HI, NO_x, SO_2, VOC's, PCDD/F, PCB, and heavy metal meld form. The proportion of solid residue in MSW incinerators varies based on waste type and process design [78]. Bottom ash makes up 25–30% of solid atrophy input. Cure can boost bottom ash penchant, like vitrification, for construction materials. However, this energy-intensive process requires high-quality fabrication. Legislation and contemporary emission requirements have pushed the incineration business to make major technological advances in the last 10–15 years. This has decreased

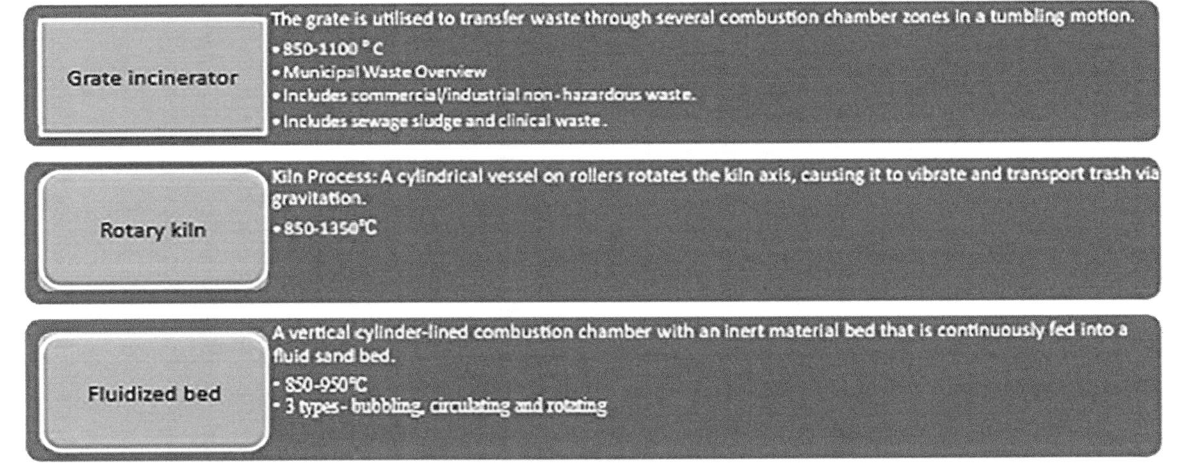

FIGURE 5.14 Classification of incineration.

air pollution hazards from waste incinerators to extremely low levels. The industry is always researching strategies to reduce operational costs while enhancing environmental performance. AEB's fourth-generation WtE plant in Amsterdam displays significant vitality and material convalescences, with net electrical efficacy predicted to exceed 30%. Meeting these lofty expectations, however, requires ongoing process development [81]. The Classification of incineration is illustrated in Figure 5.14.

5.4.1.2 Gasification

Gasification is the process of partially oxidizing organic compounds at high temperatures to generate synthesis gas, which might be used as a chemical feedstock or a fuel for electricity and heat generation. Synthesis gas includes CO, CO_2, H_2, H_2O, CH_4, trace hydrocarbons, inert gases, and impurities. It is a high-temperature process that converts organic compounds, particularly biomass, into synthetic gas. It converts carbon dioxide and water into carbon monoxide and hydrogen, producing producer gas for electricity and heat generation. Gasification uses various agents like oxygen, steam, and carbon dioxide to facilitate chemical reactions. A gasifier employs air, oxygen, steam, carbon dioxide, or a combination of these as gasification agents to generate low-energy gas (4–7 MJ Nm^{-3} NCV) and medium-energy gas (10–18 MJ Nm^{-3} NCV). Gasification containers are constructed with fixed bed, fluidized bed, and entrained flow gasifiers [82]. Figure 5.15 summarizes key qualities. For reactor applications, particularly with MSW, raw materials must be finely pulverized. Gasification of harmful waste is possible if it is fluid or finely grained. Gasifiers are divided into three categories based on their link to solids, which include petrol. Lower producer gas, tar and CO_2 eradication, char pollutants, heating worth, and gas restoration procedures are some of the challenges. To solve these issues, several reactants, adsorbents, and adsorption methods have been investigated. Dolomite, alkali, or alkaline earth minerals, notably those containing Ni, are widely employed in the gasification process [83–86].

5.4.1.3 Pyrolysis

Pyrolysis is a thermal ignominy process that produces three products: pyrolysis gas, pyrolysis liquid, and valid coke. The extents of these products pivot on the pyrolysis tacks and reactor process specifications. The calorific values of pyrolysis gas typically range from 5 to 15 MJ/m^{-3} based on MSW and 15 to 30 MJ/m^{-3} based on RDF [87].

Fixed Bed	Fluidized Bed	Entrained Flow
• Downdraft: solid and fumes transfers downcast • Updraft: solid passages downward, fumes go upright.	• Bubbling: truncated gas velocity, inactive substance carries with in the reactor • Distributing: the inactive substance elutriated, divided and recirculated.	• Category of fluidized bed • Typically, no inactive compact, high gas velocity. • May work as cyclonic gas fire.
1,000°C • Facile also robust structure • superbly granulated raw material essential	800-850°C • better tolerance to element dimension range compared to fixed beds • advanced tar levels in syngas	1,200-1,500°C • finely granulated raw material essential • less tar and CH4 content in produce gas • possible slagging of ash

FIGURE 5.15   Classification of gasification.

Plants for Hydrolysis of Waste Materials

- Preparation and grinding: Enhances heat transmission and waste quality.

- Drying: A separate drying stage improves the reactor efficiency and net calorific value of raw process gases.

- Waste pyrolysis: Produces minerals, metallics, and solid carbon-containing residue.

- Secondary treatment: burning to destroy organic ingredients and simultaneously use energy or condensing of gases for energy-usable oil combinations.

The method of pyrolysis is heating garbage, such as municipal solid atrophy (waste; MSW) and sewage mire, to create a range of products. The procedure is intricate and necessitates careful waste preparation. Hundreds of distinct chemicals are produced by it, many of which are unidentified. The existence of possible contaminants affects how useful pyrolysis is; if present, the products may become difficult or useless to employ. In addition, soil purification, cable tail treatment, synthetic waste treatment, and the recovery of metal and plastic compound materials are all part of the pyrolysis process. It has benefits for recovering materials and energy from the feed, like recovering organic fractions for fuel or material. Pretreatment is required in order to recover char for external usage. Gas engines and gas turbines, which produce energy more efficiently than traditional steam boilers, can be powered by pyrolysis gas. Smaller flue gas volumes from pyrolysis technologies result in lower treatment capital expenses. One waste treatment facility intended to lower the usage of main fuel is the Con Therm pyrolysis plant, created by RWE Power AG. The plant comprises of two 50k tons capacity drum-type kilns that run at 450–550°C in the truancy of oxygen. A cyclone is used to eradicate dust from the gas after the solid residue has been divided into coarse and fine fractions. There is no emission data available [88, 89].

As the reaction progresses, carbon becomes less amenable and evolves stable chemical adaptations, leading to the conversion stage from biomass. Cellulose and hemicellulose decay at a limited heat scale, with the rate and amount depending on method parameters. Hemicellulose breaks downward initially at 470–530 K, while cellulose tracks within the 510–620 K range [90].

5.4.1.4 Plasma Technologies

Due to the presence of charged gaseous species, plasma, the fourth state of matter, is extremely reactive. It is not like other liquids, solids, or gases. High-energy collisions between charged electrons and gaseous molecules that can occur thermally, electrically, or through electromagnetic radiation create plasma. There are two primary types of plasmas: low temperature plasmas, often known as gas discharges, and high temperature plasmas, in which every race is in a state of thermodynamic evenness. Thermal and cold plasmas, which have non-equilibrium states and quasi-equilibrium states, respectively, are further classifications for low temperature plasmas. Usually, an electric arc produced by a radio-frequency induction discharge or a plasma torch produces thermal plasmas. The relocated torch and the non-relocated torch are the two different kinds of plasma arc torches [91].

The use of plasma-based waste management systems has many benefits, such as melting materials at high temperatures, high energy densities and temperatures, high heat and reactant transmit rates, and mini-installation sort. By separating heat generation from process chemistry, electricity serves as the energy source, improving controllability and flexibility. But since electricity is a costly energy vector, financial concerns present a challenge. When the value of the products—such as syngas, hydrogen, or electricity—outweighs the actual expenses, plasmas become appealing [92].

Gasification, compaction, vitrification, and plasma pyrolysis of solid wastes are examples of plasma technologies for waste treatment, as well as combinations of these. The best course of action for treating a waste will rely on its content; for high-concentration organic compounds, plasma procedures provide an option for steam generation and full combustion. Halogen-rich waste streams, such as plastics, need to be treated at high temperatures and quenched in order to control the composition of the final product and minimize harmful emissions. Streams of inorganic solid waste can be reduced by melting, processed to remove valuable components, or oxidized and riveted in vitrified non-leaching slag [93].

5.4.1.4.1 Plasma Pyrolysis   A well-researched method of treating trash, plasma pyrolysis turns organic waste into a carbonaceous residue and combustible gas. Waste of many kinds, including plastic, old tires, agricultural residue, and medical waste, have been processed using this method. While the recovery of carbon black from used tires has shown promise, establishing and improving plasma pyrolysis techniques for industrial purposes still presents considerable technical obstacles. To reach its full potential, more research is required. Hazardous liquid and gas pyrolysis using plasma is becoming more and more popular and is a technique with a track record in industry. Ten PLASCON reactors in Australia, Japan, the USA, and Mexico use plasma pyrolysis to treat fluid wastes that contain hazardous materials. PLASCON was developed by CSIRO and SRL Plasma Ltd. Low-level radioactive waste, polymers, and medical waste are all treated with plasma pyrolysis. Nevertheless, there is no information available regarding industrial MSW or RDF facilities. The most widely used plasma-based technologies for treating solid waste are plasma gasification and vitrification; further information is provided in the next sections [94–99].

5.4.1.4.2 Plasma Gasification and Vitrification in Biomass Conversion   High temperatures are used in plasma gasification to break down organic meld into their component parts, producing a high-energy concoction gas mainly made of carbon monoxide and hydrogen. Low-energy fuels, including garbage from homes and businesses, can be made with this energy. Inorganic fractions such as glass, metals, silicates, and heavy metals are transformed into a dense, inert, non-leaching vitrified slag, and the conglomeration gas is cleaner than in traditional procedures. This gas can be converted into liquid biofuels of the second generation, power, or heat. Synthesis gas is produced via plasma gasification, which also creates steam, nitrogen, carbon dioxide, and air. For economic reasons, air is frequently used, and oxygen is utilized in reactions with organic materials. Because it lowers the amount of nitrogen and the overall gas flow, oxygen can be beneficial [100]. Higher arc voltages and jet power are produced by nitrogen and carbon dioxide-fueled plasma torches. Because of their high temperature, arc power, and plasma enthalpy, steam plasmas are advantageous for the processing of waste. Strong electrode erosion can be brought on by radicals containing hydrogen, oxygen, and hydroxide. Although argon can also be utilized as a plasma gas, enthalpy fluxes and torch power levels are reduced due to its low specific heat. Electricity is the energy source used in gas plasma technologies for waste treatment, which gives the system flexibility and control. As a stand-alone heat source, the plasma torch regulates temperature without affecting the quality of the feed or the availability of steam, oxygen, or air. Because thermal arc plasmas are insensitive to changes in process conditions, they are the dominant technology in waste treatment. Because they have large heat fluxes, transferred arc torches are perfect for decontaminating, immobilizing inorganic pollutants, and treating solid waste [101].

5.4.1.5 Torrefaction

The process of thermally converting biomass into a homogenous, energy-dense product ready for additional thermochemical reactions is known as torrefaction. It can grind readily, loses its visco-plastic qualities, gets brittle, becomes hydrophobic, and can hold up to 96% of its chemical energy. It also resists biodegradation. Although most torrefaction experiments are conducted in an inert atmosphere, other groups are experimenting with pressure, oxygenated environments, or hot compressed water (shown in Figure 5.16). The breakdown of hemicellulose is the primary cause of the bulk loss of biomass. Products of the torrefaction reaction, primarily solids, can be categorized according to ambient temperature and atmospheric pressure conditions. These include both volatile substances emitted during the process and solids such as ash, sugar structures, and char. Process variables like residence duration and torrefaction temperature affect how these products are distributed [102, 103].

Chipped biomass is used in an integrated torrefaction plant for effective drying and torrefaction as shown in Figure 5.17. Before being sized, biomass is inspected for contaminants. It is then dried to 20% moisture content, with a tiny portion being used as fuel [104]. It is possible to heat directly or indirectly using steam, flue gas, or hot air. Torrefied biomass, which is separated into flue gases and combustible gases known as "torgas," is the result of the torrefaction process. Processing factors and the biomass feedstock determine the composition and ratio. Torgas is a difficult-to-handle mixture of water, tar, and dust that produces heat for torrefaction and drying. The "point of auto-thermal operation" is where the torrefaction process should be run in order to maximize thermal efficiency and enhance process economics. Operating beyond this threshold lowers energy consumption, while operating below it necessitates the combustion

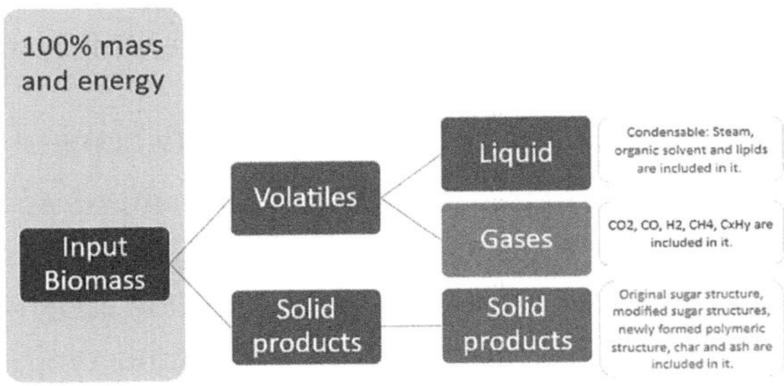

FIGURE 5.16   Torrefaction.

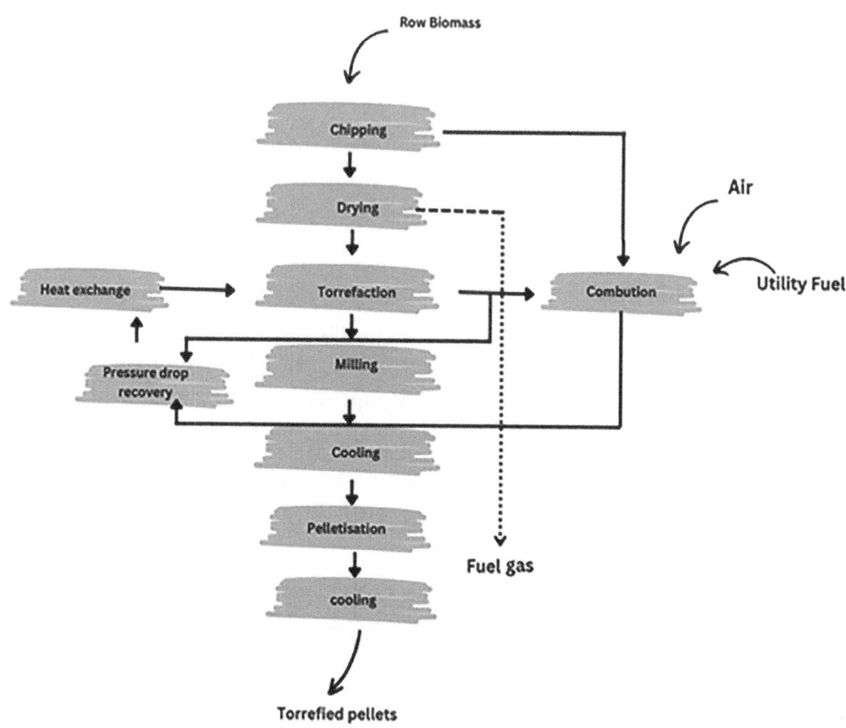

FIGURE 5.17   Filtrations process [109].

of additional biomass material. Additional biomass feedstock or utility fuel is added if there is not enough petrol. Before being pelletized in a pellet mill, torrefied biomass is ground up; cold feed is more effective in this process [105–108]. After that, the torrefied pellets are cooled down and either packaged or stored. Torrefied biomass must be fed cold to prevent overheating and lubricant melting, as hot feeding raises the temperature above the point of safe self-ignition. On the other hand, because thermally softened lignin functions as a natural binder, hot feeding is more effective.

5.4.2 Biological Conversion Technologies in Biochemical Processes

Waste-to-energy (WtE) operations utilize biochemical adaptation processes, primarily using microorganism-produced enzymes, to harness energy stored in solid fuel. These processes include anaerobic digestion for biogas production, bioethanol fermentation, and syngas generation, reducing environmental impact.

5.4.2.1 Composting

Through the aerobic process of composting, organic materials break down biologically to produce compost, carbon dioxide, water, and heat [110]. The stages of conventional composting are maturation, cooling, thermophilia, and mesophilia. Costly in-vessel systems for processing biowastes have been developed as a result of legislative pressure, enabling more accurate control over temperature and moisture. In order to comply with regulatory standards for disease control, modern composting techniques and technologies place a strong emphasis on high temperatures (> 70°C). Because compost has the ability to improve soil fertility and lower food demand in non-OECD nations, it is a standard method for fertilizers and soil conditioners. By keeping organic waste out of landfills, composting techniques improve soil fertility. Composting can be utilized as a junkyard sheathing in OECD nations to lower greenhouse gas effusion. Composting encourages good microorganisms while suppressing detrimental ones, which helps to properly hydrate organic waste and reduce soil and airborne diseases. The widespread use of bioenergy is impeded by high processing costs, despite its critical role in mitigating global warming. Extending sustainable bioenergy production from bio-wastes, such as construction, industrial, and municipal atrophy and repurposing composting heat as a renewable energy source are the challenges [111].

Previous studies have limited information on the dormant vitality mollification of fertilizer. A neoteric swotting found that during high-temperature junctures of municipal atrophy fertilizers, 1,136 kJ kg^{-1} of heat was liberated. The compost from municipal waste has low thermal conductivity coefficients, which decrease as it ages. This allows compost to cool slowly and can be used for heat exchanges. A study on compost heat reuse found that 73% of theoretical torridity vitality was relocated to water, but the placement of pipes within the compost mass was a limitation. The Deerdykes composting facility near Glasgow has estimated that compost can generate between 7,000 and 10,000 kJ kg^{-1} of energy for a 15-day composting period. A bespoke air-water heat exchanger was designed to extract the heat from the compost, providing adequate levels of heated water for domiciliary hot water and spatial heating endowing. The system was found to be the most reliable and competitively priced, providing 78% and 91% of the time, respectively. This suggests that accumulating the atrophy heat of fertilizer through a heat swapper is a realistic way out to contributing to vitality demand [112, 113].

5.4.2.2 Bioethanol Fermentation

Fermentable raw resources such as sugar-rich crops like sugarcane, plants, and fruits are used to make biofuel—or ethanol (CH_3CH_2OH). Alcohol-agitating microbes, which function at the root of co-metabolism, transform it into CH_3CH_2OH. While ethanologenic bacteria convert complicated polymeric carbohydrates into ethanol, saccharolytic bacteria break down complex polymeric carbohydrates into simpler types. As a workable substitute for ignition generators, this technique does not enhance CO_2 emissions. By using hydrolysis, fermentation, and distillation to create ethanol from banana fruit and organic waste, environmental concerns can be minimized (shown in Figure 5.18). With mass performance ranging from 346.5 to 388.7 L/t, a net energy value (NEV) between 9.86 and 9.94 MJ/L, and an energy ratio of 1.9 MJ/MJ, amylaceous materials were determined to have the best performance. The four production pathways under review have a positive energy balance, making them renewable energy sources according to the energy evaluation results [114, 115].

5.4.2.3 Anaerobic Digestion

By using microorganisms to break down organic matter in the truancy of oxygen, anaerobic digestion (AD) produces nitrogen-rich organic residue and biogas. It is utilized to produce renewable energy and reduce the chemical and biological oxygen demands in the treatment of wastewater, food, and agricultural wastes. Over 10% of organic waste in various European nations is treated with AD, which is widely utilized in Asia and Europe [116] It is categorized according to reactor design and operating parameters; a high solid content distinguishes solid-state anaerobic digestion (SS-AD). Because of its smaller reactor space, fewer energy requirements, less material handling, and reduced parasitic energy loss, SS-AD technology has advantages over liquid AD. Because of its low moisture level, it can be utilized as pelletized fuel or fertilizer. It creates biogas that is comparable to liquid AD output. As of right now, SS-AD supplies 54% of the AD capacity in Europe; there are no commercially operated facilities in the US. The benefits to the economy and environment are likely to lead to a rise in its commercial application. While there are benefits to SS-AD technology, it still has to be improved in terms of economics, production efficiency, and retention time. Compared to liquid AD, it can be retained for up to three times longer. For commercialization, advancements in feedstock preprocessing, digestate utilization, stability control, and reactor

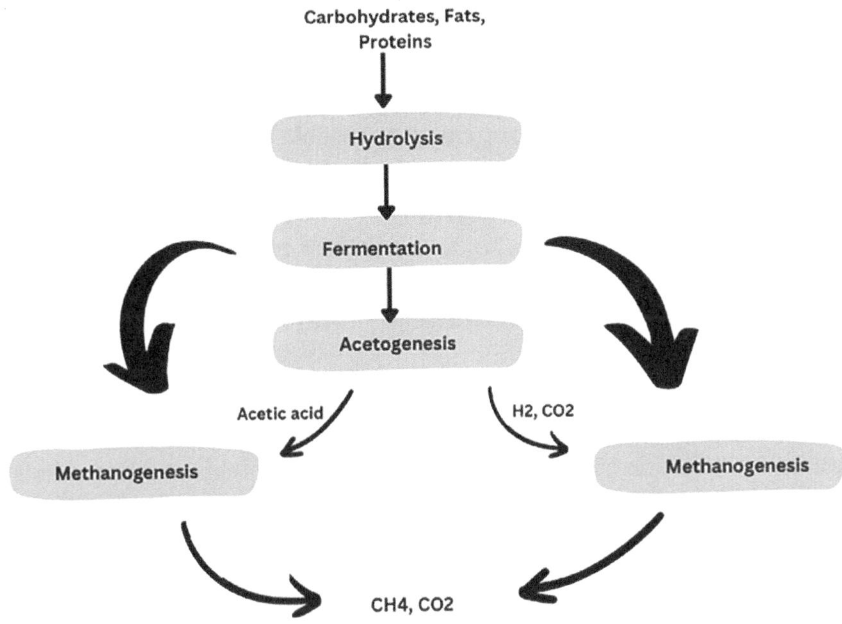

FIGURE 5.18　Bioethanol fermentation.

design are required. A biogas manufactory involves the outright trip of staples, biogas gubbins, and digester storage, followed by four steps: hydrolysis, acidogenesis/fermentation, acetogenesis, and methanogenesis [117].

5.4.2.3.1　Step 1 Hydrolysis　One of the primary phases in the multi-stage process of anaerobic digestion is hydrolysis. Methane and carbon dioxide can be produced from volatile fatty acids (VFAs), which are created when bacteria hydrolyze complicated substrates into simpler intermediates. Enzymes release these molecules through the cell membrane, facilitating the synthesis of cellular constituents and supplying energy. Starch, cellulose, monosaccharides, and polysaccharides are a few examples. Anaerobic digestion in two stages is a more efficient method of turning solid substrates into biogas than the traditional one-stage procedure. In one reactor, a vital hydrolytic and acid-forming microbial inhabitants is sustained, while in another, a methane-forming microbial community is maintained [118]. The physical, metabolic, and environmental needs of the two types of microorganisms are different. Complex insoluble substrates are hydrolyzed by bacteria into simpler, more soluble intermediates, which result in volatile fatty acids (VFAs), which can subsequently be used to produce carbon dioxide and methane. Biopolymers such as proteins, lipids, and polysaccharides hydrolyze to form solvable monomer mixes that include sugars, amino acids, and fatty acids. These biopolymers, which can either be soluble or insoluble in aqueous solutions, are widely found in biological waste. These biopolymeric organics must be hydrolyzed into smaller pieces and solubilized, which encourages metabolic deterioration, to be used. Microbes that are thermophilic and mesophilic collaborate in this process. The study highlights the importance of an efficient first hydrolytic stage in two-stage anaerobic digestion, as waste solubilization can limit treatment rates. Inoculating atrophy with microorganisms fabricating specific hydrolytic ferment can improve hydrolysis. Ferment activity levels can be used as surveillance guidelines, during anaerobic digestion of solid atrophy. The study found that both extracellular and cell-bound ferment fabricated by microorganisms led to hydrolysis of solid potato atrophy, with amylase being particularly high. The two-stage system with chaff as a biofouling carrier demean potato atrophy quickly, with high methane yields favoring its application [119].

5.4.2.3.2 Step 2 Acidogenesis   An essential stage of anaerobic digestion is called acidogenesis, during which bacteria convert complex organic materials into simpler substances, especially volatile fatty acids (VFAs). Through fermentation, organic substrates are converted into acids such as propionic, butyric, and acetic acids. These acids serve as building blocks for later stages of anaerobic digestion and increase the efficiency of biogas production [120].

5.4.2.3.3 Step 3 Acetogenesis   Acetogenesis converts gaseous fatty acids into CH_3COO^-, CO_2, and H_2 via acetogenins, crucial for biogas growth. Standard methods cannot achieve acetogenic effects due to their energy-consuming nature. Syntrophic microbial connectivity is needed for advancement impacts, as acetogenic effects cannot be achieved directly by methanogens [121].

5.4.2.3.4 Step 4 Methanogenesis   Methanogenesis Overview

- Biomethanization converts organic supports, such as acetate, H_2/CO_2, CH_4OH, and $HCOO-$, into CH_4.

- Archaea are the microorganisms that cause methane.

- CH_4 is created by the acetotrophic and hydrogenotrophic methanogenic pathways.

- The acetotrophic route produces about 70% of CH_4.

- Only a few recognized variables promote acetotrophic methanogenesis.

- The most well-known methanogenic classes are hydrogenotrophic methanogens [122].

5.4.2.4 Methane Recovery from Landfills: Sustainable Gas Capture and Utilization

In landfills, organic atrophy is reduced through micro-organisms, leading to the fabrication of landfill gas (LFG), an eruptive compound with CH_4 and CO_2 levels. LFG recovery schemes aim to limit emissions by landfills by coating them with earth, allowing fumes to be obtained by reservoirs and pipeline operations. The produced CH_4 can be ignited or used for electricity fabrication. Unrestricted and inadequately maintained dumping in metropolitan towns contributes to environmental degeneration [123]. In India, over 90% of municipal solid waste (MSW) is deployed on the ground, causing heavy metals to drain into coastal rivers. Accessibility for land for dumping is minimal in larger cities like Delhi. Most MSW is used in low-lying regions outside towns without understanding sterile landfilling principles. Most major demolition sites in India lack effective waste management systems, including leachate collection and landfill gas control. This causes all rubbish, especially toxic clinic trash, to enter the demolition site. Manufacturing scrap is frequently deposited in landfills for residential usage. Hygienic landfilling is an allowed method of managing MSW, but it will likely become India's principal work in the next years. Peculiar enlargement is required to achieve clean landfilling [124].

5.4.3 Bio-Electro Chemical Method

Bio-electro chemical system (BES) is a viable alternative to existing wastewater treatment and bioenergy conversion technologies that combine microbial metabolism with energy production (shown in Figure 5.19). It can be used in wastewater treatment, bioremediation, and synchronous bioenergy and biopolymer reproduction. BES employs electrochemical ways to convert biological energy into recyclable organic molecules, integrating bio- and electrochemical processes. Microorganism-catalyzed

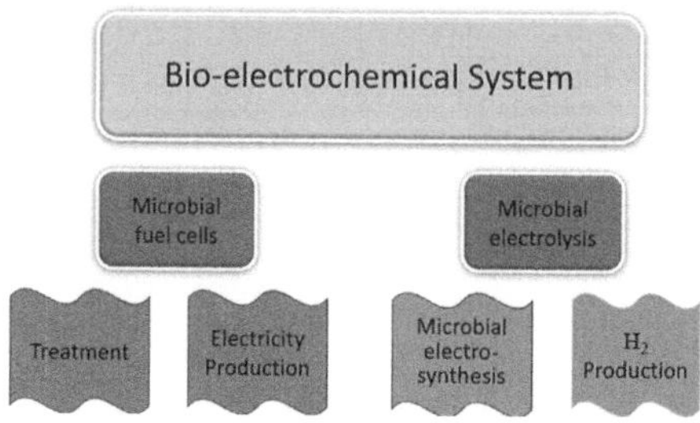

FIGURE 5.19   Basic classification of bio-electrochemical system.

redox reactions can recover bioenergy sources such as heavy alloys, nutrients, ores, and transitional industrial chemicals. BES has been widely researched for its characteristics, which include electrode substances, materials, microorganisms, and contaminant extraction. Some research, however, focuses on wastewater's potential for byproduct enhancement. BES provides environmentally friendly and encouraging solutions for renewable energy generation [125–127].

5.4.4 Economic and Environmental Perspectives of Waste-to-Energy Technologies

The growing atrophy in our biosphere is a significant concern, indicating a formidable sign for our home. Atrophy should be valued as a resource, not as vermin or scum, for economic, environmental, and social aspects. The United Nations sustainable development program has already considered atrophy management, but human activities' exploitation and environmental waste could hinder human rights and a healthy environment. The economic dimension of waste reduction involves producing biofuels and valuable products from atrophy, achieving both WtE and economic gain. However, no technology exists to reduce waste volume ultimately. Waste treatment plants should set emission standards and environmental regulations, and residue can be used as fertilizer [128, 129].

5.5 CHALLENGES AND FUTURE SCOPE

The main strength of the new civilization lies in reducing fossil fuels like petroleum and coal. At the same time, renewable vitality sources like solar, wind, hydropower, and biomass-derived electricity are rapidly being replaced, promoting alternative fossil fuels in both established and emerging nations.

Biomass Energy Potential and Tree Plantation

- Biomass waste can initiate clean vitality and essential metal nutrients.

- Increased tree planting is crucial for biomass energy production.

- Forest expansion offers long-term benefits.

- Trees like Leucaena leucocephala and bio-diesel trees like Jatropha, castor, and soybean can increase bioenergy production.

Plasma gasification is a new candidate for MSW or RDF processing, guiding substance renewal within the WtP (waste-to-product) theory. Improvements are being made during treatment for plasma automation towards debris elutriating, ameliorating its approval as a doable option for traditional elutriating rights. However, more data from research or method developers must be collected regarding emissions, performance, cost, and scientific endurance. The need for more information on plasma technology in scrap processing makes it difficult to analyze using typical approaches. However, plasma technology can help with energy efficiency and waste management. Plasma automation's commercial feasibility is predicted to increase as administrative, financial, and socio-political operators promote innovative thermo-chemical transformation apparatus. This innovation could result in major improvements to trash management and reduction.

Torrefaction is a potential biomass vitality pre- elutriating method that might help to commercialize biomass as an in-exhaustible vitality source. Its capacity to swiftly recover coal creates enormous business prospects. However, hurdles include lowering product prices while offering homogenous, hydrophobic, and durable materials with fuel flexibility. Key problems include efficient heat combination and appropriate tor gas use for auto thermal service. Torrefied beads may improve biomass commerce in a different generation, depending on raw material supply prices. Torrefaction devices must be optimized to suit end-user expectations and standardize output before they can be commercialized. Accelerated dissemination and scientific understanding are required to reduce generating costs and make torrefied bits competitive. The benefits of torrefied pellets must be proven to achieve financial success.

AD is a promising approach for extracting biomethane from various types of biomass waste. However, it faces challenges because of the unique properties of waste biomass, such as humidity content, C/N ratio, footprint components, and the constitutional prohibition on AD. Mono-digestion performance is limited, and the long-term viability of high-speed conversion is unpredictable. Co-conversion of several feedstocks is only sometimes practical. Extracellular polymeric polymers and other components remain in waste biomass due to their stable cell makeup, which affects AD degeneration performance. This type of biomass utilization demands extra processes. Fertilization may convert dewatered waste into compost, but digestate residue increases the expense of maintaining a biogas plant.

5.6 KEY HURDLES IN IMPLEMENTING RESOURCE RECOVERY FROM BIOMASS AND WASTE

The widespread use of waste to energy (WtE) technology is limited by operational, social, environmental, economic, and regulatory restrictions. These include low yield, efficiency, high transportation costs, and various biomass sizes. Separation, rectifying, agitation, pyrolysis, aerobic digestion, end-composting, and landfilling require interconnected setups or pilot plants. A standard feedstock process and a standard system model are required. Assessments are required shortly, and thermochemical, biochemical, and water-to-energy technologies appear promising. National and international research laboratories should collaborate to create a system convalescence strategy and process layout to maximize economic gains with local assistance.

5.7 CONCLUSION

Globally, there is a growing demand for safer ways of amendment of soil followed by biological stabilization due to the increased use of animal manure. More research is required to effectively inhibit pathogen recurrence in composted matter after aerobic stabilization treatment. Restoring vitality from animal debris and feedstock can help cut atrophy and generate green energy. This entails handling and breaking down atrophy, which releases greenhouse gases into the atmosphere. Zero carbon emissions and broadly accessible, environmentally benign green vitality radix are the goals of equitable growth. Generating vitality from debris and biomass might help achieve this ambition. An in-depth discussion of animal waste to energy (WtE) technologies will be provided in this chapter, emphasizing their classifications and fundamental and current applications. It highlights the necessity of a methodical approach that considers socioeconomic and environmental issues, which might range from feedstock to the layout of the product's design. The assessment emphasizes the necessity for thorough surveys while highlighting the potential of biomass atrophy for converting bio-vitality and bio-products into sustainable development.

REFERENCES

1. Kang, K., Qiu, L., Sun, G., Zhu, M., Yang, X., Yao, Y., & Sun, R. (2019). Codensification technology as a critical strategy for energy recovery from biomass and other resources-A review. *Renewable and Sustainable Energy Reviews, 116*, 109414.
2. Singh, Y., & Abd Rahim, E. (2020). Michelia Champaca: Sustainable novel non-edible oil as nano-based bio-lubricant with tribological investigation. *Fuel, 282*, 118830.
3. Fargione, J., Hill, J., Tilman, D., Polasky, S., & Hawthorne, P. (2008). Land clearing and the biofuel carbon debt. *Science, 319*(5867), 1235–1238.
4. Ge, S., Foong, S. Y., Ma, N. L., Liew, R. K., Mahari, W. A. W., Xia, C., . . . Lam, S. S. (2020). Vacuum pyrolysis incorporating microwave heating and base mixture modification: An integrated approach to transform biowaste into eco-friendly bioenergy products. *Renewable and Sustainable Energy Reviews, 127*, 109871.
5. Demirbas, M. F. (2009). Biorefineries for biofuel upgrading: A critical review. *Applied Energy, 86*, S151–S161.
6. Zareanshahraki, F., Lu, J., Yu, S., Kiamanesh, A., Shabani, B., & Mannari, V. (2020). Development of sustainable polyols with high bio-renewable content and their applications in thermoset coatings. *Progress in Organic Coatings, 147*, 105725.
7. Pappu, A., Pickering, K. L., & Thakur, V. K. (2019). Manufacturing and characterization of sustainable hybrid composites using sisal and hemp fibres as reinforcement of poly (lactic acid) via injection moulding. *Industrial Crops and Products, 137*, 260–269.
8. Sharma, B., Thakur, S., Mamba, G., Gupta, R. K., Gupta, V. K., & Thakur, V. K. (2021). Titania modified gum tragacanth-based hydrogel nanocomposite for water remediation. *Journal of Environmental Chemical Engineering, 9*(1), 104608.
9. Ma, H., Guo, Y., Qin, Y., & Li, Y. Y. (2018). Nutrient recovery technologies integrated with energy recovery by waste biomass anaerobic digestion. *Bioresource Technology, 269*, 520–531.
10. Rulkens, W. (2008). Sewage sludge as a biomass resource for the production of energy: Overview and assessment of the various options. *Energy & Fuels, 22*(1), 9–15.
11. Kaunda, C. S., Kimambo, C. Z., & Nielsen, T. K. (2012). Hydropower in the context of sustainable energy supply: A review of technologies and challenges. *International Scholarly Research Notices, 2012*.
12. Aryan, V., & Kraft, A. (2021). The crude tall oil value chain: Global availability and the influence of regional energy policies. *Journal of Cleaner Production, 280*, 124616.

13. Singh, S., & Hussain, C. M. (2021). Zero waste hierarchy for sustainable development. In *Concepts of Advanced Zero Waste Tools* (pp. 123–142). Elsevier.

14. Ates, B., Koytepe, S., Ulu, A., Gurses, C., & Thakur, V. K. (2020). Chemistry, structures, and advanced applications of nanocomposites from biorenewable resources. *Chemical Reviews, 120*(17), 9304–9362.

15. Holden, E., Linnerud, K., & Banister, D. (2014). Sustainable development: Our common future revisited. *Global Environmental Change, 26*, 130–139.

16. Kalaiselvam, S., & Parameshwaran, R. (2014). *Thermal Energy Storage Technologies for Sustainability: Systems Design, Assessment and Applications.* Elsevier.

17. Kaygusuz, K. (2002). Sustainable development of hydropower and biomass energy in Turkey. *Energy Conversion and Management, 43*(8), 1099–1120.

18. Moret, S., Peduzzi, E., Gerber, L., & Maréchal, F. (2016). Integration of deep geothermal energy and woody biomass conversion pathways in urban systems. *Energy Conversion and Management, 129*, 305–318.

19. Parlevliet, D., & Moheimani, N. R. (2014). Efficient conversion of solar energy to biomass and electricity. *Aquatic Biosystems, 10*(1), 4.

20. Bunn, D. W., Redondo-Martin, J., Muñoz-Hernandez, J. I., & Diaz-Cachinero, P. (2019). Analysis of coal conversion to biomass as a transitional technology. *Renewable Energy, 132*, 752–760.

21. Dincer, I., & Rosen, M. A. (1999). Energy, environment and sustainable development. *Applied Energy, 64*, 427–440.

22. Zhang, B., Cadotte, M. W., Chen, S., Tan, X., You, C., Ren, T., . . . Han, X. (2019). Plants alter their vertical root distribution rather than biomass allocation in response to changing precipitation. *Ecology, 100*(11), e02828.

23. Lucas, R. M., Norval, M., Neale, R. E., Young, A. R., De Gruijl, F. R., Takizawa, Y., & Van der Leun, J. C. (2015). The consequences for human health of stratospheric ozone depletion in association with other environmental factors. *Photochemical & Photobiological Sciences, 14*(1), 53–87.

24. Egbendewe-Mondzozo, A., Swinton, S. M., Bals, B. D., & Dale, B. E. (2013). Can dispersed biomass processing protect the environment and cover the bottom line for biofuel? *Environmental Science & Technology, 47*(3), 1695–1703.

25. Pandit, A., van Stokkum, I. H., van Amerongen, H., & Croce, R. (2018). Introduction: Light harvesting for photosynthesis. *Photosynthesis Research, 135*, 1–2.

26. Dincer, I., & Rosen, M. A. (1998). A worldwide perspective on energy, environment and sustainable development. *International Journal of Energy Research, 22*(15), 1305–1321.

27. Rana, A. K., Frollini, E., & Thakur, V. K. (2021). Cellulose nanocrystals: Pretreatments, preparation strategies, and surface functionalization. *International Journal of Biological Macromolecules, 182*, 1554–1581.

28. Platnieks, O., Sereda, A., Gaidukovs, S., Thakur, V. K., Barkane, A., Gaidukova, G., . . . Fridrihsone, V. (2021). Adding value to poly (butylene

succinate) and nanofibrillated cellulose-based sustainable nanocomposites by applying masterbatch process. *Industrial Crops and Products, 169,* 113669.

29. Thakur, V. K., Singha, A. S., & Thakur, M. K. (2012). In-air graft copolymerization of ethyl acrylate onto natural cellulosic polymers. *International Journal of Polymer Analysis and Characterization, 17*(1), 48–60.

30. Nikku, M., Deb, A., Sermyagina, E., & Puro, L. (2019). Reactivity characterization of municipal solid waste and biomass. *Fuel, 254,* 115690.

31. Liu, Z., Liu, Z., & Li, X. (2006). Status and prospect of the application of municipal solid waste incineration in China. *Applied Thermal Engineering, 26*(11–12), 1193–1197.

32. Zhou, H., Meng, A., Long, Y., Li, Q., & Zhang, Y. (2014). An overview of characteristics of municipal solid waste fuel in China: Physical, chemical composition and heating value. *Renewable and Sustainable Energy Reviews, 36,* 107–122.

33. Sawatdeenarunat, C., Surendra, K. C., Takara, D., Oechsner, H., & Khanal, S. K. (2015). Anaerobic digestion of lignocellulosic biomass: Challenges and opportunities. *Bioresource Technology, 178,* 178–186.

34. Zheng, W., Phoungthong, K., Lü, F., Shao, L. M., & He, P. J. (2013). Evaluation of a classification method for biodegradable solid wastes using anaerobic degradation parameters. *Waste Management, 33*(12), 2632–2640.

35. Appels, L., Baeyens, J., Degrève, J., & Dewil, R. (2008). Principles and potential of the anaerobic digestion of waste-activated sludge. *Progress in Energy and Combustion Science, 34*(6), 755–781.

36. Zhen, G., Lu, X., Kato, H., Zhao, Y., & Li, Y.-Y. (2017). Overview of pretreatment strategies for enhancing sewage sludge disintegration and subsequent anaerobic digestion: Current advances, full-scale application and future perspectives. *Renewable and Sustainable Energy Reviews, 69,* 559–577.

37. Hospido, A., Carballa, M., Moreira, M., Omil, F., Lema, J. M., & Feijoo, G. (2010). Environmental assessment of anaerobically digested sludge reuse in agriculture: Potential impacts of emerging micropollutants. *Water Research, 44*(10), 3225–3233.

38. Cordell, D., Rosemarin, A., Schröder, J. J., & Smit, A. L. (2011). Towards global phosphorus security: A systems framework for phosphorus recovery and reuse options. *Chemosphere, 84*(6), 747–758.

39. Clini, C., Musu, I., & Gullino, M. L. (2008). *Sustainable Development and Environmental Management.* Springer.

40. Gao, N., Kamran, K., Quan, C., & Williams, P. T. (2020). Thermochemical conversion of sewage sludge: A critical review. *Progress in Energy and Combustion Science, 79,* 100843.

41. Niu, Q., & Li, Y. Y. (2016). Recycling of livestock manure into bioenergy. *Recycling of Solid Waste for Biofuels and Bio-Chemicals,* 165–186.

42. Bora, R. R., Tao, Y., Lehmann, J., Tester, J. W., Richardson, R. E., & You, F. (2020). Techno-economic feasibility and spatial analysis of thermochemical conversion pathways for regional poultry waste valorization. *ACS Sustainable Chemistry & Engineering, 8*(14), 5763–5775.

43. Li, Y. Y., & Kobayashi, T. (2010). Applications and new developments of biogas technology in Japan. In *Environmental Anaerobic Technology* (pp. 35–58). Imperial College Press. https://doi.org/10.1142/9781848165434_0003.

44. Yao, Y., Huang, G., An, C., Chen, X., Zhang, P., Xin, X., . . . Agnew, J. (2020). Anaerobic digestion of livestock manure in cold regions: Technological advancements and global impacts. *Renewable and Sustainable Energy Reviews, 119*, 109494.

45. van Schalkwyk, D. L., Mandegari, M., Farzad, S., & Görgens, J. F. (2020). Techno-economic and environmental analysis of bio-oil production from forest residues via non-catalytic and catalytic pyrolysis processes. *Energy Conversion and Management, 213*, 112815.

46. Mandegari, M. A., Farzad, S., & Görgens, J. F. (2017). Recent trends on techno-economic assessment (TEA) of sugarcane biorefineries. *Biofuel Research Journal, 4*(3), 704–712.

47. Demirbas, M. F., & Balat, M. (2006). Recent advances on the production and utilization trends of bio-fuels: A global perspective. *Energy Conversion and Management, 47*(15–16), 2371–2381.

48. Yang, Y., Brammer, J. G., Wright, D. G., Scott, J. A., Serrano, C., & Bridgwater, A. V. (2017). Combined heat and power from the intermediate pyrolysis of biomass materials: Performance, economics and environmental impact. *Applied Energy, 191*, 639–652.

49. Puy, N., Murillo, R., Navarro, M. V., López, J. M., Rieradevall, J., Fowler, G., . . . Mastral, A. M. (2011). Valorisation of forestry waste by pyrolysis in an auger reactor. *Waste Management, 31*(6), 1339–1349.

50. Serrano-Ruiz, J. C., & Dumesic, J. A. (2012). Catalytic production of liquid hydrocarbon transportation fuels. In Guczi, L., & Erdôhelyi, A. (eds) *Catalysis for Alternative Energy Generation*. Springer. https://doi.org/10.1007/978-1-4614-0344-9_2

51. Stefanidis, S. D., Karakoulia, S. A., Kalogiannis, K. G., Iliopoulou, E. F., Delimitis, A., Yiannoulakis, H., . . . Triantafyllidis, K. S. (2016). Natural magnesium oxide (MgO) catalysts: A cost-effective sustainable alternative to acid zeolites for the in situ upgrading of biomass fast pyrolysis oil. *Applied Catalysis B: Environmental, 196*, 155–173.

52. Dyjakon, A., & Noszczyk, T. (2020). Alternative fuels from forestry biomass residue: Torrefaction process of horse chestnuts, oak acorns, and spruce cones. *Energies, 13*(10), 2468.

53. da Costa, T. P., Quinteiro, P., Arroja, L., & Dias, A. C. (2020). Environmental comparison of forest biomass residues application in Portugal: Electricity, heat and biofuel. *Renewable and Sustainable Energy Reviews, 134*, 110302.

54. Shi, H., Wang, Y., Chen, J., & Huisingh, D. (2016). Preventing smog crises in China and globally. *Journal of Cleaner Production, 112*, 1261–1271.

55. Liu, H., Huang, Y., Yuan, H., Yin, X., & Wu, C. (2018). Life cycle assessment of biofuels in China: Status and challenges. *Renewable and Sustainable Energy Reviews, 97*, 301–322.

56. Gomes, D., Rodrigues, A. C., Domingues, L., & Gama, M. (2015). Cellulase recycling in biorefineries—is it possible? *Applied Microbiology and Biotechnology, 99*, 4131–4143.
57. Gomes, D. G., Michelin, M., Romaní, A., Domingues, L., & Teixeira, J. A. (2021). Co-production of biofuels and value-added compounds from industrial Eucalyptus globulus bark residues using hydrothermal treatment. *Fuel, 285*, 119265.
58. Dou, B., Dupont, V., Williams, P. T., Chen, H., & Ding, Y. (2009). Thermogravimetric kinetics of crude glycerol. *Bioresource Technology, 100*(9), 2613–2620.
59. Feng, Y., Yang, Q., Wang, X., Liu, Y., Lee, H., & Ren, N. (2011). Treatment of biodiesel production wastes with simultaneous electricity generation using a single-chamber microbial fuel cell. *Bioresource Technology, 102*(1), 411–415.
60. Wijesekara, R. G., Nomura, N., Sato, S., & Matsumura, M. (2008). Pre-treatment and utilization of raw glycerol from sunflower oil biodiesel for growth and 1, 3-propanediol production by Clostridium butyricum. *Journal of Chemical Technology & Biotechnology: International Research in Process, Environmental & Clean Technology, 83*(7), 1072–1080.
61. Wolfson, A., Litvak, G., Dlugy, C., Shotland, Y., & Tavor, D. (2009). Employing crude glycerol from biodiesel production as an alternative green reaction medium. *Industrial Crops and Products, 30*(1), 78–81.
62. Nawaz, A., Li, E., Irshad, S., Xiong, Z., Xiong, H., Shahbaz, H. M., & Siddique, F. (2020). Valorization of fisheries by-products: Challenges and technical concerns to food industry. *Trends in Food Science & Technology, 99*, 34–43.
63. Choi, H.-J. (2020). Acid-fermented fish by-products broth: An influence to sludge reduction and biogas production in an anaerobic co-digestion. *Journal of Environmental Management, 262*, 110305. ISSN 0301-4797; https://doi.org/10.1016/j.jenvman.2020.110305.
64. Monsiváis-Alonso, R., Mansouri, S. S., & Román-Martínez, A. (2020). Life cycle assessment of intensified processes towards circular economy: Omega-3 production from waste fish oil. *Chemical Engineering and Processing-Process Intensification, 158*, 108171.
65. Kim, S. K., & Mendis, E. (2006). Bioactive compounds from marine processing byproducts–a review. *Food Research International, 39*(4), 383–393.
66. Chohan, N. A., Aruwajoye, G. S., Sewsynker-Sukai, Y., & Kana, E. G. (2020). Valorisation of potato peel wastes for bioethanol production using simultaneous saccharification and fermentation: Process optimization and kinetic assessment. *Renewable Energy, 146*, 1031–1040.
67. Singh, J. K., Vyas, P., Dubey, A., Upadhyaya, C. P., Kothari, R., Tyagi, V. V., & Kumar, A. (2018). Assessment of different pretreatment technologies for efficient bioconversion of lignocellulose to ethanol. *Frontiers in Bioscience, 10*(10), 2741.

68. Baiano, A. (2014). Recovery of biomolecules from food wastes—a review. *Molecules, 19*(9), 14821–14842.

69. Nayak, A., & Bhushan, B. (2019). An overview of the recent trends on the waste valorization techniques for food wastes. *Journal of Environmental Management, 233*, 352–370.

70. Schieber, A., Stintzing, F. C., & Carle, R. (2001). By-products of plant food processing as a source of functional compounds—recent developments. *Trends in Food Science & Technology, 12*(11), 401–413.

71. Han, S. K., & Shin, H. S. (2004). Biohydrogen production by anaerobic fermentation of food waste. *International Journal of Hydrogen Energy, 29*(6), 569–577.

72. Wang, Q., Wang, X., Wang, X., Ma, H., & Ren, N. (2005). Bioconversion of kitchen garbage to lactic acid by two wild strains of Lactobacillus species. *Journal of Environmental Science and Health, 40*(10), 1951–1962.

73. Zhang, C., Xiao, G., Peng, L., Su, H., & Tan, T. (2013). The anaerobic co-digestion of food waste and cattle manure. *Bioresource Technology, 129*, 170–176.

74. Zagorski, J., Reyes, G. A., Prescott, M. P., & Stasiewicz, M. J. (2021). Literature review investigating intersections between US foodservice food recovery and safety. *Resources, Conservation and Recycling, 168*, 105304.

75. Bouallagui, H., Cheikh, R. B., Marouani, L., & Hamdi, M. (2003). Mesophilic biogas production from fruit and vegetable waste in a tubular digester. *Bioresource Technology, 86*(1), 85–89.

76. Feng, S., Leung, A. K., Liu, H. W., Ng, C. W. W., Zhan, L. T., & Chen, R. (2019). Effects of thermal boundary condition on methane oxidation in landfill cover soil at different ambient temperatures. *Science of the Total Environment, 692*, 490–502.

77. Bosmans, A., Vanderreydt, I., Geysen, D., & Helsen, L. (2013). The crucial role of waste-to-energy technologies in enhanced landfill mining: A technology review. *Journal of Cleaner Production, 55*, 10–23.

78. Scarlat, N., Motola, V., Dallemand, J. F., Monforti-Ferrario, F., & Mofor, L. (2015). Evaluation of energy potential of municipal solid waste from African urban areas. *Renewable and Sustainable Energy Reviews, 50*, 1269–1286.

79. Gumisiriza, R., Hawumba, J. F., Okure, M., & Hensel, O. (2017). Biomass waste-to-energy valorisation technologies: A review case for banana processing in Uganda. *Biotechnology for Biofuels, 10*, 1–29.

80. Belgiorno, V., De Feo, G., Della Rocca, C., & Napoli, D. R. (2003). Energy from gasification of solid wastes. *Waste Management, 23*(1), 1–15.

81. Bridgwater, A. V. (1994). Catalysis in thermal biomass conversion. *Applied Catalysis A: General, 116*(1–2), 5–47.

82. Johnke, B., & Gamer, M. (2001). Draft of a German report with basic information for a BREF-document 'waste incineration'. http://files.gamta.lt/aaa/Tipk/tipk/4_kiti%20GPGB/63.pdf

83. Siwal, S. S., Zhang, Q., Sun, C., Thakur, S., Gupta, V. K., & Thakur, V. K. (2020). Energy production from steam gasification processes and parameters

that contemplate in biomass gasifier–a review. *Bioresource Technology, 297,* 122481.

84. Colmenares, J. C., Colmenares Quintero, R. F., & Pieta, I. S. (2016). Catalytic dry reforming for biomass-based fuels processing: Progress and future perspectives. *Energy Technology, 4*(8), 881–890.

85. Bridgwater, A. V. (1995). The technical and economic feasibility of biomass gasification for power generation. *Fuel, 74*(5), 631–653.

86. Tian, Y., Zhou, X., Lin, S., Ji, X., Bai, J., & Xu, M. (2018). Syngas production from air-steam gasification of biomass with natural catalysts. *Science of the Total Environment, 645,* 518–523.

87. Anthony, D. B., Howard, J. B., Hottel, H. C., & Meissner, H. P. (1976). Rapid devolatilization and hydrogasification of bituminous coal. *Fuel, 55*(2), 121–128.

88. Demirbas, A., & Arin, G. (2002). An overview of biomass pyrolysis. *Energy Sources, 24*(5), 471–482.

89. Barth, T. (1999). Similarities and differences in hydrous pyrolysis of biomass and source rocks. *Organic Geochemistry, 30*(12), 1495–1507.

90. Kucuk, M. H., & Demirbas, A. (1993). Delignification of ailanthus-altissima and spruce-orientalis with glycerol or alkaline glycerol at atmospheric-pressure. *Cellulose Chemistry and Technology, 27*(6).

91. Huang, H., & Tang, L. (2007). Treatment of organic waste using thermal plasma pyrolysis technology. *Energy Conversion and Management, 48*(4), 1331–1337.

92. Heberlein, J., & Murphy, A. B. (2008). Thermal plasma waste treatment. *Journal of Physics D: Applied Physics, 41*(5), 053001.

93. Tendero, C., Tixier, C., Tristant, P., Desmaison, J., & Leprince, P. (2006). Atmospheric pressure plasmas: A review. *Spectrochimica Acta Part B: Atomic Spectroscopy, 61*(1), 2–30.

94. Murphy, A. B., & Kovitya, P. (1993). Mathematical model and laser-scattering temperature measurements of a direct-current plasma torch discharging into air. *Journal of Applied Physics, 73*(10), 4759–4769.

95. Murphy, A. B., & McAllister, T. (1998). Destruction of ozone-depleting substances in a thermal plasma reactor. *Applied Physics Letters, 73*(4), 459–461.

96. Murphy, A. B., & McAllister, T. (2001). Modeling of the physics and chemistry of thermal plasma waste destruction. *Physics of Plasmas, 8*(5), 2565–2571.

97. Tang, L., & Huang, H. (2005). Thermal plasma pyrolysis of used tires for carbon black recovery. *Journal of Materials Science, 40*(14), 3817–3819.

98. Guddeti, R. R., Knight, R., & Grossmann, E. D. (2000). Depolymerization of polyethylene using induction-coupled plasma technology. *Plasma Chemistry and Plasma Processing, 20,* 37–64.

99. Nema, S. K., & Ganeshprasad, K. S. (2002). Plasma pyrolysis of medical waste. *Current Science,* 271–278.

100. Rutberg, P. G., Bratsev, A. N., Kuznetsov, V. A., Popov, V. E., & Ufimtsev, A. A. (2011). On efficiency of plasma gasification of wood residues. *Biomass and Bioenergy, 35*(1), 495–504.

101. Rafiq, M. H., & Hustad, J. E. (2011). Biosyngas production by autothermal reforming of waste cooking oil with propane using a plasma-assisted gliding arc reactor. *International Journal of Hydrogen Energy, 36*(14), 8221–8233.

102. Batidzirai, B., Mignot, A. P. R., Schakel, W. B., Junginger, H. M., & Faaij, A. P. C. (2013). Biomass torrefaction technology: Techno-economic status and future prospects. *Energy, 62,* 196–214.

103. Yan, W., Hastings, J. T., Acharjee, T. C., Coronella, C. J., & Vásquez, V. R. (2010). Mass and energy balances of wet torrefaction of lignocellulosic biomass. *Energy & Fuels, 24*(9), 4738–4742.

104. Medic, D., Darr, M., Potter, B., & Shah, A. (2010). Effect of torrefaction process parameters on biomass feedstock upgrading. In *2010 Pittsburgh, Pennsylvania, June 20-June 23, 2010* (p. 1). American Society of Agricultural and Biological Engineers.

105. Alamia, A., Ström, H., & Thunman, H. (2015). Design of an integrated dryer and conveyor belt for woody biofuels. *Biomass and Bioenergy, 77,* 92–109.

106. Agar, D., & Wihersaari, M. (2012). Bio-coal, torrefied lignocellulosic resources–key properties for its use in co-firing with fossil coal–their status. *Biomass and Bioenergy, 44,* 107–111.

107. Phanphanich, M., & Mani, S. (2011). Impact of torrefaction on the grindability and fuel characteristics of forest biomass. *Bioresource Technology, 102*(2), 1246–1253.

108. Prins, M. J., Ptasinski, K. J., & Janssen, F. J. (2006). More efficient biomass gasification via torrefaction. *Energy, 31*(15), 3458–3470.

109. Singh, R., Bhatia, A., & Srivastava, M. (2015). Biofuels as alternate fuel from biomass—the Indian scenario. *Energy Sustainability Through Green Energy,* 287–313.

110. Irvine, G., Lamont, E. R., & Antizar-Ladislao, B. (2010). Energy from waste: Reuse of compost heat as a source of renewable energy. *International Journal of Chemical Engineering, 2010.*

111. Kalyani, K. A., & Pandey, K. K. (2014). Waste to energy status in India: A short review. *Renewable and Sustainable Energy Reviews, 31,* 113–120.

112. Smith, M., & Aber, J. (2017). Heat recovery from composting: A step-by-step guide to building an aerated static pile heat recovery composting facility. file:///C:/Users/KU500781/Downloads/SmithandAber2017HeatRecovery fromComposting.pdf

113. Fan, L. T., Lee, Y. H., & Beardmore, D. R. (1981). Influence of major structural features of cellulose on rate of enzymatic hydrolysis. *Biotechnology and Bioengineering (United States), 23*(2).

114. Velásquez-Arredondo, H. I., & Ruiz-Colorado, A. A. (2010). Ethanol production process from banana fruit and its lignocellulosic residues: Energy analysis. *Energy, 35*(7), 3081–3087.

115. Hsieh, W. D., Chen, R. H., Wu, T. L., & Lin, T. H. (2002). Engine performance and pollutant emission of an SI engine using ethanol–gasoline blended fuels. *Atmospheric Environment, 36*(3), 403–410.

116. Li, Y., Park, S. Y., & Zhu, J. (2011). Solid-state anaerobic digestion for methane production from organic waste. *Renewable and Sustainable Energy Reviews, 15*(1), 821–826.
117. Khalid, A., Arshad, M., Anjum, M., Mahmood, T., & Dawson, L. (2011). The anaerobic digestion of solid organic waste. *Waste Management, 31*(8), 1737–1744.
118. Morgenroth, E., Kommedal, R., & Harremoës, P. (2002). Processes and modeling of hydrolysis of particulate organic matter in aerobic wastewater treatment–a review. *Water Science and Technology, 45*(6), 25–40.
119. Parawira, W., Murto, M., Read, J. S., & Mattiasson, B. (2005). Profile of hydrolases and biogas production during two-stage mesophilic anaerobic digestion of solid potato waste. *Process Biochemistry, 40*(9), 2945–2952.
120. Gujer, W., & Zehnder, A. J. (1983). Conversion processes in anaerobic digestion. *Water Science and Technology, 15*(8–9), 127–167.
121. Lalman, J. A., & Bagley, D. M. (2001). Anaerobic degradation and methanogenic inhibitory effects of oleic and stearic acids. *Water Research, 35*(12), 2975–2983.
122. Demirel, B., & Scherer, P. (2008). The roles of acetotrophic and hydrogenotrophic methanogens during anaerobic conversion of biomass to methane: A review. *Reviews in Environmental Science and Bio/Technology, 7*, 173–190.
123. Sharholy, M., Ahmad, K., Vaishya, R. C., & Gupta, R. D. (2007). Municipal solid waste characteristics and management in Allahabad, India. *Waste Management, 27*(4), 490–496.
124. Mor, S., Ravindra, K., De Visscher, A., Dahiya, R. P., & Chandra, A. (2006). Municipal solid waste characterization and its assessment for potential methane generation: A case study. *Science of the Total Environment, 371*(1–3), 1–10.
125. Zhang, J., Liu, Y., Sun, Y., Wang, H., Cao, X., & Li, X. (2020). Effect of soil type on heavy metals removal in bioelectrochemical system. *Bioelectrochemistry, 136*, 107596.
126. Jadhav, D. A., Ray, S. G., & Ghangrekar, M. M. (2017). Third generation in bio-electrochemical system research–a systematic review on mechanisms for recovery of valuable by-products from wastewater. *Renewable and Sustainable Energy Reviews, 76*, 1022–1031.
127. Wang, H., & Ren, Z. J. (2014). Bioelectrochemical metal recovery from wastewater: A review. *Water Research, 66*, 219–232.
128. Gupta, S., Mohan, K., Prasad, R., Gupta, S., & Kansal, A. (1998). Solid waste management in India: Options and opportunities. *Resources, Conservation and Recycling, 24*(2), 137–154.
129. Velenturf, A. P., & Purnell, P. (2017). Resource recovery from waste: Restoring the balance between resource scarcity and waste overload. *Sustainability, 9*(9), 1603.

CHAPTER 6

Biomass-Based Industrial Wastes

Fundamentals, Classification, and Application

Shveta Sharma and Ashish Kumar

6.1 INTRODUCTION

Biomass is the term used to describe recyclable organic material from wildlife and plants. Biomass consists of the sun's stored chemical energy, produced by plants via photosynthesis. Biomass may be burned to provide heat energy or converted into gaseous and liquid fuels using various techniques. Biomass is a primary source for generating heat and energy and acts as a fuel for conveyance. Biomass is a significant fuel source in many countries, especially for cooking and heating in underdeveloped nations. Different biomass sources used for energy production are described in Figure 6.1.

Hence, biomass usage has been increasing in industries because of the benefits attached to biomass. Further, industry is transitioning from a linear economy to a circular economy due to heightened societal awareness and stricter enforcement of environmental rules. Environmental regulations and legislation inspire companies and researchers to develop process models that facilitate the use of waste in a closed-loop system. For example, the processing of crops into final goods results in significant quantities of residual biomass, mainly composed of lignocellulosic components. This trash is barely utilized as a food source and is classified as a high-volume waste. The drawback of converting agro-waste into

DOI: 10.1201/9781003466833-8

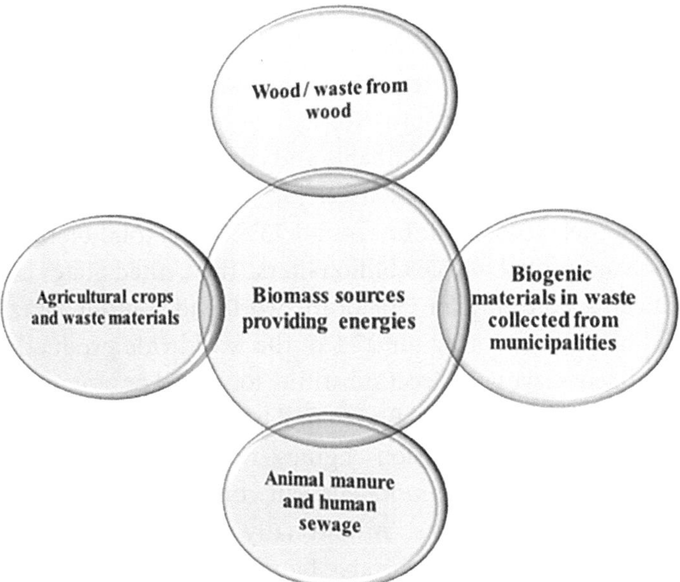

FIGURE 6.1 Different sources of biomass.

compost is the need to establish suitable well-engineered and managed facilities for the disposal of waste and the substantial duration required to get the outcome, which is neither highly esteemed nor ecologically sustainable [1]. The expansion of bio-mass-based industries is also due to the worldwide pandemic and interruptions to global industrial centers, which have significantly impacted the established chemical industry. The sector must undergo re-engineering, re-positioning, and innovation to remain significant. It is essential to adhere to and execute sustainability principles, decarbonization, process intensification, and circularity [2, 3]. To advance and enhance the chemical industry, it is necessary to transform renewable resources into biobased chemicals and biofuels. However, this can only be achieved if sustainable finance and company value are considered [4]. The themes of transportation fuels and climate change are widely discussed in both political and scientific circles worldwide. The primary example where biomass in industries is used is biofuels. The imminent depletion of global petroleum supplies necessitates the exploration of viable alternatives to fossil fuels. Furthermore, escalating conflicts in most oil-producing nations have exacerbated the fuel issue, especially for countries that do not produce oil. The existing methods of producing bio-based fuels and chemicals utilizing feedstocks derived from food sources must be revised to replace petroleum products without impacting the worldwide food supply. Non-edible or non-feed plant or plant-based materials provide a sustainable and cost-effective substitute for the current feedstocks from food sources [5]. In 2016, bioethanol, the most common biofuel, accounted for around 73% of the total biofuel production, which amounted to 135.3 billion liters. The United States is the largest producer, accounting for 59% of the worldwide output. Brazil comes in second place, accounting for 27% of the worldwide production [6, 7]. Bioethanol may serve as a direct substitute for petrol or be combined with petrol in various proportions. Bioethanol offers several benefits over petrol when used in spark ignition engines. Ethanol has a more excellent oxygen content, facilitating more efficient combustion and reducing the emissions released via exhaust. Additionally, ethanol has a higher octane number. In addition, using vegetable biomass as a source for the manufacture of bioethanol allows for the reuse of the carbon dioxide released during the combustion process, reducing carbon dioxide emissions [8, 9]. The primary drawback of first-generation bioethanol is the rivalry for cropland between biofuel feedstocks and food crops, leading to a rise in food costs [10, 11]. It is possible to produce bioethanol as a fuel from residual biomass, such as waste from the forest, industrial processes, or municipal garbage, as an alternative to the initial generation of bioethanol. These feedstocks do not pose any difficulties with food sustainability, have a consistently low price, and need little additional land [12]. There are many more uses of biomass industrial waste, such as producing usable chemicals, manufacturing adsorbents, energy production, etc. The rice industry's primary categories of agricultural waste (husk and straw) have also been utilized to produce energy [13]. Enormous quantities of biomass waste are produced globally, with rice straw (about 731.3 million tons), wheat straw (354.34 million tons), sugarcane bagasse (180.73 million tons) and maize stover (128.02 million tons) being the highest yearly producers, which can also be used for energy production [14]. Regrettably, in developing nations, several farmers incinerate agricultural remnants in their fields, therefore squandering valuable energy that might otherwise be harnessed to meet the energy needs of many others [15]. This chapter describes the essentials of biomass and the usage of biomass-based industrial waste with suitable examples.

6.2 APPLICATIONS OF BIOMASS INDUSTRIAL WASTE

6.2.1 Application of Biomass-Based Industrial Waste in the Preparation of Biofuels

The article by Branco et al. [7] investigates the research that has been conducted before that substantiates the feasibility of producing ethanol from Kraft pulp, sulfite liquor that has been wasted, and pulp and paper sludge. It also discusses the efforts to deploy biorefineries in the pulp and paper industry. The preparation of bioethanol from vegetable biomass feedstock is known as first-generation bioethanol. The most significant drawback of the first generation of bioethanol is that it creates rivalry among biofuel feedstocks for the utilization of arable land for the development of food crops, ultimately leading to a rise in the cost of food. The production of bioethanol from leftover biomass, such as trash from forests, factories, or municipalities, is an alternative to bioethanol from the first generation: the second generation. These feedstocks do not give rise to worries about long-term presence, have a cheap and steady price, and do not need more land. The pulp and paper sector is the largest consumer of woody biomass worldwide. As a result, pulp and paper mills are accountable for supplying the necessary facilities and procedures required to manage lignocellulosic biomass (LCB) and produce a significant quantity of trash. As a result of technological advancements made over the last 150 years, chemical mills can convert and fractionate LCB. Considering this, integrative biorefineries may be created inside the pulp and paper mills already in operation to produce bioethanol from the wastes and by-products. In addition to boosting the profitability of the pulp and paper sector, this strategy has the potential to benefit from the utilization of wastes and by-products, as well as the diversification of goods. Moreover, considering that the high cost of capital investment is one of the most significant obstacles in the way of the production of bioethanol of the second generation, the utilization of the machinery that is already in use in the pulp and paper industry has the potential to enhance the economics of the process, which in turn could lead to an increase in the likelihood of its success. Converting LCB into bioethanol usually begins with the first phase of feedstock preparation. This stage often comprises washing and minimizing the size of the material by milling, crushing, or dicing, which requires a significant amount of energy. The method then proceeds via the following four key steps: pretreatment, which involves breaking down the lignocellulosic network into its constituent parts; hydrolysis and saccharification, which include obtaining fermentable sugars; fermentation, which consists in converting sugars into ethanol; and recovery and dehydration, which involves separating and purifying the generated ethanol.

6.2.2 Application of Biomass-Based Industrial Waste in the Preparation of Energy

The world is transitioning away from fossil fuels and towards low-carbon energy sources right now. The pace at which this change occurs differs from one region of the world to another. It is contingent upon various conditions, including economic development, access to technical innovation, and the accomplishment of institutional changes for each nation [16]. Rice straw (RS) and rice husk (RH) are two types of solid trash produced by the rice industry. Illankoon et al. [13] investigate the energy potential of both waste products. A novel plan was devised using statistical data on rice production and paddy cultivation in every district on the island. This was accomplished by incorporating the data acquired into a geographic information system (GIS) to provide geo-referenced findings. At the twenty-second session of the United Nations International Convention on Climate Change, the government of Sri Lanka committed to producing power using renewable energy (RE) by the year 2050. Additionally, the government has set a small objective of meeting 80% of its energy requirements by 2030. Because of this, the government of Sri Lanka plans to construct 10,000 megawatts (MW) of capacity for energy (derived from natural sources that are replenished at a higher rate than consumed) over the next ten years. By 2025, the government of Sri Lanka intends to increase the amount of energy in the country by 104.62 megawatts (MW) using supplies from the agricultural, municipal, and industrial sectors. The technologies of hydropower, photovoltaics, wind turbines, and geothermal energy are all well-established and readily accessible in the commercial market. On the other hand, the comparatively high costs of investment and upkeep put a damper on their broad application in underdeveloped nations. Conversely, biomass power generation is a well-recognized method for creating alternative energy, and it is the renewable energy source used most often in Sri Lanka. It is possible to improve waste management techniques while also providing agricultural communities with decentralized power production that uses materials readily accessible in the area. Knowledge of the physico-chemical characteristics of RS and RH is required to estimate the potential energy accurately. As a result, thermogravimetric analysis was performed on both. An EDX analysis has been conducted to identify the chemical makeup of the RS and RH that used the samples in their original state. An estimate was made on the maximum energy that may be obtained from the by-products of the paddy industry. This problem is especially critical for undeveloped nations since they need more financing to adopt novel clean energy sources. Therefore, it is essential to provide economic incentives to a person, group or organization with a vested interest or stake in the decision-making and activities to reduce the financial burden. Biomass energy generation is a generally accepted and well-established technology in Sri Lanka, unlike other renewable technologies such as solar and wind. Thus, a viable strategy would need the construction and widespread use of biomass power facilities. Based on data collected from stakeholders in the value chain and prior research on the paddy industry, the excess amount of RS and RH varies in each district and is not uniform across the nation owing to their different applications. In addition, the installation of power plants may vary in size according to the available resources and financial feasibility. This includes the option of commercial-scale, small-scale, and off-grid rice residue power plants. The construction of a power plant has the potential to not only ensure the nation's reliable supply of electricity but also to stimulate economic activity in the surrounding region. From this perspective, the by-products of RS and RH power plants have the potential to be utilized in numerous industrial sectors of the economy. As an absorbent substance in water treatment, as a coating product, as a coloring agent, in the cement market, as a replacement of some percentage of rice husk ash with other raw materials, as an insulator, as a filler in the rubber industry, and in the electronics sector and other industries, these by-products have a wide range of potential applications (as shown in Figure 6.2).

Kalak [15] addresses several concerns associated with using discarded biomass products as potential sources of clean energy. The fact that the CO_2 released during the combustion or thermal conversion of biomass does not contribute to an increase in the amount of carbon dioxide in the atmosphere is one of the reasons why biomass is considered a renewable energy resource. Microorganisms degrade plants in the form of biomass, or they may be burnt and transformed into ashes in thermal oxidation in assemblies and the heat treatment plants associated with the main assembly in kilns. This process converts chemical energy into mechanical or electrical energy. Concerns over the increase in pollutant emissions brought on by human industrial production, the use of conventional fossil fuels and their

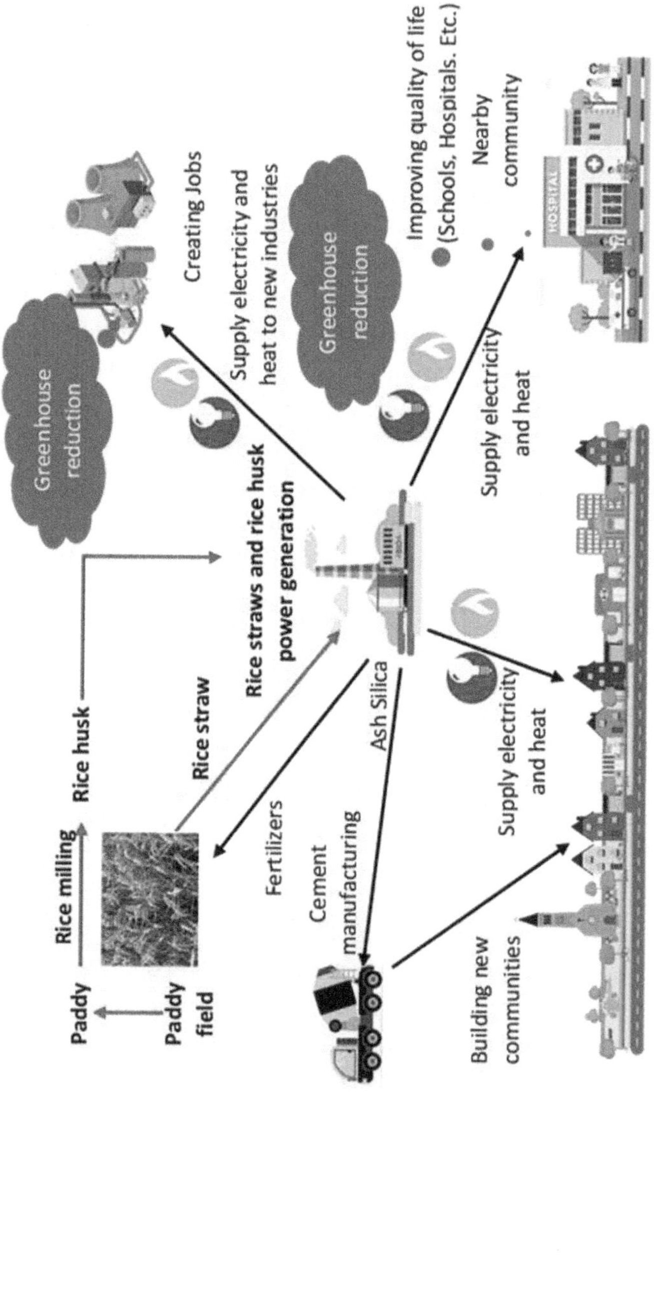

FIGURE 6.2   Usage of energy produced from RS and RH. Source: adapted from the reference [13], CCby.

eventual depletion, and the need to protect the environment all contribute to the search for renewable energy sources compatible with sustainable sustainability. Using biomass for energy purposes substantially decreases the emission of detrimental pollutants into the environment. Several governments worldwide are now using waste biomass effectively as a replacement for energy, which is a realistic option. It offers a renewable fuel source that may progressively substitute diminishing fossil fuel supplies and lower greenhouse gas emissions and the amount of solid waste. Given the diverse nature of biomass material and its dependence on local climatic conditions, different regions can seek the most suitable technological solutions to utilize their specific types of biomasses. These solutions can effectively clean polluted water bodies from contaminants, particularly heavy metals. Additionally, biomass can be converted into biofuels, which can then be further transformed into electricity and heat. The use of agricultural biomass for bioenergy generation not only yields good environmental impacts but also creates economic and social gains. Emerging technology and innovative solutions have the potential to provide more job opportunities, stimulate the local economy, and enhance the income of farmers. Nevertheless, the main hindrance remains the inadequate understanding of the energy properties of biomass materials available in various regions worldwide, as well as the lack of knowledge of the technology involved in converting biomass into energy, the necessary equipment, installation procedures, efficiency, and numerous other barriers that impede the progress of effectively utilizing biomass in the foreseeable future. Agricultural biomass is used for the generation of bioenergy. When it comes to electricity generation, studies published in academic journals have shown that the emissions of greenhouse gases that lead to the production and utilization of biomass are fewer than those that arise from coal usage.

6.2.3 Production of Chemicals from Biomass-Based Industrial Waste

Modelska et al. [1] researched whether or not it would be feasible to use waste from the sugar industry as a resource for the acid hydrolysis synthesis of furfurals. The acid hydrolysis of beetroot pulp and sugar beetroot leaves is conducted on a small scale inside a sugar mill. An example that is most often seen is biorefineries, where the fermentation of sugars derived by biotechnological or chemical procedures from the raw materials that are accessible, such as sugar cane, can result in the production of bioethanol, furfural, etc. The findings demonstrate the significant capacity of this approach for furfural synthesis. The mean furfural production rate from beet pulp is higher than that of sugar beet leaves. An assessment was conducted to determine the feasibility of establishing a large-scale industrial facility for producing furfural from sugar waste. The evaluation included the initial investment and ongoing expenses required for operating a plant of this magnitude. During the processing of sugar beet, significant quantities of lignocellulosic waste, such as beet pulp and beet leaves, are produced and were then effectively used for the synthesis of furfural in quantities (as given in Figure 6.3(a, b)) that meet the domestic demand. The further conversion of furfural into safer compounds, such as tetrahydro furfuryl alcohol, might enhance the economic feasibility of the method, as well as result in a decrease in the adverse effects of the created product on the environment. It is possible to effectively employ tetrahydrofurfuryl alcohol as a component in agricultural fertilizers, increasing the crops produced by sugar beet plants. The sugar sector is more in line with the circular economy concept due to this strategy, which decreases the amount of waste produced after production.

Coker et al. [17] reviewed the preparation of charcoal (biochar) from biomass and industrial waste. Biochar refers to charcoal that is derived from biomass and is mainly used for agricultural purposes. However, the term may also be used to describe charcoal derived from biomass and utilized for any purpose. It is possible to produce biochar using several different methods. The primary manufacturing processes are torrefaction, gasification, hydrothermal carbonization, and pyrolysis. Pyrolysis, in and of itself, uses a variety of various ways. These techniques may result in biochar's creation as the primary or secondary product; however, the most effective ways for producing biochar are pyrolysis and hydrothermal carbonization. A broad range of raw materials may be used to produce

(a)   (b)

FIGURE 6.3   (a, b) Installation of furfural production in the sugar mill. Source: adapted from the reference [1], CCby.

activated carbon. These basic ingredients should be easily accessible and affordable, with a significant amount of carbon and minimal inorganic substances. In addition, the raw materials need to be readily activated and exhibit little resistance to degradation over time. It is common practice to use thermogravimetric analysis (TGA) to characterize and assess the thermal behavior of a broad range of materials. TGA quantifies the alteration in weight (either decrease or increase) and the speed of weight alteration based on temperature, duration, and surrounding environment. The first experiments involved depositing a small amount of each feedstock sample in a cup of alumina. The sample was then heated to a temperature of 1,000°C in an environment without oxygen while closely monitoring the mass and energy flow changes using a thermal analyzer. Thermal analysis was carried out when the experimental conditions were adjusted (Figure 6.4).

Thermal analysis confirmed the mass loss in every sample; further results revealed that biochars are extensively manufactured and play a crucial role as biosorbents in the remediation of environmental contaminants. The selection of feedstocks and the use of specific processes play vital roles in enhancing the conversion efficiency in biochar manufacturing. Comprehending the processing efficiencies of biomass and industrial wastes with various features is essential to contribute to regional development and environmental conservation. Regarding applications like municipal water treatment, biomass feedstocks are preferred over hydrocarbon feedstocks due to their higher yield and greater social acceptance. Although piñon wood yielded the most favorable outcomes, it and some other biomass feedstocks possess a detrimental worth. Consequently, a dual advantage is achieved: creating a valuable and practical product,

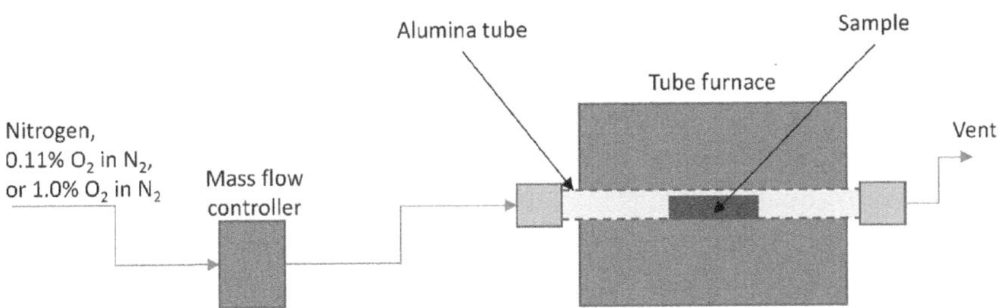

FIGURE 6.4 Diagram of pyrolysis furnace. Source: adapted from the reference [17], CCby.

activated carbon, and eliminating plant fiber waste. Yang et al. [18] investigated the biomass waste generated from the bamboo industry-prepared biochar and studied surface morphology. Utilizing charcoal kilns to convert bamboo trash into bamboo charcoal has been a prevalent method for managing waste in bamboo growing regions with steep terrain due to the effectiveness of charcoal kilns. Bamboo, a member of the Poaceae grass family, has several characteristics of woody biomass, including its relatively low ash level. This distinguishes it from other vegetative biomass products, which often have ash concentrations above 2 wt%. The research used biomass pyrolytic polygeneration technology to convert bamboo waste into charcoal. The highest amount of charcoal was produced at a temperature of 350°C. It was possible to further explore the process of pyrolysis by integrating data on the development of the char structure and the release of volatiles. During this particular kind of pyrolysis, the organized large structures in the raw biomass undergo a process of decomposition and eventual transformation into a network-like structure that is best characterized as a "three-dimensional network of benzene rings." This phenomenon takes place in the "initial breakdown stage," where the "3D arrangement of benzene rings" transforms into a "2D framework of merged rings" by breaking the functional groups that contain methyl, methylene, and oxygen. Subsequently, in the "second condensation phase," the fused rings expand and undergo dehydrogenation and cleavage of ether bonds, potentially leading to a "graphite microcrystalline framework."

6.2.4 Application of Biomass-Based Industrial Waste in the Preparation of Adsorbents

Kim et al. [19] create an effective adsorbent material for removing phosphorus (P) using industrial waste from Escherichia coli biomass. Phosphorus (P) is a vital element necessary for the proper growth and development of many organisms, such as crops and microbes. Significant phosphorus-containing waste products from highly cultivated agricultural areas are consistently released into water ecosystems, including rivers and lakes, via soil leaching, drainage, and surface runoff. An excessive influx of phosphorus into aquatic settings may result in eutrophication, which can cause significant environmental issues such as harmful algal blooms (HAB). Hence, it is crucial to effectively eliminate phosphorus from water bodies such as rivers and lakes to control and reduce its environmental consequences, particularly harmful algal blooms (HABs). Adsorption is a very efficient and dependable method for eliminating pollutants from water solutions since it does not need extra operating procedures. Nevertheless, the drawbacks of traditional adsorbents, such as their exorbitant price and inability to be reused, have prompted a growing interest in creating economical sorbents (biosorbents). Several bio-based materials, such as agricultural and industrial waste and algal, bacterial, and fungal biomasses, have shown promise as inexpensive biosorbents for removing pollutants from water solutions. The primary source of Escherichia coli biomass is the byproduct of industrial fermentation methods used for amino acid synthesis. Due to the presence of several functional groups in E. coli biomass that can form a bond with ionic contaminants, it is often used as an ecologically friendly and cost-efficient biosorbent material for removing toxic substances, precious metals, etc. Phosphorus in natural water bodies exists as phosphate ions, which are negatively charged. Depending on their pKa characteristics, the amine groups on E. coli biomass may be the primary sites for adsorbing phosphate ions. Further surface modification was performed to inhibit the presence of any remaining carboxyl groups. Ultimately, the phosphorus sorption characteristics of each produced sorbent were thoroughly evaluated using a range of sorption experiments, including pH adjustment and analysis of isotherms and kinetics. The FT-IR and XPS evaluations of the sorbents demonstrated noticeable changes in the characteristics of the functional groups present on the sorbents after chemical treatment. Additionally, the sorption capacity is influenced by the pH level. Research might provide a practical approach for recycling and using biomass waste as an effective sorbent.

6.2.5 Application of Biomass-Based Industrial Waste in the Construction Industry

One of the most significant contributors to the emission of greenhouse gases is the building sector. Carbon dioxide emissions in the atmosphere are nearing a tipping point, which may result in substantial climate change. Implementing a system that enables the collection and sequencing of carbon dioxide levels is an essential component. This technology has the potential to reduce emissions as well as the carbon footprint that is caused by the manufacturing of Portland cement and cement-based construction materials. Efforts have been made to generate waste and, as a result, to support their creation and usage as alternatives for raw materials in the manufacturing of biocomposites. In this regard, Polak et al. [20] reviewed past studies in which various types of waste material generated from different sources were utilized to prepare cement, for example lignin waste, hemp fiber waste, bamboo fiber waste, textiles, waste from coal mining, copper ore waste, and ashes collected from various industrial biomass sources etc. According to estimates, the building industry is responsible for using around 14–50% of natural resources. As a result, it is categorized as the second greatest source of carbon dioxide emissions in the atmosphere. As a result of environmental concerns, it is of the utmost importance to locate alternative resources, which in this instance may be waste products. Materials produced with human engagement as a result of various activities are referred to as municipal solid trash. The combustion of municipal solid trash and other types of waste streams produces a substantial quantity of fly ash residue, which often contains precious metals. Coal-burning combined heat and power plants usually use fly ash, a highly utilized waste material with pozzolanic properties in concrete production. Further, Pera et al. [21] explored the potential of substituting coarse aggregates (4–20 mm) in concrete with bottom ash derived from municipal trash incineration. They discovered that the use of untreated ash resulted in the occurrence of crack development and expansion.

Nevertheless, the use of sodium hydroxide in treating bottom ash enhanced the concrete's durability. However, it also led to a drop in the strength of the concrete when compared to the natural aggregate material. Lin et al. [22] conducted an experiment where they melted fly ash produced by burning municipal garbage and substituted it with mortar cement. By incorporating 10% bottom ash into the mixture, they saw an improvement in the mechanical characteristics of the concrete.

6.3 CONCLUSION AND PROSPECTS

The ongoing expansion of biomass-based industrial waste into energy is inevitable, but its speed, direction, and impact on the natural environment and humanity are challenging to foresee. Amidst these uncertainties, it is indisputable that there will be a persistent generation of agricultural and industrial waste, necessitating suitable and beneficial management. A recent study has shown that using biomass waste for energy is a viable and creative approach. This method employs ecologically friendly methods that minimize the release of dangerous compounds and have no net carbon dioxide emissions. Biomass components, including wood, forest residues, seasonal plants, post-production waste, and gases may be valuable resources for energy, heat, and liquid fuels. An advantage of biomass is its inexhaustibility, unlike fossil fuels. This characteristic makes the use of renewable energy derived from any biomass unavoidable. Further, using an integrated biorefinery strategy enables the reintegration of surplus wastes back into the production chain. Converting waste biomass into valuable goods allows humanity to create sustainable fuels, biobased materials, and chemicals for many uses in pharmaceuticals, biotechnology, nanotechnology, agro-industry, and engineering. The forest-based industry that uses and creates LCB feedstocks has the potential to play a significant role in the future economy. Utilizing current methods and instruments in the paper industry to produce ethanol and transforming these mills into integrated biorefineries shows potential for maximizing the value of generated waste, diversifying products and enhancing the profitability and acceptance of the pulp and paper industry. Another possible benefit of this process integration is the reduction in investment costs and the increase in the likelihood of success for large-scale second-generation bioethanol production. Maximizing biomass valorization and minimizing waste formation in an integrated biorefinery requires using all components of LCB to generate numerous products. When it comes to producing bioethanol on an industrial scale from LCB, two of the most significant drawbacks are the high level of technical risk and the high level of capital expenditure. LCB may be pretreated using the equipment and technologies used in Kraft mills, which have been commercially recognized for decades. These technologies and equipment include the wood preparation process and the pulping process. Kraft mills are well-established business enterprises. The most often used technique for ethanol recovery is distillation, which is a very energy-intensive process. It accounts for about 60 to 80% of the overall separation cost of bioethanol from water. This is mostly due to the low ethanol concentration in the fermented broth. It is essential to optimize the utilization of all components of LCB, including cellulose, hemicelluloses, and lignin, to generate numerous products within an integrated biorefinery. This is necessary to maximize the utilization of biomass and minimize the formation of trash.

REFERENCES

1. Modelska, M., Binczarski, M. J., Dziugan, P., Nowak, S., Romanowska-Duda, Z., Sadowski, A., & Witońska, I. A. (2020). The potential of waste biomass from the sugar industry as a source of furfural and its derivatives for use as fuel additives in Poland. *Energies, 13*(24), 6684.
2. Guragain, Y. N., & Vadlani, P. V. (2021). Renewable biomass utilization: A way forward to establish sustainable chemical and processing industries. *Clean Technologies, 3*(1), 243–259.

3. Charpentier, J. C. (2007). In the frame of globalization and sustainability, process intensification, a path to the future of chemical and process engineering (molecules into money). *Chemical Engineering Journal, 134*(1–3), 84–92.

4. Vadlani, P. V. (2020). Financing strategies for sustainable bioenergy and the commodity chemicals industry. In *Green Energy to Sustainability: Strategies for Global Industries*, 1st ed.; Vertès, A. A., Qureshi, N., Blaschek, H. P., & Yukawa, H., Eds. Hoboken, NJ, USA: John Wiley and Sons Ltd, pp. 569–586.

5. Mousdale, D. M. (2008). *Biofuels: Biotechnology, Chemistry, and Sustainable Development*. Boca Raton, FL: CRC Press, ISBN 9781420051247.

6. REN21. (2017). *Renewables 2017 Global Status Report*. Paris, France: REN21 Secretariat, pp. 30, 48. ISBN 978-3-9818107-6-9.

7. Branco, R. H., Serafim, L. S., & Xavier, A. M. (2018). Second generation bioethanol production: On the use of pulp and paper industry wastes as feedstock. *Fermentation, 5*(1), 4.

8. Balat, M. (2011). Production of bioethanol from lignocellulosic materials via the biochemical pathway: A review. *Energy Conversion and Management, 52*(2), 858–875.

9. Sebayang, A. H., Masjuki, H. H., Ong, H. C., Dharma, S., Silitonga, A. S., Mahlia, T. M. I., & Aditiya, H. B. (2016). A perspective on bioethanol production from biomass as alternative fuel for spark ignition engine. *RSC Advances, 6*(18), 14964–14992.

10. Manochio, C., Andrade, B. R., Rodriguez, R. P., & Moraes, B. S. (2017). Ethanol from biomass: A comparative overview. *Renewable and Sustainable Energy Reviews, 80*, 743–755.

11. Dutta, K., Daverey, A., & Lin, J. G. (2014). Evolution retrospective for alternative fuels: First to fourth generation. *Renewable Energy, 69*, 114–122.

12. Zabed, H., Sahu, J. N., Boyce, A. N., & Faruq, G. (2016). Fuel ethanol production from lignocellulosic biomass: An overview on feedstocks and technological approaches. *Renewable and Sustainable Energy Reviews, 66*, 751–774.

13. Illankoon, W. A. M. A. N., Milanese, C., Girella, A., Rathnasiri, P. G., Sudesh, K. H. M., Llamas, M. M., . . . Sorlini, S. (2022). Agricultural biomass-based power generation potential in Sri Lanka: A techno-economic analysis. *Energies, 15*(23), 8984.

14. Sarkar, N., Ghosh, S. K., Bannerjee, S., & Aikat, K. (2012). Bioethanol production from agricultural wastes: An overview. *Renewable Energy, 37*(1), 19–27.

15. Kalak, T. (2023). Potential use of industrial biomass waste as a sustainable energy source in the future. *Energies, 16*(4), 1783.

16. Dutt, A. (2020). *Accelerating Renewable Energy Investments in Sri Lanka*. New Delhi, India: Centre for Energy Finance, Council on Energy, Environment and Water (CEEW).

17. Coker, E. N., Lujan-Flores, X., Donaldson, B., Yilmaz, N., & Atmanli, A. (2023). An assessment of the conversion of biomass and industrial waste products to activated carbon. *Energies, 16*(4), 1606.

18. Yang, H., Huan, B., Chen, Y., Gao, Y., Li, J., & Chen, H. (2016). Biomass-based pyrolytic polygeneration system for bamboo industry waste: Evolution of the char structure and the pyrolysis mechanism. *Energy & Fuels, 30*(8), 6430–6439.
19. Kim, S., Park, Y. H., Lee, J. B., Kim, H. S., & Choi, Y. E. (2020). Phosphorus adsorption behavior of industrial waste biomass-based adsorbent, esterified polyethylenimine-coated polysulfone-Escherichia coli biomass composite fibers in aqueous solution. *Journal of Hazardous Materials, 400,* 123217.
20. Ryłko-Polak, I., Komala, W., & Białowiec, A. (2022). The reuse of biomass and industrial waste in biocomposite construction materials for decreasing natural resource use and mitigating the environmental impact of the construction industry: A review. *Materials, 15*(12), 4078.
21. Pera, J., Coutaz, L., Ambroise, J., & Chababbet, M. (1997). Use of incinerator bottom ash in concrete. *Cement and Concrete Research, 27*(1), 1–5.
22. Lin, K. L., Wang, K. S., Tzeng, B. Y., & Lin, C. Y. (2003). The reuse of municipal solid waste incinerator fly ash slag as a cement substitute. *Resources, Conservation and Recycling, 39*(4), 315–324.

CHAPTER 7

Biomass-Based Municipal Solid Wastes

Fundamentals, Classification, Properties, and Applications

Akshima Soni, Yashika Verma, Seema R. Pathak, Kamalakanta Behera and Kamal Nayan Sharma

7.1 INTRODUCTION

Biomass-based municipal solid waste (BMSW), often called organic garbage, is any material that comes from a plant or animal and is biodegradable to produce energy. Biomass—or biogenic (plant or animal products)—contains stored chemical energy from the sun that plants produce through photosynthesis. Biomass has the potential to be utilized directly for heating purposes or transformed into liquid and gaseous fuels through a variety of processes. As the world grapples with the challenges of sustainable development and waste management, the role of BMSW becomes increasingly significant.[1]

BMSW, originating from both plant and animal sources, encompasses a wide array of materials. These organic wastes are integral to our daily lives, from the leftovers on our plates to the trimmings in our gardens, untreated wood, and food-stained papers.[1] The traditional perspective of viewing these materials as mere waste is shifting, with a growing recognition of their potential in resource management and environmental conservation. The significance of BMSW lies in its versatility and biodegradability, making it a valuable resource in various sectors.[2-4]

It has emerged as a sustainable alternative for energy production, challenging conventional energy sources and contributing to reducing carbon

DOI: 10.1201/9781003466833-9

emissions. Furthermore, the proper management and utilization of BMSW can lead to substantial environmental benefits, such as reducing landfill usage, mitigating greenhouse gas emissions, and promoting soil health through composting. It delves into the classification and properties of BMSW, differentiating between various types based on their source, composition, and potential applications. Understanding the challenges and opportunities linked with BMSW is valuable for developing effective management strategies and technologies to connect its utilization.[5] Moreover, the chapter provides a comprehensive overview of the fundamental aspects and the multifaceted applications of BMSW. From its role in bioenergy production through processes like anaerobic digestion and pyrolysis to its use in creating valuable agricultural compost, BMSW presents many opportunities for sustainable development. The chapter also discusses the challenges and limitations associated with BMSW, including technological, economic, and regulatory hurdles that need to be addressed to maximize its potential. The narrative then shifts to a global perspective, examining the role of BMSW in different regions and cultures. It highlights the varied approaches to BMSW management worldwide, influenced by local environmental policies, technological advancements, and cultural practices. This global outlook provides insights into the best practices and innovative solutions adopted in different parts of the world, fostering a deeper understanding of the universal significance of BMSW.[6] Overall, this chapter introduces BMSW as a critical component in sustainable waste management, exploring its diverse sources, properties, and applications.

7.2 CLASSIFICATION OF BIOMASS-BASED MUNICIPAL SOLID WASTES

Biomass solid waste can be categorized according to its origin, composition, characteristics, and energy content. The classifications of such biomass-based wastes are described later and outlined in Figure 7.1.

7.2.1 Origin-Based Classification

The origin-based classification of biomass-based municipal waste refers to categorizing waste materials generated in various areas based on their source or origin.

7.2.1.1 Agricultural Residues

Agricultural residues encompass a diverse range of by-products from farming and harvesting activities. These residues primarily consist of

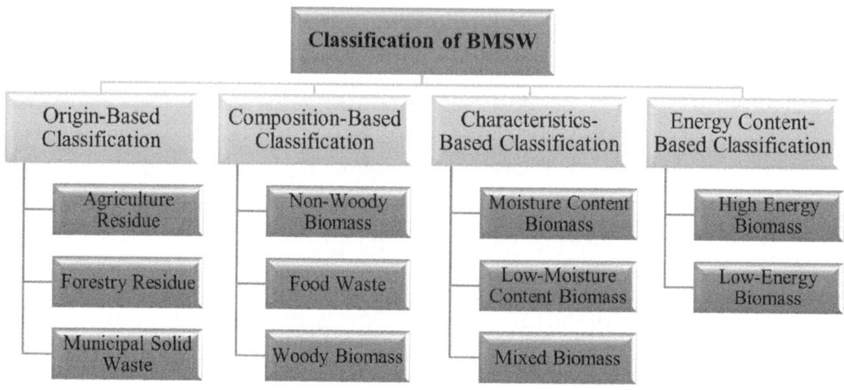

FIGURE 7.1   Classification of BMSW.

plant-based materials left over from agricultural processes. They include various crop residues such as straw, husks, and pruning residues. Crop residues like straw are the stalks left over after grain crops such as wheat, rice, or barley have been harvested. On the other hand, husks are the outer protective coverings of seeds or grains, often remaining after processing. Pruning residues typically refer to the branches, leaves, and other vegetative parts of plants that are trimmed during pruning activities in orchards or vineyards. These agricultural residues hold significant potential for utilization in various applications.

7.2.1.2 Forestry Residues

Forestry residues encompass a wide array of organic materials that result from forestry operations, encompassing both timber harvesting and forest management activities. These residues are derived from various parts of trees, including bark, branches, and wood residues, typically considered waste in forestry operations.

Bark, one of the components of forestry residues, is the protective outer layer of the tree trunk. During timber processing, bark is often removed from logs and considered a waste product. Similarly, tree branches trimmed or removed during forest management activities contribute to forestry residues. These branches and other woody materials, such as twigs and foliage, constitute a significant portion of forestry waste.

7.2.1.3 Municipal Solid Waste

Municipal Solid Waste (MSW) comprises various forms of waste generated from residential, commercial, institutional, and industrial sources within urban areas. It encompasses a heterogeneous mixture of materials, including household garbage, recyclables, organic waste, and non-recyclable materials. Municipal authorities typically collect and manage MSW through waste collection and disposal systems. Household garbage forms a significant portion of MSW and includes everyday items discarded by households, such as food waste, packaging materials, textiles, plastics, glass, and metals. Commercial waste from businesses, offices, and institutions also contributes to MSW, comprising paper, cardboard, plastics, and food waste.

7.2.2 Composition-Based Classification

Composition-based classification of BMSW involves categorizing waste according to its material types, such as organic waste, paper, and biodegradable plastics. This approach enables handmade waste management strategies like composting for organic biomass and recycling and reducing environmental impact.

7.2.2.1 Non-Woody Biomass

It encompasses a variety of organic materials that are not derived from wood. This category includes agricultural residues like straw and husks, organic waste from gardens and parks, and other plant materials that do not have a woody structure. Non-woody biomass is utilized in various applications, including biofuel production, composting for soil enhancement, and as feedstock for producing biodegradable plastics and chemicals.

7.2.2.2 Food Waste

This refers specifically to organic waste generated in kitchens, restaurants, and the food processing industry. This includes leftovers, spoiled food, and food preparation scraps. Food waste is a significant issue globally, contributing to environmental problems when not managed properly. However, it can be repurposed through composting and anaerobic digestion to produce biogas and as a resource for creating animal feed, thereby reducing its environmental impact.

7.2.2.3 Woody Biomass

This category includes materials such as wood chips, sawdust, and other primarily wood-based residues. These are typically generated from lumber processing, forestry operations, and tree maintenance tasks. Woody biomass is often used for energy production, manufacturing wood-based products, and as mulch or soil amendments in landscaping and gardening.

7.2.3 Characteristics-Based Classification

This type helps in selecting the most appropriate waste management strategies, such as recycling, composting, or thermal treatment, by identifying the nature of the waste. The characteristics-based classification of BMSW categorizes waste based on its inherent properties rather than its origin. This classification system focuses on biodegradability, combustibility, composting, and thermal treatment of waste materials, which are critical factors in determining their suitability for various waste management strategies.

7.2.3.1 Moisture Content Biomass

This category includes materials that have a high water content. Examples are food waste and green yard waste like grass clippings and leaves. The high moisture content in these materials can influence their decomposition rate and methods of processing or recycling.

7.2.3.2 Low Moisture Content Biomass

Dry materials fall under this category. Wood chips and sawdust are prime examples. Their low moisture content makes them more suitable for specific recycling or energy recovery processes, such as combustion or gasification, than their high-moisture counterparts.

7.2.3.3 Mixed Biomass

This classification encompasses heterogeneous waste streams that contain a mix of different types of organic materials, both wet and dry. Mixed biomass can be more challenging to process due to the variation in material types and moisture contents, requiring specialized treatment or separation technologies to manage effectively.

7.2.4 Energy Content-Based Classification

Energy content-based classification of biomass-based wastes involves categorizing these wastes based on their potential energy value. Different biomass materials have varying energy contents, and this classification aids in optimizing energy recovery processes, such as combustion for heat and power generation or conversion into biofuels.

7.2.4.1 High-Energy Biomass

This category includes highly calorific materials, meaning they can produce a significant amount of energy when burned. Examples of high-energy biomass are certain types of wood residues, which are dense and rich in energy, making them excellent for combustion and energy production processes.

7.2.4.2 Low-Energy Biomass

In contrast, this category encompasses materials with a lower calorific value. These materials, such as some agricultural residues (e.g., straw or husks), contain less energy per unit of weight. They are less efficient for energy production than high-energy biomass but still play a crucial role in biomass-based waste management strategies. They are often used in applications where high energy density is not a primary requirement.

7.2.5 Waste Management System-Based Classification

A waste management system is based on collecting, transporting, processing, recycling, and disposing of waste materials. Its goal is to minimize the environmental impact of waste, reduce pollution, conserve resources, and promote sustainability by efficiently managing the generation, storage, collection, transfer, processing, and disposal or recycling of waste materials.

7.2.5.1 Industrial Biomass Waste

This type of waste emerges as a byproduct of industrial activities and processes. It encompasses many organic materials from manufacturing, processing, or production operations. Examples might include wood scraps from lumber mills, pulp from paper manufacturing, or agricultural byproducts from food processing industries.

7.2.5.2 Residential Biomass Waste

Generated by households, this category includes organic waste produced in daily life. Common types of residential biomass waste are food scraps, yard trimmings, paper products, and wood-based materials. This waste stream is characterized by its diverse composition, from various items consumed and used within homes.

7.2.5.3 Commercial Biomass Waste

Arising from commercial establishments and businesses, such as restaurants, markets, and office buildings, this waste includes organic materials discarded during commercial operations. It may consist of food waste from restaurants, paper and cardboard from retail packaging, and landscaping waste from maintenance activities. Commercial biomass waste often reflects the activities and services the business generates.

7.3 IMPORTANCE OF ACCURATE CLASSIFICATION FOR EFFECTIVE WASTE MANAGEMENT STRATEGIES

Accurate waste classification plays a vital role in enhancing waste management practices. Understanding these classifications can help develop hand-crafted approaches for efficiently managing and utilizing biomass solid waste in various sectors such as energy production, agriculture, and waste-to-resource initiatives. Proper waste classification provides several benefits, including environmental protection, resource conservation, and public health improvement.[7, 8] Some significant reasons why precise categorization of waste is essential are described in the following sections.

7.3.1 Environmental Protection

Correctly classifying waste ensures that hazardous materials are properly handled, reducing the risk of soil, water, and air pollution. This is crucial for maintaining ecological balance and preventing damage to ecosystems.

7.3.2 Resource Conservation

By accurately identifying recyclable and reusable materials, waste classification contributes to resource conservation. This approach minimizes the need for raw material extraction, conserving natural resources and reducing the environmental impact of resource extraction processes.

7.3.3 Efficiency in Waste Management

Proper classification simplifies the process of waste sorting and processing. It allows for more efficient recycling and composting, reducing the volume of waste in landfills. This efficiency can lead to cost savings in waste management.

7.3.4 Public Health and Safety

Correctly identifying and segregating hazardous waste is essential for public health and safety. This prevents harmful substances from contaminating the environment and reduces the risk of exposure to toxins for the public and waste management workers.

7.3.5 Regulatory Compliance

Many regions have strict regulations regarding waste disposal. Accurate classification ensures compliance with these regulations, helping to avoid legal penalties and possible harm to the organization's reputation.

7.3.6 Promotion of Sustainable Practices

Effective waste classification encourages sustainable waste management practices. It raises awareness about the impact of waste on the environment and promotes a culture of sustainability and responsibility towards waste generation and disposal.

7.3.7 Innovation and Economic Opportunities

Precise waste classification can lead to new recycling technologies and methods. It can also create economic opportunities in the recycling and waste management sectors.

7.3.8 Data Collection and Waste Analysis

Accurate classification aids in gathering data about the types and quantities of waste produced. This information is crucial for analyzing waste trends, planning future waste management strategies, and implementing targeted waste reduction initiatives.

7.3.9 Efficient Waste-to-Energy Processes

Certain types of waste can be converted into energy through various processes, such as incineration or anaerobic digestion. Accurate classification ensures that suitable waste streams are directed towards these energy recovery methods, contributing to clean energy generation.[9]

7.4 COMPOSITION AND PROPERTIES OF BIOMASS-BASED MUNICIPAL SOLID WASTES

7.4.1 Composition of BMSW

Waste composition varies significantly worldwide, influenced by factors like socioeconomic status. This includes aspects like lifestyle—which is often income-dependent—cultural practices, and overall consumption patterns. For instance, the population ratio in developed countries is smaller than their share in global municipal solid waste (MSW) generation. This trend is reversed in countries like India and China.

Nations across the globe are categorized into regions, including but not limited to the Middle East and North Africa (MENA), East Asia and Pacific (EAP), South Asia (SA), Europe and Central Asia (ECA), North Asia (NA), Sub-Saharan Africa (SSA), and Latin America and the Caribbean (LAC). Interestingly, the MSW generation is increasing even faster than urbanization. In 2016, the most significant proportion of MSW (23%) was produced by the EAP region, followed by ECA (20%), SA (17%), NA (14%), LAC (11%), SSA (9%), and MENA (6%), according to the World Bank Group.

Projections for 2050 indicate a staggering increase in MSW production rates: SSA is expected to see a 197% increase, SA 98%, MENA 75%, LAC 60%, EAP 53%, NA 37%, and ECA 25%. Per capita waste production also varies, with NA producing the highest amount (2.21 kg/capita/day), significantly more than other regions like ECA, LAC, MENA, EAP, SA, and SSA. The composition of waste is also regionally distinct. Except for NA, where less than 30% of waste is organic and over 55% is dry recyclables, other regions like EAP, ECA, LAC, and MENA comprise about 40% of the total waste. SSA has a notable amount of inert waste, around 30%. These variations highlight how urbanization and income levels significantly influence waste composition.[10]

7.4.2 Properties of BMSW

The physical attributes of MSW can be assessed by factors such as its moisture content, particle size, density and permeability. The Central Pollution Control Board, India (CPCB) reported these properties for MSW in different urban areas across India. Various physical factors are used to identify the characteristics of solid waste, with moisture content being essential for determining the economic feasibility of waste treatment incineration due to the energy needed for water evaporation. The moisture content is also pivotal for the rate of waste decomposition. During composting and anaerobic digestion, moisture aids the process. The density of BMSW is significant as it is measured by weight per volume, which affects the design of solid waste management systems. The ability of waste to retain water, known as field capacity, is important for gauging the leachates in landfills. Another critical physical factor is the hydraulic conductivity of solid waste, which dictates how liquids and gases move through garbage in landfills.[11]

There are some important features:

Bulk density. This is an important factor in the design of a solid waste management system, and its unit is Kg/m^3. The bulk density is measured through the following steps:

- Record the weight of W1.

- Overall, the specimen is in the vessel.

- Drop the sample from 10 m height three times to settle the content.

- Fill the vessel volume with the remaining sample and measure Volume (V2).

- Record the weight (W2) and calculate bulk density = W2 − W1/V1.

Moisture content calculation. Moisture content calculation involves determining the ratio of water to the total weight of wet waste. An elevation in moisture content results in an increased weight of the solid waste, raising the overall expenses associated with transportation and collection. The measurement process comprises the following specific steps to assess moisture content accurately:

- Record the weight W1.

- Spread the specimen over the several trays.

- Place the sample in the oven for 24 hours.

- Cool the sample at room temperature and record the weight W2.

- Measure the moisture content = (W2 – W1/W2) × 100

7.4.2.1 Size

The distribution of the particles in the waste stream is the most crucial parameter due to its consequences in the design of shredders and mechanical separators.

The Central Pollution Control Board (CPCB) of India's report on the physical characteristics of BMSW generated in various Indian cities reveals diverse compositions. Paper content varies significantly, with cities like Bhopal, Calcutta, Madras, and Mumbai showing high percentages (around 10%), while Ludhiana and Visakhapatnam have lower figures (3%). Textile waste is most prevalent in Coimbatore (9%) and Nagpur (7%), whereas Calcutta reports a minimal 0.3%. Leather waste is notably present in cities like Bhopal (2%) and Patna (2%), with several cities reporting none. Plastic waste is highest in Vadodra (7%) and Pune (5%), while metal and glass contributions are relatively low across the board, with a few exceptions where they slightly peak. Ash, fine earth, and other residues show a broad range, with Delhi marking the highest at 51.5%. Compostable matter, indicating organic waste potential, varies widely, with Kochi at 58% and Nagpur at the lower end at 30.40%. The average composition across surveyed cities points towards a makeup of 5.7% paper; 3.5% textile; 0.8% leather; 3.9% plastic; 1.9% metal; 2.1% glass; 40.3% ash, fine earth, and others; and 41.8% compostable matter, showcasing the heterogeneous nature of BMSW in India.

7.5 BIOLOGICAL DEGRADATION AND THERMAL CONVERSION OF BMSW

Efficient waste management of BMSW hinges on two pivotal processes: biological degradation and thermal conversion. Within BMSW management, these dual approaches play a significant role. According to a study by Khan, biological degradation leverages microorganisms to decompose the organic components of BMSW, yielding biogas and compost. These by-products serve as renewable energy sources and soil amendments, contributing to environmental sustainability. Conversely, thermal conversion techniques such as incineration and pyrolysis subject BMSW to elevated temperatures in a controlled environment. This strategy eliminates hazardous pathogens and diminishes waste volume, mitigating potential environmental and health threats. Furthermore, thermal conversion can produce energy, either as heat or electricity, underscoring its role in sustainable waste management. By integrating biological degradation with thermal conversion, a holistic and efficient strategy for BMSW management emerges, ensuring safety and aligning with green practices and the recovery of resources.[12]

7.5.1 Thermal Conversion of BMSW

Thermal conversion technologies encompass a range of processes that transform Municipal Solid Waste (MSW) into proper energy forms. These processes include converting MSW into syngas, which can be utilized to boil water. The steam generated in this process can then produce electricity by driving turbines. This technique primarily involves the heat treatment of organic waste, resulting in either the production of heat energy or the conversion of waste into highly oxygenated fuel oil, biochar, or gas. These methods are particularly effective for treating waste with a high content of organic, non-biodegradable materials and low moisture content.

Key technologies in this category include incineration, gasification, and pyrolysis. Incineration involves the thermal decomposition of organic waste in the presence of an oxidant, producing steam, carbon dioxide, and other gases. Gasification occurs at high temperatures with limited oxygen supply, producing primarily permanent gases and some chemicals as by-products. In contrast, pyrolysis occurs without oxygen and at relatively lower temperatures, primarily yielding liquids as the end product.

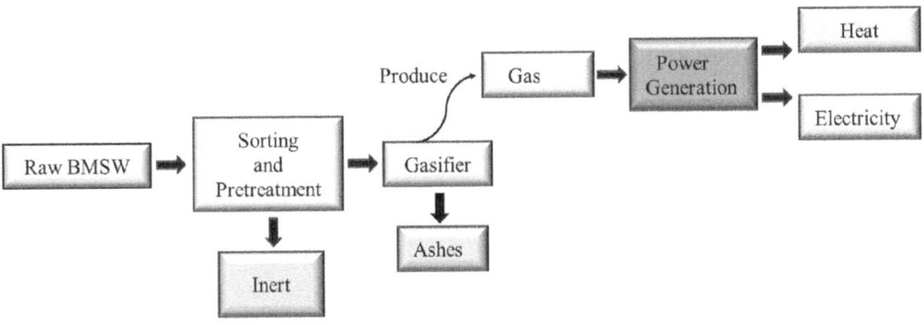

FIGURE 7.2   Gasification of BMSW.

7.5.1.1 Gasification

Gasification is an emerging waste-to-energy (WTE) conversion technology, increasingly adopted in numerous developed nations. This technology is pivotal in power generation, facilitating clean energy production at a reduced cost. This is estimated as an endothermic reaction, requiring external energy/heat to start and maintain the gasification process. In addition, sorting is an essential phase through which inert materials such as metals and glass are removed before gasification, as shown in Figure 7.2. It is versatile, allowing for the use of various feedstocks, including MSW and biomass. Gasification is a thermo-chemical process that transforms carbon-rich and organic waste materials into synthetic gas within a reactor, commonly known as syngas.[13]

The primary components of syngas are hydrogen, carbon monoxide, and methane. The energy content of syngas is roughly a third of that found in natural gas, with its energy range spanning from 4 to 50 MJ/Nm^3. An advantage of syngas is its ease of storage, transfer, and integration into existing natural gas pipeline networks. On a smaller scale, gasification systems paired with internal combustion engines can operate for extended periods, yielding high electricity efficiency with minimal emissions. There are various types of gasifiers, each with their unique benefits, drawbacks, and operational characteristics.

7.5.1.2 Pyrolysis

Pyrolysis is a unique form of thermolysis, derived from two Greek words—"pyro" (meaning fire) and "lysis" (meaning breakdown). This process plays a vital role in converting BMSW into energy. It's an endothermic procedure where heat is essential for incineration waste. Primarily used for organic materials, pyrolysis occurs in an oxygen-free setting, vaporizing the volatile components of the material being processed. This process results in three main products:

- Vapors, which can be transformed into liquid through the condensation process.

- Various gases, including carbon monoxide, carbon dioxide, and methane.

- A residue rich in carbon, comprising char and ash, useful for heating purposes.

Figure 7.3 displays the schematic representation of this technology. Pyrolysis is the initial chemical reaction in the combustion process. Its applications are diverse and significant, including the utilization in the chemical industry, conversion of biomass into syngas (synthesis gas) and employment in various cooking techniques such as baking, frying, grilling, and caramelizing.[13]

7.5.2 Biological Conversion of BMSW

Biological conversion involves the breakdown of organic material through microbial activity in an oxygen-free setting, producing biogas. This biogas primarily comprises methane and carbon dioxide. Energy can be extracted from BMSW through biological conversion methods such as composting and landfilling. Compared to thermal conversion methods, biochemical conversion techniques are more eco-friendly. These processes are particularly suitable for managing BMSW and agricultural waste, which are challenging to handle owing to their significant moisture content.

7.5.2.1 Anaerobic Digestion

Renewable energy, such as this method, offers a sustainable alternative to fossil fuels. By generating electricity within a completely enclosed system, the release of harmful emissions is minimized. This process involves using biodegradable organic matter and the combustible organic component of BMSW as feedstock in a digester tank. Digestion occurs in an oxygen-free environment, with water serving as a reactant. The raw biogas produced undergoes treatment to lower hydrogen sulfide levels, crucial for preventing corrosion and ensuring efficient operation of the combined heat and

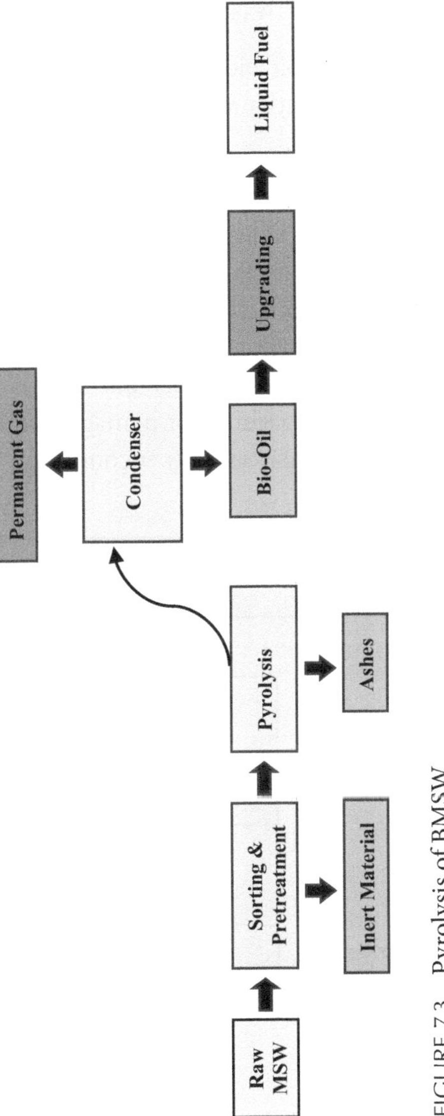

FIGURE 7.3 Pyrolysis of BMSW.

power (CHP) unit and boiler for electricity generation. The quality of the resulting biogas aligns with natural gas standards.

Anaerobic digestion comprises several steps:

- Sorting and shredding of waste.

- Feeding into an anaerobic bioreactor where micro-organisms facilitate acidogenesis and methanogenesis.

- Methane gas management.

The methane yield varies from 350 to 435 Ml/g, influenced by factors like operational conditions, reactor types, hydraulic retention time, and waste configuration. This method is particularly effective in areas with low to medium waste generation. Further enhancements can be made to reduce operational costs and increase revenue in existing incinerator plants. The procedure for managing municipal solid trash is depicted in Figure 7.4. Enhancements encompass innovative emission

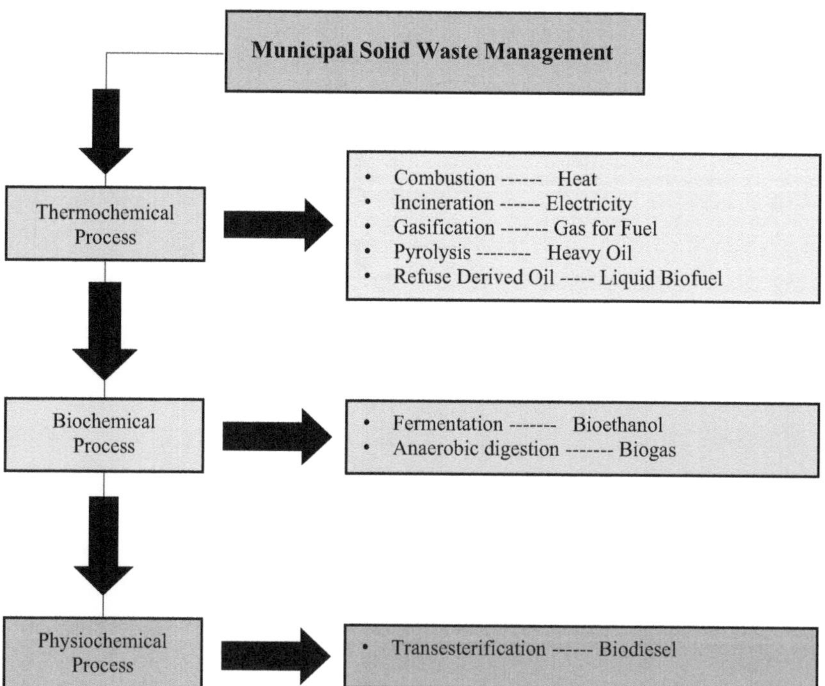

FIGURE 7.4   Municipal solid waste management.

control approaches to reduce environmental compliance costs, creating corrosion-resistant materials for minimizing maintenance expenses and utilizing advanced separation techniques to recover valuable materials[14] from ash. Improving the economic viability of existing anaerobic digestion plants is also feasible. Opportunities for this include researching co-digestion methods to boost methane production and maintain stable operation, cost-effective biogas purification to achieve pipeline-quality natural gas, innovative thermo catalytic processes for converting biogas and landfill gas into fuels and valuable co-products, and advanced reactor design and organism optimization for better conversion of gases into fuels and co-products.[14–16] The procedure for managing municipal solid trash is depicted in Figure 7.4.

7.6 ENVIRONMENTAL IMPACT AND CHALLENGES

The environmental and health implications are linked to the disposal of garbage, which is mounting in developing countries. The implementation of sustainable municipal solid waste management methods aims to reduce environmental impacts from pollutant gases associated with climate change, including greenhouse gases and ecosystem acidification due to ammonia emissions. Numerous waste management techniques have been studied to assess nitrogen and carbon losses under different environmental and operational conditions, but the comprehensive impact remains to be determined.[17] Concerns also exist regarding pollutants derived from sewage sludge in MSW, including heavy metals, polycyclic aromatic hydrocarbons (PAHs), polychlorinated biphenyls (PCBs), and dioxins and furans (PCDD/Fs). The presence of surfactants and their metabolites, which are challenging to break down under non-aerated conditions, can lead to adverse environmental effects, especially when sewage systems are overloaded and accumulate in sludge.[18]

The management of organic waste can have a significant impact on the ecosystem and the environment. When organic waste is not managed correctly, it can release greenhouse gases, such as methane and carbon dioxide, contributing to climate change. Additionally, the accumulation of organic waste can attract pests and vermin, spreading diseases and contaminating water sources. Proper management of organic waste can have positive impacts on the environment. For example, composting organic waste can produce a valuable fertilizer that can be used to improve soil health and reduce the need for chemical fertilizers. Additionally, the conversion of organic waste into biogas can provide a renewable energy source that can be used to power homes and businesses. Composts play a vital role in promoting sustainable agriculture by slowly replenishing soil nutrients, starkly contrasting the rapid nutrient release of chemical fertilizers. This slow nutrient release, particularly of nitrogen, is crucial as it reduces nitrogen loss through leaching and volatilization. Compost releases nitrogen at a gradual rate of 1–3% of its total nitrogen content per year, making it available over a more extended period and benefiting multiple crop cycles, unlike chemical fertilizers, which are typically absorbed by only one crop in a short time frame. Enriching compost with additional nutrients and beneficial microorganisms creates an environment that fosters long-term crop growth and health. Combining compost with chemical fertilizers can be particularly effective, as it ensures a steady, slow release of nutrients, leading to improved plant growth and yield.

7.6.1 Emission of Greenhouse Gases during Biomass Decomposition and Combustion

The release of greenhouse gases from the decomposition and burning of biomass is influenced by several variables, such as the kind of biomass involved, the conditions under which it is combusted, and the methods used to manage it.

7.6.1.1 Biomass Decomposition

During the decomposition of biomass, microorganisms break down organic matter, releasing greenhouse gases such as carbon dioxide and methane. The exact emissions depend on factors like the type of biomass, temperature, moisture content, and microbial activity.

7.6.1.2 Biomass Combustion

Biomass combustion releases greenhouse gases, primarily CO_2, but the environmental impact depends on various factors. Incomplete combustion can lead to the release of other pollutants, including particulate matter, carbon monoxide, and volatile organic compounds.[19]

7.6.2 Leachate Production and Soil Contamination Risks

Leachate production and soil contamination risks are critical environmental concerns associated with various human activities, particularly in the context of waste management.[20]

Leachate is the liquid that percolates through a landfill or other waste storage and extraction systems, potentially containing pollutants that can contaminate the soil and nearby water resources. Leachate is produced in two ways:

7.6.2.1 Landfill Leachate

Landfill leachate is a complex liquid substance that forms when water percolates with decomposing waste, waste materials deposited in landfills, picking up various dissolved and suspended contaminants. Landfill leachate has a high moisture content due to water percolation through waste materials. This water carries dissolved contaminants, making leachate highly concentrated and potentially hazardous. Landfill leachate often emits strong odors due to the presence of volatile organic compounds (VOCs) and sulfur-containing compounds generated during waste decomposition. These odors can be unpleasant and may pose a nuisance to surrounding communities.

7.6.2.2 Leachate from Waste Storage Facilities

Leachate can also be produced in waste storage facilities, such as lagoons and impoundments, where liquids interact with stored materials.[21] As water passes through the stored waste, it dissolves and picks up various contaminants, including organic matter, nutrients, heavy metals, and other pollutants. Typical components of such leachate include dissolved organic compounds, suspended solids, ammonia, sulfates, chlorides, and pathogens.

7.6.3 Soil Contamination Risks Caused by Leachate

Soil contamination from leachate, a liquid that percolates through waste material in landfills, poses significant environmental risks. This contamination primarily results from the introduction of harmful substances such as heavy metals and organic pollutants and the potential for groundwater contamination. Each of these contaminants poses distinct risks:

7.6.3.1 Heavy Metals

Heavy metals such as lead, mercury, cadmium, arsenic, and chromium are common in leachate. These elements are particularly concerning due to their persistence in the environment; they do not degrade over time.

Instead, they accumulate in the soil, potentially reaching toxic levels. The presence of heavy metals in soil can:

- Inhibit plant growth and microbial activity, disrupting the ecosystem's balance.

- Lead to bioaccumulation, where heavy metals are absorbed by plants and then ingested by animals and humans, posing health risks.

- Alter soil chemistry, affecting nutrient availability and soil fertility.[17]

7.6.3.2 Organic Pollutants

Organic pollutants in leachate, including pesticides, polychlorinated biphenyls (PCBs), pharmaceuticals, and petroleum hydrocarbons, are of great concern. These compounds can:

- Be toxic to soil organisms, leading to a reduction in biodiversity.

- Affect plant health and growth negatively.

- Enter the food chain, posing risks to wildlife and human health through bioaccumulation and biomagnification.

- Persist in the environment, as many organic pollutants resist degradation, leading to long-term contamination issues.[22]

7.6.3.3 Groundwater Contamination

Leachate can percolate through soil layers and contaminate groundwater, a crucial resource for drinking, agriculture, and industry. Groundwater contamination occurs when leachate, carrying heavy metals and organic pollutants, infiltrates aquifers. This process can:

- Pose direct health risks to humans and animals using the contaminated groundwater for consumption.

- Spread contamination over a wide area, far from the original landfill site, making remediation efforts more challenging.

- Affect surface water bodies connected to contaminated groundwater, spreading the impact to aquatic ecosystems.

7.7 APPLICATIONS OF BIOMASS-BASED MUNICIPAL SOLID WASTES

Transforming biodegradable MSW into compost is an effective soil conditioner when blended with nitrogen and phosphorus compounds. This not only enhances the efficiency of fertilizer usage but also benefits the soil's microbial activity. Composts inoculated with nitrogen-transforming and ammonifying nitrifying organisms and those blended with phosphorus-solubilizing microbes have significantly improved soil health and fertility. This approach, integrating compost with conventional fertilization methods, presents a sustainable and efficient strategy for nutrient management in agriculture.

The use of Municipal Solid Waste Compost (MSWC) in India, with a total application of 120,908 tons and 2406.06 kg at 16 tons per hectare, has been shown to significantly boost the grain yields of mustard and pearl millet compared to untreated soils. This improvement is particularly notable under saline conditions, where MSWC enhances soil microbial activities and provides a balanced and synchronized nutrient supply to the crop. Integrating MSWC with mineral fertilizers improves the physical condition of saline soils and enhances dehydrogenase activity and microbial biomass carbon (MBC). Studies have corroborated these positive effects on soil health conduct, demonstrating the effectiveness of MSWC in strengthening the productive potential of salt-affected soils. At the same time, MSWC can significantly improve soil's physicochemical properties, microbial health, and crop yield; careful attention must be paid to the rate and timing of its application to minimize risks such as metal toxicity.[1]

7.8 CASE STUDIES AND INNOVATIVE APPROACHES

The study focuses on a specific area within the Thessaly district, near Larisa city in Greece. This region is densely populated, leading to substantial MSW and various agricultural residues like wheat straw, maize, cotton stalks, and prunings from olive and almond trees. These types have been identified as prevalent through a Pareto analysis and are considered viable fuel sources for a proposed power plant. The intended customers for this plant are the residents of the local Ampelonas community, comprising around 1,900 households. This customer base aligns well with the energy production expected from the available MSW. These consumers rely on heating oil for warmth and electrical heat pumps for cooling. The planned facility aims to operate in a heat-match mode, adapting its output to the community's needs.

TABLE 7.1   Biomass and MSW-Related Case Study Input Data

Name of Property	MSW	Wheat Straw	Corn Stalk	Cotton Stalks	Almond Tree Prunings	Olive Tree Prunings	References
Moisture wet (%)	40	20	50	30	40	35	24
Density wet—as transported (kg/m³)	367	140	200	200	300	250	24, 25
Heating value (MJ wet kg)	8.5	14.9	12.3	15.1	13.1	13.4	24, 26
Price [for MSW gate fee]	–50	50	20	20	30	30	Prices approximate, consultation with farmer and local authorities

Financially, the waste-to-energy (WTE) facility's income would come from several streams: selling electricity to the national grid, providing heating and cooling to the local community through a district heating network, and participating in emission reduction trading. Electricity rates will be fixed through a contract with the Greek energy market. The heating cost is planned to be set at 80% of the current oil heating expenses, and similarly, cooling services will be priced at 80% of the typical costs for electrical compression chillers. This pricing strategy encourages locals to switch to the facility's services. Most agricultural biomass types considered in the study are currently unused, suggesting they could be acquired cheaply. It's important to note that the issues of biomass degradation and material loss are not a significant concern in this study. Table 7.1 assumes that these issues are minimal if the biomass is dried correctly and stored, considering the study's use of a multi-biomass approach that doesn't require long-term storage.[23]

7.9 FUTURE TRENDS AND OUTLOOK

Forecasting the quantity and composition of waste generation and comprehending the current management status is crucial for developing sustainable waste management practices. Population predictions often vary based on the underlying methodology employed for the forecast. Techniques the population's arithmetic growth during the previous few decades.[27] This computation is appropriate for large, historic cities with significant development. When applied to tiny, average, or very new cities, it will produce results below their actual worth. Using data, this method

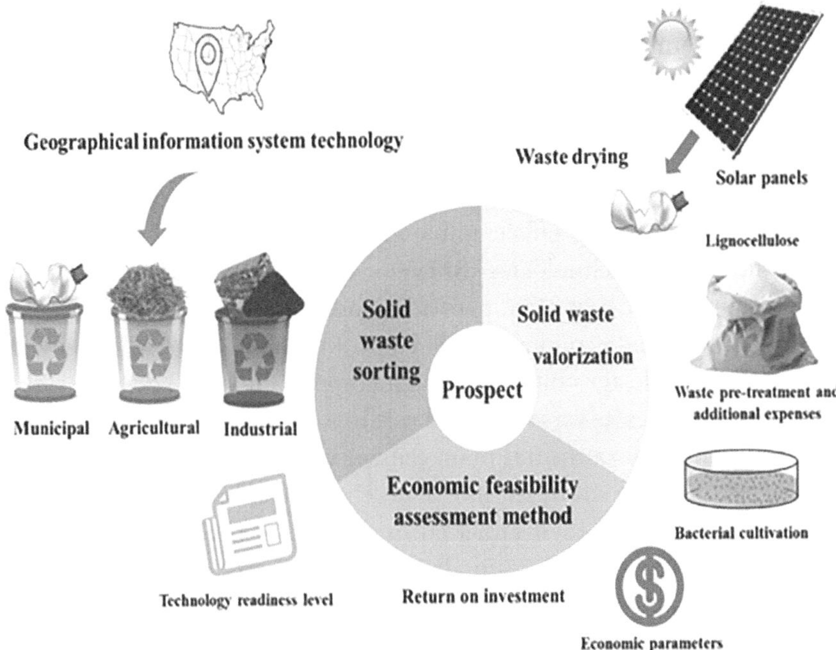

FIGURE 7.5 Improved recycling methods enhance solid waste management, supported by economic evaluations for effective utilization.[23]

determines the average population growth by decade is calculated by multiplying this increase by the current population. Therefore, it is anticipated that the population is growing steadily. In Figure 7.5, recycling and sorting represent the initial and crucial phase in enhancing and utilizing waste materials. Improving the efficiency of waste recovery and the precision of sorting processes is essential. The policy guidelines for waste recycling and sorting must be comprehensive, specifying the entities responsible for policy enforcement and the applicable standards and regulations. These strategies should be tailored to fit the specific attributes of the targeted implementation area. Additionally, the design of recycling and sorting facilities must consider the operators' age and stature to ensure accessibility, simplicity, and efficiency for all users.

7.10 CONCLUSIONS

This chapter delves into biomass-based municipal solid waste (BMSW), emphasizing its categorization, components, and pivotal role in sustainable waste management and the utilization of renewable resources. Addressing global environmental challenges necessitates effective BMSW management and exploitation. Detailed analysis of BMSW's diverse properties, such as chemical makeup, energy content, and biodegradability, is crucial for enhancing waste-to-energy technologies and recycling processes. Proper classification and comprehension of BMSW characteristics are crucial to formulating efficient management strategies and pioneering waste utilization technologies. BMSW's applications in energy production (via anaerobic digestion, pyrolysis, and gasification) offer significant environmental and economic benefits, promoting resource efficiency. Its use in sustainable agriculture, through composting and biochar production, demonstrates its versatility and role in a circular economy. The chapter highlights environmental management implications and underscores the importance of strategies like source segregation and decentralized processing to reduce environmental impacts and encourage sustainability. Understanding BMSW's fundamentals and applications is vital for advancing sustainable waste management, driving us towards a more resource-efficient and greener future through collective efforts from scientists, policymakers, industry leaders, and communities.

ACKNOWLEDGMENT

KNS thanks the Science and Engineering Research Board, New Delhi (Govt. of India), for the financial support under the SERB-SIRE award (File No. SIR/2022/001466).

REFERENCES

1. Chen, D.; Yin, L.; Wang, H.; He, P. Reprint of: Pyrolysis Technologies for Municipal Solid Waste: A Review. *Waste Manag.* **2015**, *37*, 116–136. https://doi.org/10.1016/j.wasman.2015.01.022.
2. Elfaleh, I.; Abbassi, F.; Habibi, M.; Ahmad, F.; Guedri, M.; Nasri, M.; Garnier, C. A Comprehensive Review of Natural Fibers and Their Composites: An Eco-Friendly Alternative to Conventional Materials. *Results Eng.* **2023**, *19* (April), 101271. https://doi.org/10.1016/j.rineng.2023.101271.
3. Bhatt, P.; Bhandari, G.; Turco, R. F.; Aminikhoei, Z.; Bhatt, K.; Simsek, H. Algae in Wastewater Treatment, Mechanism, and Application of Biomass for Production of Value-Added Product ☆. *Environ. Pollut.* **2022**, *309* (May), 119688. https://doi.org/10.1016/j.envpol.2022.119688.
4. Chinonso, P.; Su, H.; Wang, E. Synergistic Blending of Biomass, Sewage Sludge, and Coal for Enhanced Bioenergy Production : Exploring Residue

Combinations and Optimizing Thermal Conversion Parameters. *J. Environ. Manage.* **2024**, *352* (June 2023), 120035. https://doi.org/10.1016/j.jenvman.2024.120035.

5. Munawar, M. A.; Khoja, A. H.; Naqvi, S. R.; Mehran, M. T.; Hassan, M.; Liaquat, R.; Dawood, U. F. Challenges and Opportunities in Biomass Ash Management and Its Utilization in Novel Applications. *Renew. Sustain. Energy Rev.* **2021**, *150* (January), 111451. https://doi.org/10.1016/j.rser.2021.111451.

6. Wong, S. L.; Ngadi, N.; Abdullah, T. A. T.; Inuwa, I. M. Current State and Future Prospects of Plastic Waste as Source of Fuel: A Review. *Renew. Sustain. Energy Rev.* **2015**, *50*, 1167–1180. https://doi.org/10.1016/j.rser.2015.04.063.

7. Hameed, Z.; Aslam, M.; Khan, Z.; Maqsood, K.; Atabani, E.; Ghauri, M.; Shahzad, M.; Rehan, M.; Nizami, A. Gasification of Municipal Solid Waste Blends with Biomass for Energy Production and Resources Recovery: Current Status, Hybrid Technologies and Innovative Prospects. *Renew. Sustain. Energy Rev.* **2021**, *136* (September 2020), 110375. https://doi.org/10.1016/j.rser.2020.110375.

8. Heidari-Maleni, A.; Taheri-Garavand, A.; Rezaei, M.; Jahanbakhshi, A. Biogas Production and Electrical Power Potential, Challenges and Barriers from Municipal Solid Waste (MSW) for Developing Countries: A Review Study in Iran. *J. Agric. Food Res.* **2023**, *13* (June), 100668. https://doi.org/10.1016/j.jafr.2023.100668.

9. Christensen, T. H.; Kjeldsen, P.; Bjerg, P. L.; Jensen, D. L.; Christensen, J. B.; Baun, A.; Albrechtsen, H. J.; Heron, G. Biogeochemistry of Landfill Leachate Plumes. *Appl. Geochem.* **2001**, *16* (7–8), 659–718. https://doi.org/10.1016/S0883-2927(00)00082-2.

10. Dharmendra. Organic Waste: Generation, Composition and Valorisation. *Adv. Org. Waste Manag. Sustain. Pract. Approach.* **2022**, 3–15. https://doi.org/10.1016/B978-0-323-85792-5.00024-1.

11. Rana, R.; Ganguly, R.; Gupta, A. K. Physico-Chemical Characterization of Municipal Solid Waste from Tricity Region of Northern India: A Case Study. *J. Mater. Cycles Waste Manag.* **2018**, *20* (1), 678–689. https://doi.org/10.1007/s10163-017-0615-3.

12. Chormare, R.; Moradeeya, P. G.; Sahoo, T. P.; Seenuvasan, M.; Baskar, G.; Saravaia, H. T.; Kumar, M. A. Conversion of Solid Wastes and Natural Biomass for Deciphering the Valorization of Biochar in Pollution Abatement: A Review on the Thermo-Chemical Processes. *Chemosphere.* **2023**, *339* (July), 139760. https://doi.org/10.1016/j.chemosphere.2023.139760.

13. Chand Malav, L.; Yadav, K. K.; Gupta, N.; Kumar, S.; Sharma, G. K.; Krishnan, S.; Rezania, S.; Kamyab, H.; Pham, Q. B.; Yadav, S.; Bhattacharyya, S.; Yadav, V. K.; Bach, Q. V. A Review on Municipal Solid Waste as a Renewable Source for Waste-to-Energy Project in India: Current Practices, Challenges, and Future Opportunities. *J. Clean. Prod.* **2020**, *277.* https://doi.org/10.1016/j.jclepro.2020.123227.

14. Owens, J. M.; Chynoweth, D. P. Biochemical Methane Potential of Municipal Solid Waste (MSW) Components. *Water Sci. Technol.* **1993**, *27* (2), 1–14. https://doi.org/10.2166/wst.1993.0065.

15. Hilkiah Igoni, A.; Ayotamuno, M. J.; Eze, C. L.; Ogaji, S. O. T.; Probert, S. D. Designs of Anaerobic Digesters for Producing Biogas from Municipal Solid-Waste. *Appl. Energy.* **2008**, *85* (6), 430–438. https://doi.org/10.1016/j.apenergy.2007.07.013.

16. Yaashikaa, P. R.; Kumar, P. S.; Saravanan, A.; Varjani, S.; Ramamurthy, R. Bioconversion of Municipal Solid Waste into Bio-Based Products: A Review on Valorisation and Sustainable Approach for Circular Bioeconomy. *Sci. Total Environ.* **2020**, *748*, 141312. https://doi.org/10.1016/j.scitotenv.2020.141312.

17. Pardo, G.; Moral, R.; Aguilera, E.; del Prado, A. Gaseous Emissions from Management of Solid Waste: A Systematic Review. *Glob. Chang. Biol.* **2015**, *21* (3), 1313–1327. https://doi.org/10.1111/gcb.12806.

18. Gopinath, K. A.; Saha, S.; Mina, B. L.; Pande, H.; Kundu, S.; Gupta, H. S. Influence of Organic Amendments on Growth, Yield and Quality of Wheat and on Soil Properties during Transition to Organic Production. *Nutr. Cycl. Agroecosystems.* **2008**, *82* (1), 51–60. https://doi.org/10.1007/s10705-008-9168-0.

19. Yiping, L.; Yanxia, L.; Henley, K. B. G.; Guomo, Z. Bamboo and Climate Change Mitigation: A Comparative Analysis of Carbon Sequestration, **2010**, file:///C:/Users/KU500781/Downloads/Carbon-Publication_final_Louetal.pdf.

20. Adeolu, A. O.; Ada, O. V; Gbenga, A. A.; Adebayo, O. A. Assessment of Groundwater Contamination by Leachate near a Municipal Solid Waste Landfill. *African J. Environ. Sci. Technol.* **2011**, *5* (11), 933–940. https://doi.org/10.5897/AJEST11.272.

21. Rentizelas, A. A.; Tolis, A. I.; Tatsiopoulos, I. P. Combined Municipal Solid Waste and Biomass System Optimization for District Energy Applications. *Waste Manag.* **2014**, *34* (1), 36–48. https://doi.org/10.1016/j.wasman.2013.09.026.

22. Rayu, S. The Interactions Between Xenobiotics and Soil Microbial Communities. **2016** (January), 308. file:///C:/Users/KU500781/Downloads/Thesis_Rayu_S.pdf

23. Peng, X.; Jiang, Y.; Chen, Z.; Osman, A. I.; Farghali, M.; Rooney, D. W.; Yap, P. S. *Recycling Municipal, Agricultural and Industrial Waste into Energy, Fertilizers, Food and Construction Materials, and Economic Feasibility: A Review.* Springer International Publishing, 2023; Vol. 21. https://doi.org/10.1007/s10311-022-01551-5.

24. Voivontas, D.; Assimacopoulos, D.; Koukios, E. G. Assessment of Biomass Potential for Power Production: A GIS Based Method. *Biomass Bioenergy.* **2001**, *20* (2), 101–112. https://doi.org/10.1016/S0961-9534(00)00070-2.

25. Komilis, D.; Evangelou, A.; Giannakis, G.; Lymperis, C. Revisiting the Elemental Composition and the Calorific Value of the Organic Fraction of

Municipal Solid Wastes. *Waste Manag.* **2012**, *32* (3), 372–381. https://doi.org/10.1016/j.wasman.2011.10.034.

26. Papadopoulos, D. P.; Katsigiannis, P. A. Biomass Energy Surveying and Techno-Economic Assessment of Suitable CHP System Installations. *Biomass Bioenergy.* **2002**, *22* (2), 105–124. https://doi.org/10.1016/S0961-9534(01)00064-2.

27. Sk, M.; Ali, S. A.; Ahmad, P. A. Forecasting Municipal Solid Waste Generation in a Fast-Growing Industrial Region Using Systematic Approach. *Studies in Indian Place Names.* **2020**, *40*, 284–300.

CHAPTER 8

Food Processing Wastes

Fundamentals, Classification, Properties, and Applications

Gavirangappa Hithamani

8.1 INTRODUCTION

Food waste is a significant global issue with environmental, social and economic implications. It refers to the discarding or wastage of food that is still suitable for consumption. It can occur at various stages of the food supply chain, from production and processing to distribution, retail, and consumer levels (Beretta et al., 2013; Schuster & Torero, 2016). Food waste is a natural consequence of food processing operations such as sugar mills, dairy factories, and meat, fish, fruit, and vegetable processing plants. Food processing wastes are a source of diverse nutrients and bioactive compounds viz., carbohydrates, cellulose, hemicellulose, proteins, lignin, pectin, lipids, and antioxidants (Ng et al., 2020; Sindhu et al., 2020). The quality and quantity of wastes and wastewater generated from the food processing industry vary extensively depending on the processing method and amount of water used to wash raw materials and clean equipment. Nevertheless, dealing with food waste is difficult in many aspects due to its inadequate biological stability and the existence of pathogens, which increase microbial activity. Food wastes with high fat content are susceptive to oxidation and escalated spoilage due to incessant enzymatic activity (Russ & Meyer-Pittroff, 2004).

Food waste composition can vary depending on cultural habits, dietary patterns, and socioeconomic conditions. However, typically, food waste consists of various components, including;

DOI: 10.1201/9781003466833-10

Organic matter: this is the most significant component of food waste and includes vegetables, fruits, grains, fish, meat, and dairy products. These organic materials decompose and contribute to greenhouse gas emissions when disposed of in landfills.

Packaging materials: food packaging, such as plastics, cardboard, and aluminum foil, often accompanies food waste. While some of these materials may be recyclable, they can still end up in landfills when not adequately sorted or recycled.

Food preparation waste: peels, trimmings, and other by-products generated during food preparation contribute to food waste. These are typically high in moisture content and can decompose rapidly.

Expired or spoiled food: foods that have passed their expiration dates or have become spoiled due to improper storage or handling are a significant portion of food waste.

Leftovers: uneaten food from meals, whether at home, restaurants, or events, also contributes to food waste.

Beverages: liquid waste, such as expired beverages or unfinished drinks, can also contribute to food waste.

Non-edible parts: parts of food items that are not typically consumed, such as bones, shells, seeds, and pits, also contribute to food waste.

The process of producing food uses many resources, which results in food wastes and losses that tangentially affect the environment by causing soil erosion, deforestation, air and water pollution, and greenhouse gas emissions during the food production, storage, transportation, and disposal phases (Schanes et al., 2016, 2018). Food waste is becoming increasingly recognized by governments, corporations, NGOs, academia, and the general public as a serious issue due to these expanding environmental, social, and economic issues. Sustainable Development Goals (SDGs) were introduced by the United Nations in 2015 to promote peace and prosperity and to address poverty and environmental concerns (UN, 2015). SDG 12 aims to reduce worldwide per capita food waste by 50% by 2030 by emphasizing responsible consumption and food production. There appears to be a significant mismatch in our food systems when high rates of food waste are paired with widespread poverty and food insecurity, and the social ramifications of excessive food waste are frequently raised as a moral concern (Papargyropoulou et al., 2014). Food waste along the food value chain requires more investigation since, throughout the whole chain, little food waste is recovered, and there are only a few preventative measures (Thyberg & Tonjes, 2016). It is essential to manage food waste effectively through strategies such as composting, anaerobic digestion, food donation, and reducing food waste generation at the source to minimize its environmental impact and maximize resource efficiency. This chapter addresses the different causes of food waste, worldwide statistics, socioeconomic influences of food waste, and potential solutions for reducing the same.

8.1.1 Causes of Food Wastes

The causes of food waste are multifaceted, involving various stages of the food supply chain and consumer behavior (Figure 8.1). Causes for food waste are: i) overproduction: excessive farming and production of food beyond what is needed can lead to a surplus that may go to waste. ii) Imperfect produce: fruits and vegetables that do not meet cosmetic standards set by retailers are often discarded. iii) Losses during processing: inefficiencies during food processing can result in the loss of edible parts of food items. iv) Transportation issues: improper handling during transportation can lead to spoilage and waste. v) Expiry date policies: strict adherence to expiry dates can lead to removing perfectly edible items from shelves. vi) Over-ordering: retailers may over-order perishable goods, leading to excess that may go unsold and eventually be wasted. vii) Excess purchases: consumers often buy more food than they can consume, leading to items expiring or going bad. viii) Lack of awareness: consumers might need to know about proper food storage and handling to avoid premature spoilage. ix) Portion sizes: large portions, food that goes unsold and buffet systems contribute to restaurant and food service industry waste. x) Meal planning: lack of meal planning can lead to perishable items going unused and eventually being discarded. Also, failure to consume leftovers before they spoil contributes to household food waste.

8.2 GLOBAL STATISTICS

Food waste occurs throughout the supply chain, but the distribution is not uniform. In high-income countries, consumers contribute to a significant amount of food waste, while in low-income countries, losses are more prevalent in the production and distribution stages. (Ishangulyyev et al., 2019). Globally, it is estimated that almost one-third of all food produced

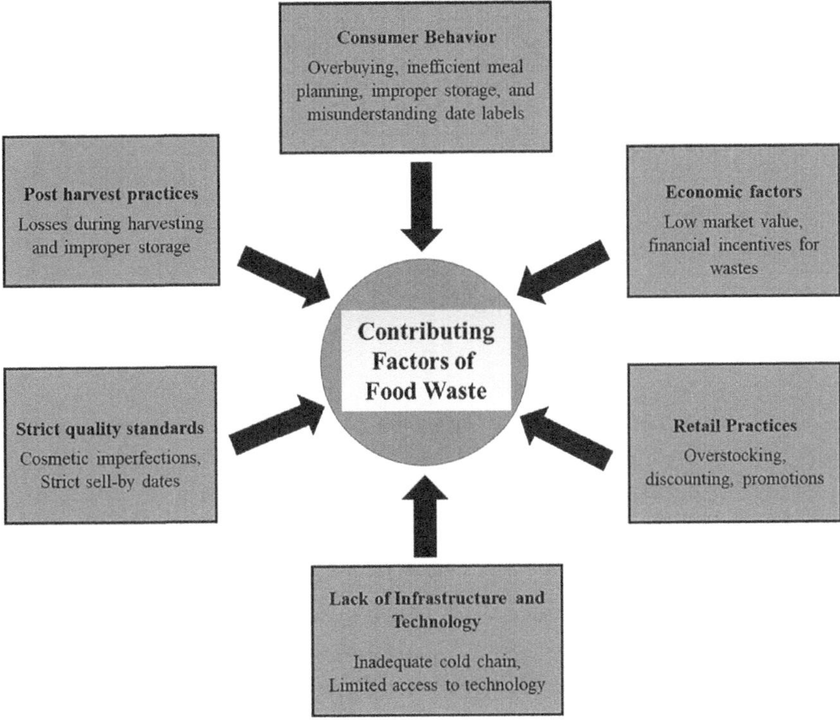

FIGURE 8.1   Major contributing factors for the generation of food waste.

for human consumption is lost or wasted. According to the United Nations Food and Agriculture Organization (FAO), the food loss amounts to about 1.3 billion tons annually. In many developed countries, households are a significant source of food waste (Parfitt et al., 2010; Beretta et al., 2013; Edjabou et al., 2016; Conrad et al., 2018; Verma et al., 2020; Conrad, 2020). On average, a considerable percentage of food waste originates from consumers discarding edible items. According to the 2021 Food Waste Index, as a part of the United Nations Environment Programme, 569 tons out of 931 million tons of food waste comes from households (McCarthy, 2021; Priya et al., 2023).

Comparing food waste amounts and composition across different countries and regions is challenging due to divergent methods, objectives, and definitions, which affect the results. However, China stands first, discarding 91.6 million tons of food annually, followed by India wasting 68.8 million tons. About 14% of food is wasted globally before it reaches the retail sector, commonly known as the "food loss," according to a report

FIGURE 8.2 Global percentage production of commodities generated as food waste.

released by the United Nations Food and Agriculture Organisation (FAO). Among these, around 33–60% of food loss is contributed by the fruits, vegetables, roots, tubers, and oil-bearing seed crops (Buchholz, 2019), as there is a significant difference in the magnitude of wastes produced globally at the commodity level (Figure 8.2), due not just to the methods used but also to agro-ecological circumstances, technological advancements, and socioeconomic settings influencing both production and postproduction. The economic cost of global food waste is substantial. It is estimated to be $1 trillion annually, considering the costs of production, transportation, and disposal. The dearth of information on the production of food waste and the dynamic character of the numbers mentioned earlier point to the challenges in understanding this phenomenon, whose importance is frequently concealed, undervalued, and misinterpreted. These statistics highlight the global scale of the food waste issue and the need for coordinated efforts across sectors to address this challenge.

8.3 CLASSIFICATION/TYPES OF FOOD WASTES

The nature of food waste encompasses various characteristics that contribute to its complexity as an environmental, social, and economic issue. Most food waste is organic, consisting of fruits, vegetables, grains, meats, and other biodegradable materials.

Food waste can be classified into different categories based on various criteria. Table 8.1 lists common classifications of food waste.

TABLE 8.1 Various Categories and Subcategories of Food Waste Are Produced

Categories of Food Wastes	Subcategories and Their Descriptions
Household food waste	Plate waste: leftovers or unconsumed portions of meals on plates. Expired or spoiled food: food that has passed its expiration date or has gone bad. Unwarranted discards: items thrown away due to overbuying, improper storage, or misinterpretation of date labels.
Retail and grocery store waste	Cosmetic limitations: fruits and vegetables that do not meet strict cosmetic standards are discarded despite being edible. Overstock and excess inventory: items that are not sold before reaching their expiration dates. Damaged or expired goods: products damaged during transportation or storage or those that have expired.
Food service industry waste	Preparation waste: trimmings, peels, and other by-products generated during food preparation. Buffet and self-serve leftovers: uneaten food on buffet lines or from self-serve stations. Expired or unsold restaurant meals: prepared meals that go unsold or exceed their serving time.
Farm and agricultural waste	Produce left in fields: unharvested or left-behind crops that do not meet market standards. Post-harvest losses: damage or spoilage occurring during storage, transportation, and processing. Rejected crops: crops deemed unsuitable for sale due to size, shape, or other factors.
Supply chain and distribution waste	Transportation losses: spoilage or damage during transit. Inadequate storage: poor storage conditions to deterioration of food quality. Order cancellations: last-minute changes in orders leading to surplus inventory.
Institutional and event waste	Schools and cafeterias: food waste generated in educational institutions and cafeteria settings. Event catering: leftovers and unused portions from events, conferences, or gatherings.
Wasted ingredients and by-products	By-products of processing: leftover materials generated during food processing that are not used or repurposed. Trimming and peeling waste: edible portions discarded during food preparation.
Non-edible food waste	Inedible parts of fruits, nuts, and meat, such as peels, seeds, pits, shells, and bones, are discarded.

Understanding these classifications can help design targeted strategies and interventions to reduce food waste at various stages of the food supply chain.

8.4 SOCIAL AND ENVIRONMENTAL IMPACTS

When the food is discarded, there is a substantial loss of resources used in food production, such as water, energy, and land. Agriculture often leads to habitat destruction and the use of pesticides, affecting biodiversity. When food is wasted, these negative environmental impacts are intensified without any corresponding benefit. The use of fertilizers and pesticides in agriculture can result in water pollution. When food is wasted, the environmental effects of these pollutants are disproportionate to the actual nutritional benefit derived from the food. Large areas of land are cleared for agriculture to meet the demand for food. Food loss and waste consume around 173 billion cubic meters of water annually, accounting for 24% of all agricultural water. This consumes 198 million hectares of cropland and 28 million tons of fertilizer annually, a significant area similar to Mexico (Kummu et al., 2012). Therefore, wasting food means squandering these resources unnecessarily. When food is wasted, the environmental costs associated with deforestation and land use change become particularly concerning. Additionally, the energy-intensive food production and transportation processes contribute to carbon emissions. Decomposing food waste in landfills produces potent greenhouse gas methane, contributing to climate change (Girotto et al., 2015). Because of microbial degradation and leachate formation, many of these biomaterials go to municipal landfills unutilized and cause significant environmental issues. The byproducts may occasionally be burnt to get rid of parasites and fungi. The expenditures associated with managing solid waste in landfills cause a negative economic effect. Furthermore, managing various degradable materials is demanding.

Consumers, businesses, and even entire nations incur financial losses due to food waste. This includes the cost of production, transportation, and disposal. Farmers may face economic challenges when their produce goes unsold or is rejected due to cosmetic imperfections. Food waste can contribute to disparities in access to food. While some regions struggle with food shortages, others discard edible food due to overproduction or inefficiencies in distribution. Given that 805 million people worldwide suffer from hunger, social repercussions can be linked to an ethical and moral component of the idea of global food security (FAO, 2014). We need more diverse nutritional sources to address the dietary issues facing modern culture. The ethical implications of wasting food, especially when so many people are in need, raise questions about responsible consumption and distribution practices. Reducing food waste in developed countries could bring down the food prices in developing countries, boost supply chain efficiency, and conserve resources for feeding the hungry, ultimately improving access to nutritious foods for vulnerable households and alleviating food insecurity.

8.5 EFFORTS TO REDUCE FOOD WASTE

Addressing food waste is a multifaceted challenge that requires concerted efforts from individuals, businesses, governments, and international organizations. Reducing food waste helps the environment and contributes to building a more sustainable and equitable global food system. Efforts to address food waste include educating consumers, raising awareness about proper food storage, understanding date labels, and encouraging responsible consumption. Improving supply chain efficiency by enhancing transportation, storage, and distribution systems minimizes losses at various supply chain stages. Technical solutions for reducing food loss and waste require other parts of the food supply chain to be effective. Postharvest losses, which account for a fifth to a third of food waste and loss in developing countries, are under-invested in agricultural research. Despite this, studies by the World Bank, FAO, and other organizations show that reducing postharvest losses can be cost-effective and yield good revenues, particularly when prices of food commodities rise.

Science about postharvest loss and waste reduction is generally less prevalent than production research, which requires multiple studies over years or seasons. Improved on-farm storage and poor forecasting techniques can't reduce food loss if farmers lack access to markets for selling their surplus. Additionally, retailers may cancel orders, negating productivity gains made by food processors. Therefore, an integrated supply chain approach is necessary for progressive food waste reduction. Initiatives to redirect surplus food via food recovery programs that collect surplus food from retailers, restaurants, and farms to those in need through food banks and other community organizations would give purpose to the produce and help alleviate hunger. However, food redistribution faces challenges due to transportation, legal, and economic factors (Mateos-Aparicio & Matias 2019; Vilas-Boas et al., 2021). Farmers and stockpiles with surplus food may not be close enough to food banks or rescue groups for economic delivery. Potential food donors may worry about legal consequences if the food is unsafe or health-threatening. Financial constraints may make it challenging for farmers to donate food, as they may incur labor, logistical, and transportation costs.

Finding creative ways to use by-products and surplus food to create new products reduces waste. Developing innovative packaging technologies is recommended to prolong the shelf life of foodstuffs without compromising quality, to reduce spoilage, and to improve storage and transportation efficiency, which can contribute to reducing food waste. For instance, fresh-cut products, introduced in the EU in the 80s, are minimally processed, which includes washing, cutting, mixing, and packing. They represent a revolutionary innovation for the industry, remodeling consumption patterns. According to the International Fresh-Cut Produce Association, the sector has expanded significantly in recent decades (Santeramo et al., 2018). Food waste can also be reduced by developing apps and technologies that help consumers track their food consumption, manage inventory, and make informed purchasing decisions. In addition, marketing strategies should be employed to encourage responsible consumer behavior, such as purchasing imperfect produce, minimizing food overstock, and utilizing various media channels to raise awareness about the impacts of food waste and promote responsible consumer behavior. It is essential to know how much consumer food waste costs daily to target reductions in food waste. With more financial freedom, people may buy better foods and lessen their influence on the environment, promoting healthy lifestyle choices. Utilizing food waste and by-products as raw materials or food additives responsibly might benefit the food industry economically, help alleviate nutritional issues, have positive health impacts, and lessen the environmental effects of improper waste management. Industries are now looking for technologies to achieve zero waste, where garbage is produced and used as a raw material for new goods. These measures can directly impact the Post 2015 Agenda, the Zero Hunger Challenge, the Sustainable Development Goals, and the Millennium Development Goals (Torres-León et al., 2018).

Implementing regulations that discourage food waste, encouraging local initiatives for composting food scraps, establishing community gardens, and encouraging responsible disposal practices can decrease the food waste, particularly organic waste, produced. Educational campaigns or school programs should incorporate education on food waste reduction in school curricula to promote accountable consumption habits from an early age. Also, implementing policies and providing economic incentives for businesses and individuals, such as tax breaks or subsidies by governments and non-governmental organizations, can bring down food waste (Schuster & Torero, 2016). Addressing food waste requires a multifaceted approach involving changes at various supply chain levels and consumer behavior. Implementing these strategies and fostering a culture of responsibility can mitigate food waste's environmental, economic, and social impacts, contributing to a more sustainable and equitable food system.

8.6 TREATMENT AND APPLICATIONS OF FOOD WASTES

Food waste valorization is a sustainable, green, and environmentally friendly method for recovering valuable nutrients from food streams, promoting a circular economy. Several conservative methods, such as combustion or incineration with proper landfilling, composting and vermicomposting, may positively correlate with sustainable environmental and economic features. An integrated waste management strategy combines conventional and traditional methods of food waste valorization, which combines various waste treatments, such as biological, chemical, and thermal processes. However, in the quest to valorize food waste sustainably, the developing enzymatic conversion of food waste into value-added chemicals has gained prominence in the world's industrialized areas.

Food waste and by-products are critical due to the abundance of proteins, fats, carbohydrates, micronutrients, bioactive substances, and dietary fibers. One of the significant issues in most developing nations is protein inadequacy and the malnutrition that goes along with it. Food fortification is essential to combat malnutrition, and significant efforts have been made in the USA and Europe to use leftovers and by-products (Mirabella et al., 2014; Girotto et al., 2015). Because these biomaterials produce many by-products, there is a larger chance of exploitation in poorer countries where hunger and malnutrition are major issues.

The food industry produces a huge amount of biodegradable waste and abandons enormous quantities of residues with high biochemical and chemical oxygen demand. Hence, appropriate management and treatment of food waste is essential to modern food processing management for environmental regulations and for recovering valuable products. Food processing procedures produce large quantities of solid and liquid phase wastes. Homogeneous and consistent wastes are converted to secondary food products or animal feed. The fruit and vegetable industry produces 25–30% of the total waste generated compared to other food processing sectors. The waste produced includes seeds, skins, peels, rinds, stones, cores, rags, pods, pomaces, and vines. Improperly disposing of fruits and vegetable waste can yield germs and invite pathogen-carrying insects, mosquitoes, and flies. In addition, the surface and groundwater get contaminated, leading to phytotoxicity and significant health-related issues. Hence, the collection approaches and storage conditions of food waste would have significant consequences for valorization. Therefore, in many research studies on food waste, the waste is frozen to reduce probable contamination and biodegradation or processed instantaneously once collected. One research group developed composite bread utilizing fruit and vegetable waste (Upadhyay et al., 2023). Cleaned and powdered fresh Indian gooseberry waste, apple waste, peels of potato and bottle gourd were used to prepare the composite bread with increased polyphenols, vitamin C, and fiber content.

Recently, interest has risen in innovative management techniques and treatments concentrating on extracting and repurposing functional components from food waste. Food processing waste is a plentiful supply of organic compounds that may be transformed into energy and then used to produce heat or electricity. Reduce, Reuse, and Recycle are the three R's of waste management, which help the food processing sector to minimize waste sent to landfills. Anaerobic digesters, which produce energy-rich digestate and biogas through the biological decomposition of organic waste, are a practical solution for handling organic waste at food processing facilities. The primary waste reduction and valorization solutions are general waste treatment techniques like composting, incineration, landfills, anaerobic fermentation, or utilizing food wastes for agricultural purposes like fertilizer or animal feed. Utilization of soybean meal produced under solid-state fermentation using a strain of Saccharomyces cerevisiae as a source of protein for the Nile tilapia (Oreochromis niloticus) diet was shown by Hassaan et al. (2015). According to the study, it is feasible to substitute fishmeal, around 37% of the typical protein supply, without hurting animal growth or health. These findings could be explained by the release of amino acids that occurs during fermentation, increasing protein availability and other nutrients.

The growth performance, feed utilization, and palatability of white shrimp as affected when dietary fishmeal was replaced with soybean meal fermented by a Saccharomyces cerevisiae strain under solid-state fermentation were documented (Sharawy et al., 2016). The findings demonstrated that fermented soybean meal may replace as much as 50% of the fish meal without negatively impacting the animals' development or behavior. The effects of the feed produced using palm kernel cake fermented by Paenibacillus polymyxa (ATCC 842) under solid-state fermentation conditions were assessed by Alshelmani et al. (2016) about the zootechnical parameters of broiler chickens. The findings show that up to 15% of the palm kernel cake fermented in the rations without impacting the animals' growth or the food's ability to be digested. Table 8.2 gives the list of compounds that are recovered from different food wastes using various production techniques.

The Berkeley technique is well-liked and tested for handling food and green waste. Food waste and leftovers are broken down into a hygienic, humus-rich material during this aerobic digestion process by various microorganisms, including fungus and bacteria, in the presence of oxygen. Finished compost may be produced in less than a month (Daud et al., 2016). The Bokashi method is a fermentation approach that uses anaerobic digestion and can be used for various food waste valorizations. The Bokashi fertilizer mainly comprises a mixture of water, sugar, rice bran, the meal of rapeseed, and essential microorganisms (Dou et al., 2012; Arumugam et al., 2021). The resultant organic fertilizer from this method helps raise the water in the soil and may also be used for soil nutrition (Kumar et al., 2004). Automated biodigesters treat food processing waste; they digest and process food waste into fertilizers enriched with nutrients. The biodigesters reduce food waste by 90% in weight while mixing and producing pre-compost in a 24-hour period (Aschonitis et al., 2014; Ma et al., 2020).

The two primary processes for converting biomass into fuel are anaerobic digestion and thermochemical treatments, such as gasification, pyrolysis, and combustion (Murugan et al., 2013). Biomass with high energy content may be converted into liquid or gaseous intermediate products via thermochemical conversion, which works best with less than 50% moisture waste. The thermal process of incineration produces heat by oxidizing the waste's combustible components. According to Murugan et al. (2013), incineration is a feasible solution for hazardous and food wastes with a relatively low water content (less than 50% by mass). A common technique for handling wastes with a high (> 50%) water content and organic value is anaerobic digestion, which uses a variety of microbes to stabilize food waste without oxygen. Because it includes phosphate, ammonia, and other minerals, the

TABLE 8.2 Compounds Recovered from Different Types of Food Waste and the Method of Recovery

Type of Food Waste	Compound Recovered	Method of Recovery/Production	References
Kitchen waste	polyunsaturated fatty acid	cultivation of Phaeodactylum strain E70	Wang et al., 2020
Food waste from restaurants and expired food products from supermarkets	Sophorolipids	cultivation of Starmerella bombicola on food waste hydrolysate	To et al., 2023
Food waste from canteens containing rice, noodles, vegetables, and meat	Glucose, free amino nitrogen, and phosphate	hydrolysis by Aspergillus awamori and Aspergillus oryzae	Pleissner et al., 2013
Food waste hydrolysate	polyunsaturated fatty acid	semi-continuous fermentation of glucose-rich food waste hydrolysate by evolved strain, ALE-Pt1 from Phaeodactylum tricornutum	Wang et al., 2022
Food waste from restaurants	saturated and unsaturated fatty acids	continuous flow culturing of the heterotrophic microalga Chlorella pyrenoidosa	Pleissner et al., 2017
Carbohydrate-rich food wastes like potato chips and oatmeal, and beverage wastes like energy drinks, fruit juices and soft drinks	Sugars	Saccharification of food and beverage mixture by glucoamylase and sucrase	Kwan et al., 2018
Seaweed solid wastes	carrageenan and bioethanol	simultaneous saccharification and fermentation with Saccharomyces cerevisiae	Tan & Lee, 2014
Waste bread	succinic acid	fermentative production by Actinobacillus succinogenes	Leung et al., 2012
Beverage waste from the supermarket	sugar and fructose	enzymatic hydrolysis, activated carbon treatment, ion exchange chromatography and ligand exchange chromatography	Haque et al., 2017
Food waste from the school cafeteria	biodiesel containing 67% medium-chain saturated fatty acids and 13% polyunsaturated fatty acid	bioconversion of waste by the production of protein and fat-rich prepupae from black soldier fly (Hermetia illucens) larvae	Surendra et al., 2016
Mixed bakery waste	L-lactic acid	non-sterilized fermentation by Thermoanaerobacterium aotearoense LA1002-G40	Yang et al., 2015

Synthetic beverage waste	5-hydroxymethylfurfural	produced by continuous-flow purification, filtration, and two-stage adsorption	Akkarawatkhoosith et al., 2024
Food waste from a student canteen	paramylon	production from Euglena gracilis and cell immobilization cultivation	Mou et al., 2024
Food waste containing mainly unconsumed and expired toast bread from a student canteen	lipid and lutein	culturing of mixotrophic Chlorella species on enzymatically hydrolyzed food waste	Wang et al., 2020
Spent tea leaves	Protein	filtration through a semipermeable membrane followed by an alkali-extraction membrane filtration	Perera et al., 2019
Restaurant pre-prepare waste, banana peel, orange bagasse, gauva bagasse	Xylooligosaccharides	Enzymatic hydrolysis followed by alkaline hydrogen peroxide solubilization	Pereira et al., 2022
Chili stalk waste	Cellulose biopolymer	Alkali (NaOH) treatment and bleaching with H_2O_2	Santhosh & Umesh, 2024
Banana peels	Xylose, xylooligosaccharides, glucose	Sequential acidic, alkaline and enzymatic treatments	Pereira et al., 2021
Pigeon pea stalks	Xylooligosaccharides	Sodium hydroxide coupled steam application followed by enzymatic hydrolysis	Samanta et al., 2013
Household food waste	Ethanol	liquefaction/saccharification	Matsakas et al., 2014
Food waste from university	Lipid and protein	Cultivation of Rhodosporidium toruloides Y2 on food waste hydrolysate	Zeng et al., 2017
Food waste hydrolysate	Lipid	Mix-culture of yeast Rhodosporidium toruloides and microalgae Chlorella vulgaris	Zeng et al., 2018
Food waste hydrolysate	Biohydrogen	dark fermentation of food waste through fungal solid-state fermentation using Aspergillus tubingensis	Manuel et al., 2022

leftover slurry from the degradation of organic substrates is employed as fertilizer (Nishio & Nakashimada, 2013). Simultaneously, biogas is created, which is utilized to minimize the usage of fossil fuels and CO_2 emissions while producing electricity through thermal energy (Pesta, 2006).

Composting is employed to biologically break down organic materials under regulated aerobic conditions and produce a solid humus-like substance to remediate garbage (Cabanillas et al., 2013). Compared to other techniques, composting technology is less expensive, easier to adopt, and better for the environment, and the resulting decomposed waste can be used as animal pellet feed or fertilizer or utilized to produce biogas (Iñiguez-Covarrubias et al., 2018). The composting site is located close to the landfill to facilitate operations and minimize material handling. Different composting methods are pit/heap, aerated static pile (ASP), compost bin or box and compost tumblers. Solar digesters (green cone technology) are also used to treat waste, where microorganisms such as bacteria, fungi, worms, and insects decompose the food dregs. With this approach, the plant receives compost water from the surrounding soil, which absorbs 90% of the waste (Mu et al., 2017).

Vermicomposting exploits several species of worms to decompose the food waste. It is an eco-friendly, eco-biotechnological, and bio-oxidative application mechanism that stabilizes the waste into a functional bioproduct (Saer et al., 2013). White worms, red wigglers, and earthworms are most commonly used for vermicomposting. According to Jayaprakash et al. (2018), earthworms play a critical role as a mediator by expanding the surface area accessible to microorganisms, enhancing enzymatic processes, and changing the physical characteristics of organic waste. Enzymes like lipase, chitinase, amylase, and cellulase are found in vermicompost. These enzymes aid in breaking organic debris in the field and releasing nutrients for plant roots.

Insect-based food waste treatment is becoming more widely acknowledged as an eco-friendly method of resource recycling, with the added benefit of being inexpensive to install (Choi et al., 2017; Leni et al., 2021). When they are produced in large quantities under regulated conditions, significant amounts of food waste can be broken down by insects. In addition, when the insects feed on this waste, they produce various expedient products from their excrement, including biofuels, fertilizer, pharmaceuticals, and insect biomass (proteins and lipids) as animal feed. In the upcoming decades, this method of treating food waste, known as "bioconversion," will be a strong competitor. The insect bioconversion field is becoming increasingly popular as a study subject in terms of commercial prospects (Fowles & Nansen, 2020; Skrivervik, 2020). Interestingly the process of bioconversion would provide value at each step, for example, the initial step of elimination of waste itself (Mutafela, 2015), generation of insect biomass as animal feed (Anankware et al., 2015), value as a means of secondary products developed (Zheng et al., 2012), and the leftover bio-converted waste can be sold as an organic soil amendment (Suantika et al., 2017). Among the few insects used for bioconversion of food waste, black soldier flies (BSF) larvae (Hermetia illucens L.) are the most common species (Wang & Shelomi, 2017) because of their capability to feed on a wide range of food sources and convert them to more useful forms. Bioconversion of food waste with BSF records a very low global warming potential score of 17.36 kg CO_2 per ton, which significantly impacts environmental sustainability (Ganesan et al., 2024). Biogas and methane production were achieved by breeding Hermetia illucens larvae on food waste, as described in a two-stage method developed by Li et al. (2023). Food waste containing vegetables, meat, cereals, oil, salt and eggs was converted into high-value-added furfural and hydroxymethyl-furfural using BSF larvae and subsequent hydrothermal catalysis with cerium-incorporated hydroxyapatite. The production method yielded 9.3 wt% of furfural and could generate a revenue of 3.14 to 584.4 USD/ton as predicted by the authors. From the perspective of the diversity of food waste generated, it is better to have selective breeding of insects and better waste-to-insect pairing to achieve the increased capability of bioconversion of food waste using insects.

8.7 FUTURE PROSPECTUS ON FOOD WASTES

Food waste is becoming more valued owing to its impact on the environment and the scarcity of resources; nevertheless, because it is diverse, it can be challenging to turn it into valuable goods. The future prospects for food waste involve a combination of technological advancements, policy initiatives, and changes in consumer behavior aimed at creating a more sustainable and efficient food system. Technological innovations include the continued development of innovative packaging technologies that extend the shelf life of food items, reducing spoilage. Advanced data analytics optimize supply chains, reduce overproduction, and minimize waste at various stages. Advancements in biotechnology such as converting food waste into valuable products, such as biofuels, enzymes, and pharmaceuticals, need to be practiced. Circular economy practices such as the implementation of closed-loop systems where food waste is recycled and by-products are reintegrated into the production process. Circular supply chain models that prioritize resource efficiency and minimize waste generation must be adopted. Policy and regulatory measures must standardize date labelling practices to reduce confusion and minimize premature discarding of edible food globally. Increased use of digital platforms and apps to educate consumers on responsible food consumption, storage, and waste reduction are the needs of the hour. Collaborative efforts by governments, NGOs, and businesses to raise awareness about the environmental and social impact of food waste have to be made. Precision agriculture techniques should be adopted to optimize farming practices, reduce losses, and enhance resource efficiency. A shift towards plant-based diets may reduce the environmental impact associated with livestock production and lower overall food waste. The future prospects of food waste hinge on the collective efforts of individuals, businesses, governments, and global organizations to embrace sustainable practices, leverage technological solutions, and foster a shift in societal attitudes towards responsible consumption and waste reduction. These developments aim to create a more resilient, equitable, and environmentally friendly food system for future generations.

8.8 CONCLUSION

Food waste is still a complicated, new, emerging, and expanding topic regarding its causes and methods. It squanders valuable resources such as water, land, and energy and contributes significantly to greenhouse gas emissions and biodiversity loss. The current book chapter presents a systematic review of food waste management, highlighting the significant issue of one-third of global food production being wasted, leading to greenhouse gas emissions, economic losses, and food divergence from the deprived population. The chapter emphasizes the need to identify the current state of food waste globally and evaluates conventional food management techniques. Addressing food waste requires a multi-faceted approach involving stakeholders at all levels of the supply chain, from producers to consumers. Furthermore, fostering a culture of mindfulness and responsibility around food consumption is essential for creating a sustainable future where food waste is minimized, and resources are used efficiently.

ACKNOWLEDGMENTS

The author GH is thankful to the Indian Council of Medical Research (ICMR), New Delhi for the award of Research Associateship.

REFERENCES

Alshelmani, M., Loh, T., Foo, H., Sazili, A., & Lau, W. 2016. "Effect of feeding different levels of palm kernel cake fermented by Paenibacillus polymyxa ATCC 842 on nutrient digestibility, intestinal morphology, and gut microflora in broiler chickens." *Animal Feed Science and Technology* 216:216–224. https://doi.org/10.1016/j.anifeedsci.2016.03.019.

Akkarawatkhoosith, Nattee, Attasak Jaree, Chotika Yoocham, Thanakorn Damrongsakul, and Tiprawee Tongtummachat. 2024. "Valorization of Beverage Waste as a Sugar Source for 5-hydroxymethylfurfural Production." *Journal of Environmental Chemical Engineering* 12 (1): 111646. https://doi.org/10.1016/j.jece.2023.111646.

Anankware P.J., Fening K.O., Osekre E., Obeng-Ofori D. 2015. "Insects as food and feed: a review." *International Journal of Agricultural Research and Review* 3: 143–151.

Arumugam, V., M.H. Ismail, T.A. Puspadaran, W. Routray, N. Ngadisih, J.N.W. Karyadi, B. Suwignyo, and H. Suryatmojo. 2021. "Food Waste Treatment Methods and its Effects on the Growth Quality of Plants: A Review." *Pertanika Journal of Tropical Agricultural Science* 45 (1): 75–101. https://doi.org/10.47836/pjtas.45.1.05.

Aschonitis, Vassilis, Dimitris Papamichail, Anastasios Lithourgidis, and Elisa Anna Fano. 2014. "Estimation of Leaf Area Index and Foliage Area Index of Rice Using an Indirect Gravimetric Method." *Communications in Soil Science and Plant Analysis* 45 (13): 1726–1740. https://doi.org/10.1080/0010 3624.2014.907917.

Beretta, Claudio, Franziska Stoessel, Urs Baier, and Stefanie Hellweg. 2013. "Quantifying Food Losses and the Potential for Reduction in Switzerland." *Waste Management* 33 (3): 764–773. https://doi.org/10.1016/j.wasman. 2012.11.007.

Buchholz, Katharina. 2019. "14 Percent of Food Goes to Waste." *Statista Daily Data*, October 16, 2019. https://www.statista.com/chart/19672/global-shares-of-different-agricultural-products-thrown-away/.

Cabanillas, Carmen, Daniel Aurelio Stobbia, and Alicia Ledesma. 2013. "Production and Income of Basil in and Out of Season with Vermicomposts from Rabbit Manure and Bovine Ruminal Contents Alternatives to Urea." *Journal of Cleaner Production* 47 (May): 77–84. https://doi.org/10.1016/j. jclepro.2013.02.012.

Choi, Byoung Deug, Nathan A.K. Wong, and Joong-Hyuck Auh. 2017. "Defatting and Sonication Enhances Protein Extraction from Edible Insects." *PubMed* 37 (6): 955–961. https://doi.org/10.5851/kosfa.2017.37.6.955.

Conrad, Zach. 2020. "Daily Cost of Consumer Food Wasted, Inedible, and Consumed in the United States, 2001–2016." *Nutrition Journal* 19 (1). https://doi.org/10.1186/s12937-020-00552-w.

Conrad, Zach, Meredith T. Niles, Deborah A. Neher, Eric D. Roy, Nicole E. Tichenor, and Lisa Jahns. 2018. "Relationship between Food Waste, Diet Quality, and Environmental Sustainability." *PLOS ONE* 13 (4): e0195405. https://doi.org/10.1371/journal.pone.0195405.

Daud, Norzaidi Mohd, Sopiah Ambong Khalid, Wan Nazriah Wan Nawawi, and Noorazlin Ramli. 2016. "Producing Fertilizer from Food Waste Recycling Using Berkeley and Bokashi Method." *PONTE International Scientific Researchs Journal* 72 (4). https://doi.org/10.21506/j.ponte.2016.4.11.

Dou, L., Komatsuzaki, M., & Nakagawa, M. 2012. "Effects of Biochar, Mokusakueki and Bokashi application on soil nutrients, yields and qualities of sweet potato." *International Research Journal of Agricultural Science and Soil Science* 2 (8): 318–327.

Edjabou, Maklawe Essonanawe, Claus Petersen, Charlotte Scheutz, and Thomas Fruergaard Astrup. 2016. "Food Waste from Danish Households: Generation and Composition." *Waste Management* 52 (June): 256–268. https://doi.org/10.1016/j.wasman.2016.03.032.

FAO. 2014. "World Hunger Falls, but 805 Million Still Chronically Undernourished." *United Nations Newsroom.* https://www.fao.org/newsroom/detail/World-hunger-falls-but-805-million-still-chronically-undernourished/en.

Fowles T and Nansen C, 2020. Insect-Based Bioconversion: Value from Food Waste. In: Närvänen, E., Mesiranta, N., Mattila, M., Heikkinen, A. (eds) *Food Waste Management* Palgrave Macmillan, Cham. https://doi.org/10.1007/978-3-030-20561-4_12.

Ganesan, Abirami Ramu, Kanwar Mohan, Sabariswaran Kandasamy, P.K. Surendran, R. Vasant Kumar, Durairaj Karthick Rajan, and Jayakumar Rajarajeswaran. 2024. "Food Waste-Derived Black Soldier Fly (*Hermetia Illucens*) Larval Resource Recovery: A Circular Bioeconomy Approach." *Process Safety and Environmental Protection* (January). https://doi.org/10.1016/j.psep.2024.01.084.

Girotto, Francesca, Luca Alibardi, and Raffaello Cossu. 2015. "Food Waste Generation and Industrial Uses: A Review." *Waste Management* 45 (November): 32–41. https://doi.org/10.1016/j.wasman.2015.06.008.

Haque, Ahteshamul, Xiaofeng Yang, Khai Lun Ong, Wentao Tang, Tsz Him Kwan, Sandeep S. Kulkarni, and Carol Sze Ki Lin. 2017. "Bioconversion of Beverage Waste to High Fructose Syrup as a Value-Added Product." *Food and Bioproducts Processing* 105 (September): 179–187. https://doi.org/10.1016/j.fbp.2017.07.007.

Hassaan, Mohamed S., Magdy A. Soltan, and Ahmed M. Abdel-Moez. 2015. "Nutritive Value of Soybean Meal after Solid State Fermentation with *Saccharomyces Cerevisiae* for Nile Tilapia, Oreochromis Niloticus." *Animal Feed Science and Technology* 201 (March): 89–98. https://doi.org/10.1016/j.anifeedsci.2015.01.007.

Iñiguez-Covarrubias, G., Rodrigo Gómez-Rizo, Walter Ramírez-Meda, and José De Jesús Bernal-Casillas. 2018. "Composting of Food and Yard Wastes Under the Static Aerated Pile Method." *Advances in Chemical Engineering and Science* 8 (4): 271–279. https://doi.org/10.4236/aces.2018.84019.

Ishangulyyev, Rovshen, Sang-Hyo Kim, and Sang Hyeon Lee. 2019. "Understanding Food Loss and Waste—Why Are We Losing and Wasting Food?" *Foods* 8 (8): 297. https://doi.org/10.3390/foods8080297.

Jayaprakash, Sachin, H.S. Lohit, and B.S. Abhilash. 2018. "Design and Development of Compost Bin for Indian Kitchen." *International Journal of Waste Resources* 8 (1). https://doi.org/10.4172/2252-5211.1000323.

Kummu, Matti, De H. Moel, Miina Porkka, Stefan Siebert, Olli Varis, and Philip J. Ward. 2012. "Lost Food, Wasted Resources: Global Food Supply Chain Losses and Their Impacts on Freshwater, Cropland, and Fertiliser Use." *Science of the Total Environment* 438 (November): 477–489. https://doi.org/10.1016/j.scitotenv.2012.08.092.

Kwan, Tsz Him, Khai Lun Ong, Ahteshamul Haque, Wing Hei Kwan, Sandeep S. Kulkarni, and Carol Sze Ki Lin. 2018. "Valorisation of Food and Beverage Waste via Saccharification for Sugars Recovery." *Bioresource Technology* 255 (May): 67–75. https://doi.org/10.1016/j.biortech.2018.01.077.

Leni, Giulia, Augusta Caligiani, and Stefano Sforza. 2021. "Bioconversion of Agri-food Waste and By-products through Insects: A New Valorization Opportunity." In *Elsevier eBooks*, 809–828. https://doi.org/10.1016/b978-0-12-824044-1.00013-1.

Leung, Cho Chark Joe, Anaxagoras Siu Yeung Cheung, Andrew Yan-Zhu Zhang, Koon Fung Lam, and Carol Sze Ki Lin. 2012. "Utilisation of Waste Bread for Fermentative Succinic Acid Production." *Biochemical Engineering Journal* 65 (June): 10–15. https://doi.org/10.1016/j.bej.2012.03.010.

Li, Ouyang, Jiaming Liang, Yundan Chen, Siqi Tang, and Li Zhen-Shan. 2023. "Exploration of Converting Food Waste into Value-Added Products via Insect Pretreatment-Assisted Hydrothermal Catalysis." *ACS Omega* 8 (21): 18760–18772. https://doi.org/10.1021/acsomega.3c00762.

Ma, Liuzheng, Duan Tie-Cheng, and Jiandong Hu. 2020. "Application of a Universal Soil Extractant for Determining the Available NPK: A Case Study of Crop Planting Zones in Central China." *Science of the Total Environment* 704 (February): 135253. https://doi.org/10.1016/j.scitotenv.2019.135253.

Manuel CR., Carlos QF., Carmen PC., Marisela VD. L., Iván MA. 2022. "Fungal solid-state fermentation of food waste for biohydrogen production by dark fermentation." *International Journal of Hydrogen Energy* 47(70): 30062–30073, https://doi.org/10.1016/j.ijhydene.2022.06.313.

Mateos-Aparicio, Inmaculada, and Ana A. Matias. 2019. "Food Industry Processing By-products in Foods." In *Elsevier eBooks*, 239–281. https://doi.org/10.1016/b978-0-12-816453-2.00009-7.

Matsakas, Leonidas, Dimitris Kekos, Maria Loizidou, and Paul Christakopoulos. 2014. "Utilization of Household Food Waste for the Production of Ethanol at High Dry Material Content." *Biotechnology for Biofuels* 7 (1). https://doi.org/10.1186/1754-6834-7-4.

McCarthy, Niall. 2021. "The Enormous Scale of Global Food Waste." *Statista Daily Data*, March 5, 2021. https://www.statista.com/chart/24350/total-annual-household-waste-produced-in-selected-countries/.

Mirabella, Nadia, Valentina Castellani, and Serenella Sala. 2014. "Current Options for the Valorization of Food Manufacturing Waste: A Review." *Journal of Cleaner Production* 65 (February): 28–41. https://doi.org/10.1016/j.jclepro.2013.10.051.

Mou, Jin-Hua, Si-Fen Liu, Lili Yang, Zi-Hao Qin, Yibo Yang, Zhenyao Wang, Hongye Li, Carol Sze Ki Lin, and Xiang Wang. 2024. "Sustainable Paramylon Production from Food Waste by Euglena Gracilis Using a Waste-Based Cell Immobilisation Technique." *Chemical Engineering Journal* 481 (February): 148594. https://doi.org/10.1016/j.cej.2024.148594.

Mu, Dongyan, Naomi Horowitz, Maeve Casey, and Kimmera Jones. 2017. "Environmental and Economic Analysis of an In-vessel Food Waste Composting System at Kean University in the U.S." *Waste Management* 59 (January): 476–486. https://doi.org/10.1016/j.wasman.2016.10.026.

Murugan, K., Chandrasekaran, V.S., Karthikeyan, P., Al-Sohaibani, S., 2013. Current state-of-the-art of food processing by-products. In: Chandrasekaran, M. (Ed.), *Valorization of Food Processing By-Products*. Taylor & Francis Group, Florida, pp. 35–62.

Mutafela, Richard. 2015. "High Value Organic Waste Treatment via Black Soldier Fly Bioconversion: Onsite Pilot Study." Master's Thesis, Royal Institute of Technology, Stockholm, Sweden, January. http://kth.diva-portal.org/smash/record.jsf?pid=diva2:868277.

Ng, Hui Suan, Phei Er Kee, Hip Seng Yim, Po-Ting Chen, Yuhong Wei, and John Chi-Wei Lan. 2020. "Recent Advances on the Sustainable Approaches for Conversion and Reutilization of Food Wastes to Valuable Bioproducts." *Bioresource Technology* 302 (April): 122889. https://doi.org/10.1016/j.biortech.2020.122889.

Nishio, Naomichi, and Yutaka Nakashimada. 2013. "Food Industry Wastes: Chapter 7." In Maria R. Kosseva and Colin Webb (eds), *Manufacture of Biogas and Fertilizer from Solid Food Wastes by Means of Anaerobic Digestion*. Academic Press, Cambridge, MA, USA: Elsevier Inc.

Papargyropoulou, Effie, Rodrigo Lozano, J. Steinberger, Nigel Wright, and Zaini Ujang. 2014. "The Food Waste Hierarchy as a Framework for the Management of Food Surplus and Food Waste." *Journal of Cleaner Production* 76 (August): 106–115. https://doi.org/10.1016/j.jclepro.2014.04.020.

Parfitt, Julian, Mark Barthel, and Sarah J. Macnaughton. 2010. "Food Waste within Food Supply Chains: Quantification and Potential for Change to 2050." *Philosophical Transactions of the Royal Society B* 365 (1554): 3065–3081. https://doi.org/10.1098/rstb.2010.0126.

Pereira M.A.F., Monteiro C.R.M., Pereira G.N., Júnior S.E.B., Zanella E., Ávila P.F., Stambuk B.U., Goldbeck R., de Oliveira D., Poletto P. 2021. "Deconstruction of banana peel for carbohydrate fractionation." *Bioprocess and Biosystems Engineering* 44 (2): 297–306, DOI: 10.1007/s00449-020-02442-1

Pereira, Beatriz Salustiano, Caroline De Freitas, Jonas Contiero, and Michel Brienzo. 2022. "Enzymatic Production of Xylooligosaccharides From Xylan Solubilized From Food and Agroindustrial Waste." *Bioenergy Research* 15 (2): 1195–1203. https://doi.org/10.1007/s12155-021-10373-2.

Perera, G.A.A.R., A.M.T. Amarakoon, D.C.K. Illeperuma, and Palavinnage Krishantha Pushpakumara Muthukumarana. 2019. "Application of Membrane Filtration Technique in Preparation of Protein-Rich Feed from Spent Tea." *Journal of Food and Agriculture* 12 (2): 13. https://doi.org/10.4038/jfa.v12i2.5220.

Pesta, G. 2006. Anaerobic digestion of organic residues and wastes. In: Oreopoulou, V., Russ, W. (Eds.), *Utilization of By-Products and Treatment of Waste in the Food Industry*. Springer Science, Business Media, New York, pp. 53–72.

Pleissner, Daniel, James C.W. Lam, Zheng Sun, and Carol Sze Ki Lin. 2013. "Food Waste as Nutrient Source in Heterotrophic Microalgae Cultivation." *Bioresource Technology* 137 (June): 139–146. https://doi.org/10.1016/j.biortech.2013.03.088.

Pleissner, Daniel, Kin Yan Lau, and Carol Sze Ki Lin. 2017. "Utilization of Food Waste in Continuous Flow Cultures of the Heterotrophic Microalga *Chlorella Pyrenoidosa* for Saturated and Unsaturated Fatty Acids Production." *Journal of Cleaner Production* 142 (January): 1417–1424. https://doi.org/10.1016/j.jclepro.2016.11.165.

Priya, Samant Shant, Sushil Kumar Dixit, Sajal Kabiraj, and Meenu Shant Priya. 2023. "Food Waste in Indian Households: Status and Potential Solutions." *Environmental Science and Pollution Research* 30 (59): 124401–124406. https://doi.org/10.1007/s11356-023-31034-1.

Russ, Winfried, and R. Meyer-Pittroff. 2004. "Utilizing Waste Products from the Food Production and Processing Industries." *Critical Reviews in Food Science and Nutrition* 44 (1): 57–62. https://doi.org/10.1080/10408690490263783.

Saer, Alex, Stephanie Lansing, Nadine H. Davitt, and Robert E. Graves. 2013. "Life Cycle Assessment of a Food Waste Composting System: Environmental Impact Hotspots." *Journal of Cleaner Production* 52 (August): 234–244. https://doi.org/10.1016/j.jclepro.2013.03.022.

Samanta, A.K., Natasha Jayapal, Atul P. Kolte, S. Senani, Manpal Sridhar, Sukriti Mishra, C.S. Prasad, and K. Suresh. 2013. "Application of Pigeon Pea (*Cajanus Cajan*) Stalks as Raw Material for Xylooligosaccharides Production." *Applied Biochemistry and Biotechnology* 169 (8): 2392–2404. https://doi.org/10.1007/s12010-013-0151-0.

Santeramo, Fabio Gaetano, Domenico Carlucci, Biagia De Devitiis, Antonio Seccia, Antonio Di Stasi, Rosaria Viscecchia, and Giorgio Nardone. 2018. "Emerging Trends in European Food, Diets and Food Industry." *Food Research International* 104 (February): 39–47. https://doi.org/10.1016/j.foodres.2017.10.039.

Santhosh, Adhithya Sankar, and Mridul Umesh. 2024. "Valorization of Waste Chili Stalks (*Capsicum Annuum*) as a Sustainable Substrate for Cellulose Extraction: Insights Into Its Thermomechanical, Film Forming and Biodegradation Properties." *Biomass Conversion and Biorefinery* (February). https://doi.org/10.1007/s13399-024-05370-2.

Schanes, Karin, Karin Dobernig, and Burcu Gözet. 2018. "Food Waste Matters – a Systematic Review of Household Food Waste Practices and Their Policy Implications." *Journal of Cleaner Production* 182 (May): 978–991. https://doi.org/10.1016/j.jclepro.2018.02.030.

Schanes, Karin, Stefan Giljum, and Edgar G. Hertwich. 2016. "Low Carbon Lifestyles: A Framework to Structure Consumption Strategies and Options to Reduce Carbon Footprints." *Journal of Cleaner Production* 139 (December): 1033–1043. https://doi.org/10.1016/j.jclepro.2016.08.154.

Schuster, Monica, and Torero Máximo. 2016. "Reducing Food Loss and Waste." *Ideas.Repec.Org.* https://ideas.repec.org/h/fpr/ifpric/9780896295827-03.html.

Sharawy, Zaki Z., Ashraf Goda, and Mohamed S. Hassaan. 2016. "Partial or Total Replacement of Fish Meal by Solid State Fermented Soybean Meal with *Saccharomyces Cerevisiae* in Diets for Indian Prawn Shrimp, *Fenneropenaeus Indicus*, Postlarvae." *Animal Feed Science and Technology* 212 (February): 90–99. https://doi.org/10.1016/j.anifeedsci.2015.12.009.

Shulin, C. 2009. *Food Waste, Food Engineering, Vol. IV. Encyclopedia of Life Support Systems (EOLSS).* Pullman, WA, USA: Department of Biological Systems Engineering, Washington State University.

Sindhu, Raveendran, Parameswaran Binod, B. Ramkumar, Sunita Varjani, Ashok Pandey, and Edgard Gnansounou. 2020. "Waste to Wealth." In *Elsevier eBooks*, 181–197. https://doi.org/10.1016/b978-0-444-64321-6.00009-4.

Skrivervik, Eili. 2020. "Insects' Contribution to the Bioeconomy and the Reduction of Food Waste." *Heliyon* 6 (5): e03934. https://doi.org/10.1016/j.heliyon.2020.e03934.

Suantika, G., Putra, R. E., Hutami, R., & Rosmiati, M. 2017. "Application of compost produced by bioconversion of coffee husk by black soldier fly larvae *(Hermetia Illucens)* as solid fertilizer to lettuce *(Lactuca Sativa Var. Crispa)."* *Proceedings of the International Conference on Green Technology* 8(1): 20–26. http://conferences.uin-malang.ac.id/index.php/ICGT/article/view/376.

Surendra, K.C., Robert Olivier, Jeffery K. Tomberlin, Rajesh Jha, and Samir Kumar Khanal. 2016. "Bioconversion of Organic Wastes Into Biodiesel and Animal Feed via Insect Farming." *Renewable Energy* 98 (December): 197–202. https://doi.org/10.1016/j.renene.2016.03.022.

Tan, Inn Shi, and Keat Teong Lee. 2014. "Enzymatic Hydrolysis and Fermentation of Seaweed Solid Wastes for Bioethanol Production: An Optimization Study." *Energy* 78 (December): 53–62. https://doi.org/10.1016/j.energy.2014.04.080.

Thyberg, Krista L., and David J. Tonjes. 2016. "Drivers of Food Waste and Their Implications for Sustainable Policy Development." *Resources, Conservation and Recycling* 106 (January): 110–123. https://doi.org/10.1016/j.resconrec.2015.11.016.

To, Ming Ho, Huaimin Wang, Yahui Miao, Guneet Kaur, Sophie Roelants, and Carol Sze Ki Lin. 2023. "Optimal Preparation of Food Waste to Increase Its Utility for Sophorolipid Production by *Starmerella Bombicola*." *Bioresource Technology* 379 (July): 128993. https://doi.org/10.1016/j.biortech.2023.128993.

Torres-León, Cristian, Nathiely Ramírez-Guzmán, Liliana Londoño-Hernández, Gloria A. Martínez-Medina, Rene Diaz-Herrera, Víctor Navarro-Macias, Olga B. Alvarez-Pérez, et al. 2018. "Food Waste and Byproducts: An Opportunity to Minimize Malnutrition and Hunger in Developing Countries." *Frontiers in Sustainable Food Systems* 2 (September). https://doi.org/10.3389/fsufs.2018.00052.

Trabold, Thomas A., and Diana Rodríguez Alberto. 2020. "Valorization of Food Processing By-products via Biofuel Production." In *Elsevier eBooks*, 53–69. https://doi.org/10.1016/b978-0-12-818293-2.00004-5.

United Nations. 2015. *Transforming Our World: The 2030 Agenda for Sustainable Development*. https://sdgs.un.org/2030agenda.

Upadhyay, Shuchi, Rajeev Tiwari, Sanjay Kumar, Shraddha Manish Gupta, Vinod Kumar, Indra Rautela, Deepika Kohli, Bhupendra Singh Rawat, and Ravinder Kaushik. 2023. "Utilization of Food Waste for the Development of Composite Bread." *Sustainability* 15 (17): 13079. https://doi.org/10.3390/su151713079.

Verma, Monika, Linda De Vreede, T.J. Achterbosch, and Martine Rutten. 2020. "Consumers Discard a Lot More Food Than Widely Believed: Estimates of Global Food Waste Using an Energy Gap Approach and Affluence Elasticity of Food Waste." *PLOS ONE* 15 (2): e0228369. https://doi.org/10.1371/journal.pone.0228369.

Vilas-Boas, Ana A., Manuela Pintado, and Ana Oliveira. 2021. "Natural Bioactive Compounds from Food Waste: Toxicity and Safety Concerns." *Foods* 10 (7): 1564. https://doi.org/10.3390/foods10071564.

Wang, Xiang, S. Balamurugan, Si-Fen Liu, Manman Zhang, Wei-Dong Yang, Jie-Sheng Liu, Hongye Li, and Carol Sze Ki Lin. 2020. "Enhanced Polyunsaturated Fatty Acid Production Using Food Wastes and Biofuels Byproducts by an Evolved Strain of *Phaeodactylum Tricornutum*." *Bioresource Technology* 296 (January): 122351. https://doi.org/10.1016/j.biortech.2019.122351.

Wang, Xiang, Si-Fen Liu, Zhenyao Wang, Ting-Bin Hao, S. Balamurugan, Dawei Li, Yuhe He, Hongye Li, and Carol Sze Ki Lin. 2022. "A Waste Upcycling Loop: Two-factor Adaptive Evolution of Microalgae to Increase Polyunsaturated Fatty Acid Production Using Food Waste." *Journal of Cleaner Production* 331 (January): 130018. https://doi.org/10.1016/j.jclepro.2021.130018.

Wang, Yu-Shiang, and Matan Shelomi. 2017. "Review of Black Soldier Fly (*Hermetia Illucens*) as Animal Feed and Human Food." *Foods* 6 (10): 91. https://doi.org/10.3390/foods6100091.

Yang, Xiaofeng, Muzi Zhu, Xiongliang Huang, Carol Sze Ki Lin, Jufang Wang, and Shuang Li. 2015. "Valorisation of Mixed Bakery Waste in Non-sterilized Fermentation for L-lactic Acid Production by an Evolved *Thermoanaerobacterium* Sp. Strain." *Bioresource Technology* 198 (December): 47–54. https://doi.org/10.1016/j.biortech.2015.08.108.

Zeng, Yu, Tian Xie, Panyu Li, Banggao Jian, Xiang Li, Yi Xie, and Yongkui Zhang. 2018. "Enhanced Lipid Production and Nutrient Utilization of Food Waste Hydrolysate by Mixed Culture of Oleaginous Yeast *Rhodosporidium Toruloides* and Oleaginous Microalgae *Chlorella Vulgaris*." *Renewable Energy* 126 (October): 915–923. https://doi.org/10.1016/j.renene.2018.04.020.

Zeng, Y., Bian, D., Xie, Y., Jiang, X., Li, X., Li, P., Zhang, Y. and Xie, T. 2017. "Utilization of food waste hydrolysate for microbial lipid and protein production by *Rhodosporidium toruloides* Y2." *Journal of Chemical Technology and Biotechnology* 92: 666-673. https://doi.org/10.1002/jctb.5049.

Zheng, Longyu, Qing Li, Jibin Zhang, and Yu Zhou. 2012. "Double the Biodiesel Yield: Rearing Black Soldier Fly Larvae, *Hermetia Illucens*, on Solid Residual Fraction of Restaurant Waste After Grease Extraction for Biodiesel Production." *Renewable Energy* 41 (May): 75–79. https://doi.org/10.1016/j.renene.2011.10.004.

III

Applications

CHAPTER 9

Biomass Wastes for Adsorbent and Waste Treatment Applications

Gianluca Viscusi, Muhammad Muneeb Ahmad and Giuliana Gorrasi

9.1 INTRODUCTION

Industrialization and municipal and agricultural activities have contributed to the ever-increasing concentrations of pollutants in water bodies.[1] The dispersion of these chemical compounds is raising environmental pollution, which is a critical issue for human health[2] and has an undeniable potential to affect the lives of all living organisms negatively. For instance, the effluent dyes take a long time to degrade[3] and block sunlight in aquatic habitats, consequently reducing oxygen solubility. The azo group (–N–N–) found in most synthetic dyes is recalcitrant to bio and photodegradation, contributing significantly to the toxicity of these compounds. Besides, even heavy metals can potentially accumulate in water bodies. Apart from these, other pollutants released from pesticides, fertilizers, plasticizer oils, detergents, and pharmaceuticals are also of great concern since they are classified as carcinogenic.[4] To solve this problem, developing efficient technologies for removing these pollutants from aqueous effluents is an urgent issue. Different technologies such as adsorption, ion exchange, biological treatment, flocculation, precipitation, coagulation, reverse osmosis, electrolysis, ozonation, and ultraviolet have already been tested. Among them, the adsorption method has been recognized as the most suitable approach to remove contaminants from wastewater.[5]

DOI: 10.1201/9781003466833-12

The overuse of synthetic adsorbents, their high cost, and their impact on the environment are stimulating the development of green and bio-based materials.

Many materials, including agriculture and industry wastes, have been studied for their potential application as adsorbents, exploring alternative, sustainable, and cost-effective water treatment solutions. This has led to examining various biosorbents sourced from natural sources[6]. Generally, biosorption processes can significantly reduce capital, operational, and total treatment costs[7]. This chapter stresses the use of waste biomass and biomaterials as adsorbents. The potential drawbacks and future perspectives on using bio-based materials as adsorbents in wastewater treatment have also been discussed. Compared with wholly synthetic products, bio-wastes are renewable resources, and converting them into versatile materials for water remediation is a sustainable strategy to reduce massive waste, reduce the use of petrochemicals, and protect the environment.

9.2 BIOMASS WASTE FOR SUSTAINABLE INDUSTRIAL APPLICATIONS

The amount of biomass waste is rapidly increasing, which leads to numerous disposal problems and governance issues[8]. The disposal, utilization, and management issues of biomass wastes are a burgeoning challenge to cities, primarily in developing countries, due to the need for more feasible and efficient recycling of these biomass waste resources. The majority of biomass wastes are left in the field to naturally decompose, discarded in landfills, or incinerated in the open for cooking, drying, and charcoal production, which not only have low efficiency but also lead to severe environmental pollution like greenhouse gas emissions and air quality deterioration. The application of low-cost and waste materials as adsorbents, derived from biomass, of dissolved metal ions has been shown to provide economic solutions to this global problem.

Biosorption is a subcategory of the adsorption method, which usually uses materials that have little or no economic value and are often disposed of as waste. Biomass wastes commonly consist of forestry residues, agricultural, animal, industrial, municipal solid, food processing, and so on[9]. Due to their abundance, cheapness, and presence of functional groups, biomass-based adsorbents have been studied in wastewater treatment. Cheaper and effective adsorbents can be formed from abundant waste materials (or products) from industrial and agricultural activities.

Generally, an adsorbent that requires little processing is abundant in nature or is a by-product or waste material from another industry is called a "low-cost" adsorbent[10]. In particular, agricultural waste can be exploited as an adsorbent because of its short regeneration cycle, effectiveness, availability, stability, environmental friendliness, and cheapness[11]. The increasing number of publications on the adsorption using biomass-derived adsorbents proves a recent increase in interest in the use of low-cost adsorbents in wastewater treatment. In recent years, many plant materials have been tested as low-cost adsorbents for various xenobiotics, such as organic dyes, heavy metals, radionuclides or endocrine disruptors.[12]. Pollutants, such as heavy metals, pesticides, pharmaceuticals, and organic and inorganic impurities, have become widespread in different water supplies, presenting substantial health and environmental hazards.[13]

The main advantage of such adsorbents is that they do not need an expensive regeneration step since they can be discarded after use because of their low cost[14]. Moreover, the cost of these biomaterials is negligible compared with synthetic adsorbents. The attractive features of this method are that they are abundant, cheap, reusable, and eco-friendly, and there is no formation of harmful by-products as the pollutants are adsorbed into the biomass. Also, the materials, especially those that originated from living things, have complex compositions, making it possible for many mechanisms, such as ion exchange, complexation, coordination, and chelation, to operate in an adsorption process.

9.2.1 Agricultural and Forestry Wastes

The global production of wood biomass wastes is 4.6 Gt every year, with 20% being production loss[9]. These wastes can be exploited as adsorbents for water remediation applications.[15] Moreover, they are locally available and cost effective. Agricultural wastes from forest industries such as sawdust, sugarcane bagasse, husk, straw, fibers, and wood shavings have been used as adsorbents. Lignocellulose-rich agricultural waste biomass has received prodigious scientific consideration for waste management and decontamination of environmental contaminants because of their profusion, sustainable features, and easy availability worldwide, particularly in developing regions. Due to their physic-chemical properties, these materials are primarily available and may have potential as adsorbents. Sometimes, they are impregnated or coated with metal oxides or other materials to improve their adsorption performance. Their adsorption performance primarily depends on the nature, treatment, and activation mode. This biomass-based material can be easily functionalized, making it suitable for adsorbing metal ions.[16] Cost-effectiveness, availability, and adsorptive properties are the main criteria for choosing an adsorbent to remove organic compounds. Considering these criteria, many researchers have investigated the adsorptive properties of unconventional adsorbents. Now, a discussion will be presented regarding the use of some adsorbents.

9.2.1.1 Natural Fibers

In the last decades, the use of natural fibers in different industrial applications has been rising.[17, 18] The natural fibers, composed of cellulose, pectin, hemicellulose, lignin, waxes, and water-soluble substances, are considered sustainable, cheap, lightweight, renewable, and biodegradable materials[19]. Using raw and treated natural fibers signifies a novel strategy to remove contaminants from wastewater.[20–22] Despite the various advantages, adsorbents based on natural fibers lack high adsorption capacities compared to typical adsorbents (e.g., carbon or zeolite). Therefore, enhancement of the adsorption performances can be achieved through different chemical/physical modifications to confer novel functionalities.[23]

Selambakkannu et al. proposed banana fibers, mixed with $NaClO_2$ and modified with glycidyl methacrylate and trimethylamine to remove acid blue (AB 80) and acid red (AR 86). The maximum adsorption capacities are 292 mg/g and 319 mg/g for AB80 and AR86.[24]

Sajab et al. produced citric acid and polyethyleneimine-modified oil palm empty fruit bunch fibers for removing methylene blue (q_{max} = 103.1 mg/g) and phenol red (q_{max} = 158.7 mg/g).[25]

Cao et al. proposed an oil adsorbent system based on Calotropis gigantea fiber modified with Ni and Cu metal nanoparticles. The oil-absorbing capacities of different compounds were investigated (Figure 9.1).[26]

Du et al. proposed carboxyl-modified jute fiber for removing Pb(II), Cd(II) and Cu(II) ions from aqueous solutions. More carboxylic groups were introduced via the formation of ester functions following microwave heating in the presence of pyromellitic dianhydride (Figure 9.2).

The maximum adsorption capacities were 157.21, 88.98 and 43.98 mg/g for Pb(II), Cd(II), and Cu(II), respectively.[27]

Shin et al. modified juniper bark fiber with lanthanum to develop an inorganic/organic hybrid adsorbent. Sorption capacities of orthophosphates as

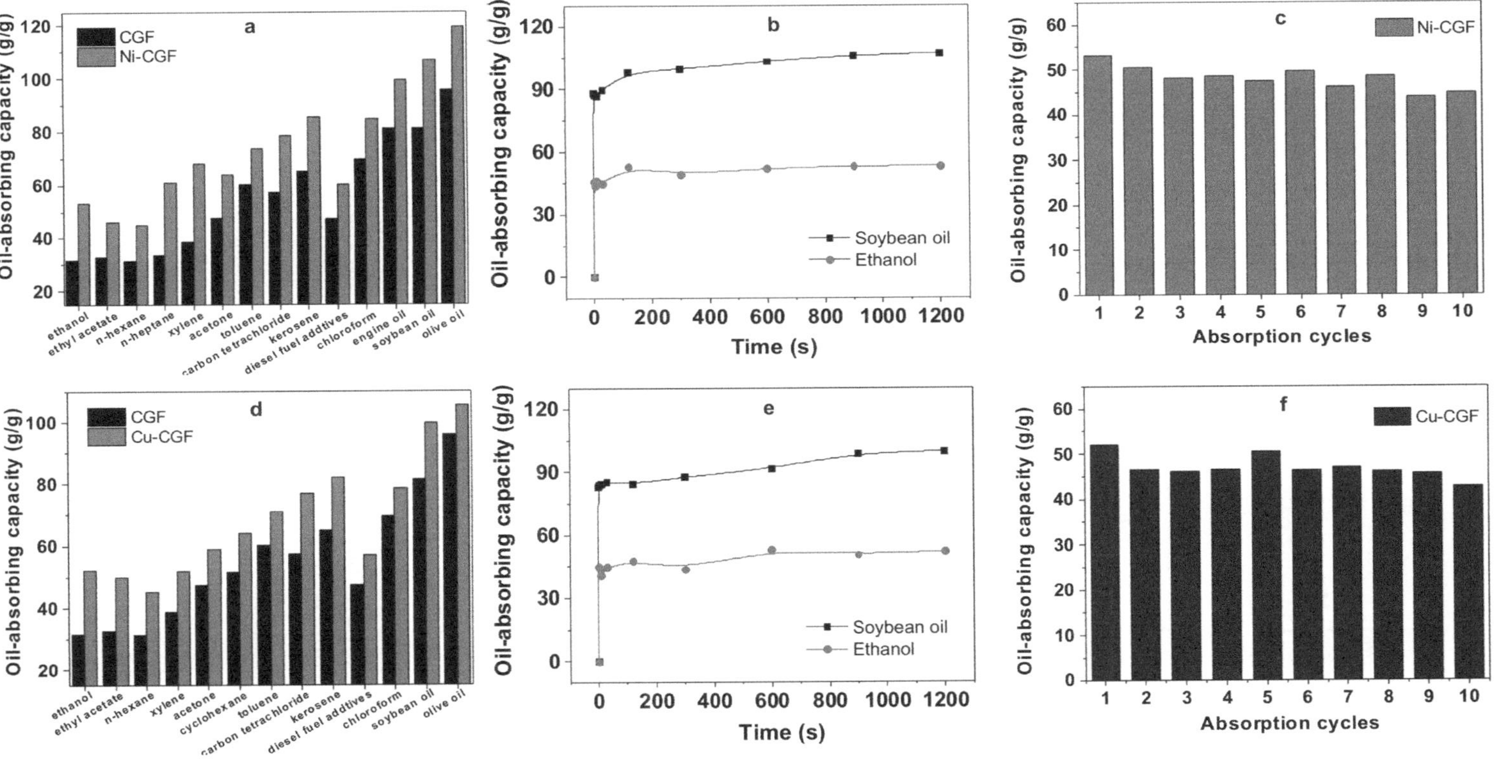

FIGURE 9.1 Oil-absorption of Ni-CGF (a, c) and Cu-CGF (d, f). (a, d) Saturated absorption capacities in various oils. (b, e) Oil-absorbing capacities over time for soybean oil and ethanol, and (c, f) reusability using ethanol (from ref. [26]).

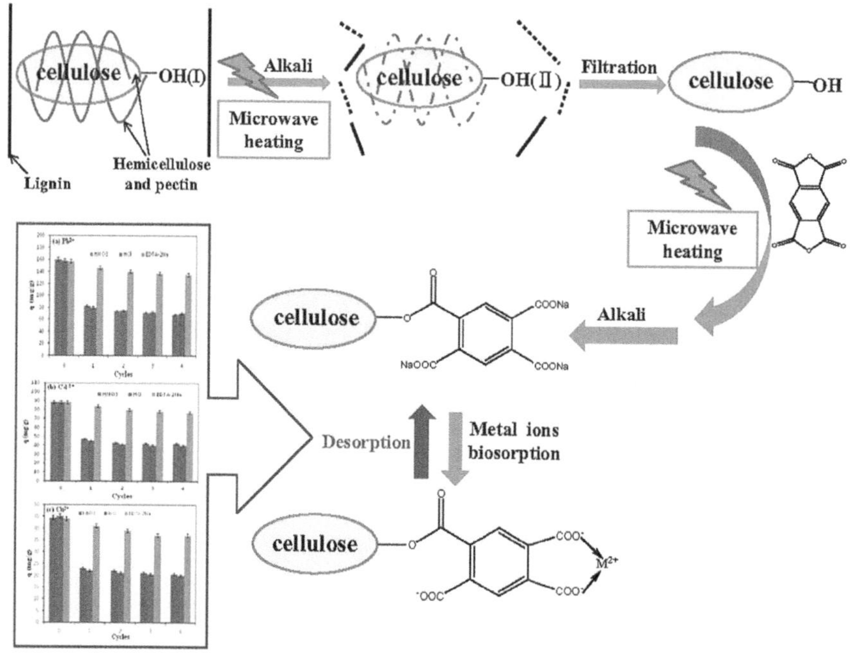

FIGURE 9.2   Scheme of jute fiber modification. [27]

a function of initial lanthanum loading varied from 0.211 to 0.351 mmol of P g^{-1}.[28]

Similar systems have been proposed in Table 9.1.

9.2.1.2 Crop Straw

Crop straw is the dry stalk or stem produced in the field after removing the grain and chaff. Nowadays, China is one of the countries with the most abundant straw resources in the world. The annual total output of straw resources in China is 600–700 million tons, and the utilization rate is 60–70%[51]. Burning is the standard and cheapest method to eliminate crop straw for preparing land after harvesting crops, leading to deteriorated air quality and other environmental problems[52]. However, crop straw is promising lignocellulosic raw materials that can be exploited to remove pollutants from water effluents. First, the structure of crop straw, rough surface, porosity, and surface area, are favorable for the physical adsorption of contaminants[53]. Second, the straw body generally comprises macromolecules such as lignin, cellulose, hemicellulose, and protein, which contain functional groups with strong coordination ability[54]. Moreover,

TABLE 9.1   Maximum Adsorption Capacity (q_{max}) of Different Natural Fiber-Based Adsorbents

Adsorbent	Pollutant	q_{max}, mg/g	Reference
Acrylic acid-modified Ficus carica fibers	Methylene blue	75.87	[29]
Luffa cylindrica fibers	Methylene blue	49	[30]
Methyl acrylate and acrylic acid graft copolymerization of Luffa cylindrica fibers	Congo red	19.24	[31]
Polyethylenimine-modified oil palm empty fruit bunch fibers	Phenol red	158.7	[25]
Citric acid-modified kenaf core fibers	Methylene blue	131.6	[32]
Glycidyl methacrylate and trimethylamine-modified banana fibers	Acid red 86	319.35	[24]
	Acid Blue 80	292.28	[24]
Ethylenediamine-modified Populus tremula fibers	Acid Blue 25	67	[33]
Corn fibers	Alcian blue	159	[34]
	Methylene blue	70	[34]
	Neutral red	50	[34]
	Brilliant blue	35	[34]
Sugarcane fiber	Crystal violet	10.44	[35]
Jute fibers	Congo red	8.12	[36]
Polyphenol tannin-modified jute fibers	Congo red	27.12	[23]
Beta-cyclodextrin and amino-terminated hyperbranched polymer cotton fibers	Methylene blue	95.8	[37]
	Congo red	236.6	[37]
Flax fibers	Basic yellow 37	512	[38]
Polyacrylonitrile-coated kapok fibers	Methyl orange	34.72	[39]
Oil palm trunk fibers	Malachite green	149.35	[40]
Hemp fiberboards	Vegetable oil	8.64	[41]
Oil palm empty fruit bunch	Crude oil	6.48	[42]
Kapok fibers	Motor oil	25.10	[43]
	Diesel oil	23.72	[43]
NaOH-modified kapok fibers	Pb^{2+}	23.40	[44]
NAClO$_2$ modified kapok fibers	Pb^{2+}	34.63	[44]
Lyocell fiber modified through xanthation	Pb^{2+}	531.29	[45]
	Cd^{2+}	505.64	[45]
	Cu^{2+}	123.08	[45]
Butyl acrylate grafting banana fiber cellulose	Pb^{2+}	209.09	[46]
Imidazole-functionalized polymer graft banana fiber	Pb^{2+}	276.24	[41]
	Zn^{2+}	101.52	[41]
	Cu^{2+}	81.97	[41]

(Continued)

TABLE 9.1 *(Continued)* Maximum Adsorption Capacity (q_{max}) of Different Natural Fiber-Based Adsorbents

Adsorbent	Pollutant	q_{max}, mg/g	Reference
Aluminum-impregnated coconut fiber ash	F⁻	3.19	[47]
N-(3-chloro-2-hydroxypropyl) trimethylammonium chloride modified kenaf fibers	F⁻	13.98	[48]
Aluminum-impregnated coconut fiber ash	F⁻	3.19	[47]
Luffa cylindrica fibers	Methyl parathion	54.88	[49]
	Coumaphos	62	[49]
PEI modified wool	Pirimiphos-methyl	625	[50]
	Monocrotophos	500	[50]
Luffa cylindrica fibers	Methyl parathion	54.88	[49]

it is known that crop straw has a strong chemical adsorption capacity towards heavy metal ions, which have electronic structures capable of forming a coordinate bond[55]. Therefore, the development and utilization of straw-based adsorbents have important practical significance, and they are the most promising way to make full use of this abundant bioresource. [56] However, the adsorption capacity of raw straw could be more efficient. Usually, chemical treatment, particularly acid modification, is one of the most universal methods to prepare straw owning the typical functional groups with high adsorption capacity[57].

Chen et al. investigated the adsorption properties of wheat straw (WS) and corn straw (CS) for Cr(VI) and Cr(III) in solution. The saturated adsorption capacity of WS for Cr(VI) and Cr(III) can reach 125.6 and 68.9 mg g⁻¹, and that of CS for Cr(VI) and Cr(III) can reach 87.4 and 62.3 mg g⁻¹, respectively [58]. Ge et al. designed citric acid functionalized magnetic graphene oxide-coated corn straw to eliminate methylene blue. The equilibrium adsorption capacity was 315.5 mg g⁻¹. The electrostatic incorporation, as well as hydrophobic interactions, determined the favorable adsorption property [59]. Feng et al. modified swede rape straw (Brassica napus L.) with tartaric acid to remove methylene blue from the aqueous solution. The maximum MB adsorption capacity was 246.4 mg g⁻¹ [57]. Baldikova et al. fabricated citric acid-NaOH-modified rye straw as a biosorbent for two organic water-soluble dyes belonging to different dye classes, namely acridine orange and methyl green. The highest values of maximum adsorption capacities were 208.3 mg/g for acridine orange and

384.6 mg/g for methyl green[12]. Pirbazari et al. designed NaOH-treated wheat straw from agriculture biomass impregnated with Fe_3O_4 magnetic nanoparticles for the removal of MB. Langmuir adsorption capacity was found to be 1,374.6 mg g^{-1} [60].

Similar results concern the use of magnetic wheat straw for arsenic adsorption[61], rice straw/Fe_3O_4 nanocomposites for adsorption of Pb(II) and Cu(II) [62], amine-crosslinked wheat straw for phosphate and Cr(VI) removal[63], wheat straw materials modified by carboxymethylation for adsorbing methylene blue [64], rice straw converted into a strong primary anion exchanger by reaction with NaOH, epichlorohydrin and trimethylamine for the removal of sulphate[65], Fe-modified steam-exploded wheat straw for the removal of Cr(VI)[66], carboxymethyl straw employed to remove MB[67], rice straw for Cr(VI) removal[15], removal of Cr(VI) utilizing Teff straw[68], adsorption of Cr(VI) on carbonized wheat straw and barley straw[69].

9.2.1.3 Husk and Pod

Husks and pods, as agricultural waste materials, possess distinct physicochemical characteristics that render them highly appropriate for water treatment. Rice husk consists of cellulose (25–35%), hemicellulose (18–21%), lignin (26–31%) and silica (15–17%). Apart from these structured components, some other nonstructural components, such as pectin, waxes, and inorganic salts, are also included.[70] Additional functional groups that can interact effectively with water contaminants via diverse adsorption mechanisms (such as H bonding, π–π interaction, electrostatic interaction, and ion exchange).[71] Their porous framework and expansive surface area amplify their adsorption performance, allowing them to trap and immobilize a broad spectrum of pollutants from aqueous solutions.[72] The adsorbents on the base of rice husks have the advantages of ecological safeness, origin from a comprehensive source of raw materials, high mechanical strength, and low costs. However, many naturally available adsorbents have low metal removal and slow process kinetics. Therefore, it is necessary to develop innovative, inexpensive adsorbents with a good affinity for metal ions, and surface functionalization technology has proven very effective. Rice husk (RH) is mainly composed of organic substances (75%) and silica (15%) and is generated in large quantities in the milling industry; it has turned out to be a severe technological and ecological waste. So, using RH as an adsorbent for removing contaminants from aquatic environments could aid in the beneficial utilization of agricultural wastes. It is known for its ecological safeness, porous structure, high hydrophobicity, high mechanical strength, and low cost.[73]

Consequently, substantial quantities of pod and husk residues are generated during the processing of these crops[74]. Unfortunately, much of this agricultural waste is often discarded or left to decompose, contributing to environmental pollution and waste management challenges. Proper waste management practices can facilitate the collection and utilization of these residues, reducing their ecological impact and transforming them into valuable resources for water contaminant remediation. The successful utilization of pods and husks as natural adsorbents in water contaminant remediation often requires pre-treatment and modification to enhance their adsorption efficiency, stability, and versatility. The chemical modification and functionalization of these adsorbent materials have the potential to significantly improve their adsorption capacity and selectivity. Various techniques have been employed to achieve these modifications, including acid treatment, alkaline treatment, composite formation, and functionalization with polymers. Treatment with sulfuric acid or hydrochloric acid can remove impurities, increase the surface area, and introduce new functional groups on the material's surface.[75] This enhances its affinity for specific contaminants. Also, alkali treatment using solutions like sodium hydroxide can remove lignin and hemicellulose, improving the materials' porosity and surface area.[76] On the other hand, coating bean pods and husk materials with polymers can introduce additional functional groups and enhance their stability in aqueous environments.[77]

Organomultiphosphonated rice husks were successfully employed to adsorb heavy metal ions, especially Au(III). The introduction of organotriphosphonic acid groups greatly enhanced the amount of gold(III) uptake. Under the optimal conditions, the maximum value of the Au(III) adsorption capacity could reach 3.25 ± 0.07 mmol/g[73].

Wheat husks (Fagopyrum esculentum) were modified by treatment with sulfuric acid to remove the 2,4-dichlorophenoxyacetic acid (2,4-D) pesticide from aqueous solutions. A Q_{max} of 161.1 mg g^{-1} at 298 K was recorded. The main interactions are hydrogen or halogen bonds, van der Waals forces, or π–π interaction, as depicted in Figure 9.3.

Examples of the use of husk and pod are reported in Table 9.2.

9.2.1.4 Sawdust

Sawdust is a possible material because it is widely produced as a solid waste at sawmills. It is a waste by-product of the timber industry that is either

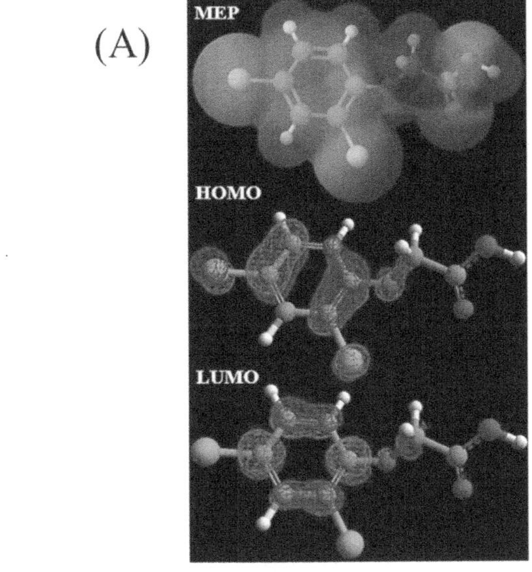

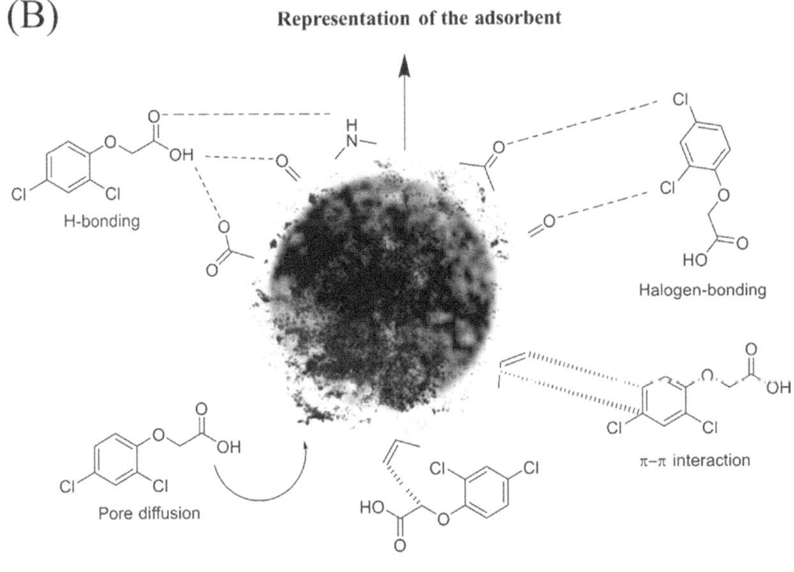

FIGURE 9.3 Mechanism of 2,4-D interaction with the adsorbent. Reproduced from [78].

TABLE 9.2   Use of Pod and Husk for Water Remediation Applications

Adsorbent	Modification	Adsorbate	Q_{max} mg/g	Reference
Bean husk	raw	Zn(II)	0.477	[79]
	raw	Ni(II)	16.89	
Bean husk	EDTA	Zn(II)	0.409	
	EDTA	Ni(II)	0.970	
Bean husk	Oxalic acid	Zn(II)	8.34	
	Oxalic acid	Ni(II)	1.29	
Bean pod	–	Crystal violet	0.38	[80]
Bean husk	Ortho-phosphoric acid	Ibuprofen	50	[81]
Bean pod	K_2CO_3	Naphtalene	300	[82]
Bean husk	HNO_3	Cd(II)	180	[83]
Bean pod	Urea	Indigo Carmine	7.42	[72]
Bean husk	Citric acid	Cr(III)	0.98	[84]
Rice husk	K_2HPO_4	Cd(II)	2000	[85]
Rice husk	–	Atrazine	5.87	[86]
		Alachlor	6.43	[86]
Rice husk treated with NaOH.		Malachite green	17.98	[87]
Rice husk		Methylene blue	40.583	[88]
		Cu(II)	11.4	
Rice husk	Tartaric acid	Cu(II)	29	[89]
		Pb(II)	108	
Green tomato husk	Formaldehyde	Mn(II)	15.22	[90]
		Fe(III)	19.83	
Peanut husk	Formalin	Pb(II)	29.14	[91]
		Cr(III)	7.67	

used as cooking fuel or a packing material[92]. Sawdust contains primarily lignin and cellulose. The good results have stimulated interest in the use of sawdust as an adsorbent. Agricultural byproducts like sawdust are generally used as a fuel. However, the bulkiness of sawdust and its tendency towards incomplete combustion are its drawbacks as an adequate fuel. Adsorption onto low-cost adsorbents offers an inexpensive option for removing different pollutants[93].

Some examples concerning the use of sawdust for water remediation uses are reported in Table 9.3.

9.2.1.5 Tea Waste

Tea waste is a cost-effective adsorbent.[101] It is an oxygen-demanding pollutant. Also, biodegradation takes a long time. When the tea waste used as

TABLE 9.3 Use of Sawdust-Based Adsorbent for Water Remediation

Adsorbent	Pollutant	Q_{max} mg/g	Reference
Beech sawdust	Cr(VI)	16.13	[94]
Meranti sawdust	Cu(II)	32.051	[95]
	Cr(III)	37.878	
	Ni(II)	35.971	
	Pb(II)	34.246	
HCl-modified oak sawdust	Cu(II)	3.60	[96]
	Ni(II)	3.37	
	Cr(VI)	1.74	
Sawdust from the poplar tree	Pb(II)	21.05	[91]
	Cr(III)	5.52	
	Cu(II)	6.585	
Juniperus procera sawdust	Cr(VI)	16.03	[97]
Mansonia wood sawdust	Methyl violet	24.6	[98]
Cedar sawdust	Methylene blue	142.36	[99]
Sawdust from Malaysian teak wood	Reactive blue	111.11	[100]

adsorbent becomes saturated, it is incinerated. The waste tea ash obtained by incineration is not a pollutant. Instead, it could be used as an adsorbent. Tea waste consists of cellulose, lignin and structural proteins, and tannin with specific functional groups that are effective in forming physicochemical reactions with heavy metals and other contaminants and, thus, can be used to remove harmful substances from solutions and wastewater.[102] The presence of carboxylate, aromatic, phenolic, hydroxyl, and oxyl groups may be responsible for its ion-exchange behavior.

Tea waste has already been used to remove methylene blue. Adsorption equilibrium was reached within five hours with a q_{max} equal to 85.16 mg/g.[101]

Tea waste was used to remove hexavalent chromium from real tannery wastewater in a batch system. Almost 97% removal of Cr(VI) was achieved under the optimized conditions (viz. adsorbent dosages of 6 g/L, solution pH of 3.9, contact time of 240 min and temperature of 303 K). At the optimum conditions, the adsorption process with TW as an adsorbent also caused the removal of 74.8% COD, 55% TDS, and 71% TS from tannery effluent.[103]

Alkali-treated tea waste (ATTW) was investigated as a novel alternative cost-effective natural adsorbent to decontaminate the pollutant of chromium (Cr(VI)). The maximum adsorption ability of ATTW was 158.73 mg/g. These results demonstrate that Cr(VI) adsorption was a complex process involving a physicochemical spontaneous monolayer and multiple

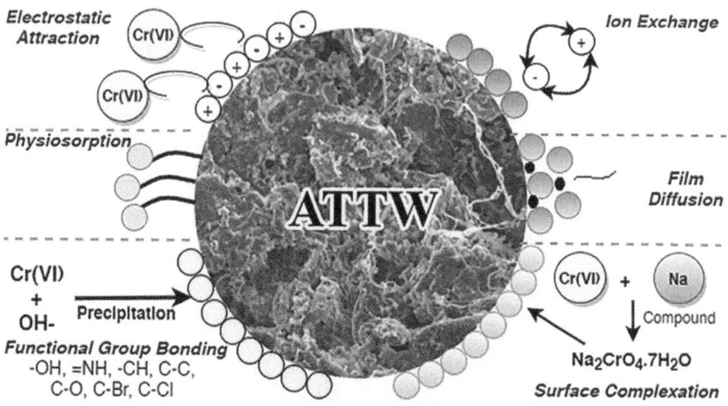

FIGURE 9.4   Adsorption mechanisms of Cr(VI) by ATTW.

rate-limiting states, which was predominantly chemical or chemisorption due to ion-exchange or electrostatic attraction/surface complexation formation, film diffusion, and intra-particle diffusion (Figure 9.4). Furthermore, the substrates exhibited an excellent regeneration capacity upon using 0.10 M HNO_3 as effluent. The results conclude that ATTW could be a promising cheap bio-adsorbent for Cr(VI) remediation[16].

Other examples are reported in Table 9.4.

9.2.1.6 Weed

The algal cell wall comprises a fiber-like structure and an amorphous embedding matrix of various polysaccharides.[118] Several functional groups on the algal cell surface can attract and sequester heavy metal ions such as amino, amido, sulfate and carboxyl.[119] Biosorption mechanisms of weed include ionic interactions and the formation of complexes between metal ions and the functional groups.

A new biosorbent-based alginate encapsulated with Myriophyllum spicatum was investigated for lead ion removal. FT-IR analysis demonstrated that the sequestration mechanism of lead ions included ion exchange and lead complexation with the adsorbent's carboxyl, carbonyl, and hydroxyl groups. The q_{max} was shown to be 238.4 mg/g.[120]

Ajmal et al. studied the efficiency of parthenium weed to remove Cd(II). The maximum adsorption efficiency of Cd(II) ions was 99.7%.[14]

Saeed investigated the use of Cicer arientinum (chickpea var. black gram) to remove heavy metals.

TABLE 9.4   Use of Tea Waste-Derived Adsorbents for Water Remediation

Adsorbent	Pollutant	Q_{max}, mg/g	Reference
Tea waste	Cu(II)	48	[104]
	Pb(II)	65	
	Congo red	43.48	[105]
Waste black tea powder	Methylene blue	302.63	[106]
Tea stem	Methylene blue	103.09	[107]
Tea waste	Cr(VI)	199.523	[108]
	Phenol	9.487	
Brewed tea waste	Pb(II)	1.197	[109]
	Zn(II)	1.457	
	Ni(II)	1.163	
	Cd(II)	2.468	
Green tea waste modified by Ca(OH)$_2$	As(III)	0.4212	[110]
	Ni(II)	0.3116	
Tea (*Camellia sinensis*) and ginger waste	Cr(VI)	7.29	[111]
Phosphoric-activated tea waste	p-nitrophenol	142.85	[112]
Tea waste	Reactive Blue 25 dye	28.99	[113]
Black tea leaves	17 β-estradiol	3.46	[114]
	17 α-ethinylestradiol	2.44	
	Bisphenol A	18.35	
Peppermint tea	Malachite green	69.7	[115]
Black tea	Nichel	90.91	[116]
H$_3$PO$_4$-modified tea leaves	Phenol	63.13	[117]

The maximum adsorption capacities were 49.97, 39.99, 33.81, 25.73, and 19.56 mg/g biomass for Pb, Cd, Zn, Cu, and Ni, respectively[121].

Arora et al. studied the performances of Chenopodium album ash (wildly growing weed) as an effective adsorbent for removing crystal violet, showing an adsorption capacity of 9.42 mg/g.[122]

Garvasis et al. synthesized superparamagnetic magnetite NPs using Siam weed flower extract. The maximum adsorption capacity recorded was 315.43 mg g^{-1} for Pb(II) ions[123].

The study of the heavy metal sorption capacity of alginate extracted from Laminaria digitata in the form of calcium beads was carried out. Pronounced differences between sorption capacities of the alginate beads concerning the different metals examined followed the order: $Pb^{2+} > Cu^{2+} > Cd^{2+}$ [124]

The potential use of the immobilized microalgae (in Ca-alginate) of Chlamydomonas reinhardtii to remove Hg(II), Cd(II) and Pb(II) ions was evaluated. The experimental results show that algal biomass immobilized alginate beads yielded high biosorption capacity for all the tested metal ions showing a maximum adsorption capacity of 116.8, 88.6, and 384.4 mg/g for Hg(II), Cd(II), and Pb(II), respectively[125].

Keshinkan et al. studied the adsorption features of Myriophyllum spicatum and Ceratophyllum demersum for lead, zinc, and copper removal. The maximum adsorption capacities (q_{max}) achieved with M. spicatum were 10.37 mg/g for Cu^{2+}, 15.59 mg/g for Zn^{2+} and 46.49 mg/g for Pb^{2+} while with C. demersum they were 6.17 mg/g for Cu^{2+}, 13.98 mg/g for Zn^{2+} and 44.8 mg/g for Pb^{2+}[126].

Pb(II) biosorption by Myriophyllum spicatum and its compost was investigated. Lead binding capacities for M. spicatum and its compost were 0.234 mmol g^{-1} and 0.287 mmol g^{-1} at pH 5.0, respectively.[127]

The dead dried alga, Chlorella vulgaris, was used for metal ion sequestering. Introducing mixed ethanol/water (50% v/v) metal ion solutions in batch systems enhanced the metal uptake of the exhausted biomass by 90% for iron, 40% for tin, and only 14% for cadmium[128].

Other studies concern the use of the residue from the extraction of agar-agar from Gracilaria debilis for linear alkyl benzene sulphonate remova [129]; pre-treated biomass of Padina sp. for Cu(II)[130]; Caulerpa racemosa var. cylindracea for the removal of malachite green[131]; biosorption of Acid Black 1 onto Nizamuddina zanardinii[132]; removal of methylene blue by Caulerpa racemosa var. cylindracea[133]; Sargassum muticum[134]; Parthenium hysterophorus[135]; Sargassum muticum[136]; lead, zinc, and copper adsorption using Ceratophyllum demersum (Coontail or hornwort)[137]; and Laminaria japonica by crosslinking with epichlorohydrin for Pb(II), Cd(II), and Fe(III) removal[138].

9.2.1.7 Shell

Shell has little economic value and is unsuitable to be used as an animal feedstock due to its toughness. For example, a Walnut shell represents an agro waste resource generated at high volume as the kernel is the only edible walnut portion. Untreated WS contained tannin, which possessed single-bond OH groups that can interact with metals[139]. Some examples are reported in Table 9.5.

TABLE 9.5 Use of Shell Biomass for Water Remediation

Adsorbent	Pollutant	Q_{max} mg/g	Reference
Walnut shell	Malachite green	90.8	[139]
Walnut shell	Cr(VI)	8.01	[140]
Hazelnut shell		8.28	
Almond		3.040	
Walnut shell	Pb(II)	32	[141]
	Cd(II)	11.6	
Bengal gram fruit shell	Acid Blue25	29.41	[142]
Peach shell	Methylene blue	183.6	[143]
Walnut shell	Cr(VI)	3.5	[144]
Hazelnut shell		3.5	
Almond shell		3.2	
Walnut shell	Pb(II)	9.912	[145]
Sunflower shell	Cd(II)	83.3	[146]
Potato		90.0	
Canola		71.4	
Walnut		76.9	
Sulfuric- and phosphoric-acid-activated Strychnine tree fruit shells	Cr(VI)	142.85	[147]
Fluted pumpkin (Telfairia occidentalis) seed shell	Pb(II)	14.286	[148]
Wheat shell	Methylene blue	21.50	[149]
Cranberry (Cornus mas) kernel shell	Cr(VI)	6.81	[150]
Rosehip (Rosa canina) seed shell	Cr(VI)	15.17	[150]
Banana peel	Cr(VI)	10.42	[150]
Horse chestnut shell	Cr(VI)	142.85	[151]

9.2.1.8 Peel

In addition to the agricultural and forest wastes mentioned earlier, many other wastes in nature, such as rattan, bagasse, and peels, have tremendous prospects for development. Fruit peel waste (FPW) has the highest potential as an active adsorbent for wastewater treatment. The main environmental problem is the generation of high amounts of agroindustrial wastes that are directly discharged into landfills without any added use-value[152]. For example, orange peel, as a typical agricultural waste, is mainly composed of cellulose, hemicellulose, pectin, and lignin components with multiple O/N-containing functional groups and pyranose/benzene rings.

Because of this, raw orange peel has been used as a candidate adsorbent for the potential removal of heavy metals[153].

Orange peel (OP) without any treatment was applied to effectively eliminate methylene blue (MB) and cadmium ions (Cd^{2+}). Batch experiments showed that in monocomponent systems, the maximum adsorption capacities were 0.7824 mmol g^{-1} for MB and 0.2884 mmol g^{-1} for Cd^{2+}. In contrast, in multicomponent systems (Cd^{2+} and MB), both contaminants competed for the adsorption sites on OP. The low cost of this material and its capacity for the individual or simultaneous removal of Cd^{2+} and MB in aqueous solutions make it a potential adsorbent for polluted water treatment processes.[154]. A further example concerns the removal of Cr(VI) from aqueous solutions investigated by acrylonitrile-grafted banana peels. Cr(VI) adsorption onto grafted banana peels was recorded to be 96%[155].

Some further examples are reported in Table 9.6.

9.2.1.9 Seed

Fruit seeds can even be used to remove a broad class of pollutants. These low-cost sorbents have generated high solid waste, which poses disposal problems. Therefore, there is a need to explore their sorptive capacity.

Mango seed kernel particles were used for removing methylene blue, showing a maximum adsorption capacity of 142.857 mg/g at 303 K.[176]

Other research focused on the use of avocado kernel seeds for Cr(VI) removal[97], dried Moringa stenopetela seed as bio-adsorbent for the removal of Cr(III)[177], removal of two chlorophenols (2-chlorophenol and 2,4-dichlorophenol) from water using surface modified mango seed waste[178], guava seeds for acid dyes removal[179], and avocado kernel seeds for basic blue 41 removal[180].

9.2.2 Carbon-Based Adsorbents

In addition, they can be used as precursors or bases to make functional carbon-based materials through pyrolysis and hydrothermal, physical, and chemical activations, which have attracted much attention in the last five years. Till now, various forms of carbon-based materials, including biochar, activated carbon (AC), graphitic carbon, and so on, have been developed. Activated carbon is widely used for wastewater treatment, especially for incredibly porous carbon from agricultural waste. Recently, several materials from agricultural waste, such as coconut coir, rice husk, wheat straw, sawdust, nutshells, etc., have been used for the preparation of AC and as adsorbents for removing pollutants from wastewater systems.

TABLE 9.6 Use of Peels for Water Remediation Applications

Adsorbent	Pollutant	Q_{max} mg/g	Reference
Banana peel	Rhodamine B	20.6	[156]
Orange peel	Rhodamine B	14.3	
Broad bean peel	Methylene blue	192.7	[157]
Banana peel	Cu(II)	27.78	[158]
Banana peel	Co(II)	9.02	[159]
	Ni(II)	8.91	
Pea peels	Cr(VI)	4.33	[111]
Banana peel	Cr(VI)	10	[111]
Papaya peels	Cr(VI)	7.16	[97]
Banana peel	Methyl orange	21	[160]
	Methylene blue	20.8	
	Methyl violet	12.2	
	Congo red	18.2	
	Rhodamine B	20.6	
	Amino black 10B	6.5	
Orange peel	Methyl orange	20.5	[160]
	Methylene blue	18.6	
	Methyl violet	11.5	
	Congo red	14.0	
	Rhodamine B	14.3	
	Amino black 10B	7.9	
Orange peel-modified phosphoric acid	Methylene blue	307.63 mg/g	[161]
K+ type orange peel	Cu(II)	59.77	[162]
Modified orange peel powder	Congo red	163	[163]
Sulfuric acid-treated orange peel	Methylene blue	50	[164]
Sulfuric acid crosslinked mangosteen peel	Cr(VI)	2 (mol/kg)	[165]
Orange peel	Reactive Gray BF-2R	11.4	[166]
Anchote peel	Methyl orange	103.03	[167]
Orange peel immobilized on calcium alginate	Thorium	29.58	[167]
Dried orange peel	Pb(II)	32.507	[168]
Orange peels by a physical-chemical process	Cu(II)	163	[169]
Orange peel adsorbent	Remazol brilliant blue	11.62	[170]
Artocarpus heterophyllus peel	Cr(VI)	64.47	[171]
Sulfured orange peel	Pb(II)	160	[172]
Cross-linked orange peel.	Cu(II)	289	[173]
Cempedak durian peel	Methyl violet 2B	238.5	
Orange peel cellulose	Cr(VI)	4.90	[174]
	Pb(II)	50.10	
	Cd(II)	29	
Pomegranate peel powder	Pb(II)	707.424	[175]

Activated carbons have shown great potential for dye removal due to properties such as large surface area, microporous structure, high adsorption capacity, as well as the presence of a broad spectrum of surface functional groups, such as carboxyls, phenols, lactones, aldehydes, ketones, quinones, hydroquinones, and anhydrides (Chingombe et al., 2005). The electrical charge of the surface groups may also enhance or decrease the adsorption of the target molecules on the carbon surface[181]. The high cost of activated carbon promotes the search for cheap materials mainly derived from biological origin. It has been proved that lignocellulosic biomasses are attractive resources for the preparation of carbonaceous materials implemented in adsorption processes. The eco-friendly nature of lignocellulosic biomasses, wide availability, and low cost are the main advantages of these resources, which makes them a suitable precursor for activated carbon preparation. Various lignocellulosic biomasses have been used naturally or in the form of activated carbon[182].

For example, many studies have used different types of activated carbon to remove Cr(VI) by adsorption[183]. However, most commercial activated carbons are usually expensive due to their high-cost sources and regeneration and reactivation procedures, which restrict their extensive application. In recent years, growing interest in the production of low-cost and highly efficient activated carbon has been focused on relatively cheap and effective raw materials with high carbon content and low inorganic content for agricultural residues, biomass, and various solid substances. Generally, biosorptive processes could reduce capital costs by 20%, operational costs by 36%, and total treatment costs by 28%, compared with the conventional systems[184].

The agroindustrial waste material, i.e., sugar beet pulp, was converted into low-cost carbon-rich material by reacting with sulfuric acid. The carbonaceous material derived from sugar beet pulp could be effectively used for the treatment of Cr(VI). The maximum adsorption capacity of chromium is about 24 mg g^{-1} for 25°C[185].

Five activated carbons from coffee residue were prepared at 600°C by chemical activation using $ZnCl_2$. The effects of the impregnation ratio ($ZnCl_2$/coffee residue) varying from 0 to 100% on the physical and chemical characteristics were studied. Results showed that the microporosity presented by the activated carbon at 25% was more critical at 95.70%. All of the prepared carbons were subjected to the adsorption of salicylic acid. The results showed that the adsorption of salicylic acid is more effective on activated carbon prepared with 25% of $ZnCl_2$ (AC 25%)[186].

Palm shell-derived activated carbon was utilized as a potential adsorbent to remove rhodamine B. A maximum dye removal efficiency of 95% was achieved at an initial dye concentration of 62.6 $\mu mol \cdot L^{-1}$, pH of 3 and temperature of 50°C[182].

The pistachio shell carbon is used to remove Pb(II) from the aqueous solution. The effects of concentration, pH, dose, and temperature have been studied. The highest maximum adsorption capacity of 7.19 mg g^{-1} of Pb(II) was recorded at pH 6. The adsorbent could desorb 82.9% of Pb(II)[187]. Other examples are reported in Table 9.7.

TABLE 9.7   Use of Carbon-Based Adsorbents for Removal of Different Pollutants

Source	Pollutant	Q_{max} mg/g	Reference
AC from sugar beet bagasse	Cr(VI)	52.87	[188]
AC from paper mill sludge	Cr(VI)	23.18	[189]
AC from apple peels	Cr(VI)	36.01	[190]
AC from watermelon shell	Pb(II)	40.984	[191]
	Zn(II)	11.312	
AC from hazelnut shell	Cr(VI)	170	[192]
AC from longan seed	Cr(VI)	35.02	[193]
AC from coconut tree	Cr(VI)	3.46	[194]
Acid-activated carbon prepared from leaves of Juniperus procera	Pb(II)	30.3	[195]
Coir pith carbon	Congo red	6.7	[196]
Activated carbon derived from Leucaena leucocephala seed pod	Cr(VI)	26.94	[197]
AC form apple peel	Cr(VI)	36.01	[198]
Acidically prepared rice husk carbon	Cr(VI)	47.62	[199]
Activated carbon derived from Leucaena leucocephala	Cr(VI)	13.85	[200]
KOH-activated carbon prepared from bamboo (Oxytenanthera abyssinica) waste	Cr(VI)	59.23	[201]
Ricinus communis seed shell active carbon	Cr(VI)	7.761	[184]
Fe-modified bamboo carbon	Cr(VI)	35.7	[202]
Palm shell-activated carbon	Cr(VI)	12.6	[203]
PEI/palm shell-activated carbon	Cr(VI)	20.5	[204]
Activated carbon resorcinol formaldehyde xerogels	Cr(VI)	241.9	[205]
H_3PO_4 activated carbon obtained from eucalyptus sawdust	Cr(VI)	12.6	[206]
	Pb(II)	7.82	

(Continued)

TABLE 9.7 *(Continued)* Use of Carbon-Based Adsorbents for Removal of Different Pollutants

Source	Pollutant	Q_{max} mg/g	Reference
Nitric acid-modified activated carbon stems from hyacinth plant	Cr(VI)	175.63	[207]
Nitric acid-modified coconut shell charcoal	Cr(VI)	10.88	[208]
Ammonium hydroxide-modified AC	Cr(VI)	48.51	[209]
Hevea Brasilinesis sawdust-activated carbon	Cr(VI)	44.05	[210]
Acidically prepared rice husk carbon	Cr(VI)	47.62	[199]
Coconut charcoal	Cr(VI)	5.257	[211]
	As(III)	0.042	
	Ni(II)	1.748	
Acid activated carbon from Juniperus procera Leaves	Pb(II)	30.3	[212]
	Cr(VI)	T23	
Porous carbon from tea waste	Methylene blue	402.25	[213]
	Eosin yellow	400	
Activated carbon from coconut tree	Cr(VI)	3.46	[214]
Activated carbon from Pinus sp.	Reactive Blue 5G	162.62	[215]
AC from apricot stone	Methylene blue	4.11	[216]
	Phenol	126	
AC coffee residue by chemical activation using $ZnCl_2$.	Salicylic acid	29.03	[217]
Fruit peel-based biochar	Cr(VI)	44.64	[218]
	chlortetracycline	10.48	
Activated carbon from rattan sawdust	Methylene blue	291.14	[219]

9.2.3 Bacteria, Fungi, and Yeast

An essential aspect of biosorption is that it can be carried out with metabolically active or inactive cells. Many microorganisms have been intensively examined for their abilities to be applied in the biosorption of heavy metals, such as bacteria, fungi, yeast, and algae. Some examples are reported in this section.

Live and heat-inactivated immobilized Trametes versicolor mycelia on the carboxymethylcellulose beads were applied to adsorb heavy metals. The maximum biosorption capacities for live and heat-inactivated Trametes versicolor were 1.51 and 1.84 mmol Cu^{2+}, 0.85 and 1.11 mmol Pb^{2+}, and 1.33 and 1.67 mmol Zn^{2+} per gram of biosorbents, respectively.[220]

Alkali pretreated (NaOH) macrophyte Lemna minor L. was used to remove heavy metals. The maximum adsorption capacities were

TABLE 9.8   Removal of Pollutants Using Bacteria, Fungi, or Yeast

Adsorbent	Pollutant	Q_{max} mg/g	Reference
Thiobacillus thiooxidans	Zn(II)	172.4	[223]
	Cu(II)	39.84	
Desulfovibrio desulfuricans	Zn(II)	49.6	[224]
	Cu(II)	16.7	
Carboxylic acid modified marine fungus Aspergillus wentii	Brilliant Blue G	384.6	[225]
Amino propyl trimethoxy Silane modified Rhizopus nigricans biomass	Cr(VI)	212	[226]
Mucor hiemalis	Cr(VI)	53.5	[227]

shown to be 83, 69, and 59 mg g^{-1} for the Cd(II), Cu(II), and Ni(II) ions, respectively[221].

Biosorption of Pb(II) and Cu(II) ions from aqueous solutions has been studied in a batch system by using Bacillus sp showing maximum biosorption capacities of Pb(II) and Cu(II) ions of 92.27 ± 1.17 mg g^{-1} and 16.25 ± 1.64 mg g^{-1}, respectively[222].

Other adsorbents based on the investigated adsorbents are reported in Table 9.8.

9.2.4 Other Biomass-Derived Adsorbents

Apart from the wide classes of adsorbents already investigated in the previous sections, other systems concern the use of adsorbents organophosphonic acid functionalized spent buckwheat to adsorb Au(III)[228], hen feathers[229], degreased coffee beans[230], maize cob[231], treated ginger waste,[232] and epicarp of Ricinus communis for removing malachite green dye[233]; peat for the removal of malachite green and methylene blue[234]; removal of basic fuchsin dye using mussel shell biomass[235]; pre-treated arca shell biomass for removing lead, copper, nickel, cobalt, and cesium [236]; modified Euterpe oleracea endocarp for the removal of Cd^{2+}, Pb^{2+}, and Cr^{3+} from water[237]; neem (Azadirachta indica) leaf powder for removing methylene blue[238] and brilliant green[239]; methylene blue adsorption by dehydrated peanut hull[240]; roots of water hyacinth to remove methylene blue and Victoria blue[241]; wheat bran for heavy metals removal[242]; coffee residues binding with clay for the removal of heavy metal ions in solution[243]; cassava waste biomass (untreated and acid treated) to remove heavy metals[244]; removal of Cr(VI) using treated waste newspaper[245];

removal of Ni ions using pine tree (Pinus nigra) materials modified with HCl[118]; removal of copper(II) from aqueous solutions by biosorption on the cone biomass of Thuja orientalis[246]; natural wool from sheep for the removal of Cr(VI)[247]; removal of chromium from industrial waste by using eucalyptus bark [248]; pinus radiata bark and tannins, chemically modified with an acidified formaldehyde solution, used for removing metal ions from aqueous solutions[249]; cactus leaves for Cr(VI) removal[250]; sugar beet pulp for Cr(III) and Cr(VI) removal[251]; sphagnum moss peat for Pb(II) removal[252]; removal of chromium(III), copper(II), and zinc(II) by adsorption on carrot residues[253]; sorption of lead ions on tree fern[254] waste; Moringa oleifera seed pods for removal of Cr(VI) and naphthol blue black[255]; modified soda lignin extracted from oil palm empty fruit bunches for removing lead(II) ions[256]; biosorption of Cr(VI) on cone biomass of Pinus sylvestris[257]; wool for Cr removal[258]; pine needles for chromium removal[258]; sugar beet pulp for removing Pb2+, Cu2+, Zn2+, Cd2+, and Ni2+ cations [259]; sulfuric acid dehydrated sesame (Sesamum indicum) stems for Cr(VI) removal[260]; barks of Acacia albida and leaves of Euclea schimperi for Cr(VI) removal[261]; Eucalyptus bark for Cr(VI) removal[262]; neem bark, hyacinth roots, and neel leaves for Cr(VI) removal[263]; garlic stem for chromium removal[151]; stem and leaves of waste Portulaca Oleracea biomass for Cr removal[264].

9.3 BIOMASS-DERIVED BIOPOLYMERS FOR WATER REMEDIATION

From sustainable biomass materials, especially agricultural plants, the most biologically renewable resources that may be identified and exploited include raw materials from terrestrial and marine animals, agricultural plants, microbes, and their wastes [265]. These raw materials and their leftovers are plentiful renewable resources that can be used all over the world to produce sustainable biomass-based biopolymers. Furthermore, using biomass helps the environment in ways like reducing greenhouse gas emissions. It can produce biomass-based biopolymers from renewable raw materials and their utilization for water remediation to minimize water pollution.

Since biopolymers are renewable and environmentally benign, a large number of them have demonstrated notable efficacy in the adsorption of contaminants from wastewater[266]. Because biopolymers are biodegradable, environmental microbes can readily break them down. Therefore, they are less likely to accumulate garbage and are environmentally friendly. In general, they are non-toxic and renewable, which decreases their reliance on non-renewable resources and makes them sustainable[266]. Because of their adaptability, biopolymer-based adsorbents can remove many harmful contaminants from wastewater[266].

Polysaccharides are biopolymers consisting of repeated monosaccharide units that are naturally occurring, sustainable, and environmentally benign. Among the sustainable and ecologically beneficial natural biopolymers are chitin/chitosan, cellulose, starch, pectin, and alginate. Numerous methods, including adsorption, reduction/degradation, and coagulation/flocculation, have shown how well polysaccharide derivatives remove inorganic pollutants like heavy metal ions and organic contaminants like dyes, nitroarenes, and pesticides[267]. Because polysaccharides are inexpensive, naturally occurring, and biocompatible, their application as adsorbents in wastewater treatment has grown in popularity. These biopolymers can also be altered to improve their adsorption capabilities, which increases their potency in eliminating contaminants from wastewater.

9.3.1 Cellulose

As the most prevalent polysaccharide on Earth, cellulose makes up around half of all plant biomass. It is a naturally occurring, renewable organic polymer that serves as the framework for plant cell walls. Cellulose's exceptional qualities, including its high strength, low toxicity, and biodegradability, have led to its widespread use in a variety of industries. It is a linear carbohydrate polymer composed of glucose monomers connected by β-1,4-glycosidic linkages. These monomers are organized into a linear chain, with cellobiose serving as the repeating unit of cellulose. All green plants and algae make cellulose, which is a significant part of their cell walls. It is used extensively in a variety of industries, including paper, food, textiles, cosmetics, pharmaceuticals, and buildings.

In recent years, cellulose has drawn much interest as a possible adsorbent for the treatment of wastewater. Due to its availability, affordability, biodegradability, and lack of toxicity, it is a desirable adsorbent material. Numerous hydroxyl groups on the surface of cellulose can interact with contaminants through a variety of methods, including adsorption, chelation, and ion exchange. Additionally, cellulose-based adsorbents can be altered to increase their capacity for adsorption and selectivity for particular contaminants. Adsorbents based on cellulose have been used to remove a variety of pollutants from wastewater, including pesticide residue, dyes, heavy metals, oil and organic solvents, and pharmaceutical residue[268].

9.3.2 Chitosan

Chitosan (CS) is one of the more exciting biopolymers and has drawn more and more attention due to its unique qualities and possible uses in a range of industries, such as food packaging, pharmaceutical, agricultural, and environmental protection. It comes from chitin, which, after cellulose, is the second most common biopolymer found in nature. Chitin is a linear polysaccharide consisting of N-acetylglucosamine (GlcNAc) residues connected by β-1,4 linkages. Chitin can be found in the cell walls of fungi, insect shells, and the exoskeletons of crustaceans like crabs and shrimp. By removing the acetyl group from GlcNAc residues in chitin, a polymer that is soluble in acidic solutions is created, known as chitosan.

Chitosan is available in a variety of forms for wastewater treatment, including films, beads, membranes, and nanoparticles[268]. Several parameters, including pH, contact time, chitosan content and the initial concentration, affect the removal of contaminants using chitosan. Because it influences both the speciation of wastewater pollutants and the surface charge of the chitosan, the pH of the wastewater is an important parameter. Because chitosan is amphoteric, the effect of pH on its adsorption capacity can be explained. Depending on the pH of the solution, the amine and hydroxyl functional groups in chitosan can be protonated or deprotonated, influencing its adsorption capacity.

The amino groups on the chitosan molecule get protonated when it is in a low pH solution, creating a positively charged surface. Electrostatic attraction causes negatively charged contaminants, like anionic dyes, to be drawn to the positively charged surface. Hydrogen bonding between the functional groups on the chitosan molecule and the pollutant can strengthen the attraction forces between the chitosan surface and the pollutant. As a result, at low pH levels, chitosan's ability to adsorb anionic contaminants rises. Pollutants can be trapped by physical or chemical means after being drawn to chitosan's surface. Nevertheless, chitosan deprotonates, and its surface charge decreases at high pH levels, which increases its ability to adsorb cationic substances at high pH values. There are various benefits of using chitosan as an adsorbent for wastewater treatment, including its low cost, biodegradability, and non-toxicity. Chitosan can be used; however, there are certain drawbacks as well, like low reusability and poor stability at high pH levels. Several modification techniques, including grafting, mixing, and crosslinking, have been developed to get around these restrictions and enhance the chitosan's qualities for use in wastewater treatment applications.

9.3.3 Alginate

Alginate is a biopolymer derived from brown algae that consist of linear chains of α-L-guluronic acid (G) and β-D-mannuronic acid (M) residues. Alginate is a naturally occurring, biodegradable, and non-toxic substance with a notable ability to adsorb a variety of contaminants. Alginate can also be strengthened mechanically and chemically by cross-linking it with different substances like calcium, magnesium, and barium ions. Additionally, cross-linking boosts alginate's ability to adsorb specific contaminants.

Alginate is a complex polysaccharide that finds widespread application. It is also utilized as a thickening, stabilizer, and gelling agent. Heavy metals, organic compounds, and dyes are among the contaminants that have been removed from wastewater using alginate-based adsorbents. Alginate and sodium hydroxide react to produce sodium alginate, which is the sodium salt of alginate. It is a soluble powder that ranges in color from white to yellowish-brown and, when hydrated, becomes viscous. Sodium alginate is a commonly used material for bio-adsorption in the removal of many kinds of inorganic and organic contaminants due to its high degree of biodegradability, sustainability, and biocompatibility[269].

9.3.4 Starch

A complex carbohydrate, starch is important to both the food and non-food sectors. Starch is said to have come from primitive plants. For development and survival, the plants employed starch as a sort of stored energy. These days, starch is taken from a variety of plants, such as cassava, wheat, potatoes, and corn. Every source of starch has a distinct physical and chemical makeup. Among the many sugar molecules that make up starch is a polysaccharide. Its main components are amylose and amylopectin, where amylopectin forms a branched chain while amylose is a straight chain of glucose molecules. Both the degree of branching and the ratio of amylose to amylopectin affect the properties of starch-based products.

Products made from starch have grown in importance since they are inexpensive, renewable, and biodegradable. Starch can be altered chemically, physically, or enzymatically to enhance its qualities and customize it for use in particular applications. Applications for it are numerous and include water treatment, food, paper, textiles, and biomedicine. Because starch is abundant, inexpensive, and renewable, there has been a surge in interest in using it as a source of biofuels in recent years. Bioethanol can be produced from starch by a process known as fermentation. Utilizing biofuels based on starch can help cut down on fossil fuel consumption and greenhouse gas emissions. Nevertheless, there are certain restrictions on the usage of products based on starch. Because of their excellent adsorption capability, affordability, and accessibility, starch-based materials have been thoroughly investigated as adsorbents for the decontamination of water over the past ten years[270]. Furthermore, starch's hydrophilicity improves the adsorption of hydrophilic contaminants. Nevertheless, there are several difficulties in using starch-based polymers as adsorbents for wastewater treatment. Because of their low water resistance and limited mechanical strength, starch-based materials may not be suitable for use in all wastewater treatment systems. By changing the surface chemistry of starch-based compounds or mixing them with other materials, their adsorption capability can be increased. The selectivity, stability, and reusability of the adsorbent can all be enhanced by combining starch-based compounds with additional substances.

9.3.5 Pectin

All terrestrial plants have large amounts of the linear heteropolymer pectin in their cell walls. It is an anionic biopolymer generated from plants that has a range of characteristics. It is extracted to a range of 0.1%–30% from plant cell walls, depending on the plant's kind and growth stage as well as the extraction technique. It consists of D-galacturonic acid (GaIA) residues connected by α-(1→4) bonds. Pectin has been extracted from the peel and pulp of many different fruits; however, some fruits—passion fruit, grapefruit, soy husk, mango, citrus fruits, pumpkin, apple, and carrot—show surprisingly high amounts.

Pectin has unique qualities that make it a desirable choice for many scientific processes, including flexibility, biodegradability, non-toxicity, and availability of OH groups[271]. Pectin does, however, have some inherent restrictions when utilized in particular industries. This inhibits the ability to manage and operate pectin solutions because of the rapid hydration, dissolution, and desaturation of pectin (due to its high water solubility) and the fast tendency of the biopolymer to form lumps and agglomerate. Because pectin is a sustainable and efficient support for metal catalysts, it has been thoroughly studied for use in catalytic systems. Several single –OH and –COOH functional groups can be found in the backbone of pectin. Particular solitary –COOH groups are found in nature as methyl esters, whereas others can be produced commercially by treating them with ammonia to create carboxamide groups[271]. These functional groups allow pectin to bind to metal ions and form complexes with them or reduce them to produce metal nanoparticles without needing a chemical reductant or stabilizer.

Pectin has distinct properties that make it helpful in removing both organic cations and ions from heavy metals. The electrostatic contact and the Van der Waals force it produces are essential for the cationic dye. When it comes to heavy metal ions, both coordination and electrostatic attraction can have an impact on the adsorption process. Specifically, the electron-rich groups, such as hydroxyl and dissociated carboxyl groups, can function as the coordination sites to generate a coordination complex. The heavy metal ions are initially drawn to the dissociated carboxyl groups through electrostatic attraction. The coordination ability towards heavy metal ions was thus dictated by the quantity of hydroxyl and carboxyl groups; the radius of the outermost occupied orbital of the targeted heavy metal ions also impacted the creation of the coordination complex.

9.4 LIMITS AND PERSPECTIVES

Without a doubt, massive progress has been made in the area of water treatment, especially in the area of adsorbent development. So far, biomass-derived adsorbents are used at the laboratory scale since they are not fully effective with natural wastewater which is known to contain a cocktail of different contaminants. Besides, the use of powdered bio-based adsorbents poses serious problems related to pressure drop, recovery and regeneration, and the high energy required post-treatment technologies. As such, incorporating bio-based powdered adsorbents into polymers, leading to the production of stable composite adsorbents, will improve its prospects for large-scale application. Future studies should, therefore, focus on facilitating the practical use of bio-based adsorbents in real wastewater treatment as well as on evaluating the quality and safety of biomass-derived products to develop highly feasible and cost-effective methods for the conversion of biomass wastes to enable industrial-scale production. One of the main drawbacks is related to the large-scale production. Besides, some issues, such as regeneration and recyclability, should be addressed in future studies.

9.5 CONCLUSION

This chapter summarizes the use of biomass-based materials for water treatment environmental applications. Performances and properties have been investigated and compared by reporting scientific works. The wide range of biosorbents presented have indicated outstanding capacities in the removal of various pollutants from the water effluents. The efficiency of these materials as adsorbents is based mainly on their structure and the presence of surface functional groups, which different functionalizations and modifications can further enhance. Overall, the unique characteristics of these materials, including their biodegradability, nontoxicity, availability, and relatively lower cost, prove their great potential to be used as efficient, economical, and environmentally friendly materials in industrial wastewater treatment.

REFERENCES

1. Goodman, S. M.; Bura, R.; Dichiara, A. B. Facile Impregnation of Graphene into Porous Wood Filters for the Dynamic Removal and Recovery of Dyes from Aqueous Solutions. *ACS Appl Nano Mater,* **2018,** *1* (10), 5682–5690. https://doi.org/10.1021/ACSANM.8B01275/ASSET/IMAGES/MEDIUM/AN-2018-01275B_0008.GIF.

2. Molinari, A.; Mayacela Rojas, C. M.; Beneduci, A.; Tavolaro, A.; Rivera Velasquez, M. F.; Fallico, C. Adsorption Performance Analysis of Alternative Reactive Media for Remediation of Aquifers Affected by Heavy Metal Contamination. *Int J Environ Res Public Health,* **2018,** *15* (5), 980. https://doi.org/10.3390/IJERPH15050980.

3. Bello, O. S.; Adegoke, K. A.; Olaniyan, A. A.; Abdulazeez, H. Dye Adsorption Using Biomass Wastes and Natural Adsorbents: Overview and Future Prospects. *Desalination Water Treat,* **2015,** *53* (5), 1292–1315. https://doi.org/10.1080/19443994.2013.862028.

4. Boakye, P.; Ohemeng-Boahen, G.; Darkwah, L.; Sokama-Neuyam, Y. A.; Appiah-Effah, E.; Oduro-Kwarteng, S.; Asamoah Osei, B.; Asilevi, P. J.; Woo, S. H. Waste Biomass and Biomaterials Adsorbents for Wastewater Treatment, **2022,** *2022,* 1–25. https://doi.org/10.5772/GEET.05; https://www.intechopen.com/journals/7/articles/22.

5. Baldikova, E.; Safarikova, M.; Safarik, I. Organic Dyes Removal Using Magnetically Modified Rye Straw. *J Magn Magn Mater,* **2015,** *380,* 181–185. https://doi.org/10.1016/J.JMMM.2014.09.003.

6. Amalina, F.; Razak, A. S. A.; Krishnan, S.; Zularisam, A. W.; Nasrullah, M. Dyes Removal from Textile Wastewater by Agricultural Waste as an Absorbent—A Review. *Clean Waste Sys,* **2022,** *3,* 100051. https://doi.org/10.1016/J.CLWAS.2022.100051.

7. Bulut, Y.; Tez, Z. Removal of Heavy Metals from Aqueous Solution by Sawdust Adsorption. *J Environ Sci,* **2007,** *19* (2), 160–166. https://doi.org/10.1016/S1001-0742(07)60026-6.

8. Zhou, C.; Wang, Y. Recent Progress in the Conversion of Biomass Wastes into Functional Materials for Value-Added Applications. *Sci Technol Adv Mater,* **2020,** *21* (1), 787–804. https://doi.org/10.1080/14686996.2020.1848213.

9. Tripathi, N.; Hills, C. D.; Singh, R. S.; Atkinson, C. J. Biomass Waste Utilisation in Low-Carbon Products: Harnessing a Major Potential

Resource. *NPJ Clim Atmos Sci*, **2019**, *2* (1), 1–10. https://doi.org/10.1038/s41612-019-0093-5.

10. Gupta, V. K. Application of Low-Cost Adsorbents for Dye Removal—A Review. *J Environ Manage*, **2009**, *90* (8), 2313–2342. https://doi.org/10.1016/J.JENVMAN.2008.11.017.

11. Mo, J.; Yang, Q.; Zhang, N.; Zhang, W.; Zheng, Y.; Zhang, Z. A Review on Agro-Industrial Waste (AIW) Derived Adsorbents for Water and Wastewater Treatment. *J Environ Manage*, **2018**, *227*, 395–405. https://doi.org/10.1016/J.JENVMAN.2018.08.069.

12. Baldikova, E.; Safarikova, M.; Safarik, I. Organic Dyes Removal Using Magnetically Modified Rye Straw. *J Magn Magn Mater*, **2015**, *380*, 181–185. https://doi.org/10.1016/J.JMMM.2014.09.003.

13. Okoro, H. K.; Orosun, M. M.; Oriade, F. A.; Momoh-Salami, T. M.; Ogunkunle, C. O.; Adeniyi, A. G.; Zvinowanda, C.; Ngila, J. C. Potentially Toxic Elements in Pharmaceutical Industrial Effluents: A Review on Risk Assessment, Treatment, and Management for Human Health. *Sustainability*, **2023**, *15* (8), 6974. https://doi.org/10.3390/SU15086974.

14. Ajmal, M.; Rao, R. A. K.; Ahmad, R.; Khan, M. A. Adsorption Studies on Parthenium Hysterophorous Weed: Removal and Recovery of Cd(II) from Wastewater. *J Hazard Mater*, **2006**, *135* (1–3), 242–248. https://doi.org/10.1016/J.JHAZMAT.2005.11.054.

15. Gao, H.; Liu, Y.; Zeng, G.; Xu, W.; Li, T.; Xia, W. Characterization of Cr(VI) Removal from Aqueous Solutions by a Surplus Agricultural Waste—Rice Straw. *J Hazard Mater*, **2008**, *150* (2), 446–452. https://doi.org/10.1016/J.JHAZMAT.2007.04.126.

16. Kabir, M. M.; Mouna, S. S. P.; Akter, S.; Khandaker, S.; Didar-ul-Alam, M.; Bahadur, N. M.; Mohinuzzaman, M.; Islam, M. A.; Shenashen, M. A. Tea Waste Based Natural Adsorbent for Toxic Pollutant Removal from Waste Samples. *J Mol Liq*, **2021**, *322*, 115012. https://doi.org/10.1016/J.MOLLIQ.2020.115012.

17. Viscusi, G.; Barra, G.; Gorrasi, G. Modification of Hemp Fibers through Alkaline Attack Assisted by Mechanical Milling: Effect of Processing Time on the Morphology of the System. *Cellulose*, **2020**, *27* (15), 8653–8665. https://doi.org/10.1007/s10570-020-03406-0.

18. Viscusi, G. Advanced Fibre Materials for Environmental Applications. *Fiber Materials: Design, Fabrication and Applications*, **2023**, 103–148. https://doi.org/10.1515/9783110992892-005/MACHINEREADABLECITATION/RIS.

19. Thyavihalli Girijappa, Y. G.; Mavinkere Rangappa, S.; Parameswaranpillai, J.; Siengchin, S. Natural Fibers as Sustainable and Renewable Resource for Development of Eco-Friendly Composites: A Comprehensive Review. *Front Mater*, **2019**, *6*, 226. https://doi.org/10.3389/FMATS.2019.00226/BIBTEX.

20. Nine, M. J.; Kabiri, S.; Sumona, A. K.; Tung, T. T.; Moussa, M. M.; Losic, D. Superhydrophobic/Superoleophilic Natural Fibres for Continuous Oil-Water Separation and Interfacial Dye-Adsorption. *Sep Purif Technol*, **2020**, *233*, 116062. https://doi.org/10.1016/J.SEPPUR.2019.116062.

21. Viscusi, G.; Lamberti, E.; Gorrasi, G. Hemp Fibers Modified with Graphite Oxide as Green and Efficient Solution for Water Remediation: Application to Methylene Blue. *Chemosphere*, **2021**, 132614. https://doi.org/10.1016/J.CHEMOSPHERE.2021.132614.

22. Viscusi, G.; Lamberti, E.; Gorrasi, G. Design of Sodium Alginate/Soybean Extract Beads Loaded with Hemp Hurd and Halloysite as Novel and Sustainable Systems for Methylene Blue Adsorption. *Polym Eng Sci*, **2022**, *62* (1), 129–144. https://doi.org/10.1002/PEN.25839.

23. Roy, A.; Adhikari, B.; Majumder, S. B. Equilibrium, Kinetic, and Thermodynamic Studies of Azo Dye Adsorption from Aqueous Solution by Chemically Modified Lignocellulosic Jute Fiber. *Ind Eng Chem Res*, **2013**. https://doi.org/10.1021/ie400236s.

24. Selambakkannu, S.; Othman, N. A. F.; Bakar, K. A.; Ming, T. T.; Segar, R. D.; Karim, Z. A. Modification of Radiation Grafted Banana Trunk Fibers for Adsorption of Anionic Dyes. *Fibers Polym*, **2019**, *20* (12), 2556–2569. https://doi.org/10.1007/S12221-019-9409-7.

25. Sajab, M. S.; Chia, C. H.; Zakaria, S.; Khiew, P. S. Cationic and Anionic Modifications of Oil Palm Empty Fruit Bunch Fibers for the Removal of Dyes from Aqueous Solutions. *Bioresour Technol*, **2013**, *128*, 571–577. https://doi.org/10.1016/J.BIORTECH.2012.11.010.

26. Cao, E.; Xiao, W.; Duan, W.; Wang, N.; Wang, A.; Zheng, Y. Metallic Nanoparticles Roughened Calotropis Gigantea Fiber Enables Efficient Absorption of Oils and Organic Solvents. *Ind Crops Prod*, **2018**, *115*, 272–279. https://doi.org/10.1016/J.INDCROP.2018.02.052.

27. Du, Z.; Zheng, T.; Wang, P.; Hao, L.; Wang, Y. Fast Microwave-Assisted Preparation of a Low-Cost and Recyclable Carboxyl Modified Lignocellulose-Biomass Jute Fiber for Enhanced Heavy Metal Removal from Water. *Bioresour Technol*, **2016**, *201*, 41–49. https://doi.org/10.1016/J.BIORTECH.2015.11.009.

28. Shin, E. W.; Karthikeyan, K. G.; Tshabalala, M. A. Orthophosphate Sorption onto Lanthanum-Treated Lignocellulosic Sorbents. *Environ Sci Technol*, **2005**, *39* (16), 6273–6279. https://doi.org/10.1021/ES048018N/SUPPL_FILE/ES048018NSI20050605_033415.PDF.

29. Gupta, V. K.; Pathania, D.; Agarwal, S.; Sharma, S. De-Coloration of Hazardous Dye from Water System Using Chemically Modified Ficus Carica Adsorbent. *J Mol Liq*, **2012**, *174*, 86–94. https://doi.org/10.1016/J.MOLLIQ.2012.07.017.

30. Demir, H.; Top, A.; Balköse, D.; Ülkü, S. Dye Adsorption Behavior of Luffa Cylindrica Fibers. *J Hazard Mater*, **2008**, *153* (1–2), 389–394. https://doi.org/10.1016/J.JHAZMAT.2007.08.070.

31. Pathania, D.; Sharma, A.; Sethi, V. Microwave Induced Graft Copolymerization of Binary Monomers onto Luffa Cylindrica Fiber: Removal of Congo Red. *Procedia Eng*, **2017**, *200*, 408–415. https://doi.org/10.1016/J.PROENG.2017.07.057.

32. Shaiful Sajab, M.; Hua Chia, C.; Zakaria, S.; Mohd Jani, S.; Khan Ayob, M.; Leong Chee, K.; Sim Khiew, P.; Siong Chiu, W. Citric Acid Modified

Kenaf Core Fibres for Removal of Methylene Blue from Aqueous Solution. *Bioresour Technol*, **2011**. https://doi.org/10.1016/j.biortech.2011.05.011.

33. Tka, N.; Jabli, M.; Saleh, T. A.; Salman, G. A. Amines Modified Fibers Obtained from Natural Populus Tremula and Their Rapid Biosorption of Acid Blue 25. *J Mol Liq*, **2018**, *250*, 423–432. https://doi.org/10.1016/j.molliq.2017.12.026.

34. Mallampati, R.; Tan, K. S.; Valiyaveettil, S. Utilization of Corn Fibers and Luffa Peels for Extraction of Pollutants from Water. *Int Biodeterior Biodegradation*, **2015**, *103*, 8–15. https://doi.org/10.1016/J.IBIOD.2015.03.027.

35. Parab, H.; Sudersanan, M.; Shenoy, N.; Pathare, T.; Vaze, B. Use of Agro-Industrial Wastes for Removal of Basic Dyes from Aqueous Solutions. *Clean (Weinh)*, **2009**, *37* (12), 963–969. https://doi.org/10.1002/CLEN.200900158.

36. Roy, A.; Chakraborty, S.; Kundu, S. P.; Adhikari, B.; Majumder, S. B. Lignocellulosic Jute Fiber as a Bioadsorbent for the Removal of Azo Dye from Its Aqueous Solution: Batch and Column Studies. *J Appl Polym Sci*, **2013**, *129* (1), 15–27. https://doi.org/10.1002/APP.38222.

37. Yue, X.; Jiang, F.; Zhang, D.; Lin, H.; Chen, Y. Preparation of Adsorbent Based on Cotton Fiber for Removal of Dyes. *Fibers Polym*, **2017**, *18* (11), 2102–2110. https://doi.org/10.1007/S12221-017-1211-9.

38. Kyzas, G. Z.; Christodoulou, E.; Bikiaris, D. N. Basic Dye Removal with Sorption onto Low-Cost Natural Textile Fibers. *Processes*, **2018**, *6* (9), 166. https://doi.org/10.3390/PR6090166.

39. Agcaoili, A. R.; Herrera, M. U.; Futalan, C. M.; Balela, M. D. L. Fabrication of Polyacrylonitrile-Coated Kapok Hollow Microtubes for Adsorption of Methyl Orange and Cu(II) Ions in Aqueous Solution. *J Taiwan Inst Chem Eng*, **2017**, *78*, 359–369. https://doi.org/10.1016/J.JTICE.2017.06.038.

40. Hameed, B. H.; El-Khaiary, M. I. Batch Removal of Malachite Green from Aqueous Solutions by Adsorption on Oil Palm Trunk Fibre: Equilibrium Isotherms and Kinetic Studies. *J Hazard Mater*, **2008**, *154* (1–3), 237–244. https://doi.org/10.1016/J.JHAZMAT.2007.10.017.

41. Selambakkannu, S.; Othman, N. A. F.; Bakar, K. A.; Shukor, S. A.; Karim, Z. A. A Kinetic and Mechanistic Study of Adsorptive Removal of Metal Ions by Imidazole-Functionalized Polymer Graft Banana Fiber. *Radiat Phys Chem*, **2018**, *153*, 58–69. https://doi.org/10.1016/J.RADPHYSCHEM.2018.09.012.

42. Onwuka, J. C.; Agbaji, E. B.; Ajibola, V. O.; Okibe, F. G. Treatment of Crude Oil-Contaminated Water with Chemically Modified Natural Fiber. *Appl Water Sci*, **2018**, *8* (3), 1–10. https://doi.org/10.1007/S13201-018-0727-5.

43. Dong, T.; Wang, F.; Xu, G. Theoretical and Experimental Study on the Oil Sorption Behavior of Kapok Assemblies. *Ind Crops Prod*, **2014**, *61*, 325–330. https://doi.org/10.1016/J.INDCROP.2014.07.020.

44. Wang, R.; Shin, C. H.; Park, S.; Park, J. S.; Kim, D.; Cui, L.; Ryu, M. Removal of Lead (II) from Aqueous Stream by Chemically Enhanced Kapok Fiber Adsorption. *Environ Earth Sci*, **2014**, *72* (12), 5221–5227. https://doi.org/10.1007/S12665-014-3804-6.

45. Bediako, J. K.; Wei, W.; Kim, S.; Yun, Y. S. Removal of Heavy Metals from Aqueous Phases Using Chemically Modified Waste Lyocell Fiber. *J Hazard Mater,* **2015**, *299*, 550–561. https://doi.org/10.1016/J.JHAZMAT.2015.07.033.

46. Rani, K.; Gomathi, T.; Vijayalakshmi, K.; Saranya, M.; Sudha, P. N. Banana Fiber Cellulose Nano Crystals Grafted with Butyl Acrylate for Heavy Metal Lead (II) Removal. *Int J Biol Macromol,* **2019**, *131*, 461–472. https://doi.org/10.1016/J.IJBIOMAC.2019.03.064.

47. Mondal, N. K.; Bhaumik, R.; Datta, J. K. Removal of Fluoride by Aluminum Impregnated Coconut Fiber from Synthetic Fluoride Solution and Natural Water. *Alex Eng J,* **2015**, *54* (4), 1273–1284. https://doi.org/10.1016/J.AEJ.2015.08.006.

48. Yusof, S. R. M.; Zahri, N. A. M.; Koay, Y. S.; Nourouzi, M. M.; Chuah, L. A.; Choong, T. S. Y. Removal of Fluoride Using Modified Kenaf as Adsorbent. *J Eng Sci Technol Special Issue SOMCHE,* **2014**, 11–22.

49. Moreno-Medina, D. A.; Sánchez-Salinas, E.; Ortiz-Hernández, M. L. Removal of Methyl Parathion and Coumaphos Pesticides by a Bacterial Consortium Immobilized in Luffa Cylindrica. *Revista internacional de contaminación ambiental,* **2014**, *30* (1), 51–63.

50. Abdelhameed, R. M.; El-Zawahry, M.; Emam, H. E. Efficient Removal of Organophosphorus Pesticides from Wastewater Using Polyethylenimine-Modified Fabrics. *Polymer (Guildf),* **2018**, *155*, 225–234. https://doi.org/10.1016/J.POLYMER.2018.09.030.

51. Zhang, S.; Deng, M.; Shan, M.; Zhou, C.; Liu, W.; Xu, X.; Yang, X. Energy and Environmental Impact Assessment of Straw Return and Substitution of Straw Briquettes for Heating Coal in Rural China. *Energy Policy,* **2019**, *128*, 654–664. https://doi.org/10.1016/J.ENPOL.2019.01.038.

52. Seglah, P. A.; Wang, Y.; Wang, H.; Bi, Y.; Zhou, K.; Wang, Y.; Wang, H.; Feng, X. Crop Straw Utilization and Field Burning in Northern Region of Ghana. *J Clean Prod,* **2020**, *261*, 121191. https://doi.org/10.1016/J.JCLEPRO.2020.121191.

53. Liu, Q.; Li, Y.; Chen, H.; Lu, J.; Yu, G.; Möslang, M.; Zhou, Y. Superior Adsorption Capacity of Functionalised Straw Adsorbent for Dyes and Heavy-Metal Ions. *J Hazard Mater,* **2020**, *382*, 121040. https://doi.org/10.1016/J.JHAZMAT.2019.121040.

54. Guo, W.; Lu, S.; Shi, J.; Zhao, X. Effect of Corn Straw Biochar Application to Sediments on the Adsorption of 17α-Ethinyl Estradiol and Perfluorooctane Sulfonate at Sediment-Water Interface. *Ecotoxicol Environ Saf,* **2019**, *174*, 363–369. https://doi.org/10.1016/J.ECOENV.2019.01.128.

55. Deng, Y.; Huang, S.; Laird, D. A.; Wang, X.; Meng, Z. Adsorption Behaviour and Mechanisms of Cadmium and Nickel on Rice Straw Biochars in Single- and Binary-Metal Systems. *Chemosphere,* **2019**, *218*, 308–318. https://doi.org/10.1016/J.CHEMOSPHERE.2018.11.081.

56. El-Bindary, A. A.; El-Sonbati, A. Z.; Al-Sarawy, A. A.; Mohamed, K. S.; Farid, M. A. Removal of Hazardous Azopyrazole Dye from an Aqueous

Solution Using Rice Straw as a Waste Adsorbent: Kinetic, Equilibrium and Thermodynamic Studies. *Spectrochim Acta A Mol Biomol Spectrosc*, **2015**, *136* (PC), 1842–1849. https://doi.org/10.1016/J.SAA.2014.10.094.

57. Feng, Y.; Zhou, H.; Liu, G.; Qiao, J.; Wang, J.; Lu, H.; Yang, L.; Wu, Y. Methylene Blue Adsorption onto Swede Rape Straw (Brassica Napus L.) Modified by Tartaric Acid: Equilibrium, Kinetic and Adsorption Mechanisms. *Bioresour Technol*, **2012**, *125*, 138–144. https://doi.org/10.1016/J.BIORTECH.2012.08.128.

58. Chen, Y.; Chen, Q.; Zhao, H.; Dang, J.; Jin, R.; Zhao, W.; Li, Y. Wheat Straws and Corn Straws as Adsorbents for the Removal of Cr(VI) and Cr(III) from Aqueous Solution: Kinetics, Isotherm, and Mechanism. *ACS Omega*, **2020**, *5* (11), 6003. https://doi.org/10.1021/ACSOMEGA.9B04356.

59. Ge, H.; Wang, C.; Liu, S.; Huang, Z. Synthesis of Citric Acid Functionalized Magnetic Graphene Oxide Coated Corn Straw for Methylene Blue Adsorption. *Bioresour Technol*, **2016**, *221*, 419–429. https://doi.org/10.1016/J.BIORTECH.2016.09.060.

60. Ebrahimian Pirbazari, A.; Saberikhah, E.; Habibzadeh Kozani, S. S. Fe3O4–Wheat Straw: Preparation, Characterization and Its Application for Methylene Blue Adsorption. *Water Resour Ind*, **2014**, *7–8*, 23–37. https://doi.org/10.1016/J.WRI.2014.09.001.

61. Tian, Y.; Wu, M.; Lin, X.; Huang, P.; Huang, Y. Synthesis of Magnetic Wheat Straw for Arsenic Adsorption. *J Hazard Mater*, **2011**, *193*, 10–16. https://doi.org/10.1016/J.JHAZMAT.2011.04.093.

62. Khandanlou, R.; Fard Masoumi, H. R.; Ahmad, M. B.; Shameli, K.; Basri, M.; Kalantari, K. Enhancement of Heavy Metals Sorption via Nanocomposites of Rice Straw and Fe3O4 Nanoparticles Using Artificial Neural Network (ANN). *Ecol Eng*, **2016**, *91*, 249–256. https://doi.org/10.1016/J.ECOLENG.2016.03.012.

63. Xu, X.; Gao, B. Y.; Tan, X.; Yue, Q. Y.; Zhong, Q. Q.; Li, Q. Characteristics of Amine-Crosslinked Wheat Straw and Its Adsorption Mechanisms for Phosphate and Chromium (VI) Removal from Aqueous Solution. *Carbohydr Polym*, **2011**, *84* (3), 1054–1060. https://doi.org/10.1016/J.CARBPOL.2010.12.069.

64. Zhang, W.; Yan, H.; Li, H.; Jiang, Z.; Dong, L.; Kan, X.; Yang, H.; Li, A.; Cheng, R. Removal of Dyes from Aqueous Solutions by Straw Based Adsorbents: Batch and Column Studies. *Chem Eng J*, **2011**, *168* (3), 1120–1127. https://doi.org/10.1016/J.CEJ.2011.01.094.

65. Cao, W.; Dang, Z.; Zhou, X. Q.; Yi, X. Y.; Wu, P. X.; Zhu, N. W.; Lu, G. N. Removal of Sulphate from Aqueous Solution Using Modified Rice Straw: Preparation, Characterization and Adsorption Performance. *Carbohydr Polym*, **2011**, *85* (3), 571–577. https://doi.org/10.1016/J.CARBPOL.2011.03.016.

66. Chun, L.; Hongzhang, C.; Zuohu, L. Adsorptive Removal of Cr(VI) by Fe-Modified Steam Exploded Wheat Straw. *Process Biochemistry*, **2004**, *39* (5), 541–545. https://doi.org/10.1016/S0032-9592(03)00087-6.

67. Zhang, W.; Dong, L.; Yan, H.; Li, H.; Jiang, Z.; Kan, X.; Yang, H.; Li, A.; Cheng, R. Removal of Methylene Blue from Aqueous Solutions by Straw Based Adsorbent in a Fixed-Bed Column. *Chem Eng J,* **2011**, *173* (2), 429–436. https://doi.org/10.1016/J.CEJ.2011.08.001.

68. Tadesse, B.; Teju, E.; Megersa, N. The Teff Straw: A Novel Low-Cost Adsorbent for Quantitative Removal of Cr(VI) from Contaminated Aqueous Samples. *Desalination Water Treat,* **2015**, *56* (11), 2925–2936. https://doi.org/10.1080/19443994.2014.968214.

69. Chand, R.; Watari, T.; Inoue, K.; Torikai, T.; Yada, M. Evaluation of Wheat Straw and Barley Straw Carbon for Cr(VI) Adsorption. *Sep Purif Technol,* **2009**, *65* (3), 331–336. https://doi.org/10.1016/J.SEPPUR.2008.11.002.

70. Ludueña, L. N.; Fasce, D. P.; Alvarez, V. A.; Stefani, P. M. Nanocellulose from Rice Husk Following Alkaline Treatment to Remove Silica. *Bioresources,* **2017**, *6* (2), 1440–1453.

71. Banu, H. T.; Karthikeyan, P.; Meenakshi, S. Zr4+ Ions Embedded Chitosan-Soya Bean Husk Activated Bio-Char Composite Beads for the Recovery of Nitrate and Phosphate Ions from Aqueous Solution. *Int J Biol Macromol,* **2019**, *130*, 573–583. https://doi.org/10.1016/J.IJBIOMAC.2019.02.100.

72. Nindjio, G. F. K.; Tagne, R. F. T.; Jiokeng, S. L. Z.; Fotsop, C. G.; Bopda, A.; Doungmo, G.; Temgoua, R. C. T.; Doench, I.; Njoyim, E. T.; Tamo, A. K.; et al. Lignocellulosic-Based Materials from Bean and Pistachio Pod Wastes for Dye-Contaminated Water Treatment: Optimization and Modeling of Indigo Carmine Sorption. *Polymers (Basel),* **2022**, *14* (18), 3776. https://doi.org/10.3390/POLYM14183776/S1.

73. Xu, M.; Yin, P.; Liu, X.; Tang, Q.; Qu, R.; Xu, Q. Utilization of Rice Husks Modified by Organomultiphosphonic Acids as Low-Cost Biosorbents for Enhanced Adsorption of Heavy Metal Ions. *Bioresour Technol,* **2013**, *149*, 420–424. https://doi.org/10.1016/J.BIORTECH.2013.09.075.

74. Meenu, M.; Chen, P.; Mradula, M.; Chang, S. K. C.; Xu, B. New Insights into Chemical Compositions and Health-Promoting Effects of Black Beans (Phaseolus Vulgaris L.). *Food Front,* **2023**, *4* (3), 1019–1038. https://doi.org/10.1002/FFT2.246.

75. Wang, B.; Lan, J.; Bo, C.; Gong, B.; Ou, J. Adsorption of Heavy Metal onto Biomass-Derived Activated Carbon: Review. *RSC Adv,* **2023**, *13* (7), 4275–4302. https://doi.org/10.1039/D2RA07911A.

76. Hassan, M. M.; Carr, C. M. Biomass-Derived Porous Carbonaceous Materials and Their Composites as Adsorbents for Cationic and Anionic Dyes: A Review. *Chemosphere,* **2021**, *265*, 129087. https://doi.org/10.1016/J.CHEMOSPHERE.2020.129087.

77. Ghadikolaei, N. F.; Kowsari, E.; Balou, S.; Moradi, A.; Taromi, F. A. Preparation of Porous Biomass-Derived Hydrothermal Carbon Modified with Terminal Amino Hyperbranched Polymer for Prominent Cr(VI) Removal from Water. *Bioresour Technol,* **2019**, *288*, 121545. https://doi.org/10.1016/J.BIORTECH.2019.121545.

78. Martínez-Sabando, J.; Coin, F.; Melillo, J. H.; Goyanes, S.; Cerveny, S. A Review of Pectin-Based Material for Applications in Water Treatment. *Materials,* **2023**, *16* (6), 2207. https://doi.org/10.3390/MA16062207.

79. Nsi, E.; Abdullahi, M.; Nsi, E. W.; Onoja, P. K. The Use of Unmodified, EDTA and Oxalic Acid Modified Beans Husk as Hazardous Remediation Agent. *IOSR J Environ Sci*, **2015**, *9* (6), 1–9. https://doi.org/10.9790/2402-09610109.

80. Akinola, L.; Umar, A. M. Adsorption of Crystal Violet onto Adsorbents Derived from Agricultural Wastes: Kinetic and Equilibrium Studies. *J Appl Sci Environ Manag*, **2015**, *19* (2), 279–288. https://doi.org/10.4314/jasem.v19i2.15.

81. Bello, O. S.; Alao, O. C.; Alagbada, T. C.; Olatunde, A. M. Biosorption of Ibuprofen Using Functionalized Bean Husks. *Sustain Chem Pharm*, **2019**, *13*, 100151. https://doi.org/10.1016/J.SCP.2019.100151.

82. Cabal, B.; Budinova, T.; Ania, C. O.; Tsyntsarski, B.; Parra, J. B.; Petrova, B. Adsorption of Naphthalene from Aqueous Solution on Activated Carbons Obtained from Bean Pods. *J Hazard Mater*, **2009**, *161* (2–3), 1150–1156. https://doi.org/10.1016/J.JHAZMAT.2008.04.108.

83. Chávez-Guerrero, L.; Rangel-Méndez, R.; Muñoz-Sandoval, E.; Cullen, D. A.; Smith, D. J.; Terrones, H.; Terrones, M. Production and Detailed Characterization of Bean Husk-Based Carbon: Efficient Cadmium (II) Removal from Aqueous Solutions. *Water Res*, **2008**, *42* (13), 3473–3479. https://doi.org/10.1016/J.WATRES.2008.04.022.

84. Pehlivan, E.; Deveci, H. Removal of Cr (VI) Ion by Modified Bean Husk. https://citeseerx.ist.psu.edu/document?repid=rep1&type=pdf&doi=3d704b 1fe005078e95a6e73a47708c1c52d06072.

85. Ajmal, M.; Ali, R.; Rao, K.; Anwar, S.; Ahmad, J.; Ahmad, R. Adsorption Studies on Rice Husk: Removal and Recovery of Cd(II) from Wastewater. *Bioresour Technol*, **2003**, *86*, 147–149.

86. Rojas, R.; Vanderlinden, E.; Morillo, J.; Usero, J.; El Bakouri, H. Characterization of Sorption Processes for the Development of Low-Cost Pesticide Decontamination Techniques. *Sci Total Environ*, **2014**, *488–489* (1), 124–135. https://doi.org/10.1016/J.SCITOTENV.2014.04.079.

87. Chowdhury, S.; Mishra, R.; Saha, P.; Kushwaha, P. Adsorption Thermodynamics, Kinetics and Isosteric Heat of Adsorption of Malachite Green onto Chemically Modified Rice Husk. *Desalination*, **2011**, *265* (1–3), 159–168. https://doi.org/10.1016/J.DESAL.2010.07.047.

88. Vadivelan, V.; Vasanth Kumar, K. Equilibrium, Kinetics, Mechanism, and Process Design for the Sorption of Methylene Blue onto Rice Husk. *J Colloid Interface Sci*, **2005**, *286* (1), 90–100. https://doi.org/10.1016/J. JCIS.2005.01.007.

89. Wong, K. K.; Lee, C. K.; Low, K. S.; Haron, M. J. Removal of Cu and Pb by Tartaric Acid Modified Rice Husk from Aqueous Solutions. *Chemosphere*, **2003**, *50* (1), 23–28. https://doi.org/10.1016/S0045-6535(02)00598-2.

90. García-Mendieta, A.; Olguín, M. T.; Solache-Ríos, M. Biosorption Properties of Green Tomato Husk (Physalis Philadelphica Lam) for Iron, Manganese and Iron–Manganese from Aqueous Systems. *Desalination*, **2012**, *284*, 167–174. https://doi.org/10.1016/J.DESAL.2011.08.052.

91. Li, Q.; Zhai, J.; Zhang, W.; Wang, M.; Zhou, J. Kinetic Studies of Adsorption of Pb(II), Cr(III) and Cu(II) from Aqueous Solution by Sawdust and Modified Peanut Husk. *J Hazard Mater*, **2007**, *141* (1), 163–167. https://doi. org/10.1016/J.JHAZMAT.2006.06.109.

92. Garg, V. K.; Gupta, R.; Kumar, R.; Gupta, R. K. Adsorption of Chromium from Aqueous Solution on Treated Sawdust. *Bioresour Technol*, **2004**, *92* (1), 79–81. https://doi.org/10.1016/J.BIORTECH.2003.07.004.

93. Karthikeyan, T.; Rajgopal, S.; Miranda, L. R. Chromium(VI) Adsorption from Aqueous Solution by Hevea Brasilinesis Sawdust Activated Carbon. *J Hazard Mater*, **2005**, *124* (1–3), 192–199. https://doi.org/10.1016/J.JHAZMAT.2005.05.003.

94. Acar, F. N.; Malkoc, E. The Removal of Chromium(VI) from Aqueous Solutions by Fagus Orientalis L. *Bioresour Technol*, **2004**, *94* (1), 13–15. https://doi.org/10.1016/J.BIORTECH.2003.10.032.

95. Rafatullah, M.; Sulaiman, O.; Hashim, R.; Ahmad, A. Adsorption of Copper (II), Chromium (III), Nickel (II) and Lead (II) Ions from Aqueous Solutions by Meranti Sawdust. *J Hazard Mater*, **2009**, *170* (2–3), 969–977. https://doi.org/10.1016/J.JHAZMAT.2009.05.066.

96. Argun, M. E.; Dursun, S.; Ozdemir, C.; Karatas, M. Heavy Metal Adsorption by Modified Oak Sawdust: Thermodynamics and Kinetics. *J Hazard Mater*, **2007**, *141* (1), 77–85. https://doi.org/10.1016/J.JHAZMAT.2006.06.095.

97. Mekonnen, E.; Yitbarek, M.; Soreta, T. R. Kinetic and Thermodynamic Studies of the Adsorption of Cr(VI) onto Some Selected Local Adsorbents. *S Afr J Chem*, **2015**, *68*, 45–52. https://doi.org/10.17159/0379-4350/2015/V68A7.

98. Ofomaja, A. E.; Ho, Y. S. Effect of Temperatures and PH on Methyl Violet Biosorption by Mansonia Wood Sawdust. *Bioresour Technol*, **2008**, *99* (13), 5411–5417. https://doi.org/10.1016/J.BIORTECH.2007.11.018.

99. Hamdaoui, O. Batch Study of Liquid-Phase Adsorption of Methylene Blue Using Cedar Sawdust and Crushed Brick. *J Hazard Mater*, **2006**, *135* (1–3), 264–273. https://doi.org/10.1016/J.JHAZMAT.2005.11.062.

100. Ratnamala, G. M.; Deshannavar, U. B.; Munyal, S.; Tashildar, K.; Patil, S.; Shinde, A. Adsorption of Reactive Blue Dye from Aqueous Solutions Using Sawdust as Adsorbent: Optimization, Kinetic, and Equilibrium Studies. *Arab J Sci Eng*, **2016**, *41* (2), 333–344. https://doi.org/10.1007/S13369-015-1666-1/METRICS.

101. Uddin, M. T.; Islam, M. A.; Mahmud, S.; Rukanuzzaman, M. Adsorptive Removal of Methylene Blue by Tea Waste. *J Hazard Mater*, **2009**, *164* (1), 53–60. https://doi.org/10.1016/J.JHAZMAT.2008.07.131.

102. Yang, S.; Wu, Y.; Aierken, A.; Zhang, M.; Fang, P.; Fan, Y.; Ming, Z. Mono/Competitive Adsorption of Arsenic(III) and Nickel(II) Using Modified Green Tea Waste. *J Taiwan Inst Chem Eng*, **2016**, *60*, 213–221. https://doi.org/10.1016/J.JTICE.2015.07.007.

103. Nigam, M.; Rajoriya, S.; Rani Singh, S.; Kumar, P. Adsorption of Cr (VI) Ion from Tannery Wastewater on Tea Waste: Kinetics, Equilibrium and Thermodynamics Studies. *J Environ Chem Eng*, **2019**, *7* (3), 103188. https://doi.org/10.1016/J.JECE.2019.103188.

104. Amarasinghe, B. M. W. P. K.; Williams, R. A. Tea Waste as a Low Cost Adsorbent for the Removal of Cu and Pb from Wastewater. *Chem Eng J*, **2007**, *132* (1–3), 299–309. https://doi.org/10.1016/J.CEJ.2007.01.016.

105. Foroughi-Dahr, M.; Abolghasemi, H.; Esmaili, M.; Shojamoradi, A.; Fatoorehchi, H. Adsorption Characteristics of Congo Red from Aqueous Solution onto Tea Waste. *Chem Eng Commun*, **2015**, *202* (2), 181–193. https://doi.org/10.1080/00986445.2013.836633.

106. Lin, D.; Wu, F.; Hu, Y.; Zhang, T.; Liu, C.; Hu, Q.; Hu, Y.; Xue, Z.; Han, H.; Ko, T. H. Adsorption of Dye by Waste Black Tea Powder: Parameters, Kinetic, Equilibrium, and Thermodynamic Studies. *J Chem*, **2020**, *2020*. https://doi.org/10.1155/2020/5431046.

107. Lee, T. C.; Wang, S.; Huang, Z.; Mo, Z.; Wang, G.; Wu, Z.; Liu, C.; Han, H.; Ko, T. H. Tea Stem as a Sorbent for Removal of Methylene Blue from Aqueous Phase. *Adv Mater Sci Eng*, **2019**, *2019*. https://doi.org/10.1155/2019/9723763.

108. Gupta, A.; Balomajumder, C. Simultaneous Adsorption of Cr(VI) and Phenol onto Tea Waste Biomass from Binary Mixture: Multicomponent Adsorption, Thermodynamic and Kinetic Study. *J Environ Chem Eng*, **2015**, *3* (2), 785–796. https://doi.org/10.1016/J.JECE.2015.03.003.

109. Çelebi, H.; Gök, G.; Gök, O. Adsorption Capability of Brewed Tea Waste in Waters Containing Toxic Lead(II), Cadmium (II), Nickel (II), and Zinc(II) Heavy Metal Ions. *Sci Rep*, **2020**, *10* (1), 1–12. https://doi.org/10.1038/s41598-020-74553-4.

110. Yang, S.; Wu, Y.; Aierken, A.; Zhang, M.; Fang, P.; Fan, Y.; Ming, Z. Mono/Competitive Adsorption of Arsenic(III) and Nickel(II) Using Modified Green Tea Waste. *J Taiwan Inst Chem Eng*, **2016**, *60*, 213–221. https://doi.org/10.1016/J.JTICE.2015.07.007.

111. Sharma, P. K.; Ayub, S.; Tripathi, C. N. Isotherms Describing Physical Adsorption of Cr(VI) from Aqueous Solution Using Various Agricultural Wastes as Adsorbents. *Cogent Eng*, **2016**, *3* (1). https://doi.org/10.1080/23311916.2016.1186857.

112. Ahmaruzzaman, M.; Gayatri, S. L. Activated Tea Waste as a Potential Low-Cost Adsorbent for the Removal of p-Nitrophenol from Wastewater. *J Chem Eng Data*, **2010**, *55* (11), 4614–4623. https://doi.org/10.1021/JE100117S/ASSET/IMAGES/LARGE/JE-2010 00117S_0004.JPEG.

113. Reza, A.; Sheikh, F. A.; Kim, H.; Zargar, M. A.; Abedin, M. Z. Facile And Efficient Strategy For Removal Of Reactive Industrial Dye By Using Tea Waste. *Adv Mater Lett*, **2016**, *7* (11), 878–885. https://doi.org/10.5185/AMLETT.2016.6363.

114. Ifelebuegu, A. O.; Ukpebor, J. E.; Obidiegwu, C. C.; Kwofi, B. C. Comparative Potential of Black Tea Leaves Waste to Granular Activated Carbon in Adsorption of Endocrine Disrupting Compounds from Aqueous Solution. *Global J Environ Sci Manag*, **2015**, *1* (3), 205–214. https://doi.org/10.7508/GJESM.2015.03.003.

115. Jung, S.; Naidoo, M.; Shairzai, S.; Navarro, A. E. On the Adsorption Of A Cationic Artificial on Spent Tea Leaves. *WIT Trans Built Environ*, **2014**, *139*, 231–241. https://doi.org/10.2495/UW140201.

116. Malakahmad, A.; Tan, S.; Yavari, S. Valorization of Wasted Black Tea as a Low-Cost Adsorbent for Nickel and Zinc Removal from Aqueous Solution. *J Chem*, **2016**, *2016*. https://doi.org/10.1155/2016/5680983.

117. Okasha, A. Y.; Ibrahim, H. G. Phenol Removal from Aqueous Systems by Sorption of Using Some Local Waste Materials. *Elec J Env Agricult Food Chem*, **2010**, *9* (4), 796–807.

118. Argun, M. E.; Dursun, S.; Gur, K.; Ozdemir, C.; Karatas, M.; Dogan, S. Nickel Adsorption on the Modified Pine Tree Materials. *Environ Technol*, **2005**, *26* (5), 479–488. https://doi.org/10.1080/09593332608618532.

119. Schiewer, S.; Volesky, B. Biosorption Processes for Heavy Metal Removal. *Environ Microbe-Metal Interact*, **2014**, 329–362. https://doi.org/10.1128/9781555818098.CH14.

120. Milojković, J. V.; Lopičić, Z. R.; Anastopoulos, I. P.; Petrović, J. T.; Milićević, S. Z.; Petrović, M. S.; Stojanović, M. D. Performance of Aquatic Weed—Waste Myriophyllum Spicatum Immobilized in Alginate Beads for the Removal of Pb(II). *J Environ Manage*, **2019**, *232*, 97–109. https://doi.org/10.1016/J.JENVMAN.2018.10.075.

121. Saeed, A.; Iqbal, M.; Akhtar, M. W. Removal and Recovery of Lead(II) from Single and Multimetal (Cd, Cu, Ni, Zn) Solutions by Crop Milling Waste (Black Gram Husk). *J Hazard Mater*, **2005**, *117* (1), 65–73. https://doi.org/10.1016/J.JHAZMAT.2004.09.008.

122. Arora, C.; Sahu, D.; Bharti, D.; Tamrakar, V.; Soni, S.; Sharma, S. Adsorption of Hazardous Dye Crystal Violet from Industrial Waste Using Low-Cost Adsorbent Chenopodium Album. *Desalination Water Treat*, **2019**, *167*, 324–332. https://doi.org/10.5004/DWT.2019.24595.

123. Garvasis, J.; Prasad, A. R.; Shamsheera, K. O.; Nidheesh Roy, T. A.; Joseph, A. Weed to Nano Seeds: Ultrasonic Assisted One-Pot Fabrication of Superparamagnetic Magnetite Nano Adsorbents from Siam Weed Flower Extract for the Removal of Lead from Water. *J Hazard Mater Adv*, **2022**, *8*, 100163. https://doi.org/10.1016/J.HAZADV.2022.100163.

124. Papageorgiou, S. K.; Katsaros, F. K.; Kouvelos, E. P.; Nolan, J. W.; Le Deit, H.; Kanellopoulos, N. K. Heavy Metal Sorption by Calcium Alginate Beads from Laminaria Digitata. *J Hazard Mater*, **2006**, *137* (3), 1765–1772. https://doi.org/10.1016/J.JHAZMAT.2006.05.017.

125. Bayramoğlu, G.; Tuzun, I.; Celik, G.; Yilmaz, M.; Arica, M. Y. Biosorption of Mercury(II), Cadmium(II) and Lead(II) Ions from Aqueous System by Microalgae Chlamydomonas Reinhardtii Immobilized in Alginate Beads. *Int J Miner Process*, **2006**, *81* (1), 35–43. https://doi.org/10.1016/J.MINPRO.2006.06.002.

126. Keskinkan, O.; Goksu, M. Z. L.; Yuceer, A.; Basibuyuk, M. Comparison of the Adsorption Capabilities of Myriophyllum Spicatum and Ceratophyllum Demersum for Zinc, Copper and Lead. *Eng Life Sci*, **2007**, *7* (2), 192–196. https://doi.org/10.1002/ELSC.200620177.

127. Milojković, J. V.; Mihajlović, M. L.; Stojanović, M. D.; Lopičić, Z. R.; Petrović, M. S.; Šoštarić, T. D.; Ristić, M. D. Pb(II) Removal from Aqueous Solution by Myriophyllum Spicatum and Its Compost: Equilibrium, Kinetic and Thermodynamic Study. *J Chem Technol Biotechnol*, **2014**, *89* (5), 662–670. https://doi.org/10.1002/JCTB.4184.

128. Al-Qunaibit, M.; Khalil, M.; Al-Wassil, A. The Effect of Solvents on Metal Ion Adsorption by the Alga Chlorella Vulgaris. *Chemosphere*, **2005**, *60* (3), 412–418. https://doi.org/10.1016/J.CHEMOSPHERE.2004.12.040.

129. Fernadez, N. A.; Chacin, E.; Gutierrez, E.; Alastre, N.; Llamoza, B.; Forster, C. F. Adsorption of Lauryl Benzyl Sulphonate on Algae. *Bioresour Technol*, **1995**, *54* (2), 111–115. https://doi.org/10.1016/0960-8524(95)00000-3.

130. Kaewsarn, P. Biosorption of Copper(II) from Aqueous Solutions by Pre-Treated Biomass of Marine Algae Padina Sp. *Chemosphere*, **2002**, *47* (10), 1081–1085. https://doi.org/10.1016/S0045-6535(01)00324-1.

131. Bekç, Z.; Yoldaş Seki, Y. Y.; Cavas, L. Removal of Malachite Green by Using an Invasive Marine Alga Caulerpa Racemosa Var. Cylindracea. *J Hazard Mater*, **2009**, *161*, 1454–1460. https://doi.org/10.1016/j.jhazmat.2008.04.125.

132. Esmaeli, A.; Jokar, M.; Kousha, M.; Daneshvar, E.; Zilouei, H.; Karimi, K. Acidic Dye Wastewater Treatment onto a Marine Macroalga, Nizamuddina Zanardini (Phylum: Ochrophyta). *Chem Eng J*, **2013**, *217*, 329–336. https://doi.org/10.1016/J.CEJ.2012.11.038.

133. Cengiz, S.; Cavas, L. Removal of Methylene Blue by Invasive Marine Seaweed: Caulerpa Racemosa Var. Cylindracea. *Bioresour Technol*, **2008**, *99* (7), 2357–2363. https://doi.org/10.1016/J.BIORTECH.2007.05.011.

134. Rubin, E.; Rodriguez, P.; Herrero, R.; Cremades, J.; Barbara, I.; Sastre de Vicente, M. E. Removal of Methylene Blue from Aqueous Solutions Using as Biosorbent Sargassum Muticum: An Invasive Macroalga in Europe. *Journal of Chemical Technology & Biotechnology*, **2005**, *80* (3), 291–298. https://doi.org/10.1002/JCTB.1192.

135. Lata, H.; Garg, V. K.; Gupta, R. K. Removal of a Basic Dye from Aqueous Solution by Adsorption Using Parthenium Hysterophorus: An Agricultural Waste. *Dyes Pigm*, **2007**, *74* (3), 653–658. https://doi.org/10.1016/J.DYEPIG.2006.04.007.

136. Rubin, E.; Rodriguez, P.; Herrero, R.; Cremades, J.; Barbara, I.; Sastre de Vicente, M. E. Removal of Methylene Blue from Aqueous Solutions Using as Biosorbent Sargassum Muticum: An Invasive Macroalga in Europe. *J Chem Technol Biotechnol*, **2005**, *80* (3), 291–298. https://doi.org/10.1002/JCTB.1192.

137. Keskinkan, O.; Goksu, M. Z. L.; Basibuyuk, M.; Forster, C. F. Heavy Metal Adsorption Properties of a Submerged Aquatic Plant (Ceratophyllum Demersum). *Bioresour Technol*, **2004**, *92* (2), 197–200. https://doi.org/10.1016/J.BIORTECH.2003.07.011.

138. Ghimire, K. N.; Inoue, K.; Ohto, K.; Hayashida, T. Adsorption Study of Metal Ions onto Crosslinked Seaweed Laminaria Japonica. *Bioresour Technol*, **2008**, *99* (1), 32–37. https://doi.org/10.1016/J.BIORTECH.2006.11.057.

139. Dahri, M. K.; Kooh, M. R. R.; Lim, L. B. L. Water Remediation Using Low Cost Adsorbent Walnut Shell for Removal of Malachite Green: Equilibrium, Kinetics, Thermodynamic and Regeneration Studies. *J Environ Chem Eng*, **2014**, *2* (3), 1434–1444. https://doi.org/10.1016/J.JECE.2014.07.008.

140. Pehlivan, E.; Altun, T. Biosorption of Chromium(VI) Ion from Aqueous Solutions Using Walnut, Hazelnut and Almond Shell. *J Hazard Mater,* **2008**, *155* (1–2), 378–384. https://doi.org/10.1016/J.JHAZMAT.2007.11.071.

141. Almasi, A.; Omidi, M.; Khodadadian, M.; Khamutian, R.; Gholivand, M. B. Lead(II) and Cadmium(II) Removal from Aqueous Solution Using Processed Walnut Shell: Kinetic and Equilibrium Study. *Toxicol Environ Chem,* **2012**, *94* (4), 660–671. https://doi.org/10.1080/02772248.2012.671328.

142. Krishna, L. S.; Yuzir, A.; Yuvaraja, G.; Ashokkumar, V. Removal of Acid Blue25 from Aqueous Solutions Using Bengal Gram Fruit Shell (BGFS) Biomass. *Int J Phytoremediation,* **2017**, *19* (5), 431–438. https://doi.org/10.1080/15226514.2016.1244161.

143. Marković, S.; Stanković, A.; Lopičić, Z.; Lazarević, S.; Stojanović, M.; Uskoković, D. Application of Raw Peach Shell Particles for Removal of Methylene Blue. *J Environ Chem Eng,* **2015**, *3* (2), 716–724. https://doi.org/10.1016/J.JECE.2015.04.002.

144. Pehlivan, E.; Altun, T. Biosorption of Chromium(VI) Ion from Aqueous Solutions Using Walnut, Hazelnut and Almond Shell. *J Hazard Mater,* **2008**, *155* (1–2), 378–384. https://doi.org/10.1016/J.JHAZMAT.2007.11.071.

145. Çelebi, H.; Gök, O. Evaluation of Lead Adsorption Kinetics and Isotherms from Aqueous Solution Using Natural Walnut Shell. *Int J Environ Res,* **2017**, *11* (1), 83–90. https://doi.org/10.1007/S41742-017-0009-3/TABLES/5.

146. Feizi, M.; Jalali, M. Removal of Heavy Metals from Aqueous Solutions Using Sunflower, Potato, Canola and Walnut Shell Residues. *J Taiwan Inst Chem Eng,* **2015**, *54*, 125–136. https://doi.org/10.1016/J.JTICE.2015.03.027.

147. Nakkeeran, E.; Selvaraju, N. Biosorption of Chromium(VI) in Aqueous Solutions by Chemically Modified Strychnine Tree Fruit Shell. *Int J Phytoremediation,* **2017**, *19* (12), 1065–1076. https://doi.org/10.1080/152265 14.2017.1328386.

148. Okoye, A. I.; Ejikeme, P. M.; Onukwuli, O. D. Lead Removal from Wastewater Using Fluted Pumpkin Seed Shell Activated Carbon: Adsorption Modeling and Kinetics. *Int J Environ Sci Technol,* **2010**, *7* (4), 793–800. https://doi.org/10.1007/BF03326188/METRICS.

149. Bulut, Y.; Aydin, H. A Kinetics and Thermodynamics Study of Methylene Blue Adsorption on Wheat Shells. *Desalination,* **2006**, *194* (1–3), 259–267. https://doi.org/10.1016/J.DESAL.2005.10.032.

150. Parlayici, Ş.; Pehlivan, E. Comparative Study of Cr(VI) Removal by Bio-Waste Adsorbents: Equilibrium, Kinetics, and Thermodynamic. *J Anal Sci Technol,* **2019**, *10* (1), 1–8. https://doi.org/10.1186/S40543-019-0175-3/TABLES/4.

151. Parlayıcı, Ş.; Pehlivan, E. Natural Biosorbents (Garlic Stem and Horse Chesnut Shell) for Removal of Chromium(VI) from Aqueous Solutions. *Environ Monit Assess,* **2015**, *187* (12), 1–10. https://doi.org/10.1007/S10661-015-4984-6/TABLES/4.

152. Giraldo, S.; Robles, I.; Ramirez, A.; Flórez, E.; Acelas, N. Mercury Removal from Wastewater Using Agroindustrial Waste Adsorbents. *SN*

Appl Sci, **2020**, *2* (6), 1–17. https://doi.org/10.1007/S42452-020-2736-X/FIGURES/2.

153. Tang, H.; Zhang, Y.; Zhang, Y.; Xiao, Q.; Zhao, X.; Yang, S. Turning Waste into Adsorbent: Modification of Discarded Orange Peel for Highly Efficient Removal of Cd(II) from Aqueous Solution. *Biochem Eng J*, **2022**, *185*, 108497. https://doi.org/10.1016/J.BEJ.2022.108497.

154. Giraldo, S.; Acelas, N. Y.; Ocampo-Pérez, R.; Padilla-Ortega, E.; Flórez, E.; Franco, C. A.; Cortés, F. B.; Forgionny, A. Application of Orange Peel Waste as Adsorbent for Methylene Blue and Cd2+ Simultaneous Remediation. *Molecules*, **2022**, *27* (16). https://doi.org/10.3390/MOLECULES27165105.

155. Ali, A.; Saeed, K.; Mabood, F. Removal of Chromium (VI) from Aqueous Medium Using Chemically Modified Banana Peels as Efficient Low-Cost Adsorbent. *Alex Eng J*, **2016**, *55* (3), 2933–2942. https://doi.org/10.1016/J.AEJ.2016.05.011.

156. Annadurai, G.; Juang, R. S.; Lee, D. J. Use of Cellulose-Based Wastes for Adsorption of Dyes from Aqueous Solutions. *J Hazard Mater*, **2002**, *92* (3), 263–274. https://doi.org/10.1016/S0304-3894(02)00017-1.

157. Hameed, B. H.; El-Khaiary, M. I. Sorption Kinetics and Isotherm Studies of a Cationic Dye Using Agricultural Waste: Broad Bean Peels. *J Hazard Mater*, **2008**, *154* (1–3), 639–648. https://doi.org/10.1016/J.JHAZMAT.2007.10.081.

158. Hossain, M. A.; Ngo, H. H.; Guo, W. S.; Nguyen, T. V. Removal of Copper from Water by Adsorption onto Banana Peel as Bioadsorbent. *GEOMATE J*, **2012**, *2* (4), 227–234.

159. Abbasi, Z.; Alikarami, M.; Nezhad, E. R.; Moradi, F.; Moradi, V. Adsorptive Removal of Co 2+ and Ni 2+ by Peels of Banana from Aqueous Solution. *Univers J Chem*, **2013**, *1* (3), 90–95. https://doi.org/10.13189/ujc.2013.010303.

160. Annadurai, G.; Juang, R. S.; Lee, D. J. Use of Cellulose-Based Wastes for Adsorption of Dyes from Aqueous Solutions. *J Hazard Mater*, **2002**, *92* (3), 263–274. https://doi.org/10.1016/S0304-3894(02)00017-1.

161. Guediri, A.; Bouguettoucha, A.; Chebli, D.; Chafai, N.; Amrane, A. Molecular Dynamic Simulation and DFT Computational Studies on the Adsorption Performances of Methylene Blue in Aqueous Solutions by Orange Peel-Modified Phosphoric Acid. *J Mol Struct*, **2020**, *1202*, 127290. https://doi.org/10.1016/J.MOLSTRUC.2019.127290.

162. Liang, S.; Guo, X.; Feng, N.; Tian, Q. Isotherms, Kinetics and Thermodynamic Studies of Adsorption of Cu2+ from Aqueous Solutions by Mg2+/K+ Type Orange Peel Adsorbents. *J Hazard Mater*, **2010**, *174* (1–3), 756–762. https://doi.org/10.1016/J.JHAZMAT.2009.09.116.

163. Munagapati, V. S.; Kim, D. S. Adsorption of Anionic Azo Dye Congo Red from Aqueous Solution by Cationic Modified Orange Peel Powder. *J Mol Liq*, **2016**, *220*, 540–548. https://doi.org/10.1016/J.MOLLIQ.2016.04.119.

164. Senthil Kumar, P.; Fernando, P. S. A.; Ahmed, R. T.; Srinath, R.; Priyadharshini, M.; Vignesh, A. M.; Thanjiappan, A. Effect of Temperature on the Adsorption of Methylene Blue Dye onto Sulfuric Acid–Treated Orange Peel. *Chem Eng Commun*, **2014**, *201* (11), 1526–1547. https://doi.org/10.1080/00986445.2013.819352.

165. Huang, K.; Xiu, Y.; Zhu, H. Removal of Hexavalent Chromium from Aqueous Solution by Crosslinked Mangosteen Peel Biosorbent. *Int J Environ Sci Technol,* **2015,** *12* (8), 2485–2492. https://doi.org/10.1007/S13762-014-0650-8/TABLES/2.

166. Nascimento, G. E. D.; Duarte, M. M. M. B.; Campos, N. F.; Rocha, O. R. S. D.; Silva, V. L. D. Adsorption of Azo Dyes Using Peanut Hull and Orange Peel: A Comparative Study. *Environ Technol,* **2014,** *35* (11), 1436–1453. https://doi.org/10.1080/09593330.2013.870234.

167. Hambisa, A. A.; Regasa, M. B.; Ejigu, H. G.; Senbeto, C. B. Adsorption Studies of Methyl Orange Dye Removal from Aqueous Solution Using Anchote Peel-Based Agricultural Waste Adsorbent. *Appl Water Sci,* **2023,** *13* (1), 1–11. https://doi.org/10.1007/S13201-022-01832-Y/TABLES/4.

168. Khamseh, A. A. G.; Ghorbanian, S. A.; Amini, Y.; Shadman, M. M. Investigation of Kinetic, Isotherm and Adsorption Efficacy of Thorium by Orange Peel Immobilized on Calcium Alginate. *Sci Rep,* **2023,** *13* (1), 1–12. https://doi.org/10.1038/s41598-023-35629-z.

169. Malook, K. Orange Peel Powder: A Potential Adsorbent for Pb(II) Ions Removal from Water. *Theor Found Chem Eng,* **2021,** *55* (3), 518–526. https://doi.org/10.1134/S004057952103012X/TABLES/2.

170. Romero-Cano, L. A.; García-Rosero, H.; Gonzalez-Gutierrez, L. V.; Baldenegro-Pérez, L. A.; Carrasco-Marín, F. Functionalized Adsorbents Prepared from Fruit Peels: Equilibrium, Kinetic and Thermodynamic Studies for Copper Adsorption in Aqueous Solution. *J Clean Prod,* **2017,** *162,* 195–204. https://doi.org/10.1016/J.JCLEPRO.2017.06.032.

171. Mafra, M. R.; Igarashi-Mafra, L.; Zuim, D. R.; Vasques, É. C.; Ferreira, M. A. Adsorption of Remazol Brilliant Blue on an Orange Peel Adsorbent. *Brazilian J Chem Eng,* **2013,** *30* (3), 657–665. https://doi.org/10.1590/S0104-66322013000300022.

172. Saranya, N.; Ajmani, A.; Sivasubramanian, V.; Selvaraju, N. Hexavalent Chromium Removal from Simulated and Real Effluents Using Artocarpus Heterophyllus Peel Biosorbent—Batch and Continuous Studies. *J Mol Liq,* **2018,** *265,* 779–790. https://doi.org/10.1016/J.MOLLIQ.2018.06.094.

173. Liang, S.; Guo, X.; Tian, Q. Adsorption of Pb2+ and Zn2+ from Aqueous Solutions by Sulfured Orange Peel. *Desalination,* **2011,** *275* (1–3), 212–216. https://doi.org/10.1016/J.DESAL.2011.03.001.

174. Feng, N.; Guo, X.; Liang, S. Adsorption Study of Copper (II) by Chemically Modified Orange Peel. *J Hazard Mater,* **2009,** *164* (2–3), 1286–1292. https://doi.org/10.1016/J.JHAZMAT.2008.09.096.

175. Rahman, A.; Yoshida, K.; Islam, M. M.; Kobayashi, G. Investigation of Efficient Adsorption of Toxic Heavy Metals (Chromium, Lead, Cadmium) from Aquatic Environment Using Orange Peel Cellulose as Adsorbent. *Sustainability,* **2023,** *15* (5), 4470. https://doi.org/10.3390/SU15054470.

176. Hashem, A.; Aniagor, C. O.; Fikry, M.; Taha, G. M.; Badawy, S. M. Characterization and Adsorption of Raw Pomegranate Peel Powder for Lead (II) Ions Removal. *J Mater Cycles Waste Manag,* **2023,** *25* (4), 2087–2100. https://doi.org/10.1007/S10163-023-01655-2/TABLES/5.

177. Kumar, K. V.; Kumaran, A. Removal of Methylene Blue by Mango Seed Kernel Powder. *Biochem Eng J*, **2005**, *27* (1), 83–93. https://doi.org/10.1016/J.BEJ.2005.08.004.

178. Maisuria, K. J.; Shah, K. A.; Rana, J. K.; Jaafar, M.; Alatabe, A.; Hussein, A. A.; Islam, M. D.; Rahaman, A.; Mahdi, M. M.; Mallik, D. Removal of Cr(III) and Other Pollutants from Tannery Wastewater by Moringa Stenopetela Seed. *IOP Conf Ser Earth Environ Sci*, **2021**, *644* (1), 012025. https://doi.org/10.1088/1755-1315/644/1/012025.

179. Bhatnagar, A.; Minocha, A. K.; Kumar, E.; Sillanpää, M.; Jeon, B. H. Removal of Phenolic Pollutants from Water Utilizing Mangifera Indica (Mango) Seed Waste and Cement Fixation. *Sep Sci Technol*, **2009**, *44* (13), 3150–3169. https://doi.org/10.1080/01496390903183303.

180. Elizalde-González, M. P.; Hernández-Montoya, V. Guava Seed as an Adsorbent and as a Precursor of Carbon for the Adsorption of Acid Dyes. *Bioresour Technol*, **2009**, *100* (7), 2111–2117. https://doi.org/10.1016/J.BIORTECH.2008.10.056.

181. Elizalde-González, M. P.; Mattusch, J.; Peláez-Cid, A. A.; Wennrich, R. Characterization of Adsorbent Materials Prepared from Avocado Kernel Seeds: Natural, Activated and Carbonized Forms. *J Anal Appl Pyrolysis*, **2007**, *78* (1), 185–193. https://doi.org/10.1016/J.JAAP.2006.06.008.

182. Aygün, A.; Yenisoy-Karakaş, S.; Duman, I. Production of Granular Activated Carbon from Fruit Stones and Nutshells and Evaluation of Their Physical, Chemical and Adsorption Properties. *Microporous Mesoporous Mater*, **2003**, *66* (2–3), 189–195. https://doi.org/10.1016/J.MICROMESO.2003.08.028.

183. Mohammadi, M.; Hassani, A. J.; Mohamed, A. R.; Najafpour, G. D. Removal of Rhodamine B from Aqueous Solution Using Palm Shell-Based Activated Carbon: Adsorption and Kinetic Studies. *J Chem Eng Data*, **2010**, *55* (12), 5777–5785. https://doi.org/10.1021/JE100730A.

184. Chingombe, P.; Saha, B.; Wakeman, R. J. Surface Modification and Characterisation of a Coal-Based Activated Carbon. *Carbon N Y*, **2005**, *43* (15), 3132 3143. https://doi.org/10.1016/J.CARBON.2005.06.021.

185. Thamilarasu, P.; Karunakaran, K. Kinetic, Equilibrium and Thermodynamic Studies on Removal of Cr(VI) by Activated Carbon Prepared from Ricinus Communis Seed Shell. *Can J Chem Eng*, **2013**, *91* (1), 9–18. https://doi.org/10.1002/CJCE.20675.

186. Altundogan, H. S.; Bahar, N.; Mujde, B.; Tumen, F. The Use of Sulphuric Acid-Carbonization Products of Sugar Beet Pulp in Cr(VI) Removal. *J Hazard Mater*, **2007**, *144* (1–2), 255–264. https://doi.org/10.1016/J.JHAZMAT.2006.10.018.

187. Khenniche, L.; Aissani, F. Preparation and Characterization of Carbons from Coffee Residue: Adsorption of Salicylic Acid on the Prepared Carbons. *J Chem Eng Data*, **2009**, *55* (2), 728–734. https://doi.org/10.1021/JE900426A.

188. Siddiqui, S. H.; Ahmad, R. Pistachio Shell Carbon (PSC)—an Agricultural Adsorbent for the Removal of Pb(II) from Aqueous Solution. *Groundw Sustain Dev*, **2017**, *4*, 42–48. https://doi.org/10.1016/J.GSD.2016.12.001.

189. Ghorbani, F.; Kamari, S.; Zamani, S.; Akbari, S.; Salehi, M. Optimization and Modeling of Aqueous Cr(VI) Adsorption onto Activated Carbon Prepared from Sugar Beet Bagasse Agricultural Waste by Application of Response Surface Methodology. *Surf Interfaces*, **2020**, *18*, 100444. https://doi.org/10.1016/J.SURFIN.2020.100444.

190. Gorzin, F.; Bahri Rasht Abadi, M. M. Adsorption of Cr(VI) from Aqueous Solution by Adsorbent Prepared from Paper Mill Sludge: Kinetics and Thermodynamics Studies. *Adsorp Sci Technol*, **2018**, *36* (1–2), 149–169. https://doi.org/10.1177/0263617416686976/ASSET/IMAGES/LARGE/10.1177_0263617416686976-FIG11.JPEG.

191. Enniya, I.; Rghioui, L.; Jourani, A. Adsorption of Hexavalent Chromium in Aqueous Solution on Activated Carbon Prepared from Apple Peels. *Sustain Chem Pharm*, **2018**, *7*, 9–16. https://doi.org/10.1016/J.SCP.2017.11.003.

192. Moreno-Barbosa, J. J.; López-Velandia, C.; Maldonado, A. D. P.; Giraldo, L.; Moreno-Piraján, J. C. Removal of Lead(II) and Zinc(II) Ions from Aqueous Solutions by Adsorption onto Activated Carbon Synthesized from Watermelon Shell and Walnut Shell. *Adsorption*, **2013**, *19* (2–4), 675–685. https://doi.org/10.1007/S10450-013-9491-X/TABLES/4.

193. Kobya, M. Removal of Cr(VI) from Aqueous Solutions by Adsorption onto Hazelnut Shell Activated Carbon: Kinetic and Equilibrium Studies. *Bioresour Technol*, **2004**, *91* (3), 317–321. https://doi.org/10.1016/J.BIORTECH.2003.07.001.

194. Yang, J.; Yu, M.; Chen, W. Adsorption of Hexavalent Chromium from Aqueous Solution by Activated Carbon Prepared from Longan Seed: Kinetics, Equilibrium and Thermodynamics. *J Ind Eng Chem*, **2015**, *21*, 414–422. https://doi.org/10.1016/J.JIEC.2014.02.054.

195. Selvi, K.; Pattabhi, S.; Kadirvelu, K. Removal of Cr(VI) from Aqueous Solution by Adsorption onto Activated Carbon. *Bioresour Technol*, **2001**, *80* (1), 87–89. https://doi.org/10.1016/S0960-8524(01)00068-2.

196. Ali, I. H.; Al Mesfer, M. K.; Khan, M. I.; Danish, M.; Alghamdi, M. M. Exploring Adsorption Process of Lead (II) and Chromium (VI) Ions from Aqueous Solutions on Acid Activated Carbon Prepared from Juniperus Procera Leaves. *Processes*, **2019**, *7* (4), 217. https://doi.org/10.3390/PR7040217.

197. Namasivayam, C.; Kavitha, D. Removal of Congo Red from Water by Adsorption onto Activated Carbon Prepared from Coir Pith, an Agricultural Solid Waste. *Dyes Pigm*, **2002**, *54* (1), 47–58. https://doi.org/10.1016/S0143-7208(02)00025-6.

198. Yusuff, A. S. Adsorption of Hexavalent Chromium from Aqueous Solution by Leucaena Leucocephala Seed Pod Activated Carbon: Equilibrium, Kinetic and Thermodynamic Studies. *Arab J Basic Appl Sci*, **2019**, *26* (1), 89–102. https://doi.org/10.1080/25765299.2019.1567656.

199. Enniya, I.; Rghioui, L.; Jourani, A. Adsorption of Hexavalent Chromium in Aqueous Solution on Activated Carbon Prepared from Apple Peels. *Sustain Chem Pharm*, **2018**, *7*, 9–16. https://doi.org/10.1016/J.SCP.2017.11.003.

200. Khan, T.; Isa, M. H.; Ul Mustafa, M. R.; Yeek-Chia, H.; Baloo, L.; Binti Abd Manan, T. S.; Saeed, M. O. Cr(vi) Adsorption from Aqueous Solution by an Agricultural Waste Based Carbon. *RSC Adv*, **2016**, *6* (61), 56365–56374. https://doi.org/10.1039/C6RA05618K.

201. Malwade, K.; Lataye, D.; Mhaisalkar, V.; Kurwadkar, S.; Ramirez, D. Adsorption of Hexavalent Chromium onto Activated Carbon Derived from Leucaena Leucocephala Waste Sawdust: Kinetics, Equilibrium and Thermodynamics. *Int J Environ Sci Technol*, **2016**, *13* (9), 2107–2116. https://doi.org/10.1007/S13762-016-1042-Z/FIGURES/5.

202. Dula, T.; Siraj, K.; Kitte, S. A. Adsorption of Hexavalent Chromium from Aqueous Solution Using Chemically Activated Carbon Prepared from Locally Available Waste of Bamboo (Oxytenanthera Abyssinica). *ISRN Environ Chem*, **2014**, *2014*, 1–9. https://doi.org/10.1155/2014/438245.

203. Wang, W.; Wang, X.; Wang, X.; Yang, L.; Wu, Z.; Xia, S.; Zhao, J. Cr(VI) Removal from Aqueous Solution with Bamboo Charcoal Chemically Modified by Iron and Cobalt with the Assistance of Microwave. *J Environ Sci*, **2013**, *25* (9), 1726–1735. https://doi.org/10.1016/S1001-0742(12)60247-2.

204. Hamad, B. K.; Noor, A. M.; Rahim, A. A. Removal of 4-Chloro-2-Methoxyphenol from Aqueous Solution by Adsorption to Oil Palm Shell Activated Carbon Activated with K2 CO3. *J Phys Sci*, **2011**, *22* (1), 39–55.

205. Owlad, M.; Aroua, M. K.; Wan Daud, W. M. A. Hexavalent Chromium Adsorption on Impregnated Palm Shell Activated Carbon with Polyethyleneimine. *Bioresour Technol*, **2010**, *101* (14), 5098–5103. https://doi.org/10.1016/J.BIORTECH.2010.01.135.

206. Oyedoh, E. A.; Ekwonu, M. C. Experimental Investigation on Chromium(VI) Removal from Aqueous Solution Using Activated Carbon Resorcinol Formaldehyde Xerogels. *Acta Polytechnica*, **2016**, *56* (5), 373–378. https://doi.org/10.14311/AP.2016.56.0373.

207. Giraldo-Gutiérrez, L.; Moreno-Piraján, J. C. Pb(II) and Cr(VI) Adsorption from Aqueous Solution on Activated Carbons Obtained from Sugar Cane Husk and Sawdust. *J Anal Appl Pyrolysis*, **2008**, *81* (2), 278–284. https://doi.org/10.1016/J.JAAP.2007.12.007.

208. Awad, F. S. Removal of Lead from Aqueous Solution on Activated Carbon and Modified Activated Carbon Prepared from Dried Water Hyacinth Plant. *Art J Anal Bioanal Tech*, **2014**. https://doi.org/10.4172/2155-9872.1000187.

209. Babel, S.; Kurniawan, T. A. Cr(VI) Removal from Synthetic Wastewater Using Coconut Shell Charcoal and Commercial Activated Carbon Modified with Oxidizing Agents and/or Chitosan. *Chemosphere*, **2004**, *54* (7), 951–967. https://doi.org/10.1016/J.CHEMOSPHERE.2003.10.001.

210. Valentín-Reyes, J.; García-Reyes, R. B.; García-González, A.; Soto-Regalado, E.; Cerino-Córdova, F. Adsorption Mechanisms of Hexavalent Chromium from Aqueous Solutions on Modified Activated Carbons. *J Environ Manage*, **2019**, *236*, 815–822. https://doi.org/10.1016/J.JENVMAN.2019.02.014.

211. Karthikeyan, T.; Rajgopal, S.; Miranda, L. R. Chromium(VI) Adsorption from Aqueous Solution by Hevea Brasilinesis Sawdust Activated Carbon.

J Hazard Mater, **2005,** *124* (1–3), 192–199. https://doi.org/10.1016/J.JHAZMAT.2005.05.003.

212. Wu, Y.; Wen, Y.; Zhou, J.; Cao, J.; Jin, Y.; Wu, Y. Comparative and Competitive Adsorption of Cr(VI), As(III), and Ni(II) onto Coconut Charcoal. *Environ Sci Pollut Res,* **2013,** *20* (4), 2210–2219. https://doi.org/10.1007/S11356-012-1066-Y/FIGURES/5.

213. Ali, I. H.; Al Mesfer, M. K.; Khan, M. I.; Danish, M.; Alghamdi, M. M. Exploring Adsorption Process of Lead (II) and Chromium(VI) Ions from Aqueous Solutions on Acid Activated Carbon Prepared from Juniperus Procera Leaves. *Processes,* **2019,** *7* (4), 217. https://doi.org/10.3390/PR7040217.

214. Borah, L.; Goswami, M.; Phukan, P. Adsorption of Methylene Blue and Eosin Yellow Using Porous Carbon Prepared from Tea Waste: Adsorption Equilibrium, Kinetics and Thermodynamics Study. *J Environ Chem Eng,* **2015,** *3* (2), 1018–1028. https://doi.org/10.1016/J.JECE.2015.02.013.

215. Selvi, K.; Pattabhi, S.; Kadirvelu, K. Removal of Cr(VI) from Aqueous Solution by Adsorption onto Activated Carbon. *Bioresour Technol,* **2001,** *80* (1), 87–89. https://doi.org/10.1016/S0960-8524(01)00068-2.

216. Martins, T. D.; Schimmel, D.; Dos Santos, J. B. O.; Da Silva, E. A. Reactive Blue 5G Adsorption onto Activated Carbon: Kinetics and Equilibrium. *J Chem Eng Data,* **2013,** *58* (1), 106–114. https://doi.org/10.1021/JE300946J/ASSET/IMAGES/JE-2012-00946J_M016.GIF.

217. Aygün, A.; Yenisoy-Karakaş, S.; Duman, I. Production of Granular Activated Carbon from Fruit Stones and Nutshells and Evaluation of Their Physical, Chemical and Adsorption Properties. *Microporous Mesoporous Mater,* **2003,** *66* (2–3), 189–195. https://doi.org/10.1016/J.MICROMESO.2003.08.028.

218. Khenniche, L.; Aissani, F. Preparation and Characterization of Carbons from Coffee Residue: Adsorption of Salicylic Acid on the Prepared Carbons. *J Chem Eng Data,* **2010,** *55* (2), 728–734. https://doi.org/10.1021/JE900426A/ASSET/IMAGES/LARGE/JE-2009-00426A_0007.JPEG.

219. Liang, J. Y.; Zhang, W. X.; Yao, X. W.; Chen, M. L.; Chen, X.; Kong, L. J.; Diao, Z. H. New Insights into Co-Adsorption of Cr6+ and Chlortetracycline by a New Fruit Peel Based Biochar Composite from Water: Behavior and Mechanism. *Colloids Surf A Physicochem Eng Asp,* **2023,** *672,* 131764. https://doi.org/10.1016/J.COLSURFA.2023.131764.

220. Hameed, B. H.; Ahmad, A. L.; Latiff, K. N. A. Adsorption of Basic Dye (Methylene Blue) onto Activated Carbon Prepared from Rattan Sawdust. *Dyes Pigm,* **2007,** *75* (1), 143–149. https://doi.org/10.1016/J.DYEPIG.2006.05.039.

221. Bayramoğlu, G.; Bektaş, S.; Arica, M. Y. Biosorption of Heavy Metal Ions on Immobilized White-Rot Fungus Trametes Versicolor. *J Hazard Mater,* **2003,** *101* (3), 285–300. https://doi.org/10.1016/S0304-3894(03)00178-X.

222. Saygideger, S.; Gulnaz, O.; Istifli, E. S.; Yucel, N. Adsorption of Cd(II), Cu(II) and Ni(II) Ions by Lemna Minor L.: Effect of Physicochemical Environment. *J Hazard Mater,* **2005,** *126* (1–3), 96–104. https://doi.org/10.1016/J.JHAZMAT.2005.06.012.

223. Tunali, S.; Çabuk, A.; Akar, T. Removal of Lead and Copper Ions from Aqueous Solutions by Bacterial Strain Isolated from Soil. *Chem Eng J,* **2006,** *115* (3), 203–211. https://doi.org/10.1016/J.CEJ.2005.09.023.

224. Liu, H. L.; Chen, B. Y.; Lan, Y. W.; Cheng, Y. C. Biosorption of Zn(II) and Cu(II) by the Indigenous Thiobacillus Thiooxidans. *Chem Eng J,* **2004,** *97* (2–3), 195–201. https://doi.org/10.1016/S1385-8947(03)00210-9.

225. Chen, B. Y.; Utgikar, V. P.; Harmon, S. M.; Tabak, H. H.; Bishop, D. F.; Govind, R. Studies on Biosorption of Zinc(II) and Copper(II) on Desulfovibrio Desulfuricans. *Int Biodeterior Biodegradation,* **2000,** *46* (1), 11–18. https://doi.org/10.1016/S0964-8305(00)00054-8.

226. Khambhaty, Y.; Mody, K.; Basha, S. Efficient Removal of Brilliant Blue G (BBG) from Aqueous Solutions by Marine Aspergillus Wentii: Kinetics, Equilibrium and Process Design. *Ecol Eng,* **2012,** *41,* 74–83. https://doi.org/10.1016/J.ECOLENG.2012.01.002.

227. Bai, R. S.; Abraham, T. E. Studies on Enhancement of Cr(VI) Biosorption by Chemically Modified Biomass of Rhizopus Nigricans. *Water Res,* **2002,** *36* (5), 1224–1236. https://doi.org/10.1016/S0043-1354(01)00330-X.

228. Tewari, N.; Vasudevan, P.; Guha, B. K. Study on Biosorption of Cr(VI) by Mucor Hiemalis. *Biochem Eng J,* **2005,** *23* (2), 185–192. https://doi.org/10.1016/J.BEJ.2005.01.011.

229. Yin, P.; Xu, M.; Qu, R.; Chen, H.; Liu, X.; Zhang, J.; Xu, Q. Uptake of Gold (III) from Waste Gold Solution onto Biomass-Based Adsorbents Organophosphonic Acid Functionalized Spent Buckwheat Hulls. *Bioresour Technol,* **2013,** *128,* 36–43. https://doi.org/10.1016/J.BIORTECH.2012.10.048.

230. Mittal, A. Adsorption Kinetics of Removal of a Toxic Dye, Malachite Green, from Wastewater by Using Hen Feathers. *J Hazard Mater,* **2006,** *133* (1–3), 196–202. https://doi.org/10.1016/J.JHAZMAT.2005.10.017.

231. Vasanth Kumar, K.; Sivanesan, S.; Ramamurthi, V. Adsorption of Malachite Green onto Pithophora Sp., a Fresh Water Algae: Equilibrium and Kinetic Modelling. *Process Biochem,* **2005,** *40* (8), 2865–2872. https://doi.org/10.1016/J.PROCBIO.2005.01.007.

232. Sonawane, G. H.; Shrivastava, V. S. Kinetics of Decolourization of Malachite Green from Aqueous Medium by Maize Cob (Zea Maize): An Agricultural Solid Waste. *Desalination,* **2009,** *247* (1–3), 430–441. https://doi.org/10.1016/J.DESAL.2009.01.006.

233. Ahmad, R.; Kumar, R. Adsorption Studies of Hazardous Malachite Green onto Treated Ginger Waste. *J Environ Manage,* **2010,** *91,* 1032–1038. https://doi.org/10.1016/j.jenvman.2009.12.016.

234. Santhi, T.; Manonmani, S.; Smitha, T. Removal of Malachite Green from Aqueous Solution by Activated Carbon Prepared from the Epicarp of Ricinus Communis by Adsorption. *J Hazard Mater,* **2010,** *179,* 178–186. https://doi.org/10.1016/j.jhazmat.2010.02.076.

235. Chieng, H. I.; Zehra, T.; Lim, L. B. L.; Priyantha, N.; Tennakoon, D. T. B. Sorption Characteristics of Peat of Brunei Darussalam IV: Equilibrium, Thermodynamics and Kinetics of Adsorption of Methylene Blue and

Malachite Green Dyes from Aqueous Solution. *Environ Earth Sci*, **2014**, *72* (7), 2263–2277. https://doi.org/10.1007/S12665-014-3135-7/TABLES/12.

236. El Haddad, M. Removal of Basic Fuchsin Dye from Water Using Mussel Shell Biomass Waste as an Adsorbent: Equilibrium, Kinetics, and Thermodynamics. *J Taibah Univ Sci*, **2016**, *10* (5), 664–674. https://doi.org/10.1016/J.JTUSCI.2015.08.007.

237. Dahiya, S.; Tripathi, R. M.; Hegde, A. G. Biosorption of Heavy Metals and Radionuclide from Aqueous Solutions by Pre-Treated Arca Shell Biomass. *J Hazard Mater*, **2008**, *150* (2), 376–386. https://doi.org/10.1016/J.JHAZMAT.2007.04.134.

238. Gonçalves Junior, A. C.; Schwantes, D.; Junior, E. C.; Zimmermann, J.; Ferreira Coelho, G. Adsorption of Cd (II), Pb (II) and Cr (III) on Chemically Modified Euterpe Oleracea Biomass for the Remediation of Water Pollution. *Chem Acta Sci Technol*, **2021**, *43*. https://doi.org/10.4025/actascitechnol.v43i1.50263.

239. Bhattacharya, K. G.; Sharma, A. Kinetics and Thermodynamics of Methylene Blue Adsorption on Neem (Azadirachta Indica) Leaf Powder. *Dyes Pigm*, **2005**, *65* (1), 51–59. https://doi.org/10.1016/J.DYEPIG.2004.06.016.

240. Bhattacharyya, K. G.; Sarma, A. Adsorption Characteristics of the Dye, Brilliant Green, on Neem Leaf Powder. *Dyes Pigm*, **2003**, *57* (3), 211–222. https://doi.org/10.1016/S0143-7208(03)00009-3.

241. Özer, D.; Dursun, G.; Özer, A. Methylene Blue Adsorption from Aqueous Solution by Dehydrated Peanut Hull. *J Hazard Mater*, **2007**, *144* (1–2), 171–179. https://doi.org/10.1016/J.JHAZMAT.2006.09.092.

242. Low, K. S.; Lee, C. K.; Tan, K. K. Biosorption of Basic Dyes by Water Hyacinth Roots. *Bioresour Technol*, **1995**, *52* (1), 79–83. https://doi.org/10.1016/0960-8524(95)00007-2.

243. Farajzadeh, M. A.; Monji, A. B. Adsorption Characteristics of Wheat Bran towards Heavy Metal Cations. *Sep Purif Technol*, **2004**, *38* (3), 197–207. https://doi.org/10.1016/J.SEPPUR.2003.11.005.

244. Boonamnuayvitaya, V.; Chaiya, C.; Tanthapanichakoon, W.; Jarudilokkul, S. Removal of Heavy Metals by Adsorbent Prepared from Pyrolyzed Coffee Residues and Clay. *Sep Purif Technol*, **2004**, *35* (1), 11–22. https://doi.org/10.1016/S1383-5866(03)00110-2.

245. Horsfall, M. J.; Arbia, A.; Spiff, A. Removal of Cu (II) and Zn (II) Ions from Wastewater by Cassava (Manihot Esculenta Cranz) Waste Biomass. *Afr J Biotechnol*, **2003**, *2* (10), 360–364. https://doi.org/10.5897/AJB2003.000-1074.

246. Dehghani, M. H.; Sanaei, D.; Ali, I.; Bhatnagar, A. Removal of Chromium(VI) from Aqueous Solution Using Treated Waste Newspaper as a Low-Cost Adsorbent: Kinetic Modeling and Isotherm Studies. *J Mol Liq*, **2016**, *215*, 671–679. https://doi.org/10.1016/J.MOLLIQ.2015.12.057.

247. Nuhoglu, Y.; Oguz, E. Removal of Copper(II) from Aqueous Solutions by Biosorption on the Cone Biomass of Thuja Orientalis. *Process Biochem*, **2003**, *38* (11), 1627–1631. https://doi.org/10.1016/S0032-9592(03)00055-4.

248. Dakiky, M.; Khamis, M.; Manassra, A.; Mer'Eb, M. Selective Adsorption of Chromium(VI) in Industrial Wastewater Using Low-Cost Abundantly

Available Adsorbents. *Adv Environ Res*, **2002**, *6* (4), 533–540. https://doi. org/10.1016/S1093-0191(01)00079-X.

249. Sarin, V.; Pant, K. K. Removal of Chromium from Industrial Waste by Using Eucalyptus Bark. *Bioresour Technol*, **2006**, *97* (1), 15–20. https://doi. org/10.1016/J.BIORTECH.2005.02.010.

250. Palma, G.; Freer, J.; Baeza, J. Removal of Metal Ions by Modified Pinus Radiata Bark and Tannins from Water Solutions. *Water Res*, **2003**, *37*, 4974– 4980. https://doi.org/10.1016/j.watres.2003.08.008.

251. Dakiky, M.; Khamis, M.; Manassra, A.; Mer'Eb, M. Selective Adsorption of Chromium(VI) in Industrial Wastewater Using Low-Cost Abundantly Available Adsorbents. *Adv Environ Res*, **2002**, *6* (4), 533–540. https://doi. org/10.1016/S1093-0191(01)00079-X.

252. Reddad, Z.; Gerente, C.; Andres, Y.; Cloirec, P. Le. Mechanisms of Cr(III) and Cr(VI) Removal from Aqueous Solutions by Sugar Beet Pulp. *Environ Technol*, **2003**, *24* (2), 257–264. https://doi.org/10.1080/09593330309385557.

253. Ho, Y. S.; McKay, G. The Kinetics of Sorption of Divalent Metal Ions onto Sphagnum Moss Peat. *Water Res*, **2000**, *34* (3), 735–742. https://doi. org/10.1016/S0043-1354(99)00232-8.

254. Nasernejad, B.; Zadeh, T. E.; Pour, B. B.; Bygi, M. E.; Zamani, A. Comparison for Biosorption Modeling of Heavy Metals (Cr (III), Cu (II), Zn (II)) Adsorption from Wastewater by Carrot Residues. *Process Biochem*, **2005**, *40* (3–4), 1319–1322. https://doi.org/10.1016/J.PROCBIO.2004.06.010.

255. Ho, Y. S.; Chiu, W. T.; Hsu, C. Sen; Huang, C. T. Sorption of Lead Ions from Aqueous Solution Using Tree Fern as a Sorbent. *Hydrometallurgy*, **2004**, *73* (1–2), 55–61. https://doi.org/10.1016/J.HYDROMET.2003.07.008.

256. Shirani, Z.; Santhosh, C.; Iqbal, J.; Bhatnagar, A. Waste Moringa Oleifera Seed Pods as Green Sorbent for Efficient Removal of Toxic Aquatic Pollutants. *J Environ Manage*, **2018**, *227*, 95–106. https://doi.org/10.1016/J. JENVMAN.2018.08.077.

257. Ibrahim, M. N. M.; Ngah, W. S. W.; Norliyana, M. S.; Daud, W. R. W.; Rafatullah, M.; Sulaiman, O.; Hashim, R. A Novel Agricultural Waste Adsorbent for the Removal of Lead (II) Ions from Aqueous Solutions. *J Hazard Mater*, **2010**, *182* (1–3), 377–385. https://doi.org/10.1016/J. JHAZMAT.2010.06.044.

258. Ucun, H.; Bayhan, Y. K.; Kaya, Y.; Cakici, A.; Faruk Algur, O. Biosorption of Chromium(VI) from Aqueous Solution by Cone Biomass of Pinus Sylvestris. *Bioresour Technol*, **2002**, *85* (2), 155–158. https://doi.org/10.1016/ S0960-8524(02)00086-X.

259. Dakiky, M.; Khamis, M.; Manassra, A. Mer'Eb, M. Selective Adsorption of Chromium(VI) in Industrial Wastewater Using Low-Cost Abundantly Available Adsorbents. *Adv Environ Res*, **2002**, *6* (4), 533–540. https://doi. org/10.1016/S1093-0191(01)00079-X.

260. Reddad, Z.; Gerente, C.; Andres, Y.; Le Cloirec, P. Adsorption of Several Metal Ions onto a Low-Cost Biosorbent: Kinetic and Equilibrium Studies. *Environ Sci Technol*, **2002**, *36* (9), 2067–2073. https://doi.org/10.1021/ ES0102989/ASSET/IMAGES/LARGE/ES0102989F00008.JPEG.

261. Selen, V.; Özer, D.; Özer, A. A Study on the Removal of Cr(VI) Ions by Sesame (Sesamum Indicum) Stems Dehydrated with Sulfuric Acid. *Arab J Sci Eng*, **2014**, *39* (8), 5895–5904. https://doi.org/10.1007/S13369-014-1266-5/METRICS.

262. Gebrehawaria, G.; Hussen, A.; Rao, V. M. Removal of Hexavalent Chromium from Aqueous Solutions Using Barks of Acacia Albida and Leaves of Euclea Schimperi. *Int J Environ Sci Technol*, **2015**, *12* (5), 1569–1580. https://doi.org/10.1007/S13762-014-0530-2/TABLES/9.

263. Sarin, V.; Pant, K. K. Removal of Chromium from Industrial Waste by Using Eucalyptus Bark. *Bioresour Technol*, **2006**, *97* (1), 15–20. https://doi.org/10.1016/J.BIORTECH.2005.02.010.

264. Singha, B.; Naiya, T. K.; Bhattacharya, A.; Kumar; Das, S. K. Cr(VI) Ions Removal from Aqueous Solutions Using Natural Adsorbents—FTIR Studies. *J Environ Prot (Irvine, Calif)*, **2011**, *2* (6), 729–735. https://doi.org/10.4236/JEP.2011.26084.

265. Mishra, A.; Dubey, A.; Shinghal, S. Biosorption of Chromium(VI) from Aqueous Solutions Using Waste Plant Biomass. *Int J Environ Sci Technol*, **2015**, *12* (4), 1415–1426. https://doi.org/10.1007/S13762-014-0516-0/TABLES/2.

266. Solihat, N. N.; Hidayat, A. F.; Ilyas, R. A.; Thiagamani, S. M. K.; Azeele, N. I. W.; Sari, F. P.; Ismayati, M.; Bakshi, M. I.; Garba, Z. N.; Hussin, M. H.; et al. Recent Antibacterial Agents from Biomass Derivatives: Characteristics and Applications. *J Bioresour Bioproducts*, **2024**. https://doi.org/10.1016/J.JOBAB.2024.02.002.

267. Yaashikaa, P. R.; Senthil Kumar, P.; Karishma, S. Review on Biopolymers and Composites—Evolving Material as Adsorbents in Removal of Environmental Pollutants. *Environ Res*, **2022**, *212*, 113114. https://doi.org/10.1016/J.ENVRES.2022.113114.

268. Saravanan, A.; Kumar, P. S.; Yuvaraj, D.; Jeevanantham, S.; Aishwaria, P.; Gnanasri, P. B.; Gopinath, M.; Rangasamy, G. A Review on Extraction of Polysaccharides from Crustacean Wastes and Their Environmental Applications. *Environ Res*, **2023**, *221*. https://doi.org/10.1016/J.ENVRES.2023.115306.

269. da Silva Alves, D. C.; Healy, B.; Pinto, L. A. de A.; Cadaval, T. R. S.; Breslin, C. B. Recent Developments in Chitosan-Based Adsorbents for the Removal of Pollutants from Aqueous Environments. *Molecules*, **2021**, *26* (3), 594. https://doi.org/10.3390/MOLECULES26030594.

270. Thakur, S. An Overview on Alginate Based Bio-Composite Materials for Wastewater Remedial. *Mater Today Proc*, **2021**, *37* (Part 2), 3305–3309. https://doi.org/10.1016/J.MATPR.2020.09.120.

271. Nasrollahzadeh, M.; Sajjadi, M.; Iravani, S.; Varma, R. S. Starch, Cellulose, Pectin, Gum, Alginate, Chitin and Chitosan Derived (Nano)Materials for Sustainable Water Treatment: A Review. *Carbohydr Polym*, **2021**, *251*, 116986. https://doi.org/10.1016/J.CARBPOL.2020.116986.

CHAPTER 10

Biomass Wastes for Catalytic Application for Biodiesel Production

Nisha Yadav, Shreya Sachdeva, Anu Rathee, Divya Choudhary, Saurabh Kumar, Anirban Das, Kamal Nayan Sharma and Gyandshwar Kumar Rao

10.1 INTRODUCTION

Biodiesel has been increasingly significant in recent decades as a substitute fuel for conventional diesel engines due to its distinct characteristics, including its environmentally benign nature, reduced toxicity, and ability to biodegrade. Nevertheless, biodiesel exhibits superior lubricity compared to petro-diesel fuel due to the absence of sulfur content, improving engine durability and increasing its lifespan [1]. Biodiesel possesses distinct oil properties, such as higher cetane numbers, improved combustion efficiency, and reduced emissions [2, 3]. It is chemically known as FAME, or fatty acid methyl ester, and is produced by reacting feedstock (vegetable oils or animal fats) with alcohol (as a solvent) and catalyst (acidic or basic), depending on the nature of the feedstock. The use of alcohol significantly impacts biodiesel production by favoring transesterification. Various types of oil have been studied for biodiesel production, including first-generation oils like palm oil [4, 5], sunflower oil [6], and soybean oil. The second-generation oils include non-edible oils such as Jatropha Curas seed oil [7], neem oil [8, 9], castor oil [10], and waste cooking oil

DOI: 10.1201/9781003466833-13

[11]. Microalgae-derived oils are classified as third-generation [12–14]. Besides oil, biodiesel can be produced from the byproduct of an edible oil refinement process, specifically bleaching clay [15]. The transesterification reaction can be catalyzed by alkali, acid, or enzyme. Enzymatic transesterification is the most efficient technique for biodiesel [16, 17]. However, the catalyst exhibits certain drawbacks, namely a substantial expense and an exceedingly sluggish reaction rate. Regarding the catalyst used for biomass conversion, homogeneous and heterogeneous catalysts are the main two types. Homogeneous base catalysts are increasingly employed for large-scale generation of biodiesel. The utilization of potassium hydroxide (KOH) and sodium hydroxide (NaOH) as a homogeneous base catalyst provides numerous advantages due to their elevated catalytic activity, brief reaction durations, gentle reaction conditions, very economical raw materials, and widespread accessibility. The presence of free fatty acids and moisture levels can influence the catalytic activity of these catalysts. In addition, the separation and purification process is challenging due to the creation of a by-product (soap) from neutralization and saponification. This is further complicated by the need for a large amount of water, which increases the operational costs of the process. This circumstance renders the catalyst environmentally detrimental [18–20]. The utilization of heterogeneous or solid catalysts for biodiesel production has become highly significant. The product can be easily separated due to the insolubility of these catalysts in the reaction mixture. Additionally, these catalysts can be recycled for subsequent reaction cycles, resulting in reduced catalyst consumption [21]. A catalyst from renewable sources such as biomass is called a bio-based or green catalyst. Biomass refers to organic materials from plants and animals which can be used as an energy source. This organic matter can be used directly as fuel or processed into various forms of biofuel. Biomass includes various materials, such as wood, crops, crop residues, animal manure, and organic components of municipal and industrial wastes. The catalyst derived from natural sources like biomass is called a bio-based or green catalyst. The recent trend shows that using calcium and carbon as a natural biological source has become a preferable heterogeneous catalyst for transesterification. Solid biomass catalysts are environmentally friendly because they are non-toxic, less corrosive, and render wastewater formation [22]. Biomass, which is both cost-effective and readily abundant, serves as the primary source of these catalysts [23]. Another compelling justification for utilizing biodiesel as an alternative fuel is the ongoing depletion of petroleum reserves and the escalating environmental

concerns [24]. The study has focused on developing cost-effective catalysts for the production of biodiesel, with an emphasis on using renewable and environmentally friendly "green catalysts." These types of catalysts can be produced using household garbage or biomass. There is significant interest in renewable catalysts, namely metal oxide catalysts derived from various seashells and carbon-based catalysts. These catalysts have gained attention because they have cheap production costs, which can help reduce the overall cost of synthesizing biodiesel [25]. Despite the ongoing challenges posed by biomass-derived wastes and residual materials, such as municipal solid waste, their utilization as bio-energy feedstock, especially in comparison to energy crops, can significantly reduce greenhouse gas emissions, alleviate pressure on land usage, and address various associated impacts, conflicts, and ethical concerns linked to food crop use. This is a substantial driver for bio-energy production, contributing to the sustainability of biofuels and chemicals as feedstock [26].

10.2 METHODS OF BIODIESEL PRODUCTION

Biodiesel synthesis uses various methods, including transesterification, non-catalytic supercritical process, and thermal cracking. Among these methods, the first method is the most conventional method of biodiesel production (Figure 10.1) [27].

10.2.1 Transesterification

The traditional transesterification approach is one of the most general methods for biodiesel production, i.e., generating the catalyzed reaction encompassing alcohol and vegetable oil to synthesize soap (glycerine) and biodiesel (methyl esters). Generally, a strong base is required for the transesterification process, including sodium methylate, sodium hydroxide, and potassium hydroxide [28]. The acid catalysts are appropriate for converting FFAs to biodiesel; however, these catalysts are not practical for biodiesel production from triglycerides as this conversion is time-consuming. The highly promising method for the conversion of free fatty acids to methyl esters takes place using two steps: first, the conversion of free fatty acids to methyl esters using acid-catalyzed treatment and then the conversion of triglycerides by the use of base-catalyzed method [29]. The decrease in viscosity of the oil is one of the advantages of the transesterification method [28]. The general use of a catalyst is to boost the yield and the reaction rate. The alcohol (usually methanol has been used due to its low cost and ability to react quickly with oil feedstock) is generally used in excess amounts to

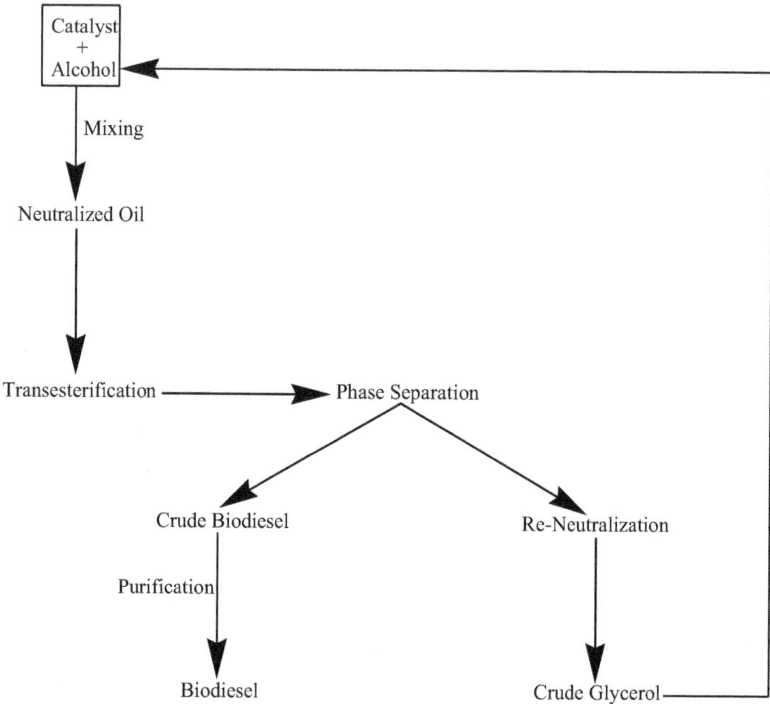

FIGURE 10.1   General scheme for biodiesel production.

shift the equilibrium towards the product side as this reaction is reversible [30]. The triglycerides react with suitable alcohol under the catalyst using constant temperature for a fixed duration [31].

10.2.2 Non-Catalytic Supercritical Process

The process known as non-catalytic supercritical transesterification displays a high rate of reaction and high conversion, which leads to a high yield in a short reaction time. The absence of a catalyst makes the purification and separation process easy. This process of biodiesel synthesis has not been affected by the presence of water and free fatty acids in fats or oils. Pre-treatment is not needed to control water content. The operating pressure ranges between 10–25 MPa, and the temperature ranges between 200–400°C. To increase the solubility of alcohol in oil, the pressure and temperature conditions should be kept above the critical values of alcohol [32]. The operating pressure and temperature of the reaction should be kept above the critical pressure (Pc) of 8.09 MPa and the critical temperature (Tc) of 239 °C of methanol. Due to the operating conditions, the process has not been of much industrial use, making biodiesel synthesis economically unfeasible [33].

10.2.3 Thermal Cracking

Thermal cracking (also known as pyrolysis) is the conversion of a substance from one form to another by using heat under the presence or absence of a catalyst. The process usually occurs by heating under the absence of oxygen or air and bond cleavage to synthesize molecules of small sizes. The pyrolysis of vegetable oil can produce different amounts of carboxylic acids, aromatics, alkadienes, alkenes, and alkanes. The thermal cracking instrument is a high cost for biodiesel production. The method needs to improve from the drawback of needing a definite distillation instrument for separation. Parawira expressed that the product produced resembles gasoline (having sulfur), which is less environmentally friendly. Pyrolytic chemistry is difficult to differentiate due to the variety of reaction products and distinct reaction pathways. Petroleum has been produced for the first time from vegetables by pyrolysis [34].

10.3 BIOMASS WASTE CATALYST CONVERSION INTO BIODIESEL

Catalysts play a crucial role in the biodiesel synthesis pathway by enhancing the rate at which triglycerides are converted into FAME. They achieve this by providing an alternative reaction pathway that requires less activation energy. These catalysts can be categorized into three main groups: homogeneous catalysts, heterogeneous catalysts, and biocatalysts. The first two groups, homogenous/heterogenous, can be further divided into acid and base catalysts (Figure 10.2) [35]. Homogeneous base catalysts are the most potent for transesterification processes and are commonly utilized in industrial applications. Nevertheless, this increased velocity of reaction is attained by sacrificing the simplicity of separating the product in large quantities [36]. Recovering homogeneous catalysts presents a hurdle due to their difficulty in being reused. In addition, the production cost of biodiesel increases due to the various stages of washing and purifying required to fulfil standard criteria [37]. Using a base catalyst, the transesterification reaction remains the most often employed method for producing biodiesel. The majority of typical catalysts used in this process are classified as homogeneous. Commonly employed base catalysts include CH_3ONa, KOH, and $NaOH$. Although these catalysts have advantages such as fast reaction rates,

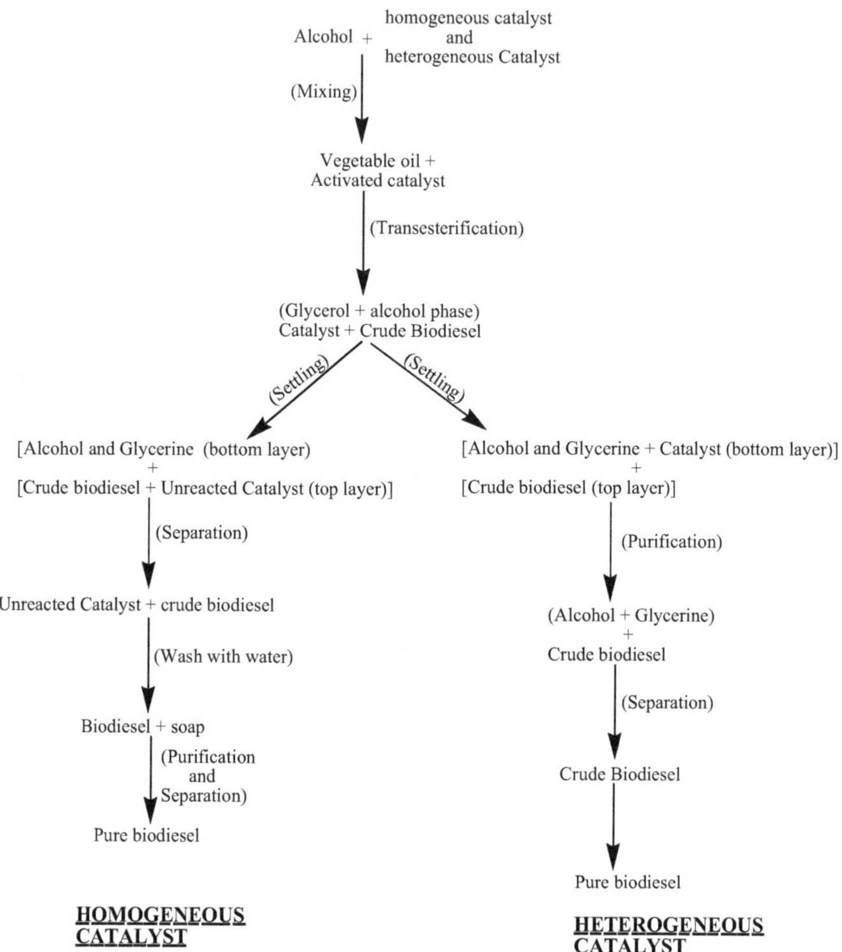

FIGURE 10.2 Flowchart of biodiesel production via the transesterification method using homogeneous or heterogenous catalysts [37].

mild reactions, and low cost, it is important to acknowledge their disadvantages [38]. According to Shahbazi et al. [39], one of the drawbacks is the saponification reaction that occurs when both free fatty acids (FFA) and a base are present, leading to the formation of soaps. Consequently, biodiesel production is significantly reduced, and there is a greater need for base catalysts due to the limited capacity to recover or reuse the catalyst. Furthermore, the process of separating the product becomes challenging and expensive due to the tendency of the esters to dissolve in the glycerol, resulting in the formation of an emulsion of ester-glycerol [40]. Thus, it can be inferred that the significance of the base catalyst in the reaction is restricted to high-quality raw materials with low levels of free fatty acids (FFA) and even water, as it facilitates the production of FFA by hydrolysis. The technique of cleansing the produced biodiesel using warm or acidified water is commonly employed in industries to purify it. As a result, a significant volume of wastewater is produced, necessitating additional purification and treatment. These variables will add to the elevated expenses associated with biodiesel production. Kulkarni and Dalai [41] discovered that enzymes are not affected by free fatty acids, in addition to the requirement of water in the process. Furthermore, the enzymes were found to have a high reusability level, which could decrease production expenses. Regrettably, the cost of producing enzymes greatly exceeds their advantages.

The slow biodiesel production process makes it impractical for large-scale industrial uses [36]. Heterogeneous catalysts have gained significant attention in recent years due to their advantages over homogeneous catalysts. This is evident from the numerous research that has emerged on the fabrication of heterogeneous catalysts for the transesterification process. In contrast to homogeneous catalysts, heterogeneous catalysts have properties that facilitate their separation from the reaction mixture and offer more excellent reusability [42].

10.3.1 Homogenous Catalyst

10.3.1.1 Homogeneous Base-Catalyzed Transesterification

The commonly used method for economic biodiesel production is the homogeneous base-catalyzed biodiesel production, which uses the selected oil feedstock, for instance, refining cooking oil, waste cooking oil, etc., to react with alcohol (like methanol) to undergo transesterification using a homogenous catalyst. Sodium methylate and sodium hydroxide are the two most generally used homogeneous catalysts, among which the latter is the most common as it is beneficial economically. The process requires various purification and filtration steps. The separation stages are necessary because of the soap formation prompted by the side reaction catalyzed by the base catalyst [43]. Base catalyst reacts with free fatty acid to synthesize soap (i.e., unwanted product). The hydrolysis of triglycerides makes the homogeneous catalysts susceptible to water. Hence, the formation and removal of soap have been significant shortcomings of the catalyst [44]. The homogeneous base catalyst serves the benefit of being inexpensive, providing a good yield of the biodiesel while using mild reaction conditions and a high reaction rate; the investment cost is also relatively low and provides validated results. However, along with various advantages, the catalyst has also been associated with constraints. For instance, the catalyst is non-recyclable, stainless steel equipment is required for its use, the cost of purification is generally high, pre-treated and refined fat or oil are required, (which are costly), and low moisture-containing feedstock is needed [43].

10.3.1.2 Homogeneous Acid-Catalyzed Transesterification

The high reaction time and the harsh reaction conditions for product formation make the acid-catalyzed biodiesel production process less significant. The harsh reaction conditions at high temperatures and high methanol-to-oil proportion have been generally needed for acid-catalyzed biodiesel production and can lead to environmental concerns due to the high corrosiveness of the catalyst [45]. Earlier, an acid-catalyzed biodiesel process was applied for the removal of free fatty acid content from the feedstock to lower the soap formation, known as pre-esterification. The cycle exhibits a few stages of alcohol recovery and reuse of the fundamentals as this process needs alcohol of high volumes in abundance [43].

The homogeneous acid catalyst is associated with a variety of advantages, including lack of soap formation, fewer reaction times with free fatty acids, transesterification and esterification taking place simultaneously, and low-cost enhancement of the branched or long-chain ester formation. Regardless of its advantages, the catalyst does suffer from drawbacks such as non-reusability, water inhibition, harsh reaction conditions, and the reaction rate being generally slow compared to the essential counterpart. Also, materials that are resistant to corrosion are required [43].

10.3.2 Heterogeneous Catalyst

Some of the drawbacks of homogeneous catalysts led to a significant focus on heterogeneous catalysts, which led to many studies on synthesizing this catalyst. These catalysts contain the properties that lead to the easy isolation of the catalyst from the reaction mixture and show high reusability. The low corrosivity, cost-effectiveness, high stability under high temperatures, and the non-toxicity of the catalysts makes them highly fascinating [37]. The heterogeneous catalysts are categorized as heterogeneous base catalysts and heterogeneous acid catalysts.

10.3.2.1 Heterogeneous Base Catalyst

There are three primary methods by which biomass can be converted into solid base catalysts for the synthesis of biodiesel. In this process, biomass can be directly converted into a catalyst, such as calcium oxide (CaO), produced through heat treatment [46–48]. Biomass can be transformed into supporting materials, such as silica support or activated carbon, to facilitate the attachment of basic active chemicals like CaO, alkali metals (KOH, NaOH), and others [49, 50]. CaO (basic earth metal oxide) serves as one of the most fascinating heterogeneous base catalysts, having the ability to catalyze transesterification with good essential strength for the production of biodiesel. CaO can be obtained efficiently from $CaCO_3$ by calcination's thermal decomposition at high temperatures. Besides, calcium oxide, strontium oxide (SrO), and magnesium oxide (MgO) are the other essential heterogeneous catalysts [51]. The benefits of a heterogeneous base catalyst lie in the fact that the catalyst is functional under mild reaction conditions, provides a high rate of reaction compared to the acid counterpart, is easier to isolate from the liquid mixture, and can be reused and regenerated. However, the drawbacks of the catalyst include the catalyst leaching, the catalyst absorbing the moisture from the surroundings, the catalyst's compassion towards the free fatty acid content present in the oil, and the high free fatty acid leading to the production of soap [43].

10.3.2.2 Heterogeneous Acid Catalyst

The conversion of several lignocellulosic biomasses into the solid acid catalyst has been explored widely, and various methods have been used, with direct sulfonation using concentrated sulfuric acid being the most likely method. This direct sulfonation pathway is straightforward and forthright. The various heterogeneous acid catalysts include sulfonic ion-exchange resins, for instance, Nafion-NR50, Amberlyst-35 and Amberlyst-15, SnO_2, TiO_2, and ZrO_2 [52].

The heterogeneous catalyst provides advantages such as insensitivity towards the free fatty acid content in the oil, easy separation and the potentiality to be reused and regenerated, and the esterification or transesterification taking place simultaneously. However, the two major drawbacks of the catalyst are the harsh reaction conditions, such as high pressure and temperature and the catalyst leaching [43].

- To summarize, homogeneous catalysts are less significant in various ways than conventional heterogeneous catalysts. The isolation of the homogeneous catalyst from product and side products (glycerine and ester) is difficult in contrast to the heterogeneous catalyst, which serves as one of the significant constraints of the homogeneous catalysts. Also, heterogeneous catalysts show higher reusability than homogeneous catalysts [37].

- Apart from being advantageous, the low surface area of the heterogeneous catalyst shows a significant drawback. Also, the base heterogeneous catalysts are easily impaired underwater [53], and the acid heterogeneous catalyst causes contamination in the biodiesel products [54].

- Homogeneous base catalyst exhibits good catalytic activity for biodiesel synthesis even though the separation of product and by-products is harsh.

- Homogeneous base catalyst gives the most effective pathway in trans-esterification reaction and has been utilized in a variety of industrial applications. Homogeneous catalysts have been difficult to recover and thus represent a challenge towards their reusability. Adding to this, the cost creation of biodiesel surges numeral stages of purification and washing of the outcome to fulfil the guidelines.

- Huge volume of water and compulsory treatment of the subsequent wastewater created from the purification technique contribute to the general expense of creation [54].

10.3.3 Solid Waste Catalyst

Currently, there is significant research interest in biomass-derived heterogeneous catalysts to explore their potential for reducing production costs and enhancing biodegradability. Biomass encompasses organic material derived from plants and animals that can be utilized for energy production through direct utilization or additional improvements. In addition to reducing costs, utilizing biomass for catalyst production offers a viable solution to the problem of disposing of biomass waste. This issue has worsened due to the continuous increase in waste production from agricultural and other human activities [55]. Research shows that only 20% of FAME yield was obtained when soyabean oil and methanol were used as feedstock in the presence of MgO. In contrast, the same experiment used CaO, which provided 93% of the biodiesel yield. This thoroughly verifies and proves the higher efficiency of CaO compared to other essential metal oxide catalysts [56]. However, heterogeneous acid catalysts display less catalytic activity than heterogeneous base catalysts for the esterification and transesterification processes. Heterogeneous acid catalysts are generally less corrosive and toxic compared to homogeneous catalysts. However, the low catalytic activity of heterogeneous catalysts means a slow reaction rate. Sulfonic ion-exchange resins (including Nafion-NR50, Amberlyst-35, and Amberlyst-15), SnO_2, TiO_2, and ZrO_2 are various heterogeneous acid catalysts [52]. Industrial waste-derived catalysts, including slag, red mud, and lime mud, have been investigated, and catalysts are derived from biological sources such as animal bones and shells. Lime mud, the waste produced in the construction industry, is composed primarily of calcium carbonate with minimal magnesium carbonate. Thus, lime mud can be used as a base catalyst as CaCO3 decomposes to CaO through calcination at high temperatures. The lime mud, which consists of CaCO3, acts as an ecologically benign and cost-effective base catalyst for transesterification, where it can provide the CaO on calcination at elevated temperatures [57]. In contrast, red mud, comprising a mixture of Fe, SiO2, and Al (Iron, silica and aluminum), has been calcinated at 200°C for five hours, exhibiting strong activity [58]. The reaction temperature and oil ratio of soybean oil were optimized for transesterification, resulting in a high biodiesel yield (94%). Dolomite rock, calcined at 800°C for two hours, has been used as the catalyst for transesterifying palm kernel oil (refined) and methanol [59].

10.4 CONCLUSIONS

The present environmental concerns resulting from the utilization of fossil fuels have spurred the need to explore fuel alternatives. Biodiesel is an excellent choice because it is a renewable substitute for diesel fuel since it is made from biomass waste materials. It also has reduced emissions of pollutants as compared to standard diesel fuel obtained from fossil sources.

Nevertheless, the present production of biodiesel encounters specific challenges, including the inability to recover the catalyst, the costly separation process, and the substantial formation of wastewater. These concerns primarily arise from the use of homogeneous catalysts. Hence, a novel catalyst or technology is required to address the issues mentioned previously. It is important to consider the advancement of diverse catalyst and reactor technologies, including batch, semi-batch/semi-continuous, or continuous process modes. In addition to conventional commercial heterogeneous catalysts like CaO available for biodiesel production, several other heterogeneous catalysts generated from renewable sources, including biomass waste, industrial waste (such as lime mud and red mud), and food waste (such as pig bones and mussel shells), have also been examined as effective catalytic systems. However, their commercialization is still pending due to the unavailability of suitable technologies. Thus, the development of appropriate technologies and catalytic systems is the primary goal to achieve in the future for industrial biodiesel production.

ACKNOWLEDGMENT

KNS thanks Science and Engineering Research Board, New Delhi (Govt. of India) for the financial support provided under the SERB-SIRE award (File No. SIR/2022/001466).

REFERENCES

1. Dehkhoda, A. M.; West, A. H.; Ellis, N. Biochar based solid acid catalyst for biodiesel production. *Appl Catal A Gen*, **2010**, 382, 197–204.
2. Demirbas, A. Comparison of transesterification methods for production of biodiesel from vegetable oils and fats. *Energy Convers Manag*, **2008**, 49, 125–130.
3. Ma, F.; Hanna, M. A. Biodiesel production: A review 1. *Biosource Tech*, **1999**, 70, 1–15.
4. Petchmala, A.; Laosiripojana, N.; Jongsomjit, B.; Goto, M.; Panpranot, J.; Mekasuwandumrong, O.; Shotipruk, A. Transesterification of palm oil and esterification of palm fatty acid in near- and super-critical methanol with SO4-ZrO2 catalysts. *Fuel*, **2010**, 89, 2387–2392.
5. Taufiq-Yap, Y. H.; Lee, H. V.; Hussein, M. Z.; Yunus, R. Calcium-based mixed oxide catalysts for methanolysis of Jatropha curcas oil to biodiesel. *Biomass Bioenergy*, **2011**, 35, 827–834.
6. Avramović, J. M.; Veličković, A. V.; Stamenković, O. S.; Rajković, K. M.; Milić, P. S.; Veljković, V. B. Optimization of sunflower oil ethanolysis catalyzed by calcium oxide: RSM versus ANN-GA. *Energy Convers Manag*, **2015**, 105, 1149–1156.
7. Vadery, V.; Narayanan, B. N.; Ramakrishnan, R. M.; Cherikkallinmel, S. K.; Sugunan, S.; Narayanan, D. P.; Sasidharan, S. Room temperature production of jatropha biodiesel over coconut husk ash. *Energy*, **2014**, 70, 588–594.
8. Betiku, E.; Omilakin, O. R.; Ajala, S. O.; Okeleye, A. A.; Taiwo, A. E.; Solomon, B. O. Mathematical modeling and process parameters optimization studies by artificial neural network and response surface methodology: A case of non-edible neem (Azadirachta indica) seed oil biodiesel synthesis. *Energy*, **2014**, 72, 266–273.
9. Selvabala, V. S.; Selvaraj, D. K.; Kalimuthu, J.; Periyaraman, P. M.; Subramanian, S. Two-step biodiesel production from Calophyllum inophyllum oil: Optimization of modified β-zeolite catalyzed pre-treatment. *Bioresour Technol*, **2011**, 102, 1066–1072.

10. Banković-Ilić, I. B.; Stamenković, O. S.; Veljković, V. B. Biodiesel production from non-edible plant oils. *Renew Sustain Energy Rev,* **2012**, 16, 3621–3647.

11. Wan Omar, W. N. N.; Amin, N. A. S. Biodiesel production from waste cooking oil over alkaline modified zirconia catalyst. *Fuel Process Technol,* **2011**, 92, 2397–2405.

12. Galadima, A.; Muraza, O. Biodiesel production from algae by using heterogeneous catalysts: A critical review. *Energy,* **2014**, 78, 72–83.

13. Teo, S. H.; Islam, A.; Yusaf, T.; Taufiq-Yap, Y. H. Transesterification of Nannochloropsis oculata microalga's oil to biodiesel using calcium methoxide catalyst. *Energy,* **2014**, 78, 63–71.

14. Tran, D. T.; Yeh, K. L.; Chen, C. L.; Chang, J. S. Enzymatic transesterification of microalgal oil from Chlorella vulgaris ESP-31 for biodiesel synthesis using immobilized Burkholderia lipase. *Bioresour Technol,* **2012**, 108, 119–127.

15. Nurfitri, I.; Maniam, G. P.; Hindryawati, N.; Yusoff, M. M.; Ganesan, S. Potential of feedstock and catalysts from waste in biodiesel preparation: A review. *Energy Convers Manag,* **2013**, 74, 395–402.

16. Christopher, L. P.; Kumar, H.; Zambare, V. P. Enzymatic biodiesel: Challenges and opportunities. *Appl Energy,* **2014**, 119, 497–520.

17. Tran, D. T.; Chen, C. L.; Chang, J. S. Effect of solvents and oil content on direct transesterification of wet oil-bearing microalgal biomass of Chlorella vulgaris ESP-31 for biodiesel synthesis using immobilized lipase as the biocatalyst. *Bioresour Technol,* **2013**, 135, 213–221.

18. Helwani, Z.; Othman, M. R.; Aziz, N.; Kim, J.; Fernando, W. J. N. Solid heterogeneous catalysts for transesterification of triglycerides with methanol: A review. *Appl Catal A Gen,* **2009**, 363, 1–10.

19. Islam, A.; Taufiq-Yap, Y. H.; Chan, E. S.; Moniruzzaman, M.; Islam, S.; Nabi, M. N. Advances in solid-catalytic and non-catalytic technologies for biodiesel production. *Energy Convers Manag,* **2014**, 88, 1200–1218.

20. Semwal, S.; Arora, A. K.; Badoni, R. P.; Tuli, D. K. Biodiesel production using heterogeneous catalysts. *Bioresour Technol,* **2011,** 102, 2151–2161.

21. Gotch, A.J. Study of heterogeneous base catalysts for biodiesel production. *J Undergrad Chem Res,* **2008**, 4, 58–62.

22. Lam, M. K.; Lee, K. T.; Mohamed, A. R. Homogeneous, heterogeneous and enzymatic catalysis for transesterification of high free fatty acid oil (waste cooking oil) to biodiesel: A review. *Biotechnol Adv,* **2010**, 28, 500–518.

23. Luque, R.; Pineda, A.; Colmenares, J. C.; Campelo, J. M.; Romero, A. A.; Serrano-Riz, J. C.; Cabeza, L. F.; Cot-Gores, J. Carbonaceous residues from biomass gasification as catalysts for biodiesel production. *J Nat Gas Chem,* **2012**, 21, 246–250.

24. Loh, Y. R.; Sujan, D.; Rahman, M. E.; Das, C. A. Review sugarcane bagasse—the future composite material: A literature review. *Resour Conserv Recycl,* **2013**, 75, 14–22.

25. Hossain, M. N.; Ullah Siddik Bhuyan, M. S.; Md Ashraful Alam, A. H.; Seo, Y. C. Biodiesel from hydrolyzed waste cooking oil using a S-ZrO$_2$/SBA-15 super acid catalyst under sub-critical conditions. *Energies,* **2018**, 11, 299.

26. Hamza, M.; Ayoub, M.; Shamsuddin, R.; Mukhtar, A.; Saqib, S.; Zahid, I.; Ameen, M.; Ullah, S.; Al-Sehemi, A. G.; Ibrahim, M. A review on the waste biomass derived catalysts for biodiesel production. *Environ Technol Innov*, **2021**, 21, 101200.

27. Sakdasri, W.; Sawangkeaw, R.; Ngamprasertsith, S. Continuous production of biofuel from refined and used palm olein oil with supercritical methanol at a low molar ratio. *Energy Convers Manag*, **2015**, 103, 934–942.

28. Demirbas, A. Comparison of transesterification methods for production of biodiesel from vegetable oils and fats. *Energy Convers Manag*, **2008**, 49, 125–130.

29. Piloto-Rodriguez, R., Melo, E. A., Goyos-Pérez, L.; Verhelst, S. Conversion of by-products from the vegetable oil industry into biodiesel and its use in internal combustion engines: A review. *Braz J Chem Eng*, **2014**, 31, 287–301.

30. Sagiroglu, A.; Isbilir, S. Ş.; Ozcan, M. H.; Paluzar, H.; Toprakkiran, N. M. Comparison of biodiesel productivities of different vegetable oils by acidic catalysis. *Chem Ind Chem Eng Q*, **2011**, 17, 53–58.

31. Leung, D. Y.; Wu, X.; Leung, M. K. H. A review on biodiesel production using catalyzed transesterification. *Appl Energy*, **2010**, 87, 1083–1095.

32. Nomanbhay, S.; Ong, M. Y. A review of microwave-assisted reactions for biodiesel production. *Bioeng*, **2017**, 4, 57.

33. Micic, R. D.; Tomić, M. D.; Kiss, F. E.; Martinovic, F. L.; Simikić, M. Đ.; Molnar, T. T. Comparative analysis of single-step and two-step biodiesel production using supercritical methanol on laboratory-scale. *Energy Convers Manag*, **2016**, 124, 377–388.

34. Parawira, W. Biodiesel production from Jatropha curcas: A review. *Sci Res Essay*, **2010**, 5, 1796–1808.

35. Soltani, S.; Rashid, U.; Al-Resayes, S. I.; Nehdi, I. A. Recent progress in synthesis and surface functionalization of mesoporous acidic heterogeneous catalysts for esterification of free fatty acid feedstocks: A review. *Energy Convers*, **2017**, 141, 183–205.

36. Semwal, S.; Arora, A. K.; Badoni, R. P.; Tuli, D. K. Biodiesel production using heterogeneous catalysts. *Bioresour Technol*, **2011**, 102, 2151–2161.

37. Konwar, L. J.; Boro, J.; Deka, D. Review on latest developments in biodiesel production using carbon-based catalysts. *Renew Sust Energy Rev*, **2014**, 29, 546–564.

38. Narowska, B.; Kułażyński, M.; Łukaszewicz, M., Burchacka, E. Use of activated carbons as catalyst supports for biodiesel production. *Renew Energy*, **2019**, 135, 176–185.

39. Shahbazi, M. R.; Khoshandam, B.; Nasiri, M.; Ghazvini, M. Biodiesel production via alkali-catalyzed transesterification of Malaysian RBD palm oil–Characterization, kinetics model. *J Taiwan Inst Chem Eng*, **2012**, 43, 504–510.

40. Avhad, M. R.; Marchetti, J. M. A review on recent advancement in catalytic materials for biodiesel production. *Renew Sust Energy Rev*, **2015,** 50, 696–718.

41. Kulkarni, M. G.; Dalai, A. K. Waste cooking oil an economical source for biodiesel: A review. *Ind Eng Chem Res*, **2006**, 45, 2901–2913.

42. Narowska, B.; Kułażyński, M.; Łukaszewicz, M., Burchacka, E. Use of activated carbons as catalyst supports for biodiesel production. *Renew Energy*, **2019**, 135, 176–185.

43. Bart, J. C.; Palmeri, N.; Cavallaro, S. *Biodiesel Science and Technology.* Elsevier, **2010**, 462.

44. Farobie, O.; Matsumura, Y. State of the art of biodiesel production under supercritical conditions. *Prog Energy Combust Sci,* **2017**, 63, 173–203.

45. Islam, A.; Taufiq-Yap, Y. H.; Chan, E. S.; Moniruzzaman, M.; Islam, S.; Nabi, M. N. Advances in solid-catalytic and non-catalytic technologies for biodiesel production. *Energy Convers Manag,* **2014**, 88, 1200–1218.

46. Piker, A.; Tabah, B.; Perkas, N.; Gedanken, A. A green and low-cost room temperature biodiesel production method from waste oil using egg shells as catalyst. *Fuel*, **2016**, 182, 34–41.

47. Muciño, G. G.; Romero, R.; Ramírez, A.; Martínez, S. L.; Baeza-Jiménez, R.; Natividad, R. Biodiesel production from used cooking oil and sea sand as heterogeneous catalyst. *Fuel*, **2014**, 138, 143–148.

48. Boro, J.; Konwar, L. J.; Thakur, A. J.; Deka, D. Ba doped CaO derived from waste shells of T striatula (TS-CaO) as heterogeneous catalyst for biodiesel production. *Fuel*, **2014**, 129, 182–187.

49. Hindryawati, N.; Maniam, G. P.; Karim, M. R.; Chong, K. F. Transesterification of used cooking oil over alkali metal (Li, Na, K) supported rice husk silica as potential solid base catalyst. *Eng Sci Technol*, **2014**, 17, 95–103.

50. Buasri, A.; Chaiyut, N.; Loryuenyong, V.; Rodklum, C.; Chaikwan, T.; Kumphan, N. Continuous process for biodiesel production in packed bed reactor from waste frying oil using potassium hydroxide supported on Jatropha curcas fruit shell as solid catalyst. *Appl Sci*, **2014**, 2, 641–653.

51. Zabeti, M.; Daud, W. M. A. W.; Aroua, M. K. Activity of solid catalysts for biodiesel production. A review. *Fuel Process Technol*, **2009**, 90, 770–777.

52. Miao, C. X.; Gao, Z. Preparation and properties of ultrafine SO_2–/ZrO_2 superacid catalysts. *Mater Chem Phys*, **1997**, 50, 15–19.

53. Tang, X.; Niu, S. Preparation of carbon-based solid acid with large surface area to catalyze esterification for biodiesel production. *Jind Eng Chem*, **2019**, 69, 187–195.

54. Mansir, N.; Taufiq-Yap, Y. H.; Rashid, U.; Lokman, I. M. Investigation of heterogeneous solid acid catalyst performance on low grade feedstocks for biodiesel production: A review. *Energy Convers Manag*, **2017**, 141, 171–182.

55. Tang, Z. E.; Lim, S.; Pang, Y. L.; Ong, H. C.; Lee, K. T. Synthesis of biomass as heterogeneous catalyst for application in biodiesel production: State of the art and fundamental review. *Renew Sust Energ Rev*, **2018**, 92, 235–253.

56. Di Serio, M.; Ledda, M.; Cozzolino, M.; Minutillo, G.; Tesser, R.; Santacesaria, E. Transesterification of soybean oil to biodiesel by using heterogeneous basic catalysts. *Ind Eng Chem*, **2006**, 45, 3009.

57. Li, H.; Niu, S.; Lu, C.; Liu, M.; Huo, M. Use of lime mud from paper mill as a heterogeneous catalyst for transesterification. *Sci China Technol Sci,* **2014**, 57, 438–444.
58. Liu, Q.; Xin, R.; Li, C.; Xu, C.; Yang, J. Application of red mud as a basic catalyst for biodiesel production. *J Environ Sci,* **2013**, 25, 823–829.
59. Ngamcharussrivichai, C.; Wiwatnimit, W.; Wangnoi, S. Modified dolomites as catalysts for palm kernel oil transesterification. *J Mol Catal A Chem,* **2007**, 276, 24–33.

CHAPTER 11

Biomass Wastes for Electrolyte and Electrode Applications

Ritika Gera, Lalit, Matthew C. Risi, Graham C. Saunders, Kamalakanta Behera and Kamal Nayan Sharma

11.1 INTRODUCTION

The current environmental and climate change issues are due to the rise of global warming and the increase in the consumption of fossil fuels. To overcome this problem, constructive initiatives and techniques have been encouraged to prevent additional climate damage. In this regard, several renewable energy sources have been researched as potential replacements for petroleum-based fuels, which help reduce environmental damage. The International Renewable Energy Agency has estimated that the worldwide consumption of renewable electricity may climb to 86% by the year 2050 due to severe policies implemented globally that promote a circular carbon economy [1]. In recent years, a global quest for sustainable energy solutions and environmentally friendly technologies has generated a growing fascination with biomass as a renewable source. Biomass, derived from organic materials such as wood, agricultural residues, and algae, offers a unique opportunity to harness waste products for energy storage and conversion by utilizing CO_2 through photosynthesis. The intricate and varied morphology of biomass-derived carbon materials (B-d-CMs)

DOI: 10.1201/9781003466833-14

and their cost-effective origin, their sustainability, and their feasibility for large-scale production make them an up-and-coming class of electrode materials for electrochemical energy storage (EES) [2]. In addition, various simple and environmentally friendly techniques have been suggested to convert biomass into beneficial carbon-based products despite the use of expensive chemical substances and complex methodologies, such as one-step pyrolysis, hydrothermal carbonization, chemical and physical activations, and molten salt carbonization [3]. The formed structure and morphology of the B-d-CMs depend primarily on the raw materials utilized during production. Also, the properties that B-d-CMs will depict rely on the temperature parameters involved during the thermal treatment of preparation. Employing biomass-based materials in our day-to-day life offers many advantages. For example, biomass waste utilization reduces the burden of landfills and waste disposal. It also minimizes the carbon footprint of energy storage and conversion technologies.

Nowadays, petrochemicals are used to create the majority of organic battery components. On the other hand, various raw compounds can now be produced from bioresources through innovative biorefinery techniques. A wide variety of carbonized and noncarbonized biomass could be used as electrode material in batteries: all types of biomass (waste) can be carbonized and utilized as conductive additives or as anodes for lithium-ion- or sodium-ion batteries, as well as cathodes for metal-oxygen or metal-sulfur batteries. However, several biomolecules with redox-active groups can be employed as redox-active agents in electrodes without chemical modification, like quinones or carboxylates. Hence, energy storage systems can become fully sustainable in the future by using biomass-based binders instead of harmful halogenated commercial binders. It is also possible to create separators and electrolytes from biomass in addition to electrodes. Thus, almost all biomass on earth may eventually be used in battery applications [4]. Similarly, many biomass precursors with economic and environmental benefits, such as banana peels, pitch and lignin, glucose, and lotus leaves, demonstrate good electrochemical performance in Na-ion batteries. Another potential biomass precursor is waste textiles, like silk fibers, which have gathered attention due to their distinct efficiency and mechanical characteristics [5]. This chapter explores the promising applications of biomass wastes in electrolytes and electrodes, focusing on their potential in energy storage, electrochemical devices, and environmental benefits.

11.2 BIOMASS AS A SUSTAINABLE RESOURCE

Biomass comprises animal and plant waste and the energy stored in them. With the ample quantity of resources, easy availability, and even reduced costs, the use of biomass in energy storage and conversion is possible. Biomass will also significantly decrease the need for traditional fossil fuels. In developing nations, the high use of biomass resources comes into view. In addition to agricultural and animal waste, a significant quantity of firewood has been used as fuel for cooking and domestic heating and to fuel steam engines for electricity production. Consequently, it will aid in air pollution. Hence, the sustainable use of biomass resources is in demand. [6].

Biomass can be of different types depending on their origin. It can be aquatic biomass, wood-derived biomass, animal and plant waste biomass, herbaceous biomass, and mixtures of several kinds [6], [7]. The current primary sources of biomass are agricultural wastes, animal residues, algae and forest leftovers (waste from manufacturing wood products, such as sawdust, splintered wood, and so on), sewage, and aquatic crops [6]. Depending on the agricultural branches in which they originated, agricultural wastes can be categorized. Waste is from food manufacturing, field cultivation, breeding processes, factory farms, and animal husbandry facilities. It is mainly composed of the following: manure and fertilizer, sludge, fruit and vegetable surplus, post-slaughter, rice straw, stalks, molasses, waste from dairy production, etc. [8]. Biomass also comprises solid waste from municipalities (MSW) and garbage streams from anthropogenic activities that cannot be reused. Vast quantities of biomass waste are created worldwide; the highest annual productions of about 731.3 million tons of rice straw, 354.34 million tons of wheat straw, 180.73 million tons of bagasse from sugarcane, and 128.02 million tons of corn stover [6].

The properties of biomass waste differ significantly because of the high heterogeneity. Hence, their conversion to proper forms requires different pretreatment processes like biochemical conversion, thermal conversion, etc. Biomass waste can be converted through biochemical conversion into sugar and starches, which can be used to form liquid biofuels like ethanol after fermentation. Based on the type of biomass used, these liquid biofuels can be first-generation (1G) liquid biofuels, second-generation (2G) liquid biofuels and third-generation (3G) liquid biofuels. Edible biomass, rich in starch, sugar, and fats, produces first-generation (1G) biofuels. Biomass from aquatic waste like cyanobacteria, algae, and fungi produces

3G biofuels. Lignocellulosic biomass or 1G biomass waste is used to produce 2G biofuels [6].

Thermochemical conversion of biomass sources converted into cellulose and hemicellulose, which on gasification formed into syngas. The biogas and biofuels obtained are used to fulfil domestic energy needs [9].

These biofuels also cover the transportation sector since they can be generated everywhere where biomass can be obtained. They offer a valuable environmentally sustainable energy source that helps grow forest-agriculture and associated businesses. The energy value of agriculturally produced biomass depends significantly on the type of crop from which it is obtained [10].

Organic matter can also produce solid fuels like pellet fuels from energy crops, industrial waste, agricultural residues, food waste, and leftover lumber. Pellet fuels can be used for pellet boilers and domestic heating furnaces. Their attributes include a low percentage of moisture, low levels of ash, and carbon-neutral greenhouse gas emissions. Consuming biomass for burning can also help lower the quantity of methane released into the atmosphere during the breakdown of biomass in open compost systems or dumps [9].

Agricultural biomass can be directly burned, vaporized, and fermented in industry. Due to the scarcity of liquid and gaseous fuels, the vaporization process is highly advantageous for producing heat. It is feasible to use agricultural biomass in warmer nations and those with more temperate climates [8]. One can use either proximate analysis, ultimate analysis, or biochemical analysis to identify critical characteristic features:

11.2.1 Analysis of Proximities

The biomass sample's moisture (M) content, amount of ash (Ash) and volatile matter (VM), and fixed carbon (FC) concentrations can be determined using proximate analysis. Thermogravimetric analysis, or TGA, is a characteristic technique used for proximate analysis.

$$FC\% = 100\% - Ash\% - VM\%$$

The elemental analysis confirms the biomass samples' elemental compositions, including C, H, N, O, S, and occasionally P. Catalytic combustion and reduction techniques are usually used to characterize it [7].

$$O\% = 100\% - C\% - H\% - N\% - S\% - P\% - Ash\%$$

Governments in developed nations support energy policies aimed at lowering greenhouse gas emissions and, in turn, preventing climate change. One noteworthy climate change initiative in the US is the "Clean Power Plan." It includes the terms that the United States Environmental Protection Agency (USEPA) declared in September 2013 to achieve a 30% reduction in emissions by 2030. For the first time, the Energy Plan established the guidelines for carbon pollution generated from utilities and power plants with the adaptability required to satisfy their requirements [11].

When biomass is burned effectively, the energy contained in the chemical bonds is released, and carbon dioxide and water are produced when atmospheric oxygen reacts with plant carbon. The process is cyclic since carbon dioxide can then be used to create new biomass. This procedure validates the acquisition of carbon credits. Therefore, using biomass instead of a plant that runs on fossil fuels can help to mitigate climate change.

Because biomass is reduced, processed, and reused, it can be considered an environmentally sustainable option. New applications for biomass are continually being identified. Making ethanol, a liquid alcohol fuel, is one method. Certain automobiles are designed to run on alcohol fuel rather than petrol. Gasoline and alcohol can also be mixed to lower our reliance on fossil fuels and oil [12].

In addition, it has additional positive environmental benefits like reduced levels of NOx compounds and sulfur emission and can aid in restoring degraded areas. Utilizing biomass waste may also reduce the effects of acid rain, soil erosion, water pollution, landfill pressure, and climate change. It can also provide habitat for wildlife and support the health of forests by improving management [13].

11.3 BIOMASS-DERIVED CARBON MATERIALS

Researchers and industrial engineers are paying close attention to energy storage as it has become an essential global concern because of the rapid growth of the economy and the rising popularity of electronic cars. On the other hand, as the need for electricity grows and fossil fuels run out, more non-conventional energy storage technologies like supercapacitors (SCs), batteries, and fuel cells are being researched and developed. Carbon and its derivates are widely used to produce energy storage devices [14].

Being the most significant and widely available materials, carbon offers many kinds of available allotropes, including mono- and bi-layered graphite slices, diamond, fullerenes, carbon nanotubes, and wonder material meta-crystalline graphene. These allotropes have distinct physical-chemical characteristics like high porosity, adjustable physico-chemical properties, high conductivity, structural flexibility, and chemical stability [15]. The arrangement of the carbon atoms mainly determines these properties and is responsible for their widespread application in lithium-ion batteries, fuel cells, and other energy storage systems (EES) [16].

Fullerenes are commonly used as electron acceptors in solar cells. As electrophiles, they can function as an e-transferring medium with mobility ranging from 10^{-4} – 10^{-3} sq. cm per V per s. Furthermore, they have been utilized in fuel cells, antibacterial coatings on water pipelines and lubricants, medicine delivery, supercapacitors, and hydrogen storage.

Carbon nanotubes (CNTs) can also be used as energy storage devices, transistors, sensors, optical devices, and in bio-applications as they offer a plethora of interesting properties and features, like chemical and mechanical stability, chirality-dependent metallic and semiconducting nature, and higher thermal and electrical conductivity.

Another C-based material, graphene, has remarkable mechanical, thermal, electrical, and electronic capabilities. Thermodynamically stable, it exhibits electron mobility of around 10^4 sq. cm per V per s in atmospheric conditions. This property makes it a promising material for energy storage and transformation, sensors, and water splitting. However, its zero bandgap nature blocks its use in electronics. To use them in energy storage devices and field-effect transistors, graphene sheets have been doped with heteroatoms such as nitrogen, phosphorus, sulfur, and boron to induce bandgap [15].

Numerous synthetic methods for preparing carbon nanotubes, C-based biomass, and carbon films have been proposed to fulfil the requirements of all these applications. Some examples include Laser ablation, direct pyrolysis, template-based synthesis, hydrothermal and solvothermal methods, combustion methods, microwave heating, etc. These techniques highlight the environmental impact and long refining time, making them incompatible with mass production [17].

Such carbon compounds are often synthesized through several processes using a costly precursor feedstock that needs to be pretreated. Certain carbon forms, such as carbon nanofibers, carbon nanotubes, and fullerenes, are produced commercially and in laboratories using costly equipment, carbonization, and plasma-enhanced chemical vapor deposition after electrospinning. To create graphene, which has a greater surface area than single-walled carbon nanotubes, several processes must be performed, including chemical and mechanical exfoliation, epitaxial growth, and chemical vapor deposition. Hence, using economical methods, the current focus is on developing C-based compounds from renewable resources.

11.3.1 Carbonization of Biomass

Considering the potential utility, biomass-derived carbon compounds are in demand because of their readily available bio precursors and more straightforward synthesizing techniques, which reduce expenses for overall development. Lignin cellulose and hemicellulose are the main constituents of biomass. Hydrothermal carbonization, microwave irradiation, laser processing, and pyrolysis are typically used to synthesize bio-derived carbon compounds [3].

11.3.1.1 Pyrolysis   In this process, anaerobic conditions are applied to bio precursors at high temperatures, decomposing at different rates and temperature ranges [3, 18]. The process occurs via free radical reactions once the -C-O- bond cleaves in the β–O–4 structure, eliminating many volatile materials along with CO, CO_2, and H_2O. Polyphenolic molecules are formed during the reaction, which can polymerize and recondense further. The solid mass rearranges and remolds into turbostratic graphene sheets by cross-linking polymers. The properties of carbon can also be enriched by accelerating the reaction rates and temperature ranges, as mentioned in Figure 11.1.

The final characteristics of the carbon material are determined by different pyrolysis and preparation factors, including the biomass's type and composition, the pyrolysis time duration, the catalyst utilized, the rate of heating, and the reaction environment. All these elements function simultaneously.

Additionally, strongly nitrogen-doped multi-layered graphene (containing 15% N), which is biomass-derived C-material, can be synthesized using silk cocoons, which can enhance the performance of supercapacitors due to the excellent electronegativity of N-dipoles. Pyrolyzing the cocoon at 400°C in an argon atmosphere and washing the pyrolysis remains with petroleum ether and acetone make the preparation method easy [19].

11.3.1.2 Hydrothermal Carbonization   In this process, biomass carbonization occurs at a temperature lower than pyrolysis. The desired reactant undergoes hydrolysis and polymerization, followed by the generation

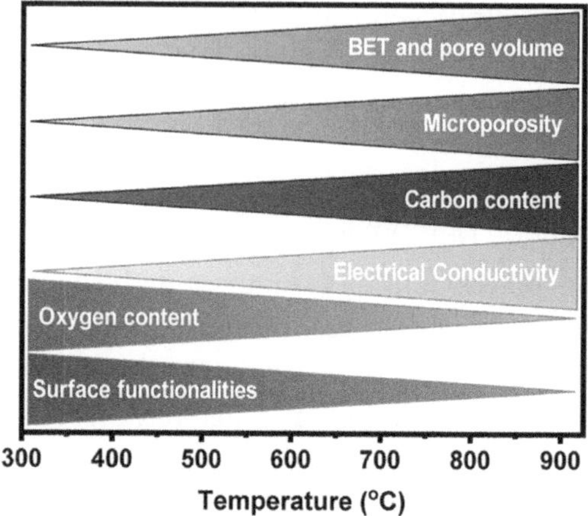

FIGURE 11.1 Shifts in the properties of biochar as the pyrolysis temperature rises.

of spherical particles. At 200°C, hydrolysis of ether bonds occurs while pyrolysis reactions involve free radical formation at high temperatures, e.g. hydrothermal carbonization of hemicellulose and lignocellulose, which undergoes hydrolysis and decomposition—

$$4\left(C_6H_{10}O_5\right)_n \rightarrow 2\left(C_{12}H_{10}O_5\right)_n + 10H_2O$$

Different characterization techniques are used to determine these biomass-derived C-materials' size and morphological structure, like scanning electron microscopy (SEM) and transmission electron microscopy (TEM). Raman spectroscopy is used to identify the graphitization extent and crystallinity disorders in particles, while XRD is used to visualize the arrangement of the particles in crystal lattices. Surface characterization by N2 adsorption and desorption isotherms, such as the size and volume of pores and surface area of particles, is evaluated through the BET method [3].

The applications of biomass-based C-materials are expanding rapidly, from micro batteries to large-scale electric cars. To meet the requirement, high-performance, smaller, lighter, more adaptable, and economically feasible energy storage devices (ESD) will be required shortly [14, 15]. They have many applications, from lithium-ion and Na-ion batteries to supercapacitors and fuel cells, because of their many characteristics like structure, thermal stability, and thermal conductivity, as depicted in Figure 11.2. The applications of biomass-based C-materials are further described as follows.

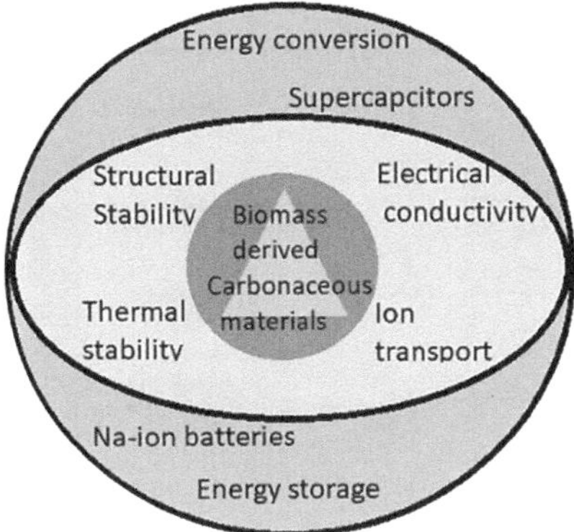

FIGURE 11.2 Properties and application of biomass-derived carbonaceous material.

11.3.2 Biomass-Derived C-Material in Li-Ion Batteries

The biomass-derived C-materials are thought to be taken as promising anode materials for the production of Li-ion batteries because they possess the desired characteristics, such as good conductivity and specific surface area, varying pore sizes (micro and meso pores), a large number of functional groups on their surfaces, tunable hydrophilicity, and economically cheap and environmentally friendly technology for production. Graphite can be used as an anode in these batteries because of its high voltage generation and moderate capacity of 372 mAh per gram with a long life cycle [16].

As the properties of biochar vary with the variation in temperature during pyrolysis treatment, the surface functionalities and the amount of oxygen in biochar decrease with the increase in temperature, as the process involved is volatile, resulting in the volatilization of H and O. Due to this, electrochemical performance as an anode material varies, and as carbon materials can increase battery capacities and power densities when utilized as anode, surface functions and oxygen content are crucial characteristics. Ultimately, homogeneous Li deposition can be facilitated by the oxygen content without requiring dendritic development. Ventosa et al. found that because of the creation of a thicker solid-electrolyte-interface (SEI) layer, the irreversible charge loss of a LIB increased with oxygen functionality. Hence, less oxygen and more carbon content are required to employ biochar as an anode material for batteries [18]. The charging and discharging process of Li-ion batteries mechanism involves the transfer of ions, i.e.,

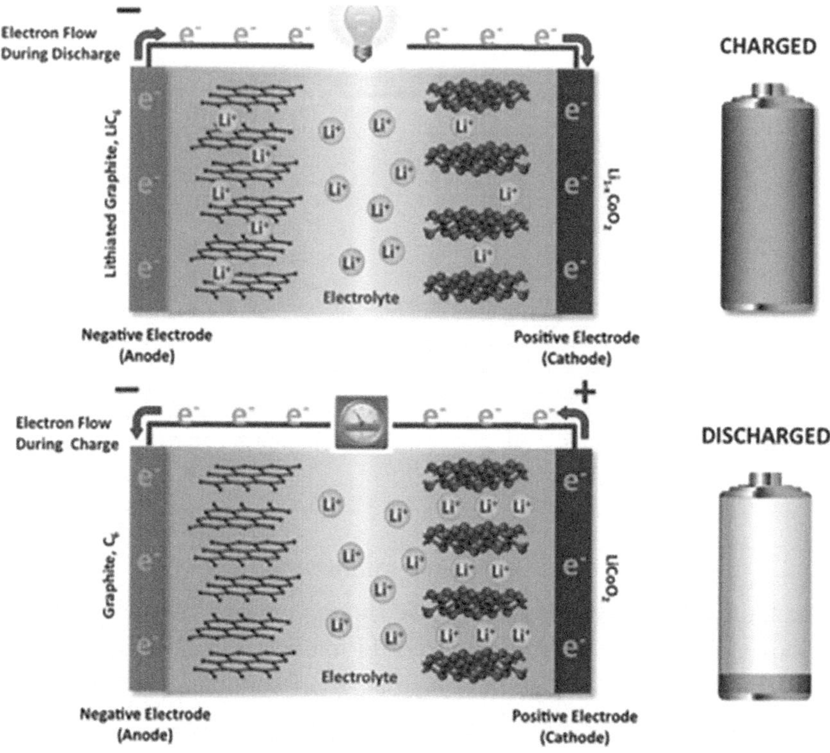

FIGURE 11.3   Charging and discharging process of Li-ion batteries.

anode and cathode, during charging/discharging [20]. While the charging anode moves to the cathode, and the discharge cathode ions move to anode ions. This charging and discharging process is shown in Figure 11.3.

11.3.3 Biomass-Derived C-Material Uses in Supercapacitors

Supercapacitors store electrical charge through reversible electroadsorption of ions from electrolytes onto electrode surfaces. Despite having a lower energy density, they may quickly produce a significant amount of power. They can run for many charge/discharge cycles with an extended lifespan compared to batteries. Supercapacitors are considered potential candidates for energy storage devices due to their fast and reversible storage techniques, which are currently used in a wide range of applications, including small devices (clocks, detectors, smartphones and tablets, speakers, etc.) to giant cells for motorized vehicles such as electric vehicle and bus batteries and for power stations [21].

Synthesizing biomass-derived carbon material through silk cocoons allows for the production of multi-layered graphene strongly doped with nitrogen (15% N content). This nitrogen enrichment enhances supercapacitor performance by leveraging the superior electronegativity of N-dipoles.

11.3.4 Biomass-Derived C-Material Uses in Fuel Cells

Fuel cells are energy-converting devices that use fuel sources to produce electricity. Directly converting chemical energy to electrical energy in fuel cells significantly increases system efficiency [22].

Silk cocoons, containing sericin and fibroin, are also excellent biomass for the biosynthesis of 3D structured biomass C-materials. The silk cocoon can retain its original morphology with numerous layers and holes after carbonization. Furthermore, the sericin in the silk cocoon can be easily removed using a degumming method. The resulting biomass C-material can be utilized as anodes for adjustable fuel cells made from microbial material with good electricity generation capability [19].

11.4 BIOMASS-DERIVED ELECTROLYTES

Numerous issues are being raised about the rapid depletion of fossil fuels [2]. Nowadays, the primary source of energy supplies is nonrenewable fossil fuels. Ideally, it must be non-fossil based and, in fact, reliable, inexpensive, and inexhaustible to create a sustainable future. To satisfy the growing energy demand, it is crucial that we investigate both organic and renewable sources of energy in place of traditional fossil fuels. This will inspire us to look for more sustainable and energy-efficient technologies. This can be achieved by energy storage and conversion. Sustainable energy sources, like biomass energy, have led to the development of several energy storage and conversion technologies, including solar cells, fuel cells, supercapacitors, and batteries [23].

Electrolytes act as a critical element in electrochemical energy storage (EES) devices, which also significantly impact energy efficiency, rate performance, durability, and safety. An electrolyte is simply a liquid or solid that facilitates the passage of ions—not electrons—through a material to conduct electricity. It often functions with a porous membrane or gel. Its identification is essential for high-performing and sustainable EES devices since it is positioned between and in close contact with the positive and negative electrodes. A suitable electrolyte should usually have excellent thermal and chemical stability and ionic solid conductivity; be economically accessible, safer, and non-toxic; have a broad electrochemical window; and chemically inert to other cell materials, like the separator and electrode substrates.

Alessandro Volta invented the first electrochemical battery, often known as the "voltaic pile," in brine, which acts as an electrolyte, whereas a solution of NH_4Cl was used as an electrolyte in Leclanche's cell, whose dry cell form is the precursor of the neutral Zinc-Ion batteries (ZIB), i.e., $Zn–MnO_2$ batteries. Many old and modern EES devices now utilize aqueous electrolytes [2]. Organic electrolytes are used to achieve high operating voltages. Many conventional rechargeable Li- or Na-ion batteries use organic electrolytes. As sodium is widely available and reasonably priced, ambient temperature sodium-ion batteries (SIBs) hold great promise for extensive grid energy storage purposes where NaClO4 or NaPF6 dissolved in carbonate ester solvents, like ethylene or propylene carbonate, are used in electrolyte formulations. When high energy is required for the device, the low working voltage of the aqueous EES devices remains a disadvantage. Organic electrolytes may provide a more extensive operating potential range than aqueous electrolytes. They serve as the foundation for the majority of traditional Li batteries. However, organic electrolytes are still hampered by safety concerns [2]. The organic solution's mobility and low flash point readily cause electrolytes to leak and evaporate, creating a chance of burning batteries. Yet, because of their high specific energy and high working voltage, they are frequently utilized in portable device batteries [24]. Numerous investigations still use standard electrolytes based on organic solvents, even though aqueous electrolytes are being researched as sustainable substitutes. Unfortunately, they typically do not meet the criteria for sustainable electrolytes since they are generated from petrochemical precursors [4].

Biomass electrodes have great potential as an alternative to these traditional multifunctional electrode materials for applications involving the conversion and storage of renewable energy [25].

The world is now wireless and rechargeable, thanks to lithium-ion batteries (LIBs), which have become essential to everyday life. The main obstacles to their practical implementation include significant electrode volume expansion, poor conductivity of electricity, the dissolution or dispersion of electrochemical components, and scarcity of sustainable anode material, which result in low Coulombic efficacy and swift capacity degradation. Nanostructured bio-based carbonaceous compounds with high carbon content have the potential to improve Li battery performance through nano spatial confinement, long- and short-range electrical conductivity, and surface and interfacial characteristics [24].

Hence, bio-based bio-carbonaceous electrolytes are ideal host material for metal ions in anode in LIBs and also serve as conductive additives due to their high degree of porosity, widespread availability, renewability, environmental sustainability, and many more properties [25, 26]. Biomass-derived carbons function differently in aqueous and non-aqueous electrolytes, depending on the conductivity and size of ions and electrode material interaction with electrolytes [27].

Sustainable biomass electrode materials offer distinct formulations, diverse microstructures, various function groups, water-soluble characteristics, and environmental benefits, making them ideal for high-energy Li-based batteries [24]. All such properties of biomass-derived materials are summarized in Figure 11.4.

Petrochemicals are currently used to generate the majority of organic battery components. Thus, the synthesis of various raw compounds from bioresources has been made possible by innovative biorefinery techniques. Organic battery materials can be obtained from biomass-derived materials through more cost-efficient methods. Apart from conventional

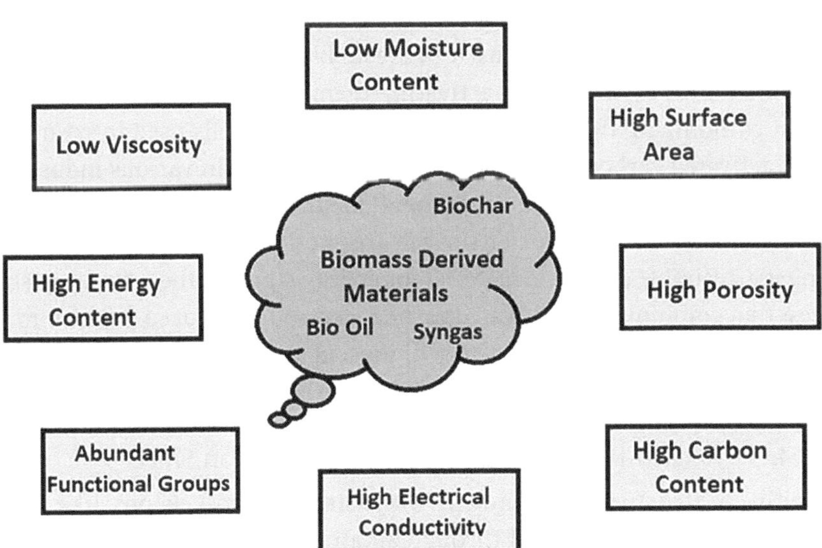

FIGURE 11.4 Characteristics of biomass-derived materials.

biopolymers like lignin and cellulose—which can be used as binding and separating agents—and solid or gel-like electrolytes and terephthalates, which are potential materials for anodes in Li/Na/ or K-ion batteries, Ande quinones used as cathode materials are also derived from bioresources. Not only this, but biomass is also used to derive ionic liquid electrolytes [4].

According to Li et al. (2020), biomass-derived activated carbon is an excellent alternative due to its ease of synthesis, low cost, abundance, and adjustable morphology and surface treatment. Biomass-derived activated carbon's porous nature and rough surface make it ideal for supercapacitor electrodes. Future energy storage devices include electrical double-layer capacitors (EDLCs), called supercapacitors. Experimental studies were conducted on EDLCs with an activated carbon electrode generated from biomass and a redox additive electrolyte of Na_2SO_4–NaI–KI. Moreover, adding KI and NaI increased the electrolyte's ionic mobility and produced a pseudo-capacitance through reduction-oxidation processes. As intended, the EDLCs with the Na_2SO_4–NaI–KI electrolyte system demonstrated energy density and specific capacitance of 22.75 W h kg^{-1} and 334.3 F g^{-1} at a maximum current density of 0.5 A g^{-1}, respectively, upon adding iodine ion (I^-) to the electrolyte.

Pomelo peels are one possible source of bio-waste, among many others. Peels from pomelo trees are a year-round crop widely available in tropical nations. Pomelo peels were used to create biomass-derived activated carbon materials by chemically activating them in potassium hydroxide and then carbonizing them. Pomelo peels are essential oils used in cosmetics and activated carbon derived from the peels is used in various industries, including agriculture and traditional medicine. Pomelo peel-activated carbon exhibited a particular surface area and porous morphology, making it a desirable electrode material for energy storage. These findings suggest that combining an activated carbon electrode produced from biomass and an electrolyte redox reagent will present a new and promising option for use in supercapacitors soon [28].

11.4.1 Chitosan-Based Electrolytes from Crustacean Shells

Chitin, a structural constituent in crustacean exoskeletons like crabs and shrimp, is deacetylated through alkaline hydrolysis to yield chitosan [29]. The stepwise methodology for the production of chitosan is depicted in Figure 11.5. It is the second-most-commonly used natural polymer. Numerous studies have focused on its benefits, including non-toxicity, biodegradability, biorenewability, odorlessness, biofunctionality, and bio-compatibility in biological tissues [29].

Chitosan has β-1,4'-glycosidic linkages and a protonated amino group in the polymeric backbone. These linkages provide structural rigidity and crystalline structure and promote the formation of intramolecular hydrogen bonds. The amino group is responsible for its polyelectrolyte behavior, making it helpful in producing high-conductivity polymers. Due to these β-1,4'-glycosidic linkages, chitosan is only soluble in diluted aqueous acidic solutions and insoluble in organic solvents [30]. The structure of chitosan is shown in Figure 11.6.

β-1,4'-Glycosidic bonds help the creation of intramolecular hydrogen bonding and give chitosan a crystalline form and rigidity. This can be resolved by changing the two functional groups in the chitosan's backbone, -NH$_2$ and -OH, to change its structure. On reaction with phthalic anhydride in dimethylformamide (DMF), organosoluble *N*-phthaloylchitosan (PhCh) is produced because chitosan has a hydrophobic phthaloyl group that prevents the formation of hydrogen bonds when its amino and hydroxyl groups interact with solvents. Hence, PhCh could be a good option for creating polymer or gel electrolytes as a polymer host. Two sets of peaks provided a confirmation of PhCh's synthesis and structure

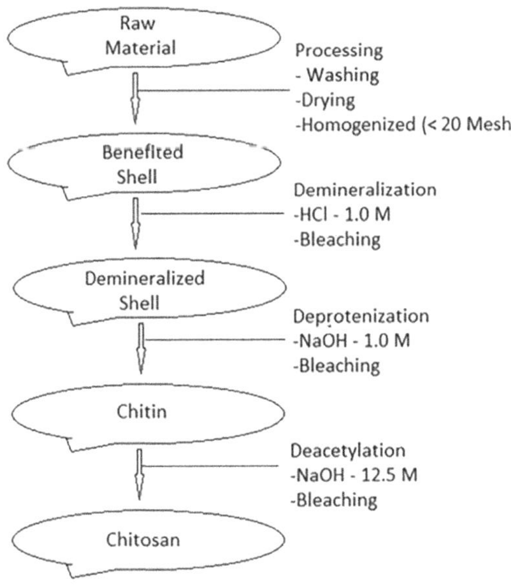

FIGURE 11.5   Production of chitosan.

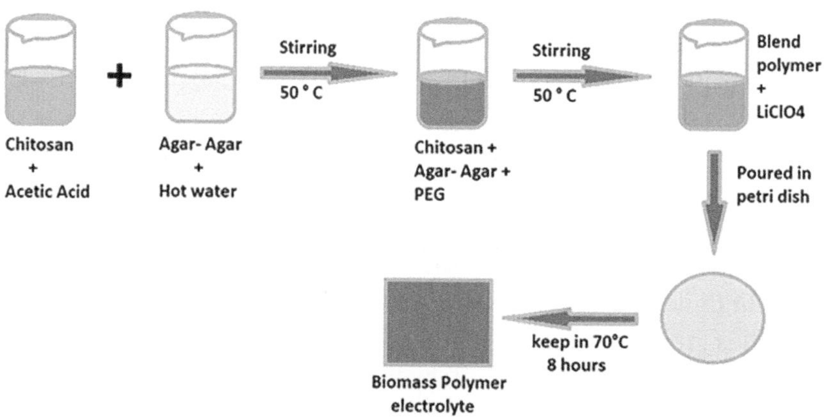

FIGURE 11.6   Structure of Chitosan.

FIGURE 11.7   Preparation of biomass polymer electrolyte.

centered at 3.0 and 7.5 ppm, respectively, obtained from 1H NMR and FT-IR studies, and the typical peaks of aromatic and phthalimido groups observed at 719, 1708, and 1,772 cm^{-1} [30].

The percentage of ions can be determined using transference number measurement (TNM), and by using linear sweep voltammetry (LSV) analysis, the electrochemical stability of the electrolyte sample can be determined [31].

Brine's technique is usually employed to extract the chitosan made from prawn shells effectively, and the solution casting method is used to create the solid polymer electrolyte, which is made up of a mixture of agar-agar and chitosan, polymerized using lithium perchlorate (LiClO$_4$) as a dopant and polyethylene glycol (PEG) as the host polymer. The chitosan blend produced in Figure 11.7 is ideal for use as an electrolyte in electrochemical devices [32].

Attenuated total reflection–Fourier transform infrared spectroscopy (ATR–FTIR), x-ray diffraction (XRD), high-resolution scanning electron microscopy (HR–SEM), and electrochemical impedance spectroscopy are used to characterize the produced polymer electrolyte.

According to the functional groups that make up its structure, chitosan has significant peaks. A stretching vibration band in chitosan corresponding to primary and secondary NH and OH is found in the 3,450–3,200 cm^{-1} range. The NHCO peak in the pyranose ring has been identified between 2,880–2,960 cm^{-1}. The peak at 1024 cm^{-1} displayed the polysaccharide bond's C–O stretching. When combined with mixed polymer electrolytes, the absorption peak at 2,871, 2,884, 2,886, 2,914, and 2,928 cm^{-1} resembles the CH$_2$ stretching of chitosan [32].

So, biomass-derived activated carbon or biochar materials are environmentally sustainable and have numerous applications because of the larger pore volumes, high specific surface area (SSA) and better-regulated microstructure. Hence, they are one of the most promising electrochemical electrode materials for various purposes, including energy storage, detectors, and wastewater treatment. Biomass-based electrolytes can also be used in biochemical species electrochemical sensing [33].

11.5 ADVANTAGES OF BIOMASS-DERIVED MATERIALS

Utilizing biomass waste in electrolyte and electrode applications provides several advantages regarding environmental sustainability, cost-effectiveness, and abundance. These advantages include biochar, which can be more effectively utilized in energy storage and conversion techniques by doping it with metal oxides, heteroatoms, and nanoparticles. Biomass-based materials are gaining popularity in oxygen electrocatalysis, supercapacitors, fuel cells, and batteries. Moreover, biomass-based carbon substances may catalyze trans-esterification, pyrolysis, energy production, and syngas synthesis.

In comparison to other uses, such as soil and water remediation, soil quality improvement (by water preservation, nutrient supplies, increased capacity for cation exchange, conductivity of electricity, and pH increase), the production of biomaterials in the chemical sector (such as the production of activated carbon and as a promoter), and the sorption of drugs and toxins in the medicine sector, it was determined that using biochar as an energy source can be preferable.

Anaerobic digestion of biomass in municipal solid waste produces energy in biogas. According to Shen et al., the biogas generated by anaerobic digestion primarily comprises 50–70% methane and 30–50% carbon dioxide, with trace amounts of ammonia, oxygen, hydrogen, nitrogen, and hydrogen sulfide. On the other hand, excessive ammonia production might be harmful. Adding biochar to soil improves microbial activity and reduces the production of ammonia and carbon dioxide.

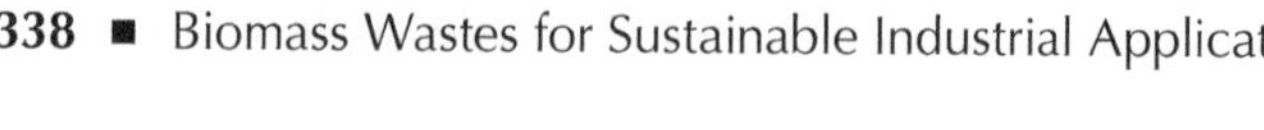

FIGURE 11.8 Applications and advantages of bioenergy sources obtained from biomass materials.

Also, bioenergy sources derived from biomass by thermal treatment can be used for waste management and pollution remediation, renewable energy, and other applications [27]. A flowchart showing the applications and advantages of bioenergy sources obtained from biomass materials is given in Figure 11.8.

11.6 CHALLENGES AND FUTURE DIRECTIONS

Biomass materials are naturally occurring macromolecules that are extensively found and renewable in nature. These materials are derived from chitosan, silk fibroin, cellulose, gelatin, collagen fiber, etc. These biomass materials have exceptional biocompatibility and biodegradability but have comparatively low mechanical and thermal stability. Fortunately, various reactive groups, such as amino-, carboxyl, or hydroxyl- groups, are present in biomass materials and can be changed to enhance their overall characteristics and features. So, to create a sustainable product, the synthesis process for making electrodes for devices to store energy should be standardized, safe, and free from toxicity. Accordingly, porous carbons with variable heteroatom doping, a large surface area, and a good morphology can have beneficial characteristics for charge storage, a high specific capacitance, environmental friendliness, a long cycle life, and compatibility with various electrolytes, such as organic, ionic, and aqueous liquids [34]. In addition, the development of environmentally friendly production processes, recyclable large-scale processing units, and a decrease in production and cost minimization are the primary obstacles that need to be addressed to enhance performance and safety in future development and the widespread use of biomass materials [33]. Hence, while the use of biomass waste in electrolyte and electrode applications is promising, several challenges must be addressed as follows:

11.6.1 Standardization

One of the significant benefits of employing biomass waste is that it is inexpensive, readily available, and has little to no greenhouse gas impact when producing value-added goods. On the contrary, poor to moderate processing yield, low specific energy content, and low biomass bulk density are the primary challenges in its production. Also, its production into valuable materials such as fuel, gases, and carbonaceous compounds is a somewhat complex process that occurs through multistep synthesis, depending on the procedure adopted [35].

11.6.2 Scale-Up

Developed biomass devices are produced on a gram or milligram scale, which will need to be scaled up in the coming years. Considerable funds and applied studies on biomass-derived carbonaceous materials or nanomaterials are required for this to happen. Using biomass waste-derived nanotechnologies, it is now possible to produce a variety of highly useful nanomaterials on a large scale for use in various applications that bulk materials could never provide, such as probing, diagnosis, catalysis, and renewable energy. However, because of their high cost and potentially harmful effects on human health, which still need to be addressed, there is still limited practical application of these applications in daily life. It is believed that if researchers and journals work towards the actual applied utilization of biomass-derived carbonaceous materials, its commercial application can be scaled up [35].

11.6.3 Performance

The outcomes of the first nanomaterials originating from the biomass were adversity of low porosity and inadequate surface area, resulting in the massive accumulation and agglomeration of molecules as the final product. However, these challenges are now mitigated, and nanomaterials extracted from biomass with extremely high surface area have introduced a reality with high potential. Carbon nanomaterials recreated from biomass mainly occur at high temperatures (ranging from 450 to

1,200°C) and are processed for a higher duration under applicable pressure conditions.

11.6.4 Safety and Toxicity

Hazardous chemicals such as H_2SO_4, H_3PO_4, $ZnCl_2$, high-potential metal catalysts, NaOH, and $KMnO_4$ are generally utilized in synthesizing carbon nanomaterials from biomass. Thus, such recycling could be inappropriate if we avoid these hurdles in the future. Several feed product hazards include fire, toxic gas generation, and dust explosion. For example, self-heating is generally produced where feed products like biomass wood are stored in large piles. Microbial activity is initiated in the presence of wood fuel and combined with moisture; storing for a longer duration may lead to heat generation and auto-ignition in several cases.

11.7 CONCLUSIONS

The depletion of fossil fuel and traditional energy sources and the need to lower CO_2 emissions make searching for renewable energy sources significant. Biomass-derived materials are regarded as one of the most adaptable, environmentally friendly, and renewable sources for creating a wide range of carbon structures. Agricultural residues, forest waste, and even wastewater sludge can be converted into valuable products for a greener future. According to reports, these biomass carbonaceous compounds have intriguing properties that make them appropriate for various energy conversion, storage, and environmental applications. Carbon materials, such as activated carbon and carbon nanotubes, have wide-ranging applications, including supercapacitors, lithium-ion batteries, and fuel cells. Biomass-derived carbon materials exhibit several advantages over their synthetic counterparts. They are cost-effective, abundant, and have a reduced environmental footprint. Utilizing biomass waste in electrolyte and electrode applications provides several advantages. Biomass waste utilization reduces the burden on landfills and waste disposal. It minimizes the carbon footprint of energy storage and conversion technologies. Also, biomass-derived materials are often more affordable than their synthetic counterparts. This can make green technologies more accessible to a broader range of users. Moreover, biomass waste is widely available, making it an abundant resource for production. This ensures a consistent and reliable supply for manufacturing processes. While the use of biomass waste in electrolyte and electrode applications is promising, several challenges must be addressed. Developing standardized procedures for biomass waste processing is essential to ensure the reproducibility and reliability of materials; scaling up the production of biomass-derived materials to meet the demands of large-scale applications remains a challenge.

ACKNOWLEDGMENT

KNS thanks Science and Engineering Research Board, New Delhi (Govt. of India) for the financial support provided under the SERB-SIRE award (File No. SIR/2022/001466).

REFERENCES

1. S. Park, J. Song, W. C. Lee, S. Jang, J. Lee, J. Kim, H. K. Kim, K. Min, *J. Chem. Eng.* 470 (2023) 144234.
2. L. Xia, L. Yu, D. Hua, G. Z. Chen, *Mater. Chem. Front.* 1 (2017) 584–618.
3. V. S. Bhat, S. Supriya, G. Hegde, *J. Electrochem. Soc.* 167 (2020) 037526.
4. C. Liedel, *Chem. Sus. Chem.* 13 (2020) 2110–2141.
5. J. Sun, D. Rakov, J. Wang, Y. Hora, M. Laghaei, N. Byrne, X. Wang, P. C. Howlett, M. Forsyth, *Chem. Electro. Chem.* 9 (2022) e202200382.
6. T. Kalak, *Energies* 16 (2023) 1783.
7. K. Wang, J. W. Tester, *Green Energy and Resour.* 1 (2023) 100005.
8. P. Drożyner, W. Rejmer, P. Starowicz, A. Klasa, K. A. Skibniewska, *Tech. Sci.* 16(3) (2013) 211–220.
9. M. Zardzewiały, M. Bajcar, C. Puchalski, J. Gorzelany, *Appl. Sci.* 13 (2023) 3195.
10. N. M. Clauser, G. González, C. M. Mendieta, J. Kruyeniski, M. C. Area, M. E. Vallejos, *Sustainability* 13 (2021) 794.
11. M. A. P. Moreno, E. S. Manzano, A. J. P. Moreno, *Sustainability* 11 (2019) 863.
12. K. O. Olaoye, *Hist. Philos. Sci.: Gen. Stud. Approach* 7 (2014) 64–79.
13. M. N. Uddin, J. Taweekun, K. Techato, M. A. Rahman, M. Mofijur, M. G. Rasul, *Energy Procedia* 160 (2019) 648–654.
14. S. Iqbal, H. Khatoon, A. H. Pandit, S. Ahmad, *Mater. Sci. Energy Technol.* 2 (2019) 417–428.
15. G. Kothandam, G. Singh, X. Guan, J. M. Lee, K. Ramadass, S. Joseph, M. Benzigar, A. Karakoti, J. Yi, P. Kumar, A. Vinu, *Adv. Sci.* 10 (2023) 2301045.
16. T. Prasankumar, S. Jose, P. M. Ajayan, M. Ashokkumar, *Mater. Res. Bull.* 142 (2021) 111425.
17. F. E. Teran, H. Perrot, O. Se, *Physchem* 3 (2023) 355–384.
18. P. Molaiyan, G. S. D. Reis, D. Karuppiah, C. M. Subramaniyam, F. G. Alvarado, U. Lassi, *Batteries* 9 (2023) 116.
19. Y. Wang, M. Zhang, X. Shen, H. Wang, H. Wang, K. Xia, Z. Yin, Y. Zhang, *Small* 17 (2021) 2008079.

20. S. Iqbal, H. Khatoon, A. H. Pandit, S. Ahmad, *Mater. Sci. Energy Technol.* 2 (2019) 417–428.
21. Y. P. Gao, Z. B. Zhai, K. J. Huang, Y. Y. Zhang, *New J. Chem.* 41 (2017) 11456-11470.
22. B. Yogeswari, I. Khan, M. S. Kumar, N. Vijayanandam, P. A. Devarani, H. Anandaram, A. Chaturvedi, W. Misganaw, *J. Nanomater.* (2022) 4949916.
23. Z. Gao, Y. Zhang, N. Song, X. Li, *Matter. Res. Lett.* 5(2) (2017) 69-88.
24. J. Liu, H. Yuan, X. Tao, Yeru Liang, S. J. Yang, J. Q. Huang, T. Q. Yuan, M. M. Titirici, Q. Zhang, *EcoMat.* 2(1) (2020) e12019.
25. M. Zhang, J. Zhang, S. Ran, W. Sun, Z. Zhu, *Electrochem. Commun.* 138 (2022) 107283.
26. A. Kumar, T. Bhattacharya, S. M. Hasnain, A. K. Nayak, M. S. Hasnain, *Mater. Sci. Energy Technol.* 3 (2020) 905–920.
27. B. Karamanova, M. Shipochka, M. Georgiev, T. Stankulov, A. Stoyanova, R. Stoyanova, *Front. Mater.* 8 (2021) 654841.
28. N. T. Nguyen, P.A. Le, V. B. T. Phung, *J. Nanopart. Res.* 22(371) (2020).
29. R. S. C. M. D. Q. Antonino, B. R. P. L. Fook, V. A. D. O. Lima, R. I. D. F. Rached, E. P. N. Lima, R. J. D. S. Lima, C. A. P. Covas, M. V. L. Fook, *Mar. Drugs* 15 (2017) 141.
30. S. N. F. Yusuf, A. D. Azzahari, R. Yahya, S. R. Majid, M. A. Careem, A. K. Arof, *RSC Adv.* 00 (2016) 1–3.
31. S. B. Aziz, M. H. Hamsan, M. M. Nofal, S. San, R. T. Abdulwahid, S. R. Saeed, M. A. Brza, M. F. Z. Kadir, S. J. Mohammed, S. A. Zangana, *Polymers* 12 (2020) 1526.
32. M. L. Edward, K. C. Dharanibalaji, K. T. Kumar, A. R. S. Chandrabose, A. M. Shanmugharaj, V. Jaisankar, *Polym. Bull.* 79 (2022) 587–604.
33. M. Mehmandoust, G. Li, N. Erk, *Ind. Eng. Chem. Res.* 62 (2023) 4628–4635.
34. P. Manasa, S. Sambasivam, F. Ran, *J. Energy Storage* 54 (2022) 105290.
35. S. K. Tiwari, M. Bystrzejewski, A. D. Adhikari, A. Huczko, N. Wang, *PECS* 92 (2022) 101023.

CHAPTER 12

Biomass Wastes for Bioenergy-Based Applications

Nafseen Ahmed, Manoj K. Banjare, Santosh Bahadur Singh, Abbul Bashar Khan, Kamal Nayan Sharma and Kamalakanta Behera

12.1 INTRODUCTION

The extensive use of fossil fuels is the leading cause of pollution and global warming. The necessity of switching to renewable energy sources is highlighted by oil supplies being limited globally and neither renewable nor infinite. The development of alternative energy sources, such as biofuels, hydrogen, wind, solar, and bioenergy, with an emphasis on sustainability and techno-economic viability, has been spurred by this realization [1]. Fossil resource extraction and processing contribute to political, economic, and environmental issues. A possible answer is biorefineries, similar to traditional refineries but using biomass as a feedstock. Across the globe, biomass is the leading renewable source of carbon. The International Energy Agency (IEA) forecast significant drops in the primary energy demand for oil (–9%), coal (–8%), and natural gas (–2%) in 2020, addressing concerns associated with the COVID-19 epidemic. On the other hand, renewable energy is projected to increase due to recent capacity increases, preferential access, and the start of new projects [2]. Bioenergy is produced from biomass by thermochemical or biochemical processes (see Figure 12.1) [3], and it can take many different forms,

DOI: 10.1201/9781003466833-15

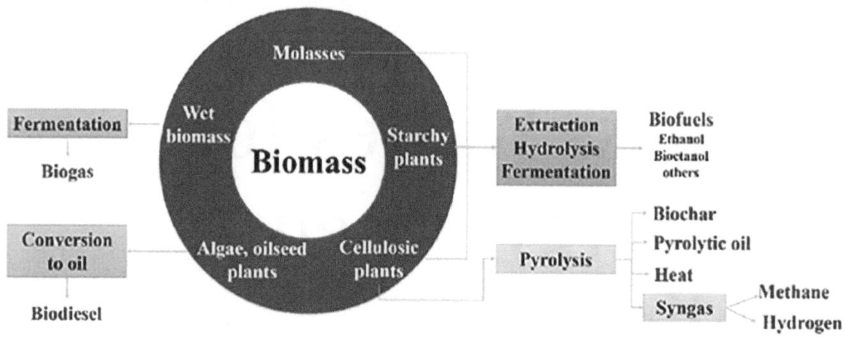

FIGURE 12.1 Various biofuel products obtained from lignocellulosic biomass [3]. Source: BCCY; Open Access Publication.

such as electricity, heat, solid, liquid, or gaseous biofuels. Direct burning of biomass is a feasible but less profitable method; nonetheless, its carbon-neutral nature is crucial. Carbohydrates, the basic building blocks of biomass, are produced via photosynthetic processes involving CO_2 from the air, water, and sunshine. Replanting compensates for the carbon dioxide emitted during biomass burning, maintaining an ongoing cycle of absorption and renewal for ongoing bioenergy production.

Biofuels are divided into three categories based on the kind of biomass that is utilized to produce them: first-generation (1G), second-generation (2G), and third-generation (3G) types. The source of first-generation biofuels is edible biomass high in lipids, carbohydrates, and sugars. Given the continued increase in the world's population, it is crucial to prioritize using such biomass for sustenance to guarantee respect for human rights. Algae and cyanobacteria are among the aquatic biomass used to create third-generation biofuels. Second-generation biofuels are obtained from lignocellulosic biomass (LCB) or the inedible residue left over from first-generation (1G) biomass. However, they provide difficulties since this biomass consists of a complex matrix of lignin, hemicelluloses, and cellulose that must be broken down to produce products with acceptable yields. Compared to its 1G and 3G cousins, 2G biofuels have advantages and need less land, water, and minerals. Low production and commercialization costs support large-scale biofuel production, preventing rivalry with food resources. One gram of ethanol is expected to cost between 0.78 and 0.97 USD/L to produce. Even while 2G biofuels are less expensive than 1G and 3G, it's important to remember that 3G ethanol production is currently only done on a pilot scale. Because of its easily accessible resources and cheap labor, Brazil produces the most economical 1G ethanol, with pricing per liter of around 0.20 USD [1]. Unquestionably, one of the most critical issues of the twenty-first century is the sustainable use of biomass obtained from agro-industrial waste to produce food, medicines, biologically active chemicals, biomaterials, and sustainable energy. Due to the large amounts of waste produced by the food and agriculture industries, there is a chance to turn these byproducts into high-value goods by using them as raw materials, which opens up the potential for sustainable manufacturing [4]. According to the Organization of the United Nations, approximately one-third of all food produced for human consumption worldwide is lost or wasted, or 1.6 billion tons annually (FAO, 2011). Notably, between 40 and 50% of all food waste comprises fruits, vegetables, roots, and tubers [5]. In this case, valuing these waste products has become a key tactic for encouraging ecologically responsible manufacturing methods. Many research projects have focused on agro-industrial wastes, such as peels, seeds, pits, pulps, press cakes, and leaves. The phenolic compounds found in these residues are especially rich in secondary plant metabolites [6], which are known to be a substantial class of bioactive chemicals with antioxidant activity in fruit tissues [7]. The anti-inflammatory, antidiabetic, antioxidant, anticancer, antimicrobial, and antiproliferative properties of these constituents have been the subject of numerous studies [8–10].

12.1.1 Sources and Classification of Biomass

Globally, biomass comes from a variety of sources, but it comes from four primary sources: "crops, agricultural waste (crop residues), forest residues, and municipal solid waste are particularly beneficial to the power sectors" [11, 12]. Table 12.1 illustrates how Abmann et al. divided these sources into three primary categories: liquid from processed waste, petrol from processed fuel, and solid in non-woody and woody biomass [13, 14].

Biomass energy development is at an all-time high, with carbon neutrality being attained. However, the amount of CO_2 accumulated over many years due to using fossil fuels has reached a point where soils and plants cannot absorb it. Thus, it becomes essential to reduce CO_2 emissions globally by developing carbon-negative energy sources. This section summarizes converting biomass to energy, focusing on a bio-refinery. The benefits and drawbacks in each area are described in the following sections, emphasizing recent advancements in the subject. Biomass energy, often known as "bio-energy," is an endless energy source used on Earth

TABLE 12.1   Sources of Biomass [14]

Processed Waste	Processed Fuels	Woody Biomass	Non-Woody Biomass
Sawmill waste	Charcoal	Shrubs	Energy crops
Plant oil cake	Briquette and densified biomass	Trees	Cereal straws
Nutshell and flesh	Plant oil from palm rape and sunflower	Forest floor sweepings	Grasses
Waste from fruits	Biogas	Bushes	Tobacco, cassava, stems, cotton, and root grasses
Bagasse	Producer gas	Palms	Banana
Bark, cereal husk, industrial wood, and logs	Ethyl alcohol and methyl alcohol	Bamboo	Soft plant stems, water plants

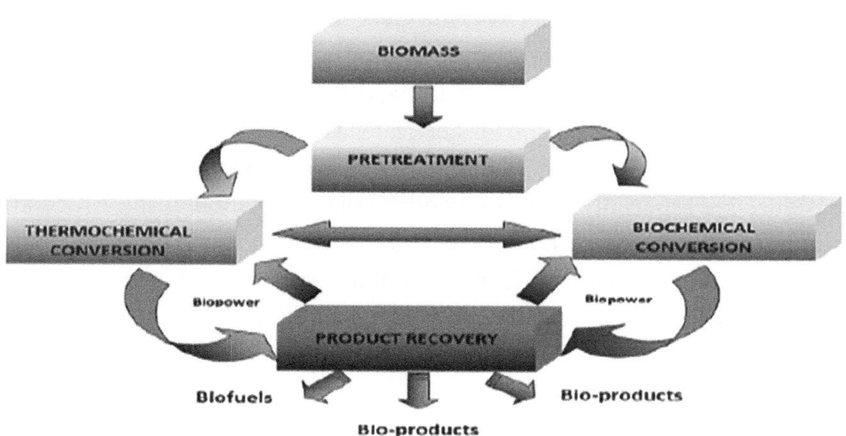

FIGURE 12.2   Illustration of biomass conversion into end products such as biofuels, bioproducts, and biopower [15]. Source: BCCY; Open Access Publication.

since the dawn of human civilization. When combined with new technology developments, biomass energy derivatives have enormous promise for resolving the energy dilemma, especially for developing countries. An adequate substitute for disposing of the massive volumes of garbage produced every day in both the urban and rural sectors is biomass innovation. Biofuels and bio-power are the two subsystems that make up bioenergy. To create biofuels, bioproducts, or biopower for integration into a grid system, pre-treatment, either thermochemically or biochemically, is crucial in these bio-energy conversions. The process of converting biomass into final goods, including biofuels, bioproducts, and biopower, is shown in Figure 12.2 [15].

12.2 AVAILABILITY OF BIOMASS AND CURRENT ENERGY SITUATION

There are abundant biomass supplies in various natural terrestrial and aquatic habitats, as well as in the form of waste from human industrial activity. The entire biomass resources in the world, both terrestrial and marine, are estimated by the World Bioenergy Association to be 1.8 trillion and 4 billion tons, respectively. Global biomass might generate around 33,000 exajoules (EJ), more than 80 times the world's yearly energy consumption [16]. Even with this potential, fossil fuels—coal, oil, and natural gas—remain the leading energy source in the world, accounting for around 81% of all primary energy supplies. However, in 2019, the percentage of renewable energy technologies in the primary energy supply was 14.1%. These technologies include solar, wind, water, biomass, and geothermal energy. 56.9 EJ of biomass were produced domestically, of which 85% came from solid biomass, which includes wood pellets, wood chips, and other biomass sources. Of the entire biomass supply, liquid biofuels accounted for 8%, industrial and municipal waste for 5%, and biogas for 2% [16].

The agriculture industry has a great deal of potential to produce more bioenergy. Cultivating various crops worldwide is possible to improve fuel and food production through bioenergy. Utilizing industrial and municipal waste is another industry that contributes to energy generation after forestry and agriculture, with an energy supply of 2.59 EJ in 2019 [16]. An estimated 27,000 terawatt-hours (TWh) of energy are produced annually worldwide, of which 2,664 TWh are made in the European Union (EU) [17–20]. An estimated 300 TWh of waste heat recovery is produced annually in the EU's industrial sector; this is divided into three categories: 25% of medium-temperature heat (200–500°C), 45% of high-temperature heat (> 500°C), and 30% of low-temperature heat (< 200°C) [17–20].

The COVID-19 pandemic's effects on fuel consumption for transport and other economic sectors resulted in a 4% (564 EJ) drop in worldwide primary energy consumption in 2020 compared to 2019. Nonetheless, energy consumption saw a resurgence in 2021, rising 5.5% (595 EJ) over 2020 levels. Global energy consumption is anticipated to continue to rise despite sporadic variations, especially in the primary energy fuels sector, which is dominated by coal and oil. With a primary energy consumption estimated to be approximately 158 EJ in 2021, China leads the world in this regard. The United States comes in second with an average use of about 93 EJ, followed by Germany, Canada, Japan, India, Russia, and other nations [21]. By 2050, annual renewable energy consumption will be predicted to increase to around 247 EJ (Figure 12.3) [22].

Based on statistical data, bioenergy output worldwide was estimated to have reached 584 terawatt hours in 2020. The energy industry's continuous transition to renewable sources from organic, biological, and waste resources is behind this growing tendency. The combustion of biomass, such as agricultural waste, waste wood, straw, sawdust, animal dung, sugar cane, sewage sludge, seaweed, other plants, organic waste (such as beetroot pulp, corn stalks, grass), animal fats, and vegetable oils, can produce bioenergy. Every year, the percentage of energy produced worldwide from renewable sources rises. A total of 2.36 terawatts of renewable energy capacity were registered in 2018. In contrast to the extensive usage of fossil fuels, renewable energy consumption is still relatively low despite this encouraging development.

12.3 BIOMASS ENERGY AND CONVERSION METHODS

In the literature, "bioenergy" is becoming increasingly common in describing energy from different biomass types. Due to ongoing technical breakthroughs, there is much potential to deploy new alternative energy sources. Using biomass resources offers a viable way to handle the large volumes of trash produced daily in rural and urban regions. When thermochemical or biochemical processes treat biomass, bioenergy can be made in heat, electricity, or biofuels (gaseous, liquid, or solid; Figure 12.4). Biomass processing is necessary to create bioenergy, bioproducts, or biofuels [21–25].

Biomass can be converted into various energy sources, including gases (such as hydrogen and synthesis gas), liquid biofuels (such as methanol and ethanol), and electricity through biochemical or thermochemical processes. While bioethanol may be made from crops like maize or sugar cane, biodiesel manufacturing can use vegetable oils like rapeseed, soybean, and waste fats [26]. Depending on the energy properties of the biomass, different amounts of biomass must be consumed to produce bioenergy. Direct thermal energy may be acquired by burning biomass, producing heat that can be used in particular product production processes or converted into energy using a steam turbine. Sometimes biomass fuel must be converted due to its low energy density to fulfil the demands of the high-energy-density fuel market, which is dominated by conventional gaseous, liquid, and solid fuels. Notably, the wood sector has significantly contributed to bioenergy production recently, providing biomass from waste wood that

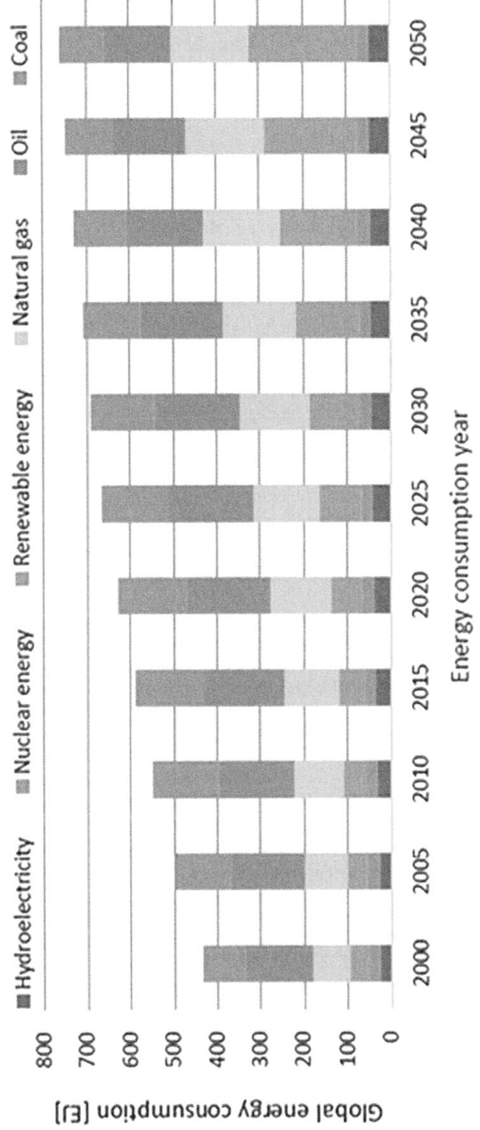

FIGURE 12.3 Global energy consumption (2000–2019) and a forecast until 2050 [22]. Source: BCCY; Open Access Publication.

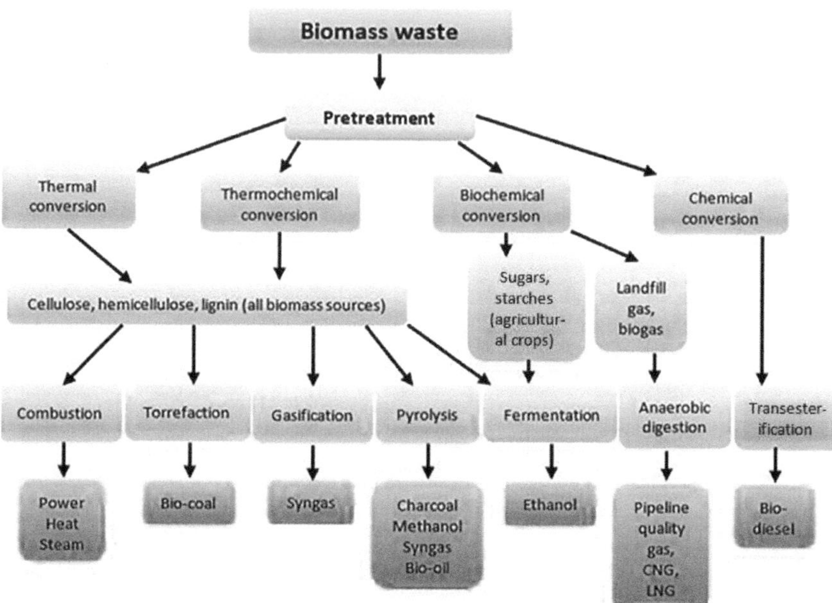

FIGURE 12.4   Biomass conversion into bioproducts [25]. Source: BCCY; Open Access Publication.

accounts for about half of the energy supplied to the market [27]. The two fuels co-combust in a furnace chamber, where coal and biomass are provided individually or as a pre-prepared combination. On the other hand, indirect co-combustion occurs in a gasifier following the gasification of biomass, and the resultant gas is then sent to a combustion chamber for burning. Burning coal and biomass separately in separate combustion chambers is called parallel combustion. Thermal carbonization of biomass produces biofuel with physicochemical properties like hydrophobicity, increased calorific value, coal-like energy, and physical characteristics like resistance to biological degradation, among others. This process is carried out at temperatures between 200 and 300°C and pressures that are similar to atmospheric pressure. Pyrolysis is another process involving biomass's thermal breakdown in anaerobic environments or with very little oxygen. After biomass is quickly pyrolyzed at high temperatures (about 500°C), char and syngas are produced, and after cooling, an oily liquid with a favorable calorific value is produced. In addition to pyrolysis oil or gas, slow pyrolysis yields charcoal with high energy density, low humidity, and enhanced stability. In a gasifier, biomass gasification entails several thermal processes, such as drying at around 150°C, oxidation at temperatures above 600°C to produce oxides, CO_2, and water vapor, and reduction of CO_2 and water vapor to CO and H_2. Gaseous components (wood gas), liquid products (acids, alcohols, steam), solids (fly ash), and tar are among the results of biomass gasification. Biochemical techniques include transesterifying vegetable and animal lipids to create biodiesel and anaerobic fermentation to create biogas or alcohol (such as methanol and ethanol). Liquid fuels like methanol, ethanol, or biogas are produced by converting biomass components like cellulose, hemicellulose, and lignin into later processes [28].

12.4 ENERGY ASPECTS OF BIOMASS

Given the present trends and growing need for renewable energy sources, biomass waste has excellent promise as an energy source [29]. "The biodegradable fraction of products, agriculture wastes (those includes animal and vegetal substances), forestry and associated industries, as well as the biodegradable fraction of municipal and industrial waste" is the definition of biomass given by Directive 2001/77/EC [30]. According to some sources, biomass refers to any organic matter originating from plants or animals, including materials that have been transformed or processed [29-32]. Several parameters, including fuel type, material type, moisture content, and calorific value, affect biomass's potential for energy [33]. Furthermore, factors including quantity, availability, chemical composition, and acquisition and processing costs contribute to its positive worth. Biomass can be burned directly or in combination with other fuels in solid, liquid, or gaseous state. Pellets, wood shavings, firewood, fruit stones, and nut shells are biofuels made from biomass. Shredded wood chips are produced by processing agricultural and forest waste. Pellets are created by crushing biofuels with binding chemicals into the shape of tiny cylinders, several millimeters in diameter and many tens of millimeters in length [34–36]. Pellets' availability in local markets, affordability, and environmental benefits have made them popular as renewable energy sources. Waste materials, plant stems, wood chips, straw, processed sawdust, post-production trash, or agricultural food waste are the sources of these clean, natural goods. A noteworthy feature of pellet production methods is using natural materials as binders during creation, such as lignin found in wood or pectin of plant fiber, without the addition of chemicals. Further benefits include cheap pelletization costs, using only natural and renewable raw materials, introducing ecological raw material cycling resulting in no additional CO_2 emissions, and efficient fuel burning.

Carbonized pellets provide better grinding qualities, a higher energy density, greater hydrophobicity, moisture resistance, and a stronger resistance to biological degradation [37]. The combustion of biomasses results in several volatile components that release organic and organochlorine compounds (polycyclic organic hydrocarbons, dioxins, etc.); high chlorine content that damages heating installations and releases HCl; elevated alkali metal oxides in fly ash that cause deposits and smearing on heating surfaces; a lower calorific value than fossil fuels; low biomass density that affects storage, transport, and feeding to heating devices; variable humidity that requires additional drying costs; and investment costs associated with buying specialized biomass combustion boilers can all be considered operational challenges of biomass combustion [38]. Co-combustion is the process of combining biomass and coal to reduce investment costs. This approach decreases the cost of purchasing conventional fuels, lowers pollution costs, and permits using more economically viable biomass. Co-combustion can be used in high- and low-power grates, pulverized beds, and fluidized bed boilers. The co-combustion process depends on the fuel composition components being chosen in the right quantities. It's important to remember that these ratios depend on the particular characteristics of the biomass, with the maximum biomass share in the composition usually going under 30% [39]. Biomass fuel is regarded as ecologically beneficial because it doesn't contribute to the excessive release of CO_2. When evaluating the total balance, CO_2 emissions are considered just for the quantity of CO_2 released during combustion that plants will take up during photosynthesis. Environmental purity is reached when CO_2 emissions from burning are equivalent to CO_2 absorption throughout the cycle of biomass renewal through photosynthesis. A thorough balance should consider the energy flow during the cycle, including any co-incineration or incineration operations [40].

12.5 ENERGY APPLICATION OF AGRO WASTES

Agro-wastes, comprising various organic residues from agricultural activities, present significant potential for various energy applications. These include biogas production, biofuel generation, biomass power generation, and thermal energy utilization, all of which contribute to sustainable energy production and waste management efforts. Through anaerobic digestion, biochemical and thermochemical conversion, and direct combustion, agro-wastes can be transformed into valuable energy resources, reducing reliance on fossil fuels and mitigating environmental impacts.

Utilizing agro-waste for energy provides renewable energy sources and helps address waste management challenges in agricultural sectors. Recent research and technological advancements have further enhanced the efficiency and feasibility of these energy applications, making agro-waste increasingly attractive as a sustainable energy source. [41] Agro-wastes, such as animal manure, crop residues, and organic by-products, can be processed through anaerobic digestion to produce biogas. Biogas is a renewable energy source primarily composed of methane and carbon dioxide, which can be used for heating, electricity generation, and cooking. Biogas production provides an alternative energy source and helps manage waste by converting organic residues into useful energy and reducing greenhouse gas emissions [42]. Agro-wastes, particularly lignocellulosic biomass like crop residues, sawdust, and straw, can be converted into biofuels such as bioethanol and biodiesel through biochemical or thermochemical processes. Bioethanol, produced via fermentation of sugars derived from biomass, can be blended with gasoline or used as a stand-alone fuel in vehicles. Biodiesel, obtained from the transesterification of vegetable oils or animal fats, is a renewable alternative to conventional diesel fuel, reducing dependence on fossil fuels and mitigating environmental impacts [43]. Agro-waste can be used as biomass power generation feedstock through combustion or gasification. Biomass power plants burn agricultural residues to produce steam, which drives turbines to generate electricity. Gasification converts biomass into syngas (synthetic gas) comprising hydrogen and carbon monoxide, which can be combusted in engines or turbines for electricity generation. Biomass power generation contributes to renewable energy production, reduces greenhouse gas emissions, and provides additional revenue streams for farmers by selling biomass feedstock [44]. Agro-wastes can undergo pyrolysis, a thermochemical process without oxygen, to produce stable carbon-rich biochar. Biochar finds applications in soil amendment, improving soil fertility, water retention, and nutrient availability, enhancing crop yields and mitigating climate change by sequestering carbon in the soil. Additionally, biochar can be used as a renewable energy source through combustion, contributing to decentralized energy production in rural areas [45]. Agro-waste, such as rice husks, coconut shells, and corn cobs, can be utilized directly as biomass fuel in stoves, boilers, and furnaces for thermal energy applications. Biomass combustion releases heat energy, which can be used for space heating, water heating, drying processes in agricultural and industrial sectors, and household cooking. Utilizing agro-wastes for thermal energy reduces reliance on fossil fuels, mitigates air pollution, and promotes sustainable resource management [46].

12.6 FUTURE SCOPE OF BIOMASS WASTE ENERGY SOURCE

The future scope of biomass waste as an energy source is promising, with ongoing research and technological advancements paving the way for increased utilization and efficiency. Biomass waste, including agricultural residues, forestry, and organic municipal waste, holds significant potential for sustainable energy production and waste management. As advancements continue in biomass conversion technologies such as anaerobic digestion, pyrolysis, and gasification, the efficiency of converting biomass waste into various forms of energy, including biogas, biofuels, and heat, is expected to improve. Furthermore, the integration of biomass waste energy systems with other renewable energy sources and innovative grid technologies offers opportunities for enhanced energy security and grid stability. With growing concerns about climate change and the need to transition to low-carbon energy sources, biomass waste energy is poised to play a crucial role in the future energy landscape. Policy support, research funding, and technological innovation will be critical drivers in realizing the full potential of biomass waste as a sustainable energy source. Prior research indicates a notable progression in applying agro-industrial wastes to create novel goods. The food production chain has historically faced difficulties due to agro-industrial waste, which is characterized by expensive removal costs and adverse environmental effects from waste disposal [47–49]. Agro-industrial waste was mainly utilized for composting or to make animal feed until recently, when it was no longer seen as a cost or an advantage [50]. However, the modern viewpoint has changed with the advent of new technologies to enhance the value of agro-industrial waste. According to researchers, state-of-the-art methods recognize that recovering added-value resources from trash may improve the food production chain's overall sustainability by addressing environmental and economic issues [51, 52].

Around the world, biorefineries are becoming commonplace, especially in nations with high agricultural output like Brazil. The proliferation of crops made possible by ideal growing circumstances, a large land area, and significant expenditures in R&D has resulted in a diversification of the productivity dynamics in Brazilian agroenergetic agriculture [53]. Crops including cotton, soy, wheat, rice, bananas, coffee, potatoes, sugar cane, eucalyptus, and maize, among others, have grown significantly across the nation in this changing environment, resulting in the production of related lignocellulosic waste. Brazilian industries have proactively implemented the biorefinery model into their industrial processes to use the potential of vegetable fiber composition and other important by-products using serial technologies [54]. An overview of the significant phases that have previously been created in conventional biorefinery models using lignocellulosic biomass generated from agricultural products is given in Figure 12.5. This strategy aims to optimize resource use and promote sustainable practices within the industrial sector.

Although the idea of a biorefinery is essential for creating a sustainable biobased economy [55], the use of food waste as a feedstock in biorefineries is still in its infancy, with few studies evaluating its commercial viability on a large scale. The survey by Cristóbal provides a techno-economic and profitability analysis of four food waste biorefineries that use feedstock made from leftovers from the processing of oranges, tomatoes, potatoes, and olives. Both good and negative features are included in the research, highlighting the importance of cautious assessment [56]. The screening study conducted at the European level reveals differences in the potential

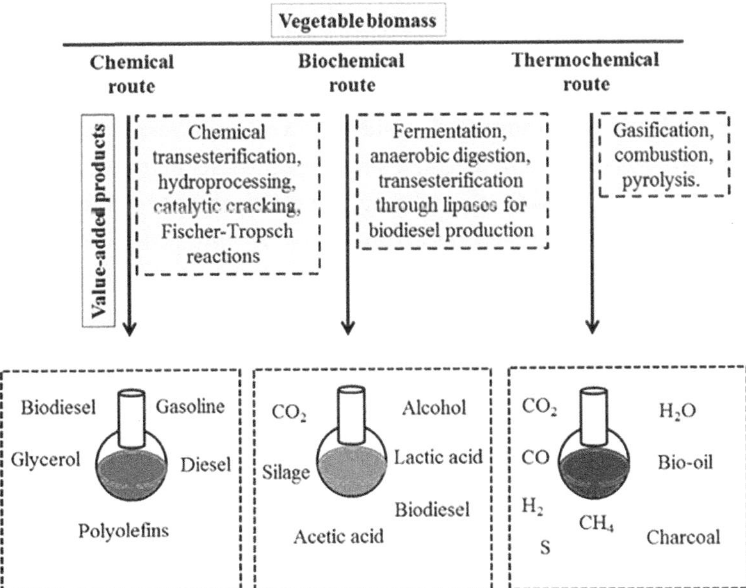

FIGURE 12.5 Summary of the main stages already consolidated in traditional models of biorefinery from lignocellulosic biomass of agricultural products [41].

of waste feedstocks. The most lucrative choices are associated with implementing fewer facilities, concentration of production, and leveraging economies of scale. Furthermore, agro-industrial waste may be used more than only the ones listed earlier. Agro-industrial waste has been investigated recently as a cost-effective adsorbent for wastewater treatment. Meseldzija et al. evaluated the effectiveness of unmodified lemon peel as an agro-industrial waste in extracting copper ions from mine effluent and aqueous solutions [57]. According to the study, lemon peel can extract up to 89% of copper from complicated mining effluent, indicating its potential as a valuable adsorbent for removing copper ions. Furthermore, biomass may substitute conventional resources in synthesizing materials for the civil construction industry. This replacement lowers energy expenses during building and makes it easier to handle agro-industrial waste effectively [58]. The authors highlight the physical-mechanical characteristics of agro-industrial waste materials, such as their strength, water absorption capacity, density, resistance to compression, and thermal conductivity, demonstrating how these features may be used to create innovative building materials. The current state of using agro-industrial waste is intimately related to developing new technologies, especially those that can extract high-value components for use as medications and dietary supplements. The following sections will provide more information about these technologies and their uses in agro-industrial waste. Green technology techniques are strongly related to advances in extracting high-added-value chemicals from agro-industrial wastes. Conventional methods can require lengthy procedures, many organic solvents, and much energy. In this discipline, the general trend is for ecologically friendly extraction techniques to replace traditional methods gradually. According to Belwal [59], green extraction techniques provide increased selectivity in product recovery, more comprehensive application, and a positive impact on sustainable sector growth. Much research has recently been done, emphasizing green extraction techniques. Supercritical fluid extraction (SFE), subcritical water extraction (SWE), ultrasound-assisted extraction (UAE), and microwave-assisted extraction (MAE) are noteworthy examples of these advancements.

12.7 AGRO-INDUSTRIAL WASTES FOR PHARMACOLOGICAL APPLICATIONS

In the field of pharmacology, there is a wealth of documentation about applying novel methods to extract essential components from agro-industrial waste [60–63]. In this regard, synthesizing pharmaceuticals using non-traditional materials and technologies offers an alternative to conventional ways, adding value and aiding in the sustainable use of large-scale biomass generated, mainly by the food sector. Focused research on agro-industrial waste has been conducted recently to apply it in a more sophisticated and environmentally responsible way than just disposing of it. The following discourse delves into several possible uses of agro-industrial waste, such as its use as biological activity sources, delivery agents, and nutritional medicine.

Agro-industrial wastes as an ingredient for food production: Agro-industrial wastes have prospects for value addition and sustainable use, making them attractive candidates as food production materials. These wastes are produced by farming and food processing and include various materials, including husks, seeds, peels from fruits and vegetables, and by-products. These agro-industrial wastes may be converted into helpful food components full of minerals, fiber, antioxidants, and bioactive substances using the proper processing and utilization methods. By including these substances in food compositions, waste creation is decreased, circular economy principles are promoted, and nutritional profiles are improved. Moreover, the use of agro-industrial wastes in food production aids in creating innovative, environmentally friendly food items that satisfy customer preferences for sustainable and health-conscious options. Using agro-industrial wastes as ingredients gives a viable path towards more resource-efficient and environmentally friendly food production processes as the food sector looks for creative ways to solve sustainability concerns. The agri-food industries produce a significant amount of trash every day. This waste comes from many points throughout the production chain, which includes harvesting, transportation, storage, industrial processing, commercialization, and end-use. These remnants, which include peels and seeds, include various elements that are helpful to health, including vitamins, minerals, fibers, carbs, and proteins, despite coming from diverse phases of development. Interestingly, these nutrients are frequently present in higher concentrations in these leftovers compared to other fruit sections. Fats, natural dyes, enzymes, pigments, antioxidants, and antimicrobials are found in large quantities in agro-industrial residues, which may be processed into high-value natural food components, supplements, and additives. These elements may be recycled for industrial manufacturing, which aids in creating novel food items; this strategy shows promise as a workable substitute that is advantageous to the environment and the economy [64].

12.8 FUTURE TRENDS AND NEW INVESTMENT OPPORTUNITIES

To establish a sustainable society, future trends and investment possibilities will depend on the explicit adoption of efficient technologies that are affordable, eco-friendly, and supportive of industrial output that can scale up. An ecologically conscientious trash reuse sector ensures the planet's well-being while advancing the economy and society. Prospects for the reuse sector seem bright, especially when considering new trends, technology, and investment possibilities. The technology used to separate value-added chemicals from garbage has undergone significant developments. Academic research has witnessed the emergence of novel, eco-friendly, and highly efficient technology. The use of supercritical and subcritical fluid technologies for biomass hydrolysis and the recovery of bioactive chemicals from waste has made these processes practical. Researchers have shown how these technologies may hydrolyze and extract biomass [61, 65]. Supercritical fluid technology, in particular, stands out as a novel and evolving technique, giving several investment prospects and options for establishing new company concepts. Future scaling-up possibilities are also promising for microwave-assisted extraction (MAE) and ultrasound-assisted extraction (UAE). Indeed, one of the most critical problems of the modern period is to recover added-value chemicals from trash. The creation of appropriate industrial plants, cost reduction for the recovery of added-value compounds, and consumer market concerns are only a few of the challenges that must be solved to build profitable industrial processes. Notwithstanding these obstacles, there is cause for hope, especially in light of recent initiatives to implement projects incorporating biorefineries and pilot plants. As noted in the study, these activities focus on the synthesis of enzymes, polymers, and edible biomass obtained from fungi and yeasts [66]. When looking at the positive sides, it's interesting to see that there may be a lot of direct and indirect job possibilities created by using agro-industrial wastes for the commercial manufacturing of value products and bioenergy generation. Furthermore, it can potentially aid in developing an innovative waste management strategy. This change creates opportunities for new investments, such as the founding of startups or new businesses directly or indirectly involved in trash reuse business models. Beyond traditional alternatives, biomass is expected to become the most practical and feasible energy source of the future. Further research and development of more accessible and effective planning, enhancement, and business models for biomass's commercial use is required to realize this promise fully. This continuing research will largely shape the future of waste management and bioenergy production.

12.9 CONCLUSIONS

Renewable energy sources are becoming more environmentally benign due to the depletion of fossil fuels and the increase in emissions of industrial pollutants. As a workable alternative, biomass efficiently lowers dangerous emissions. It is already widely used as an alternative energy source, mainly to manufacture ethanol through several conversion processes. As fossil fuel supplies decline, biomass provides a sustainable substitute that reduces greenhouse gas emissions and solid waste. Variations in biomass properties between regions drive the search for customized technical solutions for energy generation and pollution control. Furthermore, farmers have new options due to using agricultural biomass from energy crops, which benefits the economy and the environment. Technological developments in the use of biomass provide more jobs, economic revival, and higher farmer incomes. Nonetheless, issues still need to be resolved, such as understanding the properties of biomass energy, conversion methods, and related barriers that prevent its widespread use.

REFERENCES

1. S.M. Bhatt, J.S. Bal, Bioprocessing perspective in biorefineries, in: 2019: pp. 1–23. https://doi.org/10.1007/978-3-319-94797-6_1.
2. V.B. Agbor, N. Cicek, R. Sparling, A. Berlin, D.B. Levin, Biomass pretreatment: Fundamentals toward application, *Biotechnology Advances*. 29 (2011) 675–685. https://doi.org/10.1016/j.biotechadv.2011.05.005.
3. N.M. Clauser, G. González, C.M. Mendieta, J. Kruyeniski, M.C. Area, M.E. Vallejos, Biomass waste as sustainable raw material for energy and fuels, *Sustainability*. 13 (2021) 794. https://doi.org/10.3390/su13020794.
4. H.S. Ng, P.E. Kee, H.S. Yim, P.-T. Chen, Y.-H. Wei, J. Chi-Wei Lan, Recent advances on the sustainable approaches for conversion and reutilization of food wastes to valuable bioproducts, *Bioresource Technology*. 302 (2020) 122889. https://doi.org/10.1016/j.biortech.2020.122889.
5. R. Ravindran, S. Hassan, G. Williams, A. Jaiswal, A review on bioconversion of agro-industrial wastes to Industrially Important enzymes, *Bioengineering*. 5 (2018) 93. https://doi.org/10.3390/bioengineering5040093.
6. E.M.C. Alexandre, L.M.G. Castro, S.A. Moreira, M. Pintado, J.A. Saraiva, Comparison of emerging technologies to extract high-added value compounds from fruit residues: Pressure- and electro-based technologies, *Food Engineering Reviews*. 9 (2017) 190–212. https://doi.org/10.1007/s12393-016-9154-2.

7. R. Rossetto, G.M. Maciel, D.G. Bortolini, V.R. Ribeiro, C.W.I. Haminiuk, Acai pulp and seeds as emerging sources of phenolic compounds for enrichment of residual yeasts (Saccharomyces cerevisiae) through biosorption process, *LWT.* 128 (2020) 109447. https://doi.org/10.1016/j.lwt.2020.109447.

8. M. Peanparkdee, S. Iwamoto, Bioactive compounds from by-products of rice cultivation and rice processing: Extraction and application in the food and pharmaceutical industries, *Trends in Food Science & Technology.* 86 (2019) 109–117. https://doi.org/10.1016/j.tifs.2019.02.041.

9. D. Ballesteros-Vivas, G. Álvarez-Rivera, S.J. Morantes, A. del P. Sánchez-Camargo, E. Ibáñez, F. Parada-Alfonso, A. Cifuentes, An integrated approach for the valorization of mango seed kernel: Efficient extraction solvent selection, phytochemical profiling and antiproliferative activity assessment, *Food Research International.* 126 (2019) 108616. https://doi.org/10.1016/j.foodres.2019.108616.

10. T.S. Tunna, M.Z.I. Sarker, K. Ghafoor, S. Ferdosh, J.M. Jaffri, F.Y. Al-Juhaimi, M.E. Ali, M.J.H. Akanda, M.S. Awal, Q.U. Ahmed, J. Selamat, Enrichment, in vitro, and quantification study of antidiabetic compounds from neglected weed Mimosa pudica using supercritical CO 2 and CO 2 -Soxhlet, *Separation Science and Technology.* 53 (2018) 243–260. https://doi.org/10.1080/01496395.2017.1384015.

11. J. Ben-Iwo, V. Manovic, P. Longhurst, Biomass resources and biofuels potential for the production of transportation fuels in Nigeria, *Renewable and Sustainable Energy Reviews.* 63 (2016) 172–192. https://doi.org/10.1016/j.rser.2016.05.050.

12. M.N. Uddin, K. Techato, J. Taweekun, M. Mofijur, M.G. Rasul, T.M.I. Mahlia, S.M. Ashrafur, An overview of recent developments in biomass pyrolysis technologies, *Energies.* 11 (2018) 3115. https://doi.org/10.3390/en11113115.

13. T.B. Johansson, K. McCormick, L. Neij, W.C. Turkenburg, The potentials of renewable energy, in *Renewable Energy*, Routledge, n.d.: pp. 15–47.

14. B. Dukhnytskyi, World agricultural production Ekon, *APK.* 7 (2019) 59–65.

15. K. Sivabalan, S. Hassan, H. Ya, J. Pasupuleti, A review on the characteristic of biomass and classification of bioenergy through direct combustion and gasification as an alternative power supply, *Journal of Physics: Conference Series.* 1831 (2021) 012033. https://doi.org/10.1088/1742-6596/1831/1/012033.

16. B. Kummauru, World bioenergy association–WBA global bioenergy statistics 2017 (n.d.)., https://www.worldbioenergy.org/uploads/WBA%20GBS%202017_hq.pdf.

17. L. Bonaccorsi, A. Fotia, A. Malara, P. Frontera, advanced adsorbent materials for waste energy recovery, *Energies.* 13 (2020) 4299. https://doi.org/10.3390/en13174299.

18. J. Ziemele, R. Kalnins, G. Vigants, E. Vigants, I. Veidenbergs, Evaluation of the industrial waste heat potential for its recovery and integration into a fourth generation district heating system, *Energy Procedia.* 147 (2018) 315–321. https://doi.org/10.1016/j.egypro.2018.07.098.

19. M. Papapetrou, G. Kosmadakis, A. Cipollina, U. La Commare, G. Micale, Industrial waste heat: Estimation of the technically available resource in the EU per industrial sector, temperature level and country, *Applied Thermal Engineering.* 138 (2018) 207–216. https://doi.org/10.1016/j.applthermaleng.2018.04.043.

20. R. Agathokleous, G. Bianchi, G. Panayiotou, L. Aresti, M.C. Argyrou, G.S. Georgiou, S.A. Tassou, H. Jouhara, S.A. Kalogirou, G.A. Florides, P. Christodoulides, Waste heat recovery in the EU industry and proposed new technologies, *Energy Procedia.* 161 (2019) 489–496. https://doi.org/10.1016/j.egypro.2019.02.064.

21. Global primary energy consumption (n.d.), https://www.energyinst.org/statistical-review#:~:text=2023%20saw%20a%20second%20consecutive,in%20absolute%20terms)%20was%20recorded.

22. H. Durak, Comprehensive assessment of thermochemical processes for sustainable waste management and resource recovery, *Processes.* 11 (2023) 2092. https://doi.org/10.3390/pr11072092.

23. J.N. Chung, Grand challenges in bioenergy and biofuel research: Engineering and technology development, environmental impact, and sustainability, *Frontiers in Energy Research.* 1 (2013). https://doi.org/10.3389/fenrg.2013.00004.

24. M. Fiala, L. Nonini, Biomass and biofuels, *EPJ Web of Conferences.* 189 (2018) 00006. https://doi.org/10.1051/epjconf/201818900006.

25. T. Kalak, Potential use of industrial biomass waste as a sustainable energy source in the future, *Energies.* 16 (2023) 1783. https://doi.org/10.3390/en16041783.

26. S.N. Naik, V.V. Goud, P.K. Rout, A.K. Dalai, Production of first and second generation biofuels: A comprehensive review, *Renewable and Sustainable Energy Reviews.* 14 (2010) 578–597. https://doi.org/10.1016/j.rser.2009.10.003.

27. D. Gielen, F. Boshell, D. Saygin, M.D. Bazilian, N. Wagner, R. Gorini, The role of renewable energy in the global energy transformation, *Energy Strategy Reviews.* 24 (2019) 38–50. https://doi.org/10.1016/j.esr.2019.01.006.

28. N.L. Panwar, R. Kothari, V.V. Tyagi, Thermo chemical conversion of biomass—eco friendly energy routes, *Renewable and Sustainable Energy Reviews.* 16 (2012) 1801–1816. https://doi.org/10.1016/j.rser.2012.01.024.

29. M.-A. Perea-Moreno, E. Samerón-Manzano, A.-J. Perea-Moreno, Biomass as renewable energy: Worldwide research trends, *Sustainability.* 11 (2019) 863. https://doi.org/10.3390/su11030863.

30. E.U. Directive, Directive 2001/77/EC of the European parliament and of the council of 27 September 2001 on the promotion of electricity produced from renewable energy sources in the internal electricity market, *Official Journal of the European Communities L.* (n.d.) 283.

31. A. Mehedintu, M. Sterpu, G. Soava, Estimation and forecasts for the share of renewable energy consumption in final energy consumption by 2020 in the European Union, *Sustainability.* 10 (2018) 1515. https://doi.org/10.3390/su10051515.

32. C. Contescu, S. Adhikari, N. Gallego, N. Evans, B. Biss, Activated carbons derived from high-temperature pyrolysis of lignocellulosic biomass, *C.* 4 (2018) 51. https://doi.org/10.3390/c4030051.

33. I. Niedziółka, M. Szpryngiel, B. Zaklika, Possibilities of using biomass for energy purposes, *Agricultural Engineering.* 1 (2014) 155–163.

34. A. Acar, Ş. Ayanoğlu, Determination of higher heating values (HHVs) of biomass fuels, *Energy Education Science And Technology Part A-Energy Science And Research.* 28 (2012) 749–758.

35. G. Li, C. Liu, Z. Yu, M. Rao, Q. Zhong, Y. Zhang, T. Jiang, Energy saving of composite agglomeration process (CAP) by optimized distribution of pelletized feed, *Energies.* 11 (2018) 2382. https://doi.org/10.3390/en11092382.

36. O. Williams, S. Taylor, E. Lester, S. Kingman, D. Giddings, C. Eastwick, Applicability of mechanical tests for biomass pellet characterisation for bioenergy applications, *Materials.* 11 (2018) 1329. https://doi.org/10.3390/ma11081329.

37. P. Pradhan, S.M. Mahajani, A. Arora, Production and utilization of fuel pellets from biomass: A review, *Fuel Processing Technology.* 181 (2018) 215–232. https://doi.org/10.1016/j.fuproc.2018.09.021.

38. F. Rosillo-Calle, A review of biomass energy—shortcomings and concerns, *Journal of Chemical Technology & Biotechnology.* 91 (2016) 1933–1945. https://doi.org/10.1002/jctb.4918.

39. S. Munir, A review on biomass-coal co-combustion: Current state of knowledge, *Proceedings of the Pakistan Academy of Sciences.* 47 (2010) 265–287.

40. J. Chen, C. Li, Z. Ristovski, A. Milic, Y. Gu, M.S. Islam, S. Wang, J. Hao, H. Zhang, C. He, H. Guo, H. Fu, B. Miljevic, L. Morawska, P. Thai, Y.F. LAM, G. Pereira, A. Ding, X. Huang, U.C. Dumka, A review of biomass burning: Emissions and impacts on air quality, health and climate in China, *Science of The Total Environment.* 579 (2017) 1000–1034. https://doi.org/10.1016/j.scitotenv.2016.11.025.

41. L.C. Freitas, J.R. Barbosa, A.L.C. da Costa, F.W.F. Bezerra, R.H.H. Pinto, R.N. de Carvalho Junior, From waste to sustainable industry: How can agro-industrial wastes help in the development of new products?, Resources, Conservation and Recycling. 169 (2021) 105466. https://doi.org/10.1016/j.resconrec.2021.105466.

42. S.C. Bhatia, Biogas, in: *Advanced Renewable Energy Systems*, Elsevier, 2014: pp. 426–472. https://doi.org/10.1016/B978-1-78242-269-3.50017-6.

43. M. Boro, A.K. Verma, D. Chettri, V.K. Yata, A.K. Verma, Strategies involved in biofuel production from agro-based lignocellulose biomass, *Environmental Technology & Innovation.* 28 (2022) 102679. https://doi.org/10.1016/j.eti.2022.102679.

44. J. Makwana, A.D. Dhass, P.V. Ramana, D. Sapariya, D. Patel, An analysis of waste/biomass gasification producing hydrogen-rich syngas: A review, *International Journal of Thermofluids.* 20 (2023) 100492. https://doi.org/10.1016/j.ijft.2023.100492.

45. C.E. Brewer, R.C. Brown, Biochar, in: *Comprehensive Renewable Energy*, Elsevier, 2012: pp. 357–384. https://doi.org/10.1016/B978-0-08-087872-0.00524-2.

46. M. Yulianto, E. Hartulistiyoso, L.O. Nelwan, S.E. Agustina, C. Gupta, Thermal characteristics of coconut shells as boiler fuel, *International Journal of Renewable Energy Development*. 12 (2023) 227–234. https://doi.org/10.14710/ijred.2023.48349.

47. A. Cattaneo, G. Federighi, S. Vaz, The environmental impact of reducing food loss and waste: A critical assessment, *Food Policy*. 98 (2021) 101890. https://doi.org/10.1016/j.foodpol.2020.101890.

48. B. Cakar, S. Aydin, G. Varank, H.K. Ozcan, Assessment of environmental impact of FOOD waste in Turkey, *Journal of Cleaner Production*. 244 (2020) 118846. https://doi.org/10.1016/j.jclepro.2019.118846.

49. Q.D. Read, S. Brown, A.D. Cuéllar, S.M. Finn, J.A. Gephart, L.T. Marston, E. Meyer, K.A. Weitz, M.K. Muth, Assessing the environmental impacts of halving food loss and waste along the food supply chain, *Science of The Total Environment*. 712 (2020) 136255. https://doi.org/10.1016/j.scitotenv.2019.136255.

50. F. Nazzaro, F. Fratianni, M.N. Ombra, A. D'Acierno, R. Coppola, Recovery of biomolecules of high benefit from food waste, *Current Opinion in Food Science*. 22 (2018) 43–54. https://doi.org/10.1016/j.cofs.2018.01.012.

51. I.A. Udugama, L.A.H. Petersen, F.C. Falco, H. Junicke, A. Mitic, X.F. Alsina, S.S. Mansouri, K.V. Gernaey, Resource recovery from waste streams in a water-energy-food nexus perspective: Toward more sustainable food processing, *Food and Bioproducts Processing*. 119 (2020) 133–147. https://doi.org/10.1016/j.fbp.2019.10.014.

52. K.C. Lai, K.H. Yeap, S.K. Lim, P.C. Teh, H. Nisar, An Investigation on food waste recovery: A preliminary step of waste-to-energy (WtE) development, *Energy Procedia*. 138 (2017) 169–174. https://doi.org/10.1016/j.egypro.2017.10.145.

53. G.C. Fonseca, C.B.B. Costa, A.J.G. Cruz, Economic analysis of a second-generation ethanol and electricity biorefinery using superstructural optimization, *Energy*. 204 (2020) 117988. https://doi.org/10.1016/j.energy.2020.117988.

54. A.A. Longati, G. Batista, A.J.G. Cruz, Brazilian integrated sugarcane-soybean biorefinery: Trends and opportunities, *Current Opinion in Green and Sustainable Chemistry*. 26 (2020) 100400. https://doi.org/10.1016/j.cogsc.2020.100400.

55. G. Dragone, A.A.J. Kerssemakers, J.L.S.P. Driessen, C.K. Yamakawa, L.P. Brumano, S.I. Mussatto, Innovation and strategic orientations for the development of advanced biorefineries, *Bioresource Technology*. 302 (2020) 122847. https://doi.org/10.1016/j.biortech.2020.122847.

56. J. Cristóbal, C. Caldeira, S. Corrado, S. Sala, Techno-economic and profitability analysis of food waste biorefineries at European level, *Bioresource Technology*. 259 (2018) 244–252. https://doi.org/10.1016/j.biortech.2018.03.016.

57. S. Meseldzija, J. Petrovic, A. Onjia, T. Volkov-Husovic, A. Nesic, N. Vukelic, Utilization of agro-industrial waste for removal of copper ions from aqueous solutions and mining-wastewater, *Journal of Industrial and Engineering Chemistry*. 75 (2019) 246–252. https://doi.org/10.1016/j.jiec.2019.03.031.

58. N. Jannat, A. Hussien, B. Abdullah, A. Cotgrave, Application of agro and non-agro waste materials for unfired earth blocks construction: A review, *Construction and Building Materials.* 254 (2020) 119346. https://doi.org/10.1016/j.conbuildmat.2020.119346.

59. T. Belwal, F. Chemat, P.R. Venskutonis, G. Cravotto, D.K. Jaiswal, I.D. Bhatt, H.P. Devkota, Z. Luo, Recent advances in scaling-up of non-conventional extraction techniques: Learning from successes and failures, *TrAC Trends in Analytical Chemistry.* 127 (2020) 115895. https://doi.org/10.1016/j.trac.2020.115895.

60. H. Ben Mohamed, K.S. Duba, L. Fiori, H. Abdelgawed, I. Tlili, T. Tounekti, A. Zrig, Bioactive compounds and antioxidant activities of different grape (Vitis vinifera L.) seed oils extracted by supercritical CO_2 and organic solvent, *LWT.* 74 (2016) 557–562. https://doi.org/10.1016/j.lwt.2016.08.023.

61. F.W.F. Bezerra, W.A. da Costa, M.S. de Oliveira, E.H. de Aguiar Andrade, R.N. de Carvalho, Transesterification of palm pressed-fibers (Elaeis guineensis Jacq.) oil by supercritical fluid carbon dioxide with entrainer ethanol, *The Journal of Supercritical Fluids.* 136 (2018) 136–143. https://doi.org/10.1016/j.supflu.2018.02.020.

62. A. Brenes, A. Viveros, S. Chamorro, I. Arija, Use of polyphenol-rich grape by-products in monogastric nutrition. A review, *Animal Feed Science and Technology.* 211 (2016) 1–17. https://doi.org/10.1016/j.anifeedsci.2015.09.016.

63. F.A. Espinosa-Pardo, V.M. Nakajima, G.A. Macedo, J.A. Macedo, J. Martínez, Extraction of phenolic compounds from dry and fermented orange pomace using supercritical CO_2 and cosolvents, *Food and Bioproducts Processing.* 101 (2017) 1–10. https://doi.org/10.1016/j.fbp.2016.10.002.

64. M.F. Bellemare, M. Çakir, H.H. Peterson, L. Novak, J. Rudi, On the measurement of food waste, *American Journal of Agricultural Economics.* 99 (2017) 1148–1158. https://doi.org/10.1093/ajae/aax034.

65. B.M. Pedras, G. Regalin, I. Sá-Nogueira, P. Simões, A. Paiva, S. Barreiros, Fractionation of red wine grape pomace by subcritical water extraction/hydrolysis, *The Journal of Supercritical Fluids.* 160 (2020) 104793. https://doi.org/10.1016/j.supflu.2020.104793.

66. M. Narra, D.M. Rudakiya, K. Macwan, N. Patel, Black liquor: A potential moistening agent for production of cost-effective hydrolytic enzymes by a newly isolated cellulo-xylano fungal strain Aspergillus tubingensis and its role in higher saccharification efficiency, *Bioresource Technology.* 306 (2020) 123149. https://doi.org/10.1016/j.biortech.2020.123149.

Biomass Wastes for Electricity Generation, Bioenergy, and Harnessing of Nature Power Applications

Abhinay Thakur, Saurav Dixit and Ashish Kumar

13.1 INTRODUCTION

The utilization of biomass for energy production has deep historical roots, dating back to the earliest human civilizations. Biomass, derived from organic materials such as plants, agricultural residues, and animal waste, has been a fundamental energy source for cooking, heating, and various industrial processes. However, with the advent of fossil fuels in the Industrial Revolution, biomass took a backseat, leading to increased reliance on non-renewable and environmentally detrimental energy sources. In recent decades, there has been a resurgence of interest in biomass as a sustainable alternative. This renewed focus is driven by growing environmental concerns, volatile fossil fuel prices, and the imperative to transition towards cleaner and greener energy solutions (1–4). Technological advances have also played a crucial role in making biomass a more viable and efficient option for energy production. From traditional combustion methods to cutting-edge gasification and biomass power plants, the potential applications of biomass have expanded, prompting a reevaluation of

DOI: 10.1201/9781003466833-16

its role in the global energy landscape. The widespread implementation of combined heat and power (CHP) plants in Sweden has positioned the country as a frontrunner in biomass utilization. The emphasis on district heating systems powered by wood chips and pellets has been instrumental in steering Sweden towards a fossil fuel-free energy sector (5, 6). Brazil stands out globally for its substantial reliance on biomass, particularly sugarcane residues, for bioenergy production. The nation's success in integrating biofuels, notably ethanol, into its transportation fuel has enhanced energy security and significantly contributed to reducing greenhouse gas emissions. Finland has carved a sustainable niche in biomass energy, drawing from its well-established forest industry for wood-based biomass. The country's approach underscores the potential of biomass in supporting renewable energy transitions while maintaining ecological balance. In the United States, diverse applications of biomass range from traditional use in wood stoves to cutting-edge technologies like biomass power plants and biofuel production (7, 8). Various states, especially those with robust agricultural and forestry sectors, have implemented biomass initiatives to generate electricity and foster local economic development. Denmark's commitment to biomass is evident in its investment in biomass-fired power plants and integrated systems. The country has successfully combined wind, biomass, and other renewable sources to advance its carbon neutrality and sustainable energy goals. India, focusing on decentralized power generation in rural areas, has increasingly turned to biomass gasification systems. These systems provide off-grid electricity, contributing to both energy access and rural development. Germany's comprehensive approach to biomass as part of its Energiewende strategy exemplifies its commitment to renewable energy. Biomass plays a significant role in Germany's bioenergy sector, contributing to electricity generation, heating, and biofuel production. These global examples underscore the versatility of biomass and its adaptability across different contexts, showcasing its resurgence as a critical player in the worldwide shift towards cleaner and more sustainable energy solutions (9, 10).

A distinctive feature of biomass in the renewable energy portfolio is its carbon neutrality. While the combustion of biomass releases carbon dioxide, the plants from which it originates absorb an equivalent amount of carbon dioxide during their growth. This closed carbon cycle effectively mitigates the net increase in atmospheric carbon dioxide, positioning biomass as an environmentally friendly alternative to fossil fuels and supporting efforts to combat climate change. Biomass emerges as a viable solution for waste management, addressing environmental pollution concerns. Agricultural residues, forestry byproducts, and municipal solid waste, which would otherwise contribute to pollution, find a purpose as biomass feedstock for energy production, as shown in Figure 13.1.

This dual benefit of waste reduction and energy generation underscores biomass's environmental stewardship potential, aligning with sustainable resource utilization principles (12, 13). In addition to its environmental advantages, the decentralization of biomass resources enhances energy security by reducing reliance on centralized power systems. Localized biomass facilities offer the capacity to provide energy to rural and remote areas, fostering economic development and improving energy access in regions where centralized power infrastructure may be challenging to implement. For instance, in an experiment, Ezealigo et al. (14) significantly addressed the gap in Nigeria's bioenergy sector, where a comprehensive assessment of resource availability was lacking. Through a thorough investigation spanning 2008 to 2018, the research focused on evaluating the bioenergy potential inherent in agricultural residues and municipal solid and liquid waste. By employing a computational and analytical approach grounded in judicious assumptions, the study aimed to provide a robust understanding of the technical potential for cellulosic ethanol and biogas production derived from the identified biomass sources. This investigation reveals noteworthy insights into the comparative energy generation capabilities of biogas and cellulosic ethanol from the same type of residue. Biogas emerged as a more efficient performer, showcasing its potential as a superior bioenergy resource. The calculated technical potential of 84 Mt from crop residues yields substantial outputs, with cellulosic ethanol and biogas reaching 14,766 ML/yr (8 Mtoe) and 15,014 Mm^3/yr (13 Mtoe), respectively. This underscores the significant energy potential locked within biomass resources and highlights biogas's versatility, especially considering its various applications ranging from heat to electric power generation. A crucial aspect of this study is its relevance to addressing Nigeria's ongoing electricity crisis. The promising energy outputs from biogas offer a viable solution to the country's energy challenges. The versatile nature of biogas positions it as a multifaceted resource capable of contributing to various energy needs, aligning seamlessly with the nation's quest to meet the targets outlined in the 7th Sustainable Development Goal (SDG 7) on clean and affordable energy. By emphasizing the potential role of biomass valorization, this research advocates for sustainable bioenergy solutions in Nigeria, providing valuable insights that can inform policy decisions

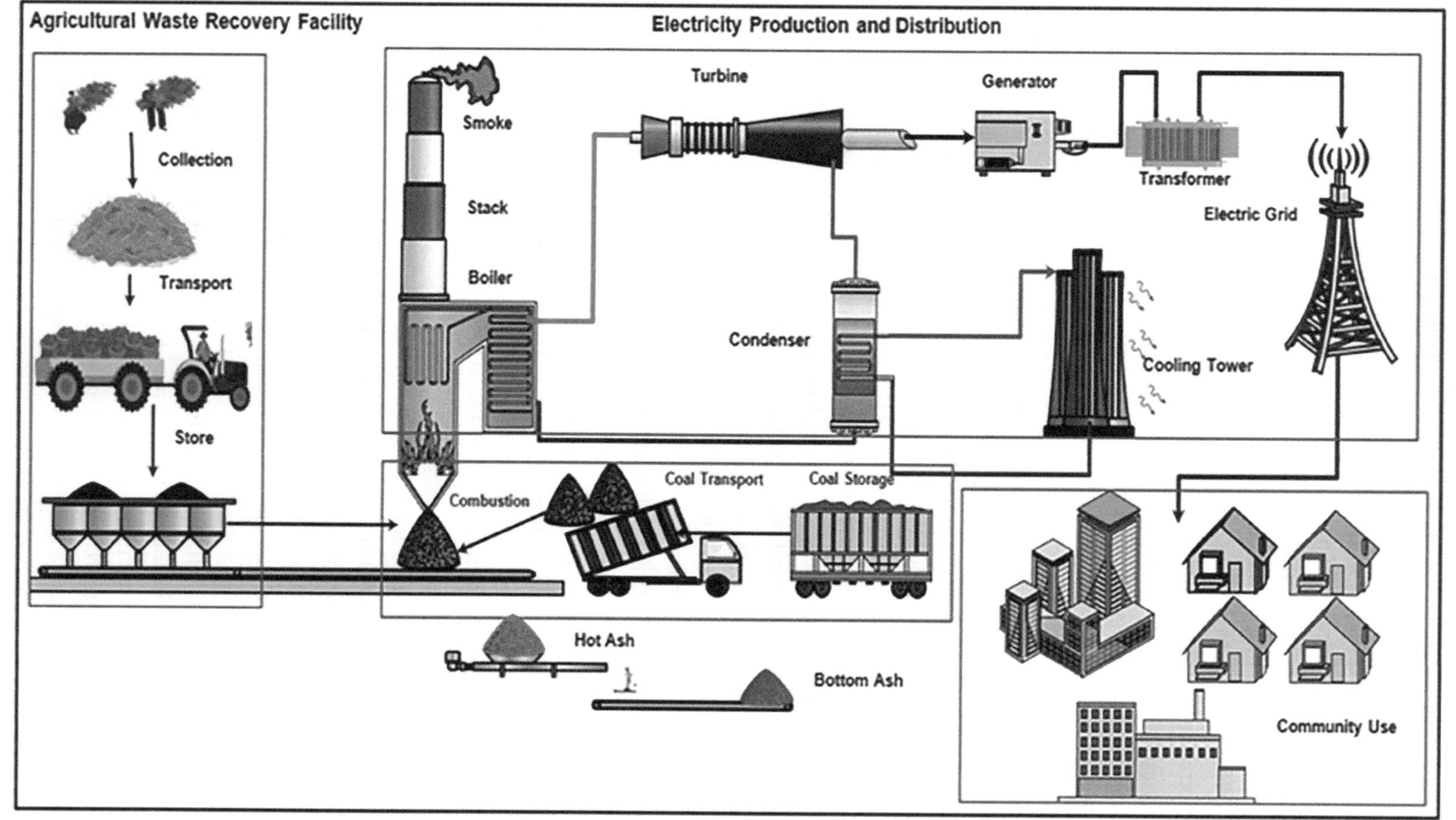

FIGURE 13.1   Illustration of the recovery of agricultural waste and power production. It is adapted from ref. (11) under CCBY 4.0.

and strategic planning for the country's energy future. This decentralized approach contributes to resilience in energy supply and aligns with principles of inclusive and sustainable development. As nations grapple with the multifaceted challenges posed by climate change and the urgent need to transition to a low-carbon economy, the significance of biomass as a renewable energy source becomes increasingly apparent. Its sustainable and versatile attributes and its potential for carbon neutrality, waste management, and decentralized energy production position biomass as a crucial player in the global pursuit of cleaner and more sustainable energy solutions.

13.2 BIOMASS COMPOSITION AND SOURCES

13.2.1 Organic Materials

The cornerstone of biomass energy lies in the rich and varied composition of organic materials obtained from living organisms. Organic materials, which form the fundamental building blocks of biomass, encompass various biological substances, from plants and crops to animal byproducts. These materials are pivotal in sustainable bioenergy production, offering a renewable alternative to traditional fossil fuels (15–17). As primary contributors to organic biomass, plants play a central role in the energy cycle. Plants capture and store solar energy through photosynthesis, converting it into complex organic compounds such as cellulose, hemicellulose, and lignin. These compounds collectively constitute the bulk of plant biomass, serving as the primary feedstock for various bioenergy production processes. The abundance of these complex carbohydrates in plant matter makes them valuable resources for generating energy from heat, electricity, or biofuels. In addition to plant-derived materials, organic biomass includes contributions from animal waste, presenting a valuable and often underutilized component. Animal waste, such as manure, is particularly rich in organic matter. This organic matter becomes a potent resource for bioenergy production when subjected to anaerobic digestion. Through this microbial process, organic materials in manure break down, releasing biogas as a byproduct. Biogas, primarily composed of methane, can be harnessed as a renewable energy source, offering an environmentally friendly power generation and heating alternative. Including organic materials in the biomass portfolio is integral to enhancing its sustainability. Unlike finite fossil fuels, organic biomass resources can be continually replenished through natural processes. The cyclic nature of this replenishment, driven by the growth and regeneration of plants and the ongoing

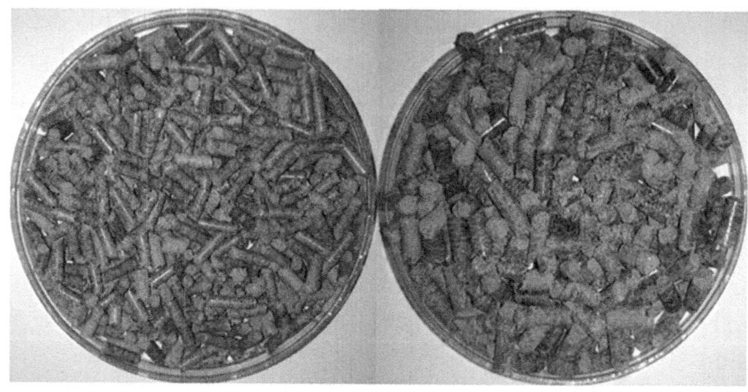

FIGURE 13.2 Prepared pellets from biomass waste. It is adapted from ref. (18) under CCBY 4.0.

production of organic waste, forms a critical part of the renewable energy cycle. Greinert et al. (18) focused on harnessing waste biomass from both the wood industry and municipal sources as a valuable resource for energy production within the realm of renewable energy sources. In response to societal concerns about burning crop plants, the research explored viable alternatives in mixtures and pellets incorporating waste materials, as shown in Figure 13.2.

The characterized mixtures, derived from a combination of wood and municipal waste, underwent an evaluation of their calorific values, revealing a considerable range from 7.4 to 18.2 MJ/kg. Notably, 47% of the analyzed mixtures exhibited calorific values exceeding the critical threshold of 15 MJ/kg. A crucial aspect highlighted in the study pertains to the stabilizing influence of wood shavings and sewage sludge on the durability of the produced pellets. This finding underscores the potential for optimizing the composition of biomass-derived mixtures to enhance the overall quality and resilience of energy pellets. Additionally, the study delves into the environmental implications of the combustion process, specifically examining emissions of acidic anhydrides (including NO, NO_2, NO_x, and H_2S) during the combustion of pellets from waste biomass. The observed lower emissions compared to willow pellets signify a positive environmental attribute of the examined waste biomass. However, the study emphasizes the need to optimize combustion process parameters further. This optimization is crucial not only for maximizing the energetic potential of the biomass but also for minimizing environmental emissions. The research thus contributes valuable insights into the utilization of waste biomass for energy production, with implications for both environmental sustainability and the efficient generation of renewable energy. This perpetual availability positions biomass as a sustainable energy source, contributing to the broader goals of reducing dependence on non-renewable resources and mitigating environmental impacts associated with conventional energy production.

13.2.2 Agricultural Residues

Agricultural residues are a substantial and highly accessible biomass reservoir, significantly contributing to the evolving global bioenergy landscape. This category encompasses diverse materials left behind post-harvest, constituting various non-edible parts, including stalks, leaves, husks, and other plant remnants (19, 20). Key examples of these agricultural residues include corn stover, wheat straw, and rice husks, all of which exhibit high cellulose and hemicellulose content, rendering them optimal feedstocks for bioenergy production. The richness of cellulose and hemicellulose in these crop residues makes them ideal candidates for advanced bioenergy processes. Cellulose, a complex carbohydrate, is a primary structural component in plant cell walls, while hemicellulose complements cellulose's structural role. Both compounds can be effectively converted into energy through various thermochemical conversion methods—such as pyrolysis and gasification—and biochemical processes like fermentation for biofuel production.

Beyond their potential as bioenergy feedstocks, the utilization of agricultural residues holds immense promise for promoting sustainable agriculture. Repurposing crop residues for bioenergy offers multifaceted benefits for farmers and the environment. First, this practice contributes to enhanced soil health. Instead of removing all crop residues, leaving some on the fields can help retain moisture, reduce soil erosion, and improve nutrient content, fostering a more sustainable and resilient agricultural ecosystem. Moreover, incorporating agricultural residues into bioenergy processes reduces the reliance on traditional practices like burning, often employed to clear fields of post-harvest residues (21, 22). The shift towards bioenergy utilization mitigates air pollution associated with burning and minimizes the release of greenhouse gases, thereby contributing to climate change mitigation efforts. Integrating agricultural residues into advanced bioenergy processes exemplifies their pivotal role in supporting a cleaner and more sustainable energy future. Thermochemical conversion methods, such as gasification, involve heating these residues in a controlled environment, producing synthesis gas or biochar that can be used for energy generation or soil improvement, respectively. Biofuel production through processes like fermentation further expands the potential applications of these residues, offering alternatives to conventional fossil fuels and reducing greenhouse gas emissions.

13.2.3 Forestry Byproducts

Forestry byproducts, derived from timber harvesting and forest management activities, stand as a vital and often overlooked reservoir within biomass resources. These byproducts encompass an array of materials, including branches, bark, wood chips, and sawdust, all of which emerge as residual elements from various forestry operations. Crucially, these residues play a pivotal role in the biomass landscape, representing a sustainable and renewable energy source, thanks to responsible forestry practices prioritizing the continual replenishment of forest resources. The inherent sustainability of forestry byproducts is closely tied to the principles of responsible forest management (23). As trees are harvested for timber, the residues left behind become valuable ecosystem components. When forestry operations adhere to sustainable practices such as reforestation, afforestation, and responsible harvesting techniques, the balance between resource extraction and environmental preservation is maintained. This sustainable approach ensures that the forestry byproducts continue to be renewable, contributing to the long-term health and resilience of forests. The rich lignocellulosic content embedded in forestry byproducts underscores their significance as a valuable feedstock for various bioenergy applications. Lignocellulose, a complex structure comprising cellulose, hemicellulose, and lignin, serves as a robust and energy-dense material. Wood pellets, a prominent example of bioenergy derived from forestry residues, have gained widespread recognition as a clean and efficient fuel source. These pellets find applications in biomass power plants and residential heating systems, offering a renewable alternative to traditional fossil fuels.

The utilization of forestry residues for energy production addresses the increasing demand for sustainable energy sources and aligns with broader environmental goals. By converting these residues into energy, the forestry sector contributes to reducing reliance on non-renewable resources and mitigating greenhouse gas emissions associated with conventional energy production (24, 25). Moreover, using wood pellets in residential heating systems provides a cleaner and more environmentally friendly alternative to traditional heating methods, improving air quality. Significantly, using forestry residues goes beyond energy production; it actively promotes and supports sustainable forest management practices. Responsible forestry operations that integrate timber extraction with the utilization of byproducts contribute to maintaining the ecological balance of forest ecosystems. By recognizing the value of forestry residues, these practices emphasize the importance of meeting present energy needs and doing so in a manner that safeguards the health and biodiversity of forests for future generations.

13.2.4 Municipal Solid Waste

Municipal solid waste (MSW) has emerged as a non-traditional yet increasingly significant source of biomass for energy production, marking a transformative shift in waste management paradigms. As global urbanization accelerates, the surge in waste generation within cities has become a pressing issue. MSW, characterized by its heterogeneous mix of organic and inorganic materials, including food waste, paper, plastics, and yard waste, presents a unique opportunity to harness energy potential while addressing the challenges of waste management in an urbanized world. Integrating MSW into the biomass energy portfolio reflects a commitment to the circular economy, where discarded materials are repurposed to extract value, mitigating environmental impact (26, 27). Thermochemical processes, such as incineration and gasification, offer efficient means to convert the organic fraction of MSW into energy, including heat and electricity. Incineration involves combusting MSW at high temperatures, generating heat that can be converted into electricity. On the other hand, gasification transforms MSW into syngas, a mixture of hydrogen and carbon monoxide, which can be further used for power generation. An alternative and environmentally friendly technology for MSW is anaerobic digestion. This process involves the breakdown of organic waste by microorganisms without oxygen, producing biogas primarily composed of methane. Biogas extracted through anaerobic digestion is a renewable and versatile energy source, offering a sustainable alternative to conventional fossil fuels.

The inclusion of MSW in the biomass energy portfolio aligns closely with resource efficiency and environmental sustainability principles. By diverting organic waste from landfills—which would contribute to methane emissions—and instead utilizing it for energy production, MSW-based bioenergy serves a dual purpose—addressing waste management challenges and reducing greenhouse gas emissions (28, 29). Moreover, transforming waste streams into valuable energy resources exemplifies the potential of MSW as a renewable source, turning a once-discarded material into a valuable asset in pursuing a sustainable energy future. The benefits extend beyond waste management; integrating MSW into biomass energy systems contributes to local energy resilience. By decentralizing energy production through smaller-scale facilities strategically located within urban centers, dependence on centralized power systems is reduced, enhancing energy security and reliability. This decentralized approach aligns with the broader goals of creating more sustainable and resilient urban environments.

13.3 ELECTRICITY GENERATION FROM BIOMASS

13.3.1 Biomass Power Plants

Biomass power plants, at the intersection of ecological sustainability and energy generation, play a vital role in diversifying the global energy mix. These plants are ingeniously designed to tap into the organic wealth of living organisms, offering a renewable and environmentally friendly alternative to conventional power sources. A prevailing method employed by these plants involves the direct combustion of biomass. This process exemplifies simplicity and reliability and underscores the potential for harnessing the energy stored in the intricate structures of plant materials. The cornerstone of biomass power generation is the combustion process, a well-established technique that transforms biomass's complex carbohydrates into a potent heat energy source (30, 31). The primary constituents of plant biomass, namely cellulose, hemicellulose, and lignin, undergo a thermal metamorphosis during combustion, breaking down into elemental compounds. This exothermic reaction liberates heat energy, setting a chain of events in motion that culminates in electricity production. The scalability of biomass power plants adds a layer of adaptability, allowing for deployment across a spectrum of sizes and settings. From compact, decentralized facilities catering to local energy needs in rural areas to large-scale, grid-connected power stations supplying electricity to urban centers, biomass power plants exhibit versatility. This adaptability positions biomass as a responsive solution to the diverse energy requirements of communities worldwide.

The global adoption of combustion-based biomass power plants is driven by regions blessed with abundant biomass resources. Agricultural residues, forestry byproducts, and other organic materials serve as readily available feedstocks, making the integration of biomass power generation a pragmatic choice in areas surrounded by nature's bounty. The straightforward conversion of biomass through combustion aligns seamlessly with established practices in traditional power generation, ensuring a smooth transition to more sustainable energy solutions. However, the journey toward sustainability is challenging, and the combustion process poses an environmental conundrum. While biomass is considered a carbon-neutral energy source over the long term, the combustion releases carbon dioxide (CO_2) into the atmosphere. The net carbon impact hinges on the sustainability of biomass sourcing and the ability of the source plants to sequester carbon during their growth phase. Addressing this challenge requires a holistic approach, considering the entire biomass lifecycle from cultivation to combustion. A multifaceted strategy is imperative to enhance the sustainability of biomass power plants. One avenue involves strategically utilizing waste biomass and diverting materials that would otherwise contribute to environmental degradation. Waste biomass, such as agricultural residues and forestry byproducts, not only becomes a valuable energy resource but also mitigates the environmental impact of improper disposal. Additionally, cultivating dedicated energy crops optimized for efficient biomass conversion contributes to a more sustainable and reliable biomass supply. Technological innovations further augment the efficiency and environmental performance of biomass power plants. Vrabie (32) delved into alternative energy solutions, drawing inspiration from existing supply chain (SC) models for MSW management and gleaning insights from successful city models such as Bergen and Tønsberg in Norway, London in the UK, and Barcelona in Spain. The objective was to craft a conceptual framework for municipal officials in Romanian municipalities and beyond, encouraging innovation by transforming MSW into biogas for utilization in public services, particularly public transportation. The findings of this research underscore that when innovation is embraced and effectively implemented by all stakeholders, the benefits accrued for citizens and municipalities far exceed those obtained through conventional MSW collection and disposal methods. The proposed approach not only aligns with EU environmental policies but also offers a pertinent strategy for implementation, addressing potential delays commonly observed, especially in the case of Romania. Integrating advanced emission control technologies, including scrubbers and filters, assists in minimizing the release of pollutants. Co-firing biomass with traditional fossil fuels in existing power plants provides a transitional strategy, gradually phasing in cleaner and more sustainable energy sources.

13.3.2 Gasification Technologies

Gasification technologies represent a cutting-edge approach to transforming the landscape of electricity generation from biomass. In a stark departure from conventional combustion, gasification intricately orchestrates the partial oxidation of biomass within a controlled environment, yielding syngas—a complex amalgamation of hydrogen, carbon monoxide, and methane. This transformative process facilitates electricity generation and underscores its adaptability, positioning gasification as a technological vanguard in sustainable energy solutions (33–35). At the nucleus of gasification lies the gasifier, where biomass undergoes a sophisticated thermochemical conversion. Governed by meticulously controlled high temperatures and a regulated oxygen or steam supply, this process induces the breakdown of intricate organic compounds within biomass, culminating in syngas production. This synthesis gas, a dynamic blend of chemical constituents, is a potent energy carrier with diverse applications. The adaptability of gasification technology enables the utilization of various biomass feedstocks, including woody biomass, agricultural residues, and municipal solid waste, showcasing its versatility in contributing to a diversified and sustainable biomass supply chain. Syngas, the progeny of gasification, assumes a pivotal role in electricity generation. Its efficient deployment in internal combustion engines or gas turbines ensures a direct and efficient conversion of chemical energy into electrical power. Distinguished by high thermodynamic efficiency and reduced environmental impact compared to traditional combustion methods, gasification contributes to the ongoing paradigm shift towards more sustainable and cleaner electricity generation practices. The scalability of gasification technologies amplifies their significance, providing flexibility across diverse deployment scales. From small-scale distributed systems meeting localized energy demands to large-scale power plants integrated into centralized grids, gasification technology accommodates a spectrum of energy needs. This adaptability fosters decentralized energy production and enhances energy resilience by reducing reliance on centralized power infrastructure. Yet, the narrative of gasification extends beyond electricity generation. The syngas, post-production, undergoes a refinement process to eliminate impurities, enabling the production of advanced biofuels. This transformation positions gasification as a pivotal player in pursuing sustainable transportation solutions (36, 37). Synthetic diesel and bio-methane emerge as viable alternatives, addressing the imperative to reduce carbon emissions in the transportation sector. Gasification, through the production of biofuels, seamlessly aligns with broader sustainability goals. By diversifying the energy landscape and providing alternatives to conventional fossil fuels, gasification significantly reduces the carbon footprint and advances a more sustainable and environmentally conscious transportation sector. This multifaceted role underscores the transformative potential of gasification technologies in shaping a cleaner and more sustainable energy future.

13.3.3 Combustion Processes

Combustion processes have served as the backbone of biomass power generation, offering a traditional and widely adopted method valued for its simplicity and scalability across diverse scales. In this method, biomass undergoes direct combustion, liberating heat that is effectively used to produce steam. The ensuing steam powers turbines connected to generators, ultimately converting thermal energy into electricity (38, 39). Despite its widespread use, combustion encounters challenges, particularly in emissions management, with a primary focus on carbon dioxide. Although biomass combustion is considered carbon-neutral over time, ensuring sustainability requires responsible sourcing and careful land management practices. While combustion efficiently generates electricity, releasing other pollutants poses environmental challenges. Carbon dioxide, despite being carbon-neutral in the long term, prompts the need for continuous advancements in emission control technologies. Integrating advanced control measures is crucial, ensuring compliance with stringent environmental standards. The sustainability of biomass combustion is contingent on responsible biomass sourcing and land management practices to mitigate potential environmental impacts.

Combustion technologies are categorized based on biomass type and specific combustion conditions. Grate-fired boilers are commonly employed for solid biomass, demonstrating efficiency in handling diverse feedstocks. Meanwhile, fluidized bed combustion showcases versatility, accommodating a broader range of biomass feedstocks, including those with high moisture content. Advanced combustion technologies, such as biomass co-firing in existing coal-fired power plants, present transitional pathways toward more sustainable energy production. Beyond carbon dioxide, releasing pollutants such as particulate matter, nitrogen oxides, and sulfur dioxide raises environmental concerns. Stringent control measures are imperative, and advanced combustion technologies integrate features like scrubbers, filters, and catalytic converters to mitigate these emissions effectively. This aligns with environmental regulations, improving air quality and fostering a cleaner environment. Grate-fired boilers, stalwarts in solid biomass combustion, employ a stationary grate for controlled and efficient burning (40, 41). This method excels in handling a spectrum of solid biomass feedstocks, showcasing adaptability from wood chips to agricultural residues. In contrast, fluidized bed combustion suspends biomass particles in an aerated bed, facilitating thorough and efficient burning. This approach accommodates solid biomass and a broader range of feedstocks, including those with high moisture content. Biomass co-firing emerges as a strategic approach within advanced combustion technologies. This involves integrating biomass combustion into existing coal-fired power plants, presenting a transitional pathway towards sustainability. Co-firing allows for a gradual shift from conventional fossil fuels to renewable biomass, leveraging existing infrastructure while reducing the overall carbon footprint. This adaptive strategy aligns seamlessly with global initiatives to decarbonize the energy sector.

13.3.4 Advanced Conversion Technologies

13.3.4.1 Pyrolysis Technology

Pyrolysis, a cutting-edge biomass conversion technology, involves the rapid heating of biomass in the absence of oxygen, resulting in the production of bio-oil, biochar, and syngas, as evidenced by Howari et al. (11). Bio-oil, a complex mixture of organic compounds, plays a crucial role in advancing renewable energy solutions. Bio-oil can be converted into transportation fuels through further refinement, providing a sustainable alternative to traditional fossil fuels. This versatility positions bio-oil as a pivotal component in diversifying the energy landscape, especially addressing the transportation sector's reliance on non-renewable resources. Biochar, another valuable byproduct of pyrolysis, extends its role beyond energy production. As a carbon-rich material, biochar is a soil amendment that enhances soil fertility and structure. Its incorporation into agricultural practices sequesters soil carbon and promotes sustainable farming by improving water retention and nutrient availability. This dual functionality underscores biochar's contribution to environmental sustainability and its potential to address challenges in modern agriculture. Syngas, comprising hydrogen, carbon monoxide, and methane, represents the gaseous pyrolysis fraction and emerges as a dynamic energy resource. With applications in energy production, chemical synthesis, and biofuel production, syngas showcases its versatility in contributing to various sectors. Its adaptability positions it as a valuable resource, offering solutions to diverse energy needs and contributing to the broader shift toward sustainable energy practices (42, 43). Pyrolysis's significance extends beyond its ability to generate multiple products and its capacity to process a wide range of biomass feedstocks. From woody biomass sourced from forests to agricultural residues left after harvest, pyrolysis provides a scalable and efficient solution for biomass utilization. This adaptability ensures the sustainable and efficient conversion of organic matter into valuable products, contributing to environmental stewardship goals and the pursuit of a more sustainable energy future.

13.3.4.2 Anaerobic Digestion

As an advanced biomass conversion technology, anaerobic digestion capitalizes on microbial processes to break down organic matter without oxygen. The paramount outcome of this anaerobic breakdown is the production of biogas, a mixture predominantly composed of methane and carbon dioxide, which holds significant promise for diverse energy applications. The versatility of feedstocks suitable for anaerobic digestion is a defining feature, encompassing a broad spectrum of organic waste streams such as agricultural residues, food waste, and wastewater sludge (44–46). The microbial consortium engaged in anaerobic digestion undertakes the metabolic transformation of complex organic compounds, culminating in biogas production. This renewable energy source can be harnessed in combined heat and power (CHP) systems, offering the dual benefits of electricity and heat. Alternatively, it can be upgraded for injection into the natural gas grid.

The environmental advantages of anaerobic digestion are manifold. Beyond providing an efficient means of managing organic waste and alleviating concerns associated with landfill disposal, it concurrently generates a renewable energy source. The symbiotic relationship between waste management and energy production positions anaerobic digestion as a sustainable solution for decentralized energy generation. This technology addresses the imperative to transition towards cleaner and greener energy alternatives and contributes to waste reduction and the promotion of a circular economy. In essence, anaerobic digestion emerges as a dynamic and environmentally conscious strategy, aligning waste management practices with renewable energy production to foster sustainable and decentralized energy solutions.

13.3.4.3 Microbial Fuel Cells (MFCs)

Microbial fuel cells (MFCs) represent a groundbreaking approach to electricity generation by leveraging the metabolic activity of microorganisms. In these devices, microorganisms play a pivotal role in directly converting organic matter into electricity. Typically residing in an anode chamber, the microbial community undergoes electrochemical reactions, liberating electrons that flow through an external circuit, thereby generating electric current. The promise of MFCs lies in their potential for decentralized and environmentally friendly electricity generation, particularly in small-scale applications (47, 48). The simplicity of the microbial electricity generation process, combined with the flexibility to utilize various organic substrates, positions MFCs as an innovative and sustainable solution for energy production. This aspect is particularly advantageous in scenarios where traditional energy infrastructure may be impractical or environmentally disruptive.

Beyond electricity generation, MFCs exhibit remarkable versatility. These systems can treat wastewater by harnessing the microbial activity involved in electrochemical processes. As microorganisms break down organic matter during this electrochemical reaction, electricity is produced, and the wastewater is also effectively treated. This dual functionality underscores the potential of MFCs as integrated systems for energy production and environmental remediation. The environmental implications of MFCs extend further, as they offer a potential means to address two critical challenges simultaneously: the need for sustainable energy sources and wastewater treatment. MFCs provide a compelling example of how innovative technologies can contribute to a more sustainable and integrated approach to energy production and environmental management by merging these functions. As research in this field progresses, MFCs may find broader applications, playing a role in decentralized energy solutions and sustainable water treatment practices.

13.4 BIOENERGY APPLICATIONS

Bioenergy applications encompass diverse technologies and processes that leverage biological materials to generate energy. This multifaceted field plays a crucial role in transitioning to sustainable and renewable energy sources, contributing to reduced reliance on fossil fuels and mitigating environmental impacts. In this comprehensive exploration, we delve into various bioenergy applications, focusing on biofuels, biogas, and the environmental benefits of biomass in bioenergy (49–53).

13.4.1 Biofuels

13.4.1.1 Introduction to Biofuels

Biofuels constitute a prominent category within bioenergy, encompassing liquid or gaseous fuels derived from biomass. These fuels serve as alternatives to traditional fossil fuels, potentially mitigating greenhouse gas emissions and promoting energy security. Key biofuels include biodiesel, bioethanol, and advanced biofuels.

13.4.1.2 Biodiesel

Biodiesel, a renewable substitute for conventional diesel, is typically produced through transesterification, converting triglycerides found in feedstocks like vegetable oils or animal fats into biodiesel and glycerol. This chemical reaction, facilitated by alcohol and a catalyst, results in fatty acid methyl esters (FAME) or fatty acid ethyl esters (FAEE), the key components of biodiesel. The choice of feedstocks is diverse and includes common options like soybean oil, canola oil, palm oil, and even waste materials such as used cooking oil. The selection of feedstocks is crucial for the sustainability of biodiesel production, impacting both environmental and economic considerations. Animal fats, like tallow or lard, are additional feedstock options, contributing to using byproducts from the meat industry. Biodiesel finds wide-ranging applications in the transportation sector, offering a renewable and accessible alternative to traditional diesel. Its compatibility with existing diesel engines, with minimal modifications, makes biodiesel adoption convenient. Blending biodiesel with petroleum diesel, known as biodiesel blends, is a common practice to balance performance, emissions, and cost considerations. The environmental benefits of biodiesel include reduced greenhouse gas emissions, as the carbon released during combustion is part of a closed carbon cycle. This cycle involves plants absorbing equivalent carbon dioxide during their growth, making biodiesel a carbon-neutral or low-carbon alternative. By diversifying fuel sources and incorporating biodiesel into the mix, nations enhance energy security by reducing reliance on finite fossil fuels. Biodiesel's lubricating properties, improved lubricity, and renewable energy integration further contribute to its significance in pursuing sustainable and resilient energy solutions. In an experiment, Patrinou et al. (54) treated poultry litter extract (PLE) using a microbial consortium predominantly led by the filamentous cyanobacterium Leptolyngbya sp. in collaboration with heterotrophic microorganisms inherent in poultry waste. The research encompassed laboratory- and pilot-scale experiments conducted in aerobic conditions, employing suspended and attached growth photobioreactors. This experiment implemented diverse dilutions of the PLE, leading to varying initial pollutant concentrations, encompassing nitrogen, phosphorus, dissolved chemical oxygen demand (d-COD), and total sugars. The outcomes revealed noteworthy nutrient removal rates, biomass productivity, and maximal lipid production across all investigated systems. Attached growth reactors, both in laboratory- and pilot-scale experiments, demonstrated elevated removal rates for d-COD, nitrogen, phosphorus, and total sugars (reaching up to 94.0%, 88.2%, 97.4%, and 79.3%, respectively, in laboratory-scale; and up to 82.0%, 69.4%, 81.0%, and 83.8%, respectively, in pilot-scale). Moreover, pilot-scale attached growth experiments showcased substantial total biomass productivity (up to 335.3 mg/L/d). The produced biomass exhibited lipid content of up to 19.6% (w/w) on a dry weight basis, with saturated and monounsaturated fatty acids constituting more than 70% of the total fatty acids, indicating significant potential for biodiesel production. The findings of this study suggest that the developed processing systems can effectively treat PLE while concurrently generating lipids suitable as feedstock for biodiesel manufacturing. Additionally, Buasri et al. (55) delved into a continuous biodiesel production process utilizing waste frying oil (WFO) and a solid catalyst consisting of potassium hydroxide supported on Jatropha curcas fruit shell activated carbon (KOH/JS). Rigorous characterization of the catalyst was carried out using x-ray diffraction (XRD), scanning electron microscopy (SEM), and the Brunauer–Emmett–Teller (BET) method. The study systematically explored various reaction parameters, such as residence time, reaction temperature, methanol/oil molar ratio, and catalyst bed height in a packed bed reactor (PBR), to assess their impact on biodiesel yield. The catalyst characterization revealed a uniform distribution of KOH on the Jatropha curcas fruit shell-activated carbon. Optimal conditions yielding an 86.7% conversion were identified, including a residence time of 2 hours, a reaction temperature of 60°C, a methanol/oil molar ratio of 16, and a catalyst bed height of 250 mm in the packed bed reactor. SEM images exhibited a favorable distribution of KOH on the catalyst support. The study demonstrated that the KOH/JS catalyst maintained its catalytic activity over five consecutive cycles without requiring additional activation treatment, indicating its reusability. The findings underscore the industrial potential of the KOH/JS catalyst for the transesterification of WFO, offering promising applications in biodiesel production. Additionally, the study determined the fuel properties of the biodiesel, contributing to a comprehensive understanding of the produced biodiesel's characteristics.

Ianda et al. (56) explored the potential of implementing a biodiesel program in sub-Saharan countries, focusing on Guinea-Bissau, where limited electricity and fuel access and fragile institutional organizations impede overall development. Recognizing the role of bioenergy in addressing these challenges, the study introduces the concept of "institutional transplant" and proposes a preliminary institutional and organizational framework. This framework draws inspiration from Brazil's successful biodiesel production experience and is a foundational blueprint for initiating a Biodiesel Production Program in Guinea-Bissau. To project the potential socioeconomic impacts, the study constructs scenarios for Guinea-Bissau's future energy requirements and conducts simulations to estimate biodiesel production volumes. Adopting a B10 blend (10% biodiesel and 90% diesel) is a promising strategy. According to the simulations, implementing this blend could generate 35,785 temporary jobs, each offering remuneration equivalent to 1.5 minimum salaries. This translates to an annual income of US$5,162,701.00. Given that a significant portion of Guinea-Bissau's population lives on less than US$1.0 a day, the biodiesel production initiative holds substantial potential for poverty alleviation. Furthermore, the study highlights the potential of the B10 blend to reduce dependency on imported diesel, contributing to Guinea-Bissau's energy self-sufficiency. Overall, the findings underscore the positive socioeconomic impacts and energy security benefits that could result from implementing a biodiesel program based on the Brazilian experience.

13.4.1.3 Bioethanol

Bioethanol, a prominent alcohol-based biofuel, derives primarily from the fermentation of sugars or starches in crops like corn and sugarcane. Bioethanol production involves two main pathways: directly fermenting sugars extracted from biomass or converting starches into sugars before fermentation. Direct fermentation is typical in regions where sugarcane is prevalent, while corn-based bioethanol production is widespread in the United States. The choice of feedstocks is critical, with corn and sugarcane being primary sources, raising concerns about potential competition with food production and environmental impacts. Additionally, technological advancements enable the use of cellulosic biomass, including agricultural residues and wood, for bioethanol production, reducing concerns related to food-versus-fuel competition. Bioethanol's environmental advantages include reduced greenhouse gas emissions compared to conventional gasoline, as the carbon released during combustion is part of a closed carbon cycle. Furthermore, bioethanol has lower emissions of certain pollutants, improving air quality, particularly in urban areas. The potential utilization of non-food biomass for cellulosic ethanol production aligns with the circular economy principles, promoting waste reduction and sustainable resource management. However, careful evaluation of land use, feedstock choices, and potential impacts on food production is essential to ensure the overall sustainability of bioethanol as a renewable energy option. As technology advances, developing more sustainable and diverse feedstock sources becomes crucial for further enhancing bioethanol's environmental profile. In an experiment, Shrivastava et al. (57) focused on evaluating the efficacy of wheat straw (WS) hydrolysate, prepared through fungal pre-treatment, for the dual production of ethanol and electricity in an electrochemical bioreactor. The process involved using three white rot fungi—Phanerochaete chrysosporium, Phlebia floridensis, and Phlebia brevispora—to degrade WS and prepare the hydrolysate. The study also monitored the production of lignocellulolytic enzymes during the pre-treatment process. Subsequently, the yeast Pichia fermentans was employed to ferret all three hydrolysates over 12 days. The yeast demonstrated optimal electrochemical responses on the twelfth day, showing an open circuit voltage of 0.672 V, a current density of 542.42 mA m^{-2}, and a power density of 65.09 mW m^{-2}, particularly in the hydrolysate prepared using Phlebia floridensis. Maximum ethanol production reached 9.2% (w/v) on the seventh day, with a fermentation efficiency of approximately 62.1%. Additionally, the coulombic efficiency displayed improvement from 0.06 to 1.46% during the 12-day experiment. These results suggest the promising potential for converting lignocellulosic biomass into bioethanol while simultaneously generating bioelectricity in an integrated and sustainable manner. Additionally, Martis et al. (58) investigated the possibility of date palm waste as an economical and abundant biomass feedstock for biofuel production in the United Arab Emirates (UAE). The study explores various conversion routes, including thermochemical processes such as pyrolysis and gasification and biochemical fermentation. The simulations conducted using Aspen Plus v.10 indicate specific outcomes for each process in terms of biofuel production. For every ton of biomass feed, gasification generates 56 kg of hydrogen, and fermentation produces 233 kg of ethanol. However, the study reveals that the energy demands associated with these processes offset the economic value derived from bioethanol production. In more detail, the net duty for pyrolysis, gasification, and fermentation for 1 ton of biomass feed are determined to be 37 kJ, 725 kJ, and 7481.5 kJ, respectively. Additionally, the economic analysis demonstrates that pyrolysis generates a profit of USD 768, gasification yields USD 166, while fermentation incurs an expenditure of USD 763, rendering it economically unviable. Further economic examination of the fermentation process suggests that reducing the system's net duty to 6,500 kJ/ton biomass, coupled with converting 30% of hemicellulose and cellulose content, leads to a breakeven bioethanol fuel price of 1.85 USD/L. This falls within the acceptable commercial feasibility range of 0.8–2.4 USD/L, making it competitive with bioethanol produced through alternative processes. In conclusion, the economic analysis underscores that pyrolysis and gasification exhibit greater economic feasibility than fermentation. The study proposes utilizing wasted hemicellulose and lignin from fermentation in thermochemical processes for additional fuel production to optimize profits.

Nenciu et al. (59) focused on an integrated approach involving alcoholic fermentation and pyrolysis of sweet sorghum biomass residues to produce high-grade chemicals and biofuels. Led by authors Florin Nenciu, Maria Paraschiv, Valentin Nicolae Vladut, Radu Kuncser, Constantin Stan, and Diana Cocarta, the study addressed the challenge of leveraging marginal lands for second-generation energy crops. This approach seeks to ensure sustainable bioenergy production without compromising resources allocated to food production. Sweet sorghum is identified as a suitable technical plant for challenging growing conditions, displaying characteristics such as high biomass and sugar yields, efficient photosynthesis, low fertilizer requirements, drought resistance, and adaptability to diverse climates. Despite these advantages, the study recognizes existing challenges in agronomy, techno-economics, and methodologies that may impact biomass quantity and profitability. The research compared sweet sorghum crops cultivated on marginal and regular lands. The assessment encompasses plant development characteristics, juice production, and the potential for bioethanol generation. Subsequently, vegetal wastes resulting from the process undergo pyrolysis to maximize the production of high-quality liquid biofuels and chemicals. Acknowledging the need for in-depth research to understand biomass productivity and advanced quality evaluation, the study emphasizes the development of efficient combined technologies. The charcoal obtained from the pyrolysis process is considered an amendment, highlighting the integrated and sustainable nature of the proposed approach. This research aligns with global initiatives aiming to utilize marginal lands for bioenergy production, offering a promising strategy to contribute to high-quality biofuels and chemicals without competing with the agri-food industry. Furthermore, Mohammed et al. (60) focused on exploring the untapped bioenergy potential of millet chaff, an abundant agro-residue in sub-Saharan Africa. The study provides a comprehensive analysis of millet chaff's physicochemical and combustion characteristics using thermogravimetric analysis and process simulation through Aspen Plus. Though overlooked in the past, millet chaff emerges as a promising feedstock for sustainable bioenergy solutions. The collected millet chaff sample underwent thorough proximate and ultimate analyses, revealing crucial parameters. Key findings include 71.25 wt% volatile matter, 15.35 wt% fixed-carbon, 13.40 wt% ash content, and a higher heating value of 13.15 MJ/kg. The material exhibited low nitrogen and sulfur content, with notable inorganic components such as potassium, aluminum, magnesium, calcium, iron, and sodium. The kinetic study, employing the distributed activation energy model (DAEM), provided valuable insights into the combustion process. Notably, an average frequency factor and activation energy of 1.41×10^{-18} (s^{-1}) and 149.39 kJ/mol, respectively, were determined. Ignition and burnout temperatures fell between 232–244°C and 430–489°C. Further analysis of combustion thermodynamic parameters, including ΔH, ΔG, and ΔS, revealed respective values of 144.75 kJ/mol, 167.12 kJ/mol, and −40.08 J/mol. Process simulation using Aspen Plus, integrated with a steam turbine cycle, showcased excellent combustion efficiency at an air-fuel ratio of 5.14 (stoichiometric air). Noteworthy results included power generation and electric efficiency of 0.7 kWh/kg and 21.07%, achieved at 24% excess air with minimal environmental impacts. In essence, this study emphasizes the considerable potential of millet chaff as a viable biomass feedstock for producing clean bioenergy. The findings contribute valuable insights into the utilization of this overlooked agro-residue, paving the way for the development of sustainable and efficient bioenergy solutions in the region.

13.4.1.4 Advanced Biofuels

Advanced biofuels stand at the forefront of innovation, utilizing diverse feedstocks and cutting-edge technologies to produce liquid fuels with enhanced sustainability characteristics. Key advancements include:

- Cellulosic biomass conversion: the biofuels landscape is evolving towards advanced solutions prioritizing non-food feedstocks, with cellulosic biomass playing a pivotal role in this paradigm shift. Unlike traditional biofuels that rely on edible crops, advanced biofuels utilize complex carbohydrates in cellulosic biomass derived from agricultural residues and forestry byproducts. The process involves advanced enzymatic hydrolysis, where enzymes act as catalysts to break cellulose and hemicellulose into sugars (61–66). Subsequent biochemical processes employ microorganisms for fermentation, converting these sugars into advanced biofuels like ethanol. This approach ensures a sustainable and environmentally friendly alternative to traditional fossil fuels. The emphasis on non-food feedstocks addresses concerns related to food-versus-fuel competition and promotes feedstock diversity, contributing to a resilient bioenergy landscape. Despite challenges, ongoing research focuses on innovations in enzymatic processes, fermentation efficiency, and developing robust microorganisms. The carbon neutrality of advanced biofuels derived from cellulosic biomass distinguishes them as crucial in pursuing a sustainable and low-carbon bioenergy future.

- Algae-based biofuels: algae emerge as a promising and versatile feedstock in advanced biofuels, offering distinctive advantages for sustainable energy production. Notable for their high oil content and rapid growth rates, algae present an appealing solution to sourcing renewable fuels. The capacity of algae to thrive in various environments, including non-arable land and wastewater, is a crucial advantage, mitigating concerns related to land competition with food crops and providing flexibility in cultivation practices (67–69). The inherent adaptability of algae to diverse growth conditions contributes to resource efficiency and scalability in large-scale biofuel production. The appeal of algae-based biofuels lies in their potential to address environmental and resource challenges. With their lipid-rich composition, algae can be harnessed for biodiesel production, serving as a sustainable alternative to conventional diesel derived from fossil fuels. The rapid growth rates of algae enable frequent harvests and shorter cultivation cycles, enhancing the efficiency of biomass accumulation. Furthermore, the versatility of algae cultivation methods, ranging from ponds to photobioreactors, allows for customized approaches based on specific requirements and constraints.

- Cutting-edge technologies, such as algal biotechnology and genetic engineering, play a pivotal role in optimizing the lipid production of algae. By manipulating the genetic makeup of algae, researchers aim to enhance the efficiency of lipid synthesis pathways, ultimately increasing the overall yield of biofuels. This innovative approach aligns with the broader goal of improving algae-based biofuel technologies' economic viability and competitiveness. Beyond the financial considerations, algae-based biofuels offer environmental benefits. Algae can capture carbon dioxide during photosynthesis, contributing to carbon sequestration and mitigating the environmental impact of traditional fossil fuels. This dual advantage of providing a renewable fuel source while actively participating in carbon capture positions algae-based biofuels as a sustainable and eco-friendly option in transitioning to cleaner energy sources. Overall, the unique combination of high oil content, rapid growth rates, and adaptability to diverse environments positions algae as a promising and environmentally conscious feedstock for producing advanced biofuels, playing a crucial role in advancing the agenda for a more sustainable bioenergy landscape.

- Synthetic biology: in pursuing advanced biofuels, synthetic biology is a transformative force, offering unprecedented capabilities to engineer microorganisms for enhanced efficiency and specificity in biofuel production. This field harnesses genetic engineering techniques to modify the genetic makeup of microorganisms, tailoring them to perform specific tasks in the complex process of converting diverse feedstocks into biofuels (70–73). Genetic engineering empowers scientists to manipulate the genetic code of microorganisms, such as bacteria or yeast, to optimize their metabolic pathways for biofuel synthesis. This approach involves introducing or modifying specific genes responsible for vital enzymatic reactions in biomass conversion into biofuels. By doing so, researchers can enhance the efficiency of these microorganisms in breaking down complex organic compounds, such as cellulose or hemicellulose, into fermentable sugars. One of the primary advantages of synthetic biology in advanced biofuels is its ability to customize microorganisms for specific tasks and conditions. For instance, certain microorganisms may excel in the breakdown of lignocellulosic biomass, while others may be better suited for the fermentation of sugars into biofuels. Synthetic biology allows for the precise tuning of microbial traits to optimize their performance at different stages of the biofuel production process. Synthetic biology's versatility extends to microbial consortia engineering, where multiple species of microorganisms work collaboratively to perform complementary tasks. This cooperative approach mimics natural microbial ecosystems and enhances the overall efficiency of biofuel production. Each member of the microbial consortium can be designed to contribute specialized functions, creating a synergistic system that maximizes resource utilization and biofuel yield (74–77). Moreover, synthetic biology enables the creation of microorganisms with enhanced tolerance to harsh conditions commonly encountered in biofuel production processes, such as high temperatures or inhibitors. This increased resilience contributes to the robustness and reliability of biofuel production systems, addressing challenges associated with the variability of feedstocks and processing conditions.

- The applications of synthetic biology in advanced biofuels extend beyond the laboratory as researchers work towards the scalable and commercial implementation of engineered microorganisms. The ability to design microorganisms with tailored traits not only enhances the overall efficiency of biofuel production but also opens avenues for developing novel biofuel pathways and utilizing unconventional feedstocks. As advances in synthetic biology continue to unfold, the potential for creating microorganisms with enhanced capabilities for biofuel production becomes increasingly promising (78–81). This innovative approach plays a pivotal role in shaping the future of sustainable bioenergy, offering solutions to challenges related to feedstock diversity, process efficiency, and environmental impact. Through the precise manipulation of microbial genetics, synthetic biology emerges as a powerful tool driving the evolution of advanced biofuels toward greater sustainability and economic viability.

13.4.2 Biogas

13.4.2.1 Overview of Biogas

Biogas, a versatile renewable energy source, results from microorganisms' anaerobic digestion of organic matter. The predominant component of biogas is methane, which can be utilized for electricity generation, heating, and as a vehicle fuel. This section explores the process of anaerobic digestion, feedstock options, and the applications of biogas in various sectors.

13.4.2.2 Biogas Production from Agricultural Residues

Agricultural residues, encompassing crop residues and animal manure, emerge as valuable and abundant feedstocks for biogas production. This section delves into the potential of harnessing agricultural waste for biogas generation, highlighting its multifaceted role in fostering sustainable agriculture, efficient waste management, and renewable energy production (82, 83). Agricultural residues, often considered byproducts of farming activities, possess rich organic content suitable for anaerobic digestion—a process central to biogas production. Anaerobic digestion involves the microbial breakdown of organic matter in the absence of oxygen, generating biogas as a byproduct. Crop residues, including stalks, leaves, husks, and animal manure, represent organic materials abundant in carbon and nitrogen, providing an ideal substrate for anaerobic digestion. Using agricultural residues for biogas production holds significant promise in addressing environmental and agricultural challenges. Farmers can simultaneously manage agricultural waste effectively and derive renewable energy by repurposing these residues through anaerobic digestion. This symbiotic approach aligns with the principles of circular economy and sustainable agriculture, contributing to resource efficiency and reduced environmental impact. Moreover, biogas produced from agricultural residues serves as a versatile energy source. It can be used for electricity generation, heating, and as a vehicle fuel. Integrating biogas into diversified energy applications enhances its role in providing decentralized and clean energy solutions for rural and urban settings. The byproduct of anaerobic digestion, known as digestate, further holds value as an organic fertilizer, closing the loop in sustainable agriculture practices. In an experiment, Siciliano et al. (84) introduced a comprehensive strategy for effectively utilizing agro-industrial waste, aiming to simultaneously produce biofuels (biogas and syngas) and recover phosphorus. The initial phase involves subjecting two different combinations of raw agro-industrial residues to anaerobic digestion (AD). These mixtures include asparagus, tomato wastes (mixture-1), potatoes, and kiwifruit residues (mixture-2). The investigation reveals that these mixtures' inherent characteristics significantly influence the AD process's dynamics. Despite having a lower organic load, mixture-1 demonstrates superior biogas yield, approximately 0.44 L/gCOD removed, while mixture-2 encounters challenges due to the accumulation of acidity. In the subsequent phase, the digestates derived from AD undergo supercritical water gasification (SCWG) at 450°C and 250 bar, producing syngas. Both digests exhibit efficient conversion into syngas, primarily composed of H2, CO_2, CH4, and CO. Mixture-2 demonstrates superior performance, attaining maximum values for global gasification efficiency at 56.5 g/kgCOD and gas yield at 1.8 mol/kg. The final step of the integrated treatment focuses on recovering phosphorus from the residual liquid fraction of SCWG, forming $MgKPO_4 \cdot 6H_2O$. Experimental results indicate that almost complete phosphorus removal can be achieved under specific conditions (pH = 10 and Mg/P = 1). The digestion process in mixture-1 demonstrated high efficiency, characterized by a cumulative curve devoid of a lag phase for biomass acclimation, as depicted in Figure 13.3. This observation underscores

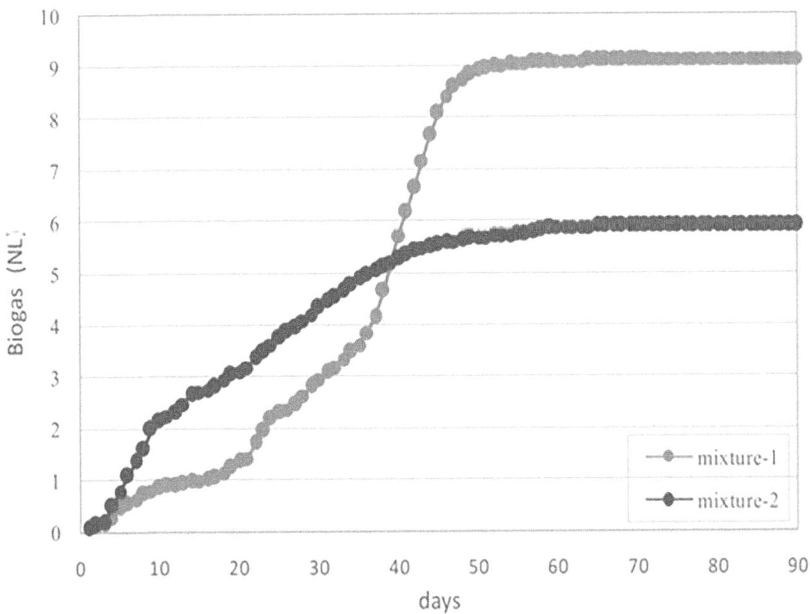

FIGURE 13.3 Production of the biogas throughout the co-digestion examination. It is adapted from ref. (84) under CCBY 4.0.

the favorable biodegradability of substrates and the absence of inhibitory compounds within the mixture. The process was initiated promptly, with a moderate volume of biogas generated in the initial two weeks, although low methane fractions, ranging from 2–14%, were recorded. During this early phase, it is likely that the hydrolysis of complex organic compounds occurred, leading to the formation of volatile fatty acids (VFAs). Consequently, the predominant components of the biogas were CO_2 and H_2. The exponential phase of biogas production commenced after 18 days, ultimately reaching the asymptotic value of 9.16 L within 50 days. The average CH4 fraction stabilized at 74% throughout this period, indicating a consistent methanation process, as illustrated in Figure 13.3.

Moreover, scanning electronic microscopy confirms the composition of the produced precipitate as magnesium potassium phosphate crystals. This innovative integrated approach showcases the potential for simultaneously deriving biofuels and recovering phosphorus from agro-industrial waste, highlighting the versatility and sustainability of the proposed treatment strategy. The study advances sustainable waste management practices with valuable bioenergy production and resource recovery applications. Ezealigo et al. (14) delve into the comprehensive assessment of the bioenergy potential inherent in biomass residues within Nigeria. The focus extends to agricultural residues and municipal solid and liquid waste, with data from 2008 to 2018 serving as the foundation for a computational and analytical exploration, incorporating conservative assumptions. The primary objective is to estimate the technical potential for producing cellulosic ethanol and biogas. The outcomes of the study highlight biogas' superior energy generation capabilities compared to cellulosic ethanol when derived from the same type of residue. The technical potential, originating from available crop residues totaling 84 million tons, yields substantial outputs of 14,766 ML/yr (8 Mtoe) for cellulosic ethanol and 15,014 Mm^3/yr (13 Mtoe) for biogas. Notably, biogas emerges as a promising solution due to its versatile applications, spanning from heat to electric power generation, and holds significant promise in addressing Nigeria's prevailing electricity crisis. Additionally, the study underscores the pivotal role of biogas in contributing substantially to achieving Sustainable Development Goal 7 (SDG 7) concerning clean and affordable energy.

13.4.2.3 Integration of Biogas into Energy Systems

The integration of biogas into energy systems represents a pivotal step toward maximizing the utility and impact of this renewable energy source. This section explores the technological dimensions, benefits, and challenges of incorporating biogas into existing energy infrastructure, emphasizing grid injection and combined heat and power (CHP) applications. Grid injection involves injecting biogas into the natural gas grid, typically upgraded to biomethane. This process facilitates the distribution of biogas over a broader geographical area and allows for its utilization in diverse sectors, including residential, commercial, and industrial applications. Biomethane, with its compatibility with existing natural gas infrastructure, provides a renewable and low-carbon alternative to traditional natural gas, contributing to the decarbonization of the gas grid (85, 86). In CHP applications, biogas is utilized for simultaneous electricity and heat generation. This combined approach enhances energy efficiency by utilizing the thermal energy produced during electricity generation. CHP systems find application in various sectors, such as wastewater treatment plants, agricultural facilities, and industrial complexes, where the demand for electricity and heat is substantial. Integrating biogas into CHP systems maximizes energy utilization and contributes to the resilience and reliability of energy supply in decentralized settings.

While integrating biogas into energy systems presents numerous benefits, challenges persist. Variability in biogas production, influenced by factors such as feedstock composition and digestion process stability, poses a technical challenge in aligning biogas supply with energy demand. Additionally, the economic viability of biogas projects, including the costs associated with upgrading biogas to biomethane and infrastructure investments, requires careful consideration.

13.4.3 Environmental Benefits of Biomass in Bioenergy

13.4.3.1 Carbon Neutrality of Biomass

Biomass's carbon neutrality is a cornerstone of its environmental benefits, representing a crucial aspect in the quest for sustainable energy solutions. This section delves into the concept of carbon neutrality within the context of biomass, elucidating how the carbon dioxide released during combustion is effectively counterbalanced by the carbon absorbed during the biomass's growth phase. The inherent carbon neutrality of biomass is intricately linked to the natural carbon cycle (87–89). During photosynthesis, plants absorb carbon dioxide from the atmosphere and convert it into organic compounds, storing solar energy in complex carbohydrates like cellulose, hemicellulose, and lignin. This stored carbon becomes part of the plant's biomass, contributing to its growth and development.

When biomass is subsequently utilized for energy through combustion, the stored carbon is released back into the atmosphere as carbon dioxide. Crucially, the carbon dioxide emitted during combustion is equivalent to the amount absorbed by the plants during their growth, creating a closed-loop carbon system. This closed carbon cycle ensures that the net carbon emissions from biomass combustion are mitigated by the carbon sequestration capacity of the source plants during their life cycle. Understanding and harnessing the carbon neutrality of biomass is essential for evaluating its environmental impact. As the global community grapples with the imperative to reduce carbon emissions and combat climate change, biomass emerges as a renewable energy source that aligns with sustainability goals by maintaining a balance in the atmospheric carbon cycle.

13.4.3.2 Biomass for Waste Management

Biomass is pivotal in addressing waste management challenges, offering an environmentally friendly solution to repurpose various organic materials as feedstock for energy production. Agricultural residues, comprising crop residues and forestry byproducts, including branches and wood chips, can be effectively repurposed as biomass feedstock. In doing so, biomass utilization addresses waste disposal challenges and fosters a circular economy by transforming waste into a valuable resource for energy generation. This dual benefit exemplifies biomass's environmental stewardship potential, contributing to waste reduction and renewable energy generation (90, 91). Municipal solid waste (MSW) further demonstrates the versatility of biomass in waste management. As urbanization accelerates globally, the volume of waste generated in cities continues to rise. Harnessing the energy potential within MSW through thermochemical processes like incineration and gasification transforms the organic fraction of waste into biogas or syngas, offering a renewable alternative to traditional fossil fuels. This integration of MSW into the biomass energy portfolio aligns with resource efficiency and environmental sustainability principles, effectively turning waste streams into valuable energy resources.

13.4.3.3 Decentralization and Energy Security

The decentralization of biomass resources emerges as a strategic element in enhancing energy security and fostering sustainable development (92–94). Unlike centralized power systems that rely on large-scale energy production and distribution, decentralized biomass facilities operate on a smaller scale, often tailored to local or regional energy needs. By utilizing locally available biomass resources, these facilities contribute to the diversification of energy sources and reduce dependence on external, centralized power infrastructure. Localized biomass facilities play a transformative role in rural and remote areas, where access to traditional energy sources may be limited. They not only provide a reliable source of energy but also contribute to job creation and community development. The decentralization of energy production through biomass aligns with the broader goals of sustainable development, ensuring communities have access to clean and affordable energy. Moreover, decentralized biomass facilities enhance energy security by reducing vulnerability to disruptions in centralized power systems. Localized energy production, powered by readily available biomass resources, creates a resilient energy infrastructure that can withstand natural disasters or supply chain disruptions.

13.5 CONTEMPORARY ENERGY CHALLENGES

13.5.1 Global Energy Landscape

The contemporary energy landscape is a dynamic interplay of challenges and opportunities shaped by the ever-growing global population, rapid urbanization, and industrialization. This scenario intensifies the energy demand, while concerns about climate change and environmental degradation underscore the urgency of transitioning to sustainable and low-carbon energy sources (95, 96). The reliance on fossil fuels, a primary contributor to the current energy mix, faces scrutiny due to its environmental impact and introduces geopolitical complexities. Navigating this complex energy landscape requires a holistic approach addressing the increasing demand for energy and the imperative to reduce environmental harm.

13.5.2 Role of Biomass in Addressing Challenges

Biomass emerges as a pivotal player in this intricate energy landscape, offering unique attributes that position it as a sustainable alternative to traditional fossil fuels. Its renewability, carbon neutrality, and versatility make biomass a valuable resource in pursuing a cleaner and more sustainable energy mix. Biomass provides a decentralized and locally available energy source, significantly diversifying the energy portfolio. Its applications in electricity generation, biofuels, and waste-to-energy processes offer a multifaceted approach to addressing the complexities of the global energy landscape. At the intersection of energy security, environmental sustainability, and economic development, biomass stands as a promising solution for the challenges faced by the global energy sector.

13.5.3 Greenhouse Gas Emissions Reduction

One of the paramount challenges in the global energy landscape is the imperative to reduce greenhouse gas (GHG) emissions to mitigate climate change. Fossil fuel combustion, a significant contributor to GHG emissions, releases carbon dioxide and other pollutants. Biomass plays a pivotal role by offering a carbon-neutral energy option. The closed carbon cycle of biomass, where the carbon released during combustion is offset by the carbon absorbed during plant growth, acts as a mitigating factor in the net increase of atmospheric carbon dioxide. Utilizing biomass for energy purposes aligns with global efforts to achieve carbon neutrality, limit global warming, and transition towards sustainable and environmentally friendly energy sources.

13.5.4 Biodiversity Conservation

In pursuing sustainable energy, it is imperative to consider the impact of energy practices on biodiversity. Unsustainable energy production, habitat destruction, and climate change threaten global biodiversity. Biomass, when sourced responsibly, can contribute positively to biodiversity conservation. Sustainable biomass practices use feedstocks that do not compete with food crops, ensuring minimal ecosystem impact (97, 98). Adopting agroforestry practices and sustainable forestry management promotes biodiversity by maintaining natural habitats. Integrating biodiversity considerations into biomass production and utilization is crucial for fostering a harmonious relationship between energy development and ecological preservation.

13.5.5 Sustainable Energy Practices

The overarching goal in addressing contemporary energy challenges is the pursuit of sustainable energy practices that foster resilience and equity. Biomass aligns with these objectives by offering a renewable and locally available resource that can be harnessed through environmentally responsible practices. Sustainable biomass production involves land-use planning, conservation of natural ecosystems, and community engagement. Additionally, integrating advanced technologies, such as gasification and advanced conversion processes, enhances biomass utilization's efficiency and environmental performance. As part of a broader strategy, biomass realizes sustainable energy practices that prioritize environmental, social, and economic dimensions. In a nutshell, Figure 13.4 depicts the several challenges for biofuel production in biorefineries using biomass waste.

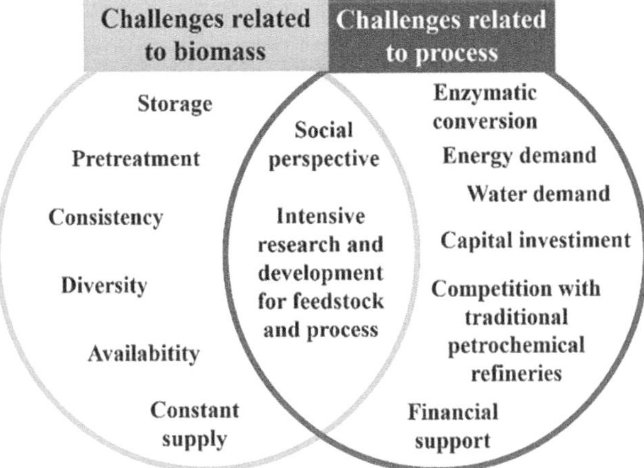

FIGURE 13.4   Numerous obstacles exist to the production of biofuels in biorefineries utilizing biomass waste. They are adapted from ref. (99) under CCBY 4.0.

13.6 CONCLUSION

In concluding this exploration of biomass as a renewable energy source, its multifaceted role in addressing contemporary energy challenges becomes evident. Biomass, deeply ingrained in human history, has evolved from traditional applications to meet demands in the modern global energy landscape. Its inherent sustainability, versatility, and carbon-neutral nature position biomass as a compelling solution to pressing energy challenges. The closed carbon cycle, where emissions from biomass combustion are offset by plant absorption, underscores its significance in reducing greenhouse gas emissions. Moreover, the diverse composition of biomass, ranging from organic materials to agricultural residues and municipal solid waste, emphasizes its local availability and potential for decentralized energy solutions. Advanced technologies, including gasification and other innovative conversion processes, amplify biomass's role, making it a key contender in shaping a cleaner and sustainable energy landscape.

Looking toward the future, the prospects of biomass depend on continued research and development initiatives. Technological advancements, feedstock diversification, and comprehensive environmental impact assessments are essential to optimize energy production efficiency and understand the broader implications of biomass utilization on ecosystems, biodiversity, and water resources. Integrating biomass with other renewable sources, formulating supportive policies, and considering socioeconomic factors are crucial aspects that demand focused research efforts. Navigating these research directions will empower biomass to play a pivotal role in fostering a sustainable energy future, contributing significantly to the ongoing global shift towards cleaner and resilient energy systems.

REFERENCES

1. Ortiz Cebolla R, Navas C. Supporting hydrogen technologies deployment in EU regions and member states: The smart specialisation platform on energy (S3PEnergy). *Int J Hydrogen Energy [Internet].* 2019;44(35):19067–19079. Available from: https://doi.org/10.1016/j.ijhydene.2018.05.041

2. Child M, Bogdanov D, Breyer C. The Baltic Sea region: Storage, grid exchange and flexible electricity generation for the transition to a 100% renewable energy system. *Energy Procedia [Internet].* 2018;155:390–402. Available from: https://doi.org/10.1016/j.egypro.2018.11.039

3. Low S, Baum CM, Sovacool BK. Rethinking net-zero systems, spaces, and societies: "Hard" versus "soft" alternatives for nature-based and engineered carbon removal. *Glob Environ Chang [Internet].* 2022;75(December 2021):102530. Available from: https://doi.org/10.1016/j.gloenvcha.2022.102530

4. Calvo-Saad MJ, Solís-Chaves JS, Murillo-Arango W. Suitable municipalities for biomass energy use in Colombia based on a multicriteria analysis from a sustainable development perspective. *Heliyon.* 2023;9(10).

5. Olabi AG, Elsaid K, Obaideen K, Abdelkareem MA, Rezk H, Wilberforce T, et al. Renewable energy systems: Comparisons, challenges and barriers, sustainability indicators, and the contribution to UN sustainable development goals. *Int J Thermofluids [Internet].* 2023;20(October):100498. Available from: https://doi.org/10.1016/j.ijft.2023.100498

6. Kudelin A, Kutcherov V. Wind ENERGY in Russia: The current state and development trends. *Energy Strateg Rev [Internet].* 2021;34(June 2020):100627. Available from: https://doi.org/10.1016/j.esr.2021.100627

7. Ugwu CO, Ozoegwu CG, Ozor PA, Agwu N, Mbohwa C. Waste reduction and utilization strategies to improve municipal solid waste management on Nigerian campuses. *Fuel Commun [Internet].* 2021;9:100025. Available from: https://doi.org/10.1016/j.jfueco.2021.100025

8. Hao HTN, Karthikeyan OP, Heimann K. Bio-refining of carbohydrate-rich food waste for biofuels. *Energies.* 2015;8(7):6350–6364.

9. Abdelaal A, Antolini D, Piazzi S, Patuzzi F, Villot A, Gerente C, et al. Steam reforming of tar using biomass gasification char in a Pilot-scale gasifier. *Fuel [Internet].* 2023;351(January):128898. Available from: https://doi.org/10.1016/j.fuel.2023.128898

10. Kataya G, Cornu D, Bechelany M, Hijazi A, Issa M. Biomass waste conversion technologies and its application for sustainable environmental development—a review. *Agronomy.* 2023;13(11):2833.

11. Howari H, Parvez M, Khan O, Alhodaib A, Mallah A, Yahya Z. Multi-objective optimization for ranking waste biomass materials based on performance and emission parameters in a pyrolysis process—an AHP–TOPSIS approach. *Sustain.* 2023;15(4).

12. Dunlap A. The green economy as counterinsurgency, or the ontological power affirming permanent ecological catastrophe. *Environ Sci Policy [Internet].* 2023;139(October 2022):39–50. Available from: https://doi.org/10.1016/j.envsci.2022.10.008

13. Sovacool BK, Baum C, Low S. The next climate war? Statecraft, security, and weaponization in the geopolitics of a low-carbon future. *Energy Strateg Rev [Internet].* 2023;45(November 2022):101031. Available from: https://doi.org/10.1016/j.esr.2022.101031

14. Ezealigo US, Ezealigo BN, Kemausuor F, Achenie LEK, Onwualu AP. Biomass valorization to bioenergy: Assessment of biomass residues' availability and bioenergy potential in Nigeria. *Sustain.* 2021;13(24).

15. Ugwu CO, Ozoegwu CG, Ozor PA, Agwu N, Mbohwa C. Waste reduction and utilization strategies to improve municipal solid waste management on Nigerian campuses. *Fuel Commun [Internet].* 2021;9:100025. Available from: https://www.sciencedirect.com/science/article/pii/S2666052021000182

16. Blair MJ, Gagnon B, Klain A, Kulišić B. Contribution of biomass supply chains for bioenergy to sustainable development goals. *Land.* 2021;10(2):1–28.

17. Sovacool BK, Baum CM, Low S. Climate protection or privilege? A whole systems justice milieu of twenty negative emissions and solar geoengineering technologies. *Polit Geogr [Internet].* 2022;97(December 2021):102702. Available from: https://doi.org/10.1016/j.polgeo.2022.102702

18. Greinert A, Mrówczyńska M, Szefner W. The use of waste biomass from thewood industry and municipal sources for energy production. *Sustain.* 2019;11(11).

19. Capellán-Pérez I, de Castro C, Miguel González LJ. Dynamic energy return on energy investment (EROI) and material requirements in scenarios of global transition to renewable energies. *Energy Strateg Rev [Internet].* 2019;26(July):100399. Available from: https://doi.org/10.1016/j.esr.2019.100399

20. Slorach PC, Jeswani HK, Cuéllar-Franca R, Azapagic A. Environmental sustainability of anaerobic digestion of household food waste. *J Environ Manage [Internet].* 2019;236(August 2018):798–814. Available from: https://doi.org/10.1016/j.jenvman.2019.02.001

21. Fetio Ngoune N, Kanouo Djousse BM, Djoukeng GH, Nguimeya CGF, Tangka KJ, Tchoffo M. Contribution of the mix renewable energy potentials in delivering parts of the electric energy needs in the west region of Cameroon. *Heliyon [Internet].* 2023;9(3):e14554. Available from: https://doi.org/10.1016/j.heliyon.2023.e14554

22. Rudek TJ, Huang HT. Flexible experimentation as a remedy for uncertainties - reflexive public reason behind the energy transition in the People's Republic of China. *Energy Res Soc Sci.* 2024;107(December 2023).

23. Das TK, Kundu D. Feasibility and sensitivity analysis of a self-sustainable hybrid system: A case study of a mountainous region in Bangladesh. *Energy Convers Manag X [Internet]*. 2023;20(July 2022):100411. Available from: https://doi.org/10.1016/j.ecmx.2023.100411

24. Mitchell J, Olaf T, Kavadis N, Wendt S. Current research in environmental sustainability green bonds and sustainable business models in Nordic energy companies. *Curr Res Environ Sustain [Internet]*. 2024;7(July 2023):100240. Available from: https://doi.org/10.1016/j.crsust.2023.100240

25. Karatayev M, Hall S. Establishing and comparing energy security trends in resource-rich exporting nations (Russia and the Caspian Sea region). *Resour Policy [Internet]*. 2020;68(February):101746. Available from: https://doi.org/10.1016/j.resourpol.2020.101746

26. Kabir Ahmad R, Anwar Sulaiman S, Yusup S, Sham Dol S, Inayat M, Aminu Umar H. Exploring the potential of coconut shell biomass for charcoal production. *Ain Shams Eng J [Internet]*. 2022;13(1):101499. Available from: https://doi.org/10.1016/j.asej.2021.05.013

27. Solakivi T, Paimander A, Ojala L. Cost competitiveness of alternative maritime fuels in the new regulatory framework. *Transp Res Part D Transp Environ [Internet]*. 2022;113(October):103500. Available from: https://doi.org/10.1016/j.trd.2022.103500

28. Blaschke T, Biberacher M, Gadocha S, Schardinger I. "Energy landscapes": Meeting energy demands andhuman aspirations. *Biomass and Bioenergy [Internet]*. 2013;55:3–16. Available from: http://dx.doi.org/10.1016/j.biombioe.2012.11.022

29. Icaza-Alvarez D, Jurado F, Tostado-Véliz M, Arevalo P. Decarbonization of the Galapagos Islands. Proposal to transform the energy system into 100% renewable by 2050. *Renew Energy*. 2022;189:199–220.

30. Ogbolumani OA, Nwulu NI. Environmental impact assessment for a meta-model-based food-energy-water-nexus system. *Energy Reports [Internet]*. 2024;11(December 2023):218–232. Available from: https://doi.org/10.1016/j.egyr.2023.11.033

31. Caetano BC, Santos NDSA, Hanriot VM, Sandoval OR, Huebner R. Energy conversion of biogas from livestock manure to electricity energy using a Stirling engine. *Energy Convers Manag X*. 2022;15(April).

32. Vrabie C. Converting municipal waste to energy through the biomass chain, a key technology for environmental issues in (smart) cities. *Sustain*. 2021;13(9).

33. Borowski PF. Management of energy enterprises in zero-emission conditions: Bamboo as an innovative biomass for the production of green energy by power plants. *Energies*. 2022;15(5):1928.

34. Full J, Shoshi A, Gamero E, Baumgarten Y, Protte K, Kiemel S, et al. Biointelligent waste-to-X systems: A novel concept for sustainable, decentralized and interconnected value creation. *Procedia CIRP [Internet]*. 2023;116:576–581. Available from: https://doi.org/10.1016/j.procir.2023.02.097

35. Okumu B, Kehbila AG, Osano P. A review of water-forest-energy-food security nexus data and assessment of studies in East Africa. *Curr Res Environ Sustain [Internet]*. 2021;3:100045. Available from: https://doi.org/10.1016/j.crsust.2021.100045

36. Nkuna SG, Olwal TO, Chowdhury SD. Assessment of thermochemical technologies for wastewater sludge-to-energy: An advance MCDM model. *Clean Eng Technol [Internet]*. 2022;9(May):100519. Available from: https://doi.org/10.1016/j.clet.2022.100519

37. Ekwenna EB, Wang Y, Roskilly A. Bioenergy production from pretreated rice straw in Nigeria: An analysis of novel three-stage anaerobic digestion for hydrogen and methane co-generation. *Appl Energy [Internet]*. 2023;348(July):121574. Available from: https://doi.org/10.1016/j.apenergy.2023.121574

38. Huarachi-Olivera R, Dueñas-Gonza A, Yapo-Pari U, Vega P, Romero-Ugarte M, Tapia J, et al. Bioelectrogenesis with microbial fuel cells (MFCs) using the microalga Chlorella vulgaris and bacterial communities. *Electron J Biotechnol [Internet]*. 2018;31:34–43. Available from: https://doi.org/10.1016/j.ejbt.2017.10.013

39. Sovacool BK, Iskandarova M, Geels FW. "Bigger than government": Exploring the social construction and contestation of net-zero industrial megaprojects in England. *Technol Forecast Soc Change [Internet]*. 2023;188(January):122332. Available from: https://doi.org/10.1016/j.techfore.2023.122332

40. Torkayesh AE, Deveci M, Karagoz S, Antucheviciene J. A state-of-the-art survey of evaluation based on distance from average solution (EDAS): Developments and applications. *Expert Syst Appl*. 2023;221(November 2022).

41. Logroño W, Ramírez G, Recalde C, Echeverría M, Cunachi A. Bioelectricity generation from vegetables and fruits wastes by using single chamber microbial fuel cells with high Andean soils. *Energy Procedia [Internet]*. 2015;75:2009–2014. Available from: http://dx.doi.org/10.1016/j.egypro.2015.07.259

42. Nunes LJR, Matias JCO. Biomass torrefaction as a key driver for the sustainable development and decarbonization of energy production. *Sustain*. 2020;12(3):1–9.

43. Chai YH, Yusup S, Kadir WNA, Wong CY, Rosli SS, Ruslan MSH, et al. Valorization of tropical biomass waste by supercritical fluid extraction technology. *Sustain*. 2021;13(1):1–24.

44. Ali L, Baloch KA, Palamanit A, Raza SA, Laohaprapanon S, Techato K. Physicochemical characterisation and the prospects of biofuel production from rubberwood sawdust and sewage sludge. *Sustain*. 2021;13(11):1–16.

45. Tshikovhi A, Motaung TE. Technologies and Innovations for biomass energy production. *Sustain*. 2023;15(16):1–21.

46. Lalhmangaihzuala S, Laldinpuii Z, Lalmuanpuia C, Vanlaldinpuia K. Glycolysis of poly(Ethylene terephthalate) using biomass-waste derived recyclable heterogeneous catalyst. *Polymers (Basel)*. 2021;13(1):1–13.

47. Sharma S, Tsai ML, Sharma V, Sun PP, Nargotra P, Bajaj BK, et al. Environment friendly pretreatment approaches for the bioconversion of lignocellulosic biomass into biofuels and value-added products. *Environ - MDPI.* 2023;10(1).

48. Duque-Acevedo M, Belmonte-Ureña LJ, Yakovleva N, Camacho-Ferre F. Analysis of the circular economic production models and their approach in agriculture and agricultural waste biomass management. *Int J Environ Res Public Health.* 2020;17(24):1–34.

49. Thakur A, Kumar A. Exploring the potential of ionic liquid-based electrochemical biosensors for real-time biomolecule monitoring in pharmaceutical applications: From lab to life. *Results Eng [Internet].* 2023;20(July):101533. Available from: https://doi.org/10.1016/j.rineng.2023.101533

50. Thakur A, Kaya S, Kumar A. Recent innovations in nano container-based self-healing coatings in the construction industry. *Curr Nanosci.* 2021;18(2):203–216.

51. Sharma D, Thakur A, Sharma MK, Jakhar K, Kumar A, Sharma AK, Hari OM. Synthesis, electrochemical, morphological, computational and corrosion inhibition studies of 3-(5-Naphthalen-2-yl-[1,3,4]oxadiazol-2-yl)-pyridine against mild steel in 1 M HCl. *Asian J Chem.* 2023;35(5): 1079–1088.

52. Kaya S, Lgaz H, Thakkur A, Kumar A, Özbakır D, Karakuş N, et al. Molecular insights into the corrosion inhibition mechanism of omeprazole and tinidazole: A theoretical investigation. *Mol Simul.* 2023;1–15.

53. Thakur A, SAVAŞ K, Kumar A. Recent trends in the characterization and application progress of nano-modified coatings in corrosion mitigation of metals and alloys. *Appl Sci.* 2023;13:730.

54. Patrinou V, Tsolcha ON, Tatoulis TI, Stefanidou N, Dourou M, Moustaka-Gouni M, et al. Biotreatment of poultry waste coupled with biodiesel production using suspended and attached growth microalgal-based systems. *Sustain.* 2020;12(12).

55. Buasri A, Chaiyut N, Loryuenyong V, Rodklum C, Chaikwan T, Kumphan N. Continuous process for biodiesel production in packed bed reactor from waste frying oil using potassium hydroxide supported on Jatropha curcas fruit shell as solid catalyst. *Appl Sci.* 2012;2(3):641–653.

56. Ianda TF, Padula AD. Exploring the Brazilian experience to design and simulate the impacts of a biodiesel program for sub-Saharan countries: The case of Guinea-Bissau. *Energy Strateg Rev [Internet].* 2020;32:100547. Available from: https://doi.org/10.1016/j.esr.2020.100547

57. Shrivastava A, Sharma RK. Conversion of lignocellulosic biomass: Production of bioethanol and bioelectricity using wheat straw hydrolysate in electrochemical bioreactor. *Heliyon [Internet].* 2023;9(1):e12951. Available from: https://doi.org/10.1016/j.heliyon.2023.e12951

58. Martis R, Al-Othman A, Tawalbeh M, Alkasrawi M. Energy and economic analysis of date palm biomass feedstock for biofuel production in UAE: Pyrolysis, gasification and fermentation. *Energies.* 2020;13(22).

59. Nenciu F, Paraschiv M, Kuncser R, Stan C, Cocarta D, Vladut VN. High-grade chemicals and biofuels produced from marginal lands using an

integrated approach of alcoholic fermentation and pyrolysis of sweet sorghum biomass residues. *Sustain.* 2022;14(1).

60. Mohammed IY, Kabir G, Abakr YA, Apasiku MAA, Kazi FK, Abubakar LG. Bioenergy potential of millet chaff via thermogravimetric analysis and combustion process simulation using Aspen Plus. *Clean Chem Eng.* 2022;3(May):100046.

61. Thakur A, Kumar A, Sharma S, Ganjoo R, Assad H. Computational and experimental studies on the efficiency of Sonchus arvensis as green corrosion inhibitor for mild steel in 0.5 M HCl solution. *Mater Today Proc [Internet].* 2022;66:609–621. Available from: https://doi.org/10.1016/j.matpr.2022.06.479

62. Thakur A, Kumar A, Kaya S, Benhiba F, Sharma S. Electrochemical and computational investigations of the Thysanolaena latifolia leaves extract: An eco-benign solution for the corrosion mitigation of mild steel. *Results Chem [Internet].* 2023;6(September):101147. Available from: https://doi.org/10.1016/j.rechem.2023.101147

63. Sharma D, Thakur A, Kumar M, Sharma R, Kumar S, Om H. Effective corrosion inhibition of mild steel using novel 1, 3, 4-oxadiazole-pyridine hybrids: Synthesis, electrochemical, morphological, and computational insights. *Environ Res [Internet].* 2023;234(July):116555. Available from: https://doi.org/10.1016/j.envres.2023.116555

64. Thakur A, Sharma S, Ganjoo R, Assad H, Kumar A. Anti-corrosive potential of the sustainable corrosion Inhibitors based on biomass waste: A review on preceding and perspective research. *J Phys Conf Ser.* 2022;2267(1):012079.

65. Kaya S, Thakur A, Kumar A. The role of in Silico/DFT investigations in analyzing dye molecules for enhanced solar cell efficiency and reduced toxicity. *J Mol Graph Model.* 2023;124(May).

66. Thakur A, Sharma S, Ganjoo R, Assad H, Kumar A. Anti-corrosive potential of the sustainable corrosion inhibitors based on biomass waste: A review on preceding and perspective research. *J Phys Conf Ser.* 2022;2267(1).

67. Grande L, Pedroarena I, Korili SA, Gil A. Hydrothermal liquefaction of biomass as one of the most promising alternatives for the synthesis of advanced liquid biofuels: A review. *Materials (Basel).* 2021;14(18).

68. Barbot YN, Al-Ghaili H, Benz R. A review on the valorization of macroalgal wastes for biomethane production. *Mar Drugs.* 2016;14(6).

69. Bhowmik D, Chetri S, Erhons K, Naha A, Deb T, Shah MP, et al. Cleaner and circular bioeconomy multitudinous approaches, challenges and opportunities of bioelectrochemical systems in conversion of waste to energy from wastewater treatment plants. *Clean Circ Bioeconomy [Internet].* 2023;4(October 2022):100040. Available from: https://doi.org/10.1016/j.clcb.2023.100040

70. Yahya SA, Iqbal T, Omar MM, Ahmad M. Techno-economic analysis of fast pyrolysis of date palm waste for adoption in Saudi Arabia. *Energies.* 2021;14(19).

71. Sulman AM, Matveeva VG, Bronstein LM. Cellulase Immobilization on nanostructured supports for biomass waste processing. *Nanomaterials.* 2022;12(21):1–20.

72. Peres S, Loureiro E, Santos H, Vanderley e Silva F, Gusmao A. The production of gaseous biofuels using biomass waste from construction sites in Recife, Brazil. *Processes*. 2020;8(4).

73. Jahirul MI, Rasul MG, Chowdhury AA, Ashwath N. Biofuels production through biomass pyrolysis- A technological review. *Energies*. 2012;5(12): 4952–5001.

74. Thakur A, Kumar A. Unraveling the multifaceted mechanisms and untapped potential of activated carbon in remediation of emerging pollutants: A comprehensive review and critical appraisal of advanced techniques. *Chemosphere [Internet]*. 2024;346(November 2023):140608. Available from: https://doi.org/10.1016/j.chemosphere.2023.140608

75. Thakur A, Kumar A. Recent trends in nanostructured carbon-based electrochemical sensors for the detection and remediation of persistent toxic substances in real-time analysis. *Mater Res Express*. 2023;10(3).

76. Dhonchak C, Agnihotri N. Computational Insights in the spectrophotometrically 4H-chromen-4-one complex using DFT method. *Biointerface Res Appl Chem*. 2023;13(4):357.

77. Verma C, Thakur A, Ganjoo R, Sharma S, Assad H, Kumar A, et al. Coordination bonding and corrosion inhibition potential of nitrogen-rich heterocycles: Azoles and triazines as specific examples. *Coord Chem Rev [Internet]*. 2023;488(April):215177. Available from: https://doi.org/10.1016/j.ccr.2023.215177

78. Thakur A, Kumar A, Singh A. Adsorptive removal of heavy metals, dyes, and pharmaceuticals: Carbon-based nanomaterials in focus. *Carbon N Y [Internet]*. 2023;217(November 2023):118621. Available from: https://doi.org/10.1016/j.carbon.2023.118621

79. Kumar A, Thakur A. Encapsulated nanoparticles in organic polymers for corrosion inhibition [Internet]. In *Corrosion Protection at the Nanoscale*. Elsevier Inc.; 2020. 345–362. Available from: http://dx.doi.org/10.1016/B978-0-12-819359-4.00018-0

80. Arifa Farzana B, Mujafarkani N, Thakur A, Kumar A, Mushira Banu A, Shifana M. Evaluating (p-Semidine-Guanidine-Formaldehyde) terpolymer resin efficiency as anti-corrosive agent for mild steel in 1 M H2SO4: An experimental and computational approach. *Inorg Chem Commun [Internet]*. 2023;158(P1):111572. Available from: https://doi.org/10.1016/j.inoche.2023.111572

81. Thakur A, Kumar A. Ecotoxicity analysis and risk assessment of nanomaterials for the environmental remediation. *Macromol Symp*. 2023;410(1):1–23.

82. Kumar T, Eswari Jujjavarappu S. A critical review on an advanced bioelectrochemical system for carbon dioxide sequestration and wastewater treatment. *Total Environ Res Themes [Internet]*. 2023;5(December 2022):100023. Available from: https://doi.org/10.1016/j.totert.2022.100023

83. Okonkwo EC, Namany S, Fouladi J, Almanassra IW, Mahmood F, Al-Ansari T. A multi-level approach to the energy-water-food nexus: From molecule to governance. *Clean Environ Syst [Internet]*. 2023;8(October 2022):100110. Available from: https://doi.org/10.1016/j.cesys.2023.100110

84. Siciliano A, Limonti C, Mehariya S, Molino A, Calabrò V. Biofuel production and phosphorus recovery through an integrated treatment of agro-industrial waste. *Sustain.* 2019;11(1):1–17.

85. Elegbede JA, Ajayi VA, Lateef A. Microbial valorization of corncob: Novel route for biotechnological products for sustainable bioeconomy. *Environ Technol Innov [Internet].* 2021;24:102073. Available from: https://doi.org/10.1016/j.eti.2021.102073

86. Briones-Hidrovo A, Uche J, Martínez-Gracia A. Hydropower and environmental sustainability: A holistic assessment using multiple biophysical indicators. *Ecol Indic [Internet].* 2021;127(April):107748. Available from: https://doi.org/10.1016/j.ecolind.2021.107748

87. Aragón-Briceño CI, Grasham O, Ross AB, Dupont V, Camargo-Valero MA. Hydrothermal carbonisation of sewage digestate at wastewater treatment works: Influence of solid loading on characteristics of hydrochar, process water and plant energetics. *Renew Energy.* 2020;157:959–973.

88. Cheng F, Li X. Preparation and application of biochar-based catalysts for biofuel production. *Catalysts.* 2018;8(9):1–35.

89. Oyewo AS, Solomon AA, Bogdanov D, Aghahosseini A, Mensah TNO, Ram M, et al. Just transition towards defossilised energy systems for developing economies: A case study of Ethiopia. *Renew Energy [Internet].* 2021;176:346–365. Available from: https://doi.org/10.1016/j.renene.2021.05.029

90. Walmsley TG, Philipp M, Picón-Núñez M, Meschede H, Taylor MT, Schlosser F, et al. Hybrid renewable energy utility systems for industrial sites: A review. *Renew Sustain Energy Rev.* 2023;188(October).

91. Khare V, Khare CJ, Nema S, Baredar P. Path towards sustainable energy development: Status of renewable energy in Indian subcontinent. *Clean Energy Syst [Internet].* 2022;3(April):100020. Available from: https://doi.org/10.1016/j.cles.2022.100020

92. Pender A, Kelleher L, Neill EO, Pender A. Regulation of the bioeconomy: Barriers, drivers and potential for innovation in the case of Ireland school of architecture, planning & environmental policy, University College Dublin, Dublin BiOrbic Bioeconomy SFI Research Centre, University College. *Clean Circ Bioeconomy [Internet].* 2023;100070. Available from: https://doi.org/10.1016/j.clcb.2023.100070

93. Malode SJ, Prabhu KK, Mascarenhas RJ, Shetti NP, Aminabhavi TM. Recent advances and viability in biofuel production. *Energy Convers Manag X [Internet].* 2021;10(December 2020):100070. Available from: https://doi.org/10.1016/j.ecmx.2020.100070

94. Srivastava RK, Sarangi PK, Vivekanand V, Pareek N, Shaik KB, Subudhi S. Microbial fuel cells for waste nutrients minimization: Recent process technologies and inputs of electrochemical active microbial system. *Microbiol Res [Internet].* 2022;265(September):127216. Available from: https://doi.org/10.1016/j.micres.2022.127216

95. Zardzewiały M, Bajcar M, Puchalski C, Gorzelany J. The possibility of using waste biomass from selected plants cultivated for industrial purposes to produce a renewable and sustainable source of energy. *Appl Sci.* 2023;13(5).

96. Mensah TNO, Oyewo AS, Breyer C. The role of biomass in sub-Saharan Africa's fully renewable power sector—the case of Ghana. *Renew Energy [Internet]*. 2021;173:297–317. Available from: https://doi.org/10.1016/j.renene.2021.03.098

97. Alzate Acevedo S, Díaz Carrillo ÁJ, Flórez-López E, Grande-Tovar CD. Recovery of banana waste-loss from production and processing: A contribution to a circular economy. *Molecules*. 2021;26(17):1–30.

98. Ramos M, Dias APS, Puna JF, Gomes J, Bordado JC. Biodiesel production processes and sustainable raw materials. *Energies*. 2019;12(23).

99. Clauser NM, González G, Mendieta CM, Kruyeniski J, Area MC, Vallejos ME. Biomass waste as sustainable raw material for energy and fuels. *Sustain*. 2021;13(2):1–21.

CHAPTER 14

Biomass Wastes for Lubricating, Adhesive, and Anticorrosive Applications

Ambrish Singh, Akhiu K. Yimchunger, Therola Sangtam, Shivani Singh, Kashif R. Ansari, Dheeraj Singh Chauhan, Ashish Kumar, Kailash Juglan, Yuanhua Lin and M. A. Quraishi

14.1 INTRODUCTION

Metals, including iron, copper, zinc, and their alloys, are used in many industries. They are of great importance, have high heat conductivity, have excellent ecological strength, and have excellent mechanical and physical qualities. The metals are often exposed to corrosive solutions during the pickling processes that deteriorate their desirable properties. Several methods have been used to protect the metals from corrosive solutions, including inhibitors [1]. Numerous variables can lead to corrosion in metals and alloys; Figure 14.1 roughly depicts a few of these primary significant ecological sources.

The more popular, straightforward, and economical approach is the application of corrosion inhibitors. As corrosion inhibitors, inorganic substances such as phosphates, nitrates, arsenates, tetraborates, chromates, and organic substances containing heteroatoms, several bonds, and

DOI: 10.1201/9781003466833-17

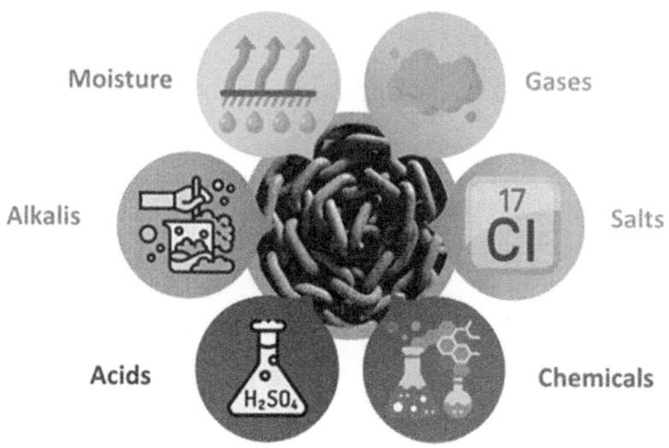

FIGURE 14.1   Environmental causes of metals corrosion.

polymers are employed. Though organic inhibitors adsorb upon metallic surfaces utilizing heteroatoms that are S, O, N, aromatic rings π electrons, and numerous bonds, traditional inorganic inhibitors oxidize the metal surface to generate an impermeable coating that prevents aggressive chemicals in the environment from reaching the metal. The acceptability of organic and inorganic inhibitors has been the subject of intense discussion recently because of concerns about their expressivity, non-biodegradability, dangerous heavy metal concentration, and biotoxicity. The usage of polymers as moderate inhibitors that are not economical has been reported in several studies. Therefore, while choosing an inhibitor, consideration should be given to the overall cost of operation, toxicity, biodegradability, and environmental dangers of the material in question, in addition to the strength of the protection offered. As a result, more research and studies are being done to develop eco-friendly inhibitors [2]. The application of corrosion mitigation research is shown in Figure 14.2.

Extracts from plants are readily recyclable, non-toxic, and reasonably priced. They are being shown to function as effective corrosion inhibitors. In science, the holistic approach and the utilization of waste products and materials have gained popularity. It has been found that straw from rice extract [3], waste from shrimp protein [4], and coconut fiber [5] can all suppress the corrosion of various metals in acidic environments. Human hair, composed of lipids, the pigment melanin, and the hard fibrous protein keratin, is another source of biowaste [6]. The near majority of municipalities across the globe have garbage systems that contain human hair. Human hair leachate can lead to eutrophication issues, and burning hair releases harmful gases and an unpleasant smell. Creating goods or systems that use human hair as a resource would be the best approach to cope with

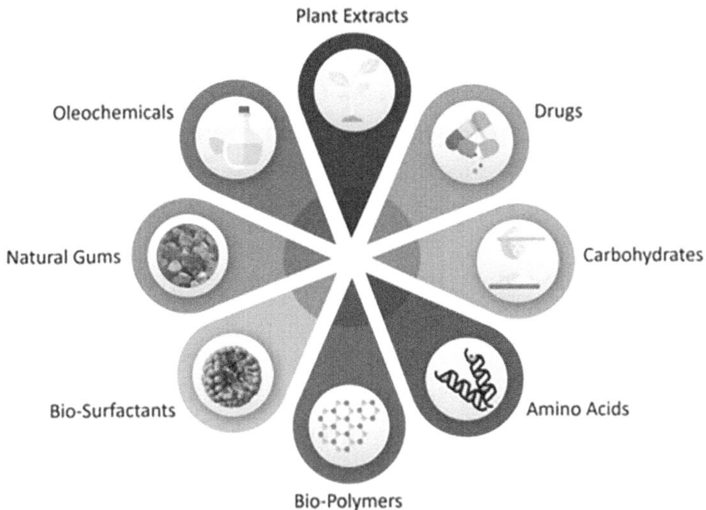

FIGURE 14.2   Natural materials used as green corrosion inhibitors.

the enduring issue of hair waste management. Hair has been utilized in applications for heavy metal removal, oil spill cleanup, construction material reinforcement, oil filtration, textile and fiber stuffing, molded furniture and items, and fertilizer up to this point [7]. Additionally, the use of human hair is being researched in areas such as concrete reinforcement [8], tissue regeneration [9], biomaterials engineering [10], superconducting system composites [11], catalytic nanoparticle platforms [12], suturing material [13], and microelectrodes [14].

It's interesting to note that some naturally existing bio-wastes originating from plant parts can be used as eco-friendly corrosion inhibitors because of their low price, ease of availability, and safety for the ecosystem. Over time, scientists and researchers have developed an interest in bio-waste valuation. Kitchen wastes contain several helpful bioactive molecules with various applications, enhance economic viability, and lessen environmental restraints, spurring interest in extracting essential bioactive compounds from them [15].

Many commercial uses related to sustainability and ecology revolve around plant matter and agricultural trash products. Because nature offers such a wide variety of matrices with different chemical compositions, there is a greater need for renewable resources. Since various forms of biomass are necessary for nourishing humans and animals, it is required to balance their use for other purposes. Ideally, the biomass known as "bio-waste," which is left behind after the nutritional components have been extracted, could be used for different processes. One area where sustainable chemistry is widely used in plant-based goods to provide innovative solutions to minimize ecological effects and wastes is metallic surface protection [16–19]. Table 14.1 represents the plant extracts application in corrosion mitigation.

Biomass is an abundant and renewable natural resource that may be processed into high-value goods. The link between biomass and biomass conversion products requires the presence of sugars such as glucose, fructose, xylose, and others. Usually, cellulose and hemicellulose undergo hydrolysis and saccharification to convert these monosaccharides. But lignocellulose's intricate and stable chemical and physical structures prevent it from breaking down or changing. Lignin prevents hydrolases from binding to cellulose and hemicellulose, which lowers the biomass's saccharification efficiency. To maximize the value of biomass, the lignocellulosic structure must be effectively destroyed. To optimize biomass utilization, pretreatment is needed to overcome the biomass's resistance to breakdown and increase its accessibility to hydrolytic enzymes.

The potential of biomass waste as organic corrosion inhibitors has received less attention than it should, even though organic plant-based sustainable corrosion inhibitors are currently the subject of various investigations. Chemicals called "green corrosion inhibitors" are used to stop metal surfaces from corroding. They are safer and more environmentally friendly than traditional inhibitors [72]. Agriculture trash, forest garbage, and feed scraps are all considered biomass wastes and can be burned or anaerobically digested to provide renewable energy [73]. Although organic green corrosion inhibitors are uncommon, one source[74] reported their accomplishment, underscoring the vast potential use of biowaste for additional research. High inhibitory efficiency has been observed in several studies. For instance, it was noted that inhibitors made from rice straw, orange peel, and agricultural waste showed encouraging results when used to prevent mild steel from corroding in an acidic medium [75]. The inhibitors' C = O, N-H, C-H, O-H, and C-transform infrared spectroscopy (FTIR) and ultraviolet/visible light spectroscopy (UV/VIS) revealed absorption bands associated with these functional groups. These bands reveal information about the specific type of adsorption that heteroatoms use to interact with the metal. This clarifies how plant-based chemicals and structural groupings can persist in biological waste, even in trace

TABLE 14.1   Plant Extracts Application in Corrosion Mitigation

S.No.	Biomass Extract	Corrosive Solution	Metal to be Protected	Techniques Used	Efficiency	References
1.	Rice husk ash in NaOH solution	0.5M NaCl	Mild steel	EIS, EDX, FTIR	88%	[20]
2.	Coconut husk fibers in ethyl acetate	0.1M HCl	Carbon steel	EIS	90%	[21]
3.	Human hair extract in water	Water	Mild steel	EIS, PDP, SEM	98.6%	[22]
4.	Elephant grass biomass in water	3.5% HCl	Mild steel	WL, EIS, PDP, LPR	81.7%	[23]
5.	Lignin polymers from elaeis guineensis	Ethanol	Mild steel	WL, EIS, FTIR, SEM	93%	[24]
6.	Pomegranate peels crude extract	Ethanol	Mild steel	EIS, SEM, FTIR	95%	[25]
7.	Sugarcane bagasse ash sand	Water	Steel	EIS, SEM	94%	[26]
8.	Barley-agro industrial waste	Water	SS–AISI 304	WL, EIS, PDP, SEM	81.6%	[27]
9.	Irvingia wombolu	Ethanol	Mild steel	WL, FTIR	97.87%	[28]
10.	Groundnut leaves	Ethanol	Mild steel	WL	86.3%	[29]
11.	Shrimp waste	H_2SO_4	Steel	WL, EIS, PDP, SEM	96.79%	[30]
12.	The hard shell of the pistachio nut	Water	Mild steel	WL, EIS	92%	[31]
13.	Punica granatum	Methanol	Mild steel	WL, EIS, PDP	80%	[32]
14.	Hen feather	H_2SO_4	Mild steel	WL, EIS	95.5%	[33]
15.	Chicken nail	Ethanol	Mild steel	WL, PDP, SEM	74.04%	[34]
16.	Chlorella sorokiniana	HCl	Mild steel	WL, PDP, EIS	94.6%	[35]
17.	Coffee waste	Water	Carbon steel	EIS, PDP, HPLC	94.83%	[36]
18.	Tomato peel waste	NaCl	Tin	WL, EIS	75.9%	[37]
19.	Allium cepa peel waste	Ethanol	X80 steel	FTIR, XRD, SEM	95.8%	[38]
20.	Honeycomb waste	Ethanol	SS 304	EIS, SEM	97.29%	[39]
21.	Egg shell waste	NaCl	SS 316	SEM, RSM	96.66%	[40]
22.	Activated sludge waste	NaCl	Carbon steel	EIS, PDP, FTIR	80%	[41]
23.	Banana stem waste	H_2SO_4	Mild steel	SEM, FTIR	92%	[42]
24.	Date palm waste	HCl	Carbon steel	EIS, PDP, FTIR	89%	[43]
25.	Vigna unguiculata waste	NaOH	Aluminum	WL	79.63%	[44]
26.	Tea waste	HCl	BQ steel	LCMS, EIS, PDP	84.53%	[45]

(Continued)

TABLE 14.1 (*Continued*) Plant Extracts Application in Corrosion Mitigation

S.No.	Biomass Extract	Corrosive Solution	Metal to be Protected	Techniques Used	Efficiency	References
27.	Coconut shell waste	H2SO4	Aluminum	WL, EIS, SEM	99%	[46]
28.	Orange peel	HCl	Carbon steel	EIS, SEM	95%	[47]
29.	Orange peel	HCl/H_2SO_4	Cu, Zn, Al	EIS, WL, SEM	82%	[48]
30.	Cockroach wings	Seawater	N80 steel	WL, EFM, SECM, SKP, SEM	96.8%	[49]
31.	*Clarias batrachus* fins	H_2SO_4	Mild steel	EFM, SEM, AFM	95.8%	[50]
32.	Passion peel	H_3PO_4	Mild steel	FTIR, EIS, DFT, XPS	90%	[51]
33.	Cashew peel	HCl	Carbon steel	WL, EIS	86%	[52]
34.	Carrot peel	HCl	Mild steel	EIS, DFT, AFM	97%	[53]
35.	Plantain peel	H_2SO_4	Mild steel	WL, EIS, SEM	70%	[54]
36.	Papaya peel	HCl	Aluminum	PDP, EIS, SEM, AFM	95.5%	[55]
37.	Water melon rind	HCl	Mild steel	UV, FTIR, SEM, PDP	89%	[56]
38.	Sweet melon peel	HCl	Mild steel	PDP, EIS, SEM	90%	[57]
39.	Pineapple peel	HCl	Mild steel	Weight loss	71%	[58]
40.	Pomelo Peel	H_3PO_5	Mild steel	UV, FTIR, SEM	95%	[59]
41.	Onion peel	NaCl	Mild steel	EIS, PDP, SEM, OCP	90%	[60]
42.	Tangerine	NaCl	J55 steel	FTIR, EIS PDP	83%	[61]
43.	Potato peel	NaCl	Carbon steel	EIS, LPR, SEM	70%	[62]
44.	Pea fruit peel	HCl	Mild steel	FTIR, EIS PDP, SEM	91%	[63]
45.	Orange peel	HCl	Mild steel	PDP, EIS	90%	[64]
46.	Chinese yam peel	HCl	Mild steel	PDP, EIS, DFT	93.9%	[65]
47.	African walnut shell (*Plukenetia Conophora*)	HCl	Pipeline steel	FTIR, EIS, PDP	83.9%	[66]
48.	Saccharum sinense waste	NaCl	J55	FTIR, EIS, PDP	90%	[67]
49.	Cocoa pod	HCl	Mild steel	SEM, WL, FTIR	95.4%	[68]
50.	Sorghum waste	HCl	Aluminum	WL, Adsorption	79%	[69]
51.	Biomass wastes	HCl, H_2SO_4	Mild steel, carbon steel	FTIR, EIS, PDP	------	[70]
52.	Plant wastes	HCl, NaCl	Mild steel, aluminum	FTIR, EIS, PDP	------	[71]

amounts [76]. Given the abundance of readily available biomass waste sources—including livestock manure, municipal solid waste, forest debris, and agricultural leftovers—it makes sense to exploit these enormous resources to create corrosion inhibitors. This problem is solved by using biomass waste, which is easily accessible. The highest inhibitory efficacy of any inhibitor derived from biomass waste can differ depending on several variables, including the kind of biomass waste utilized, the preparation technique, the type of metal being shielded, and the surrounding circumstances. This review compares the results of several studies in this field to assess how effective biomass extracts prevent metal corrosion. The review aims to provide a solid foundation to develop more potent natural inhibitors, encourage a shift from harmful to safe inhibitors, and educate metal industries about the advantages of using biomass extracts for corrosion inhibition. The study focuses on the evidence, traits, and methods for obtaining many biomass extracts in corrosion prevention. This work screens the performance of plant extracts in terms of corrosion mitigation. Its goals are to provide researchers with a solid knowledge base for developing more potent natural inhibitors and to educate the metal industries about the advantages of employing biomass extracts for corrosion inhibition. The work focused on sources, attributes, and extraction methods for several biomass extracts used in corrosion protection. The main objective is to demonstrate how biomass extracts can stop metal corrosion. Tin, carbon steel, and mild steel are just a few metal substrates used in the study to assess how well these extracts work.

14.2 TECHNIQUES FOR MANUFACTURING BIOMASS EXTRACTS

Biomass can be extracted using various techniques, including pressurized liquid extraction, ultrasound, microwave, ultrasonic assistance, solvent, and supercritical fluid extraction. Furthermore, a simplified illustration of a representative scientific process for producing a powdered chamomile extract is provided in Figure 14.3 [77].

Every technique has advantages of its own for removing desirable substances. These approaches may be combined, contingent upon the particular use case and intended results. When choosing a biomass extraction technique, it's critical to consider sustainability, affordability, purity, and yield. Confirming that the selected procedure is safe and conforms to all relevant laws is also vital. Solvent extraction is a widely employed technique that dissolves and separates desired chemicals from biomass using

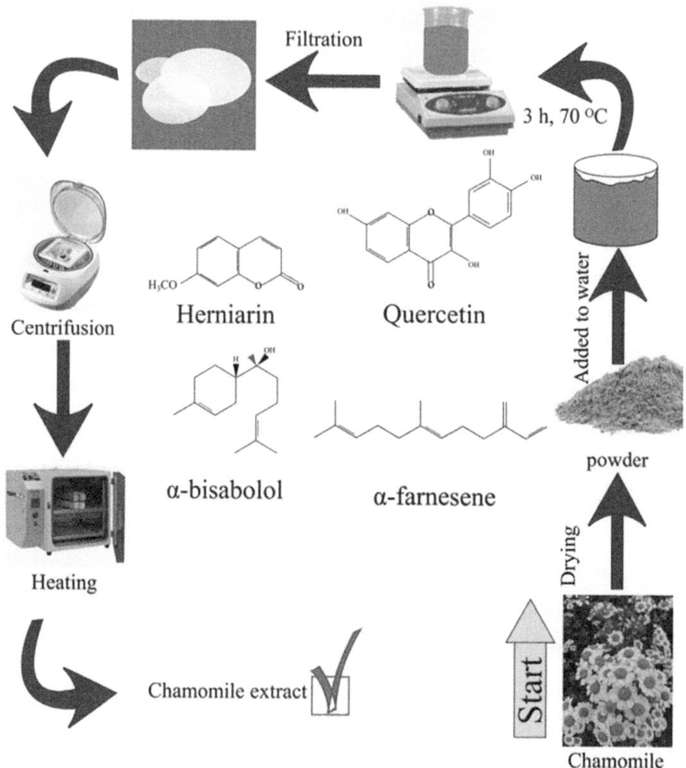

FIGURE 14.3 Steps involved in extract preparation along with the active ingredients.

a solvent such as ethanol. Nevertheless, there are disadvantages to this approach, including the possibility of undesired impurities and the final product consisting of solvent residue. Supercritical fluid extraction (SFE) is an additional technique that extracts chemicals from biomass using a high-temperature, high-pressure fluid, such as carbon dioxide. This method is much more environmentally friendly and efficient than solvent extraction. The microwave-assisted extraction (MAE) technique is a quick and efficient way to extract desirable molecules, especially phytochemicals and bioactive compounds, using microwaves. Last, high-frequency sound waves are used in ultrasonic extraction to disrupt the biomass's structure and liberate the required chemicals. The natural substance and the components that must be extracted should determine the employed method.

Standard and unorthodox methods are the two main techniques used in this extraction [78].

14.3 TRADITIONAL TECHNIQUES FOR EXTRACTING BIOMASS

The two traditional techniques for extracting plant matter are liquid-solid extraction (LSE) and liquid-liquid extraction (LLE).

14.3.1 Extractions Using Liquids

A critical step in extraction using liquids is choosing the right combination of solvents. Choosing which solvents to employ requires careful consideration of several criteria. The solute must, above all, dissolve more readily in a solvent compared to the aqueous medium. Thus, it's become essential to comprehend the partitioning coefficient of solute. Another prerequisite is that two solvents cannot combine to create a uniform solution in water. Third, there must be no reaction between the solvents and the solute. Furthermore, the biomass should be volatile to facilitate the solvent's simple extraction [79]. Superior production scaling specificity and low-temperature performance are just two advantages of this approach. However, it has certain drawbacks, including price, harmfulness, and combustibility [80]. Several investigations have demonstrated the effectiveness of containing liquid-liquid recovery of corrosion-preventing substances using biological extract. In acidic conditions, the derived inhibitors effectively prevented mild steel from corroding [81, 82]. While there are other methods for liquid-liquid extraction, we focus primarily on acid-base extraction because of its versatility; it can be used to extract various substances, such as organic acids, flavonoids, and alkaloids. Because of its adaptability, it can be used with different kinds of biomass. Extraction based on acid-base separates organic compounds according to whether they are essential or acidic. Changing the biomass's pH causes a change in how charged it is, depending upon its nature, which is either necessary or acidic. Since many compounds don't change from their neutral state, they are solubilized more readily in organic solvents than in an aqueous medium. However, when it takes on an ionic state, an organic component becomes more soluble in water. The success of this technique depends upon the organic component's dissolution and the notable change in dissolution between its initial phase and its salty form [83]. Raja et al. (2008) conducted a study using acid-base extraction to separate alkaloids from neem leaf extracts. It was discovered that the alkaloids significantly prevented mild steel from corroding in a solution of hydrochloric acid [84].

14.3.1.1 Liquid-Solid Extraction

Analytes can be extracted from various arrays using liquid-solid extraction (LSE). Sonication, Soxhlet extraction, maceration, percolation, and distillation using steam and reflux extraction are traditional methods. These techniques have several drawbacks, though, including the need for dangerous solvents and a lengthy processing time. Additionally, temperature fluctuations during the process may cause thermo-labile metabolites to break down [85].

14.3.1.1.1 Maceration   A menstruum is used to cover the entire drug and plant-based materials. To guarantee complete extraction, the vessel is sealed and kept in place for at least three days, with periodic agitation of the contents. Filtration or decantation separates the micelle and marc mixture following extraction. After that, the micelle is allowed to evaporate, either in an oven or a water bath. This method is practical, feasible, and works exceptionally well with heat-sensitive plant material [86]. The author Shukla and his group looked at applying neem leaf extract to mitigate mild steel corrosion in acidic environments using electrochemical impedance spectroscopy (EIS). The findings demonstrated % inhibition performance of 87% at 200 mg/L [87]. The drawbacks of this approach include its poor extraction efficiency and length of time.

14.3.1.1.2 Percolation   The percolation process uses a percolator, a small glass jar with a conical shape and apertures on both ends. This process starts with finely ground plant material immersed in an extraction solvent in a sanitized container. After that, some more solvent is added to the mixture and left to settle for four hours. The processed product is put in a percolator and unaltered over a few hours while the bottom part is closed. [88]. The extracted solvent is poured over the medicinal material via the top portion until fully saturated, and the liquid escapes through the lower percolator section. The medicinal material is extracted by continuously adding new solvents made more accessible by gravity. Once 75% of the entire volume meant for processing is administered, solvent replenishment stops. The extract is separated by filtering and decanting, and the final solvent dose is added once the leftover material has been expressed to get the required volume. This method is significantly more effective than maceration since it is a process that continuously adds additional solvent to the mixture that has already reached saturation [89]. It looked at how corrosion inhibitors may be extracted using an extract of the olive leaf by percolation. In solutions containing chloride, the isolated compounds significantly inhibited the corrosion of aluminum alloys. It was determined that phenolic acids and flavonoids are present in the extract [90].

14.3.1.1.3 Decoction   The extraction process needs water in an ongoing heat-driven extraction. Plant material dried, powdered, and pulverized is put into a sanitized container. Water is then added and combined with the substance. After that, heat is used to speed up the extraction procedure. This process usually ends in a short time—a few minutes or even up to 15 minutes. Typically, a 4:1 or 16:1 solvent-to-crude-medicine ratio is used. Plant material that is soluble in heat and water can be extracted using this technique. The decoction extract contains a variety of various water-soluble pollutants. A decoction is not a suitable method for removing volatile or thermo-labile compounds [68]. Bayarri et al. (2020) used rosemary leaf extract for copper mitigation and exhibited encouraging capabilities for inhibiting corrosion, demonstrating a mitigation value of 65% [91].

14.3.1.1.4 Reflux   This process requires the evaporation and condensation of water with no solvent loss. The procedure is widely utilized in the herbal industry due to its affordability, ease of usage, and efficiency. Reflux extraction utilizes less solvent and is less environmentally hazardous than percolation or maceration. Using it will prevent the extraction of thermo-labile compounds [91].

14.3.1.1.5 Soxhlet Extraction   The Soxhlet extraction process incorporates the benefits of percolation and reflux extraction by fusing this phenomenon with continuous extraction of the plant substance using a fresh solvent. Soxhlet extraction is an ongoing mechanically extracted method that has excellent extraction effectiveness when compared with maceration or percolation. However, there is a higher chance of thermal damage because of the Soxhlet approach's longer extraction time and higher temperature [92].

14.3.2 Unconventional Biomass Extraction Methods

These are alternate processes of extracting essential elements through biomass without depending on traditional physical or chemical methods [93]. Among them are:

14.3.2.1 Pressurized Extraction of Liquid

Pressurized fluid, high-pressure solvent, and rapid solvent extractions are some of the extraction techniques referred to as "pressurized liquid extraction" (PLE). Accelerated solvent extraction (ASE) uses higher temperatures and pressures than conventional methods to extract materials. A significant amount of pressure is needed to keep the solvent liquid at this high temperature. PLE is a practical and dependable approach for numerous extractions, especially the first extraction of many samples. The flask containing the extract and the solvent is heated and kept at a specific pressure for a predetermined period. Nitrogen gas is used to purify the material while the procedure is repeated with a new solvent. Elevated temperatures and pressures accelerate biomass's solubility and solvent penetration, hence increasing yield. In a 2019 study, Pan et al. examined the PLE of lignin from sugarcane bagasse. In a sodium chloride solution, the produced lignin extract effectively inhibited the corrosion of aluminum alloy [94]. Because PLE requires less solvent, it is an environmentally friendly and cost-effective form of cleaning. As researchers have shown, it is possible to carry out many extractions using the same solvent or one with a higher polarity as long as the material is completely dried after each extraction. The reproducibility of the procedure is increased by its programming potential, determining temperature, time, and appropriate solvent [95]. Similarly, some studies had trouble extracting antioxidants from grape pomace using PLE at 200°C, while others were successful [96, 97].

14.3.2.2 Supercritical Fluid Extraction (SFE)

This extraction uses supercritical fluid as the solvent in a separation process known as supercritical fluid extraction (SFE). Supercritical fluids may solubilize many chemicals because of their unique solvation capabilities, which come from their combination of gas-like diffusivity and liquid-like solubility. Their solvating capacities can be fine-tuned with tiny changes in temperature and pressure, although they fluctuate considerably around critical points. Because of its many advantageous qualities, including its non-toxic nature, selectivity, inertness, etc., supercritical carbon dioxide (S-CO_2) is a frequent solvent in SFE. S-CO_2 is the most effective solution for eliminating low-polarity materials. Nonetheless, the S-CO_2 performance can be significantly improved by including a modifier to increase its solvating capabilities [98]. The extracted tannins established an ardent film on the metal surface, demonstrating considerable corrosion prevention [99].

SFE is a new and efficient method for extracting essential oils from many plants. The resultant extracts are valuable raw materials that can be used to create pharmaceuticals. SFE enables the selective extraction of distinct compounds under a range of situations. SFE has gained popularity because of its capacity to perform extractions at nearly equal room temperature, preventing thermal denaturation of the target compounds. SFE can be pretty discriminating despite the higher expenditure needed. Supercritical fluids are involved in the mechanistic separation of the solutes added to the chromatographic system in supercritical fluid chromatography (SFC), another separation method [100]. Chemicals can be extracted from plant material using an exciting technique called supercritical fluid extraction. In industrial contexts, it could replace procedures like solvent extraction and steam distillation [101].

14.3.2.3 Ultrasound Assisted Extraction (UAE)

Ultrasonic wave energy is utilized in the UAE to facilitate the extraction process. When ultrasonic waves are applied, the solvent becomes cavitated, which accelerates the solute's movement and dissolution along with heat transfer and enhances the efficiency of extraction. UAE provides more advantages, such as lower temperatures, shorter extraction periods, and less energy and solvent consumption. Additionally, UAE is used to extract compounds that are both unstable and thermo-labile, and it usually uses a variety of procedures to extract natural products [78]. A negative pressure bubble can form, expand, oscillate, and possibly split or implode. This process is known as cavitation. The powerful waves created when these cavitation bubbles burst can be used to remove different kinds of chemicals and particles from the matrix's surface. Strong liquid microjets and shock waves produced by collapsing bubbles of cavitation at or near the surface of the sample speed up the extraction process. This implosion creates localized circumstances with elevated temperatures and pressures. Nevertheless, several problems could occur when extracting solid samples, such as the analyst disintegrating if caught in bubble cavitation [102]. Solid biomass material cells are subjected to ultrasonic mechanical stress by creating cavitation within the sample. Cell lysis increases output by encouraging metabolites to dissolve in the solvent. The device's frequency, along with the duration and temperature of the sonication, determines the extent of chemical extraction. UAE was primarily used for small-quantity extraction. It typically simplifies the extraction of intracellular metabolites from plant cell cultures [103, 104].

14.3.2.4 Enzyme Assisted Extraction (EAE)

Olive leaves are rich in phenolic chemicals, which are recognized for their ability to reduce corrosion and act as antioxidants. With an 82.5% recovery rate, EAE effectively recovered phenolic components from olive leaves using pectinase and cellulase. In acidic environments, the isolated phenolic chemicals have effectively inhibited carbon steel corrosion [105, 106].

14.4 FUTURE POTENTIAL OF BIOMASS EXTRACTS IN CORROSION MITIGATION

The number of studies on the relationship between human health and the natural world has grown significantly in the last decade. Given that corrosion is a significant financial trend, it is imperative to identify the most ecologically sound methods to prevent metallurgical products from deteriorating. Further studies ought to emphasize the possible application of corrosion mitigation manufactured using plant-based materials. Extreme deterioration primarily affects submerged metallic structures and petroleum and natural gas pipes. The improved comprehension of the inhibitory characteristics concerning the morphological and inner architecture of a coating is one of the main obstacles to applying biomass coatings; consequently, more in-depth study is required in this area [107]. The application of biomass blockers has been the subject of several academic investigations, and many articles exist in the scientific literature. Nonetheless, these undertakings have a connection to noteworthy difficulties. Dehydration is the first step in using biomass and agricultural waste extract. It is necessary to conduct more thorough research on this dehydration procedure, considering elements like ideal temperature and surrounding circumstances. The production of corrosion-inhibiting substances from biomass and agricultural waste must follow sustainable methods to keep dangerous materials out of the natural setting. Future technological advancements in versatile substances would concentrate on creating customized, bioinspired coverings to prevent corrosion. The potential to avoid corrosion is significantly increased when coatings combine with other intelligent properties like healing, cleaning themselves, preventing fouling, superhydrophobicity, etc. Researching the integration of porous materials impregnated with slippery solutions on various organic protective coatings for various metallic objects is imperative. Additional research is necessary since the anti-fouling characteristic attained after utilizing these substances is unsatisfactory. It will be exciting to examine the defensive strategy associated with bioinspired coatings in depth and to use various statistical approaches. Computerized modeling can forecast biomass extracts' effectiveness as corrosion inhibitors—the application of coating enhanced by carbon nanomaterials used as an anti-corrosion agent. Carbon black nanofillers, graphene, graphene oxide, carbon nanotubes, and carbon nanofiber may significantly impact corrosion avoidance. To lessen the consequences of corrosion, their numerous uses in organic coatings are being thoroughly investigated [108].

14.5 CONCLUSION

A viable and long-term strategy for reducing corrosion in various settings is applying biomass extracts as environmentally acceptable inhibiting agents for metallic materials. Compared to traditional manmade anti-corrosion agents, biomass extracts from natural substances like plants and microbes are more affordable and ecologically safe. Substantial inhibition of corrosion efficacy has been shown by these biomass extracts across various environments, particularly systems comprising chloride, acids, and bases. Their ability to adhere to metallic surfaces and create barrier layers is due to an abundance of reactive functional units, including oxygen, sulfur, and nitrogen. Furthermore, biomass extracts are sustainable corrosion inhibitors due to their lower toxicity and biodegradability. In many settings, using biomass extracts to prevent metallic corrosion has been scientifically proven to be effective. Biomass extracts have demonstrated efficacy in preventing metallic corrosion in acidic conditions, like ones observed in manufacturing operations, when applied to terrestrial settings. They have also been effectively used to shield metals against corrosion in maritime settings, where exposure to saltwater and inclement weather can hasten deterioration. The creation and use of corrosion inhibitors produced from biomass have bright futures. Present-day studies aim to maximize the active chemicals' production and efficiency by optimizing extracting techniques. Furthermore, studies exploring the combinatorial impacts of mixing various biomass extracts have been carried out to improve effectiveness in corrosion prevention. In addition, for biomass-based corrosion suppressants to be used practically in multiple industries, novel recipes and distribution methods must be developed.

REFERENCES

1. B. Rani, B.B.J. Basu, Green inhibitors for corrosion protection of metals and alloys: an overview, *International Journal of Corrosion*, 2012 (2012).

2. C. Verma, E.E. Ebenso, I. Bahadur, M. Quraishi, An overview on plant extracts as environmental sustainable and green corrosion inhibitors for metals and alloys in aggressive corrosive media, *Journal of Molecular Liquids*, 266 (2018) 577–590.

3. O. Oyewole, T.S. Abayomi, T.A. Oreofe, T.A. Oshin, Anti-corrosion using rice straw extract for mild steel in 1.5 M H2SO4 solution, *Results in Engineering*, 16 (2022) 100684.

4. A.A. Farag, A.S. Ismail, M. Migahed, Environmental-friendly shrimp waste protein corrosion inhibitor for carbon steel in 1 M HCl solution, *Egyptian Journal of Petroleum*, 27 (2018) 1187–1194.

5. C. Liu, Y. Liu, Z. Xia, Z. Wang, B. Wu, Coconut coir dust extract as a novel green corrosion inhibitor for carbon steel in the chloride-contaminated concrete pore solution, *Journal of Building Engineering*, 82 (2024) 108194.

6. V. Hemapriya, M. Prabakaran, S. Chitra, M. Swathika, S.-H. Kim, I.-M. Chung, Utilization of biowaste as an eco-friendly biodegradable corrosion inhibitor for mild steel in 1 mol/L HCl solution, *Arabian Journal of Chemistry*, 13 (2020) 8684–8696.

7. J.-J. Oh, J.Y. Kim, Y.J. Kim, S. Kim, G.-H. Kim, Utilization of extracellular fungal melanin as an eco-friendly biosorbent for treatment of metal-contaminated effluents, *Chemosphere*, 272 (2021) 129884.

8. N. Bheel, P. Awoyera, O. Aluko, S. Mahro, A. Viloria, C.A.S. Sierra, Sustainable composite development: novel use of human hair as fiber in concrete, *Case Studies in Construction Materials*, 13 (2020) e00412.

9. H.E. Abaci, A. Coffman, Y. Doucet, J. Chen, J. Jacków, E. Wang, Z. Guo, J.U. Shin, C.A. Jahoda, A.M. Christiano, Tissue engineering of human hair follicles using a biomimetic developmental approach, *Nature communications*, 9 (2018) 5301.

10. H. Lee, K. Noh, S.C. Lee, I.-K. Kwon, D.-W. Han, I.-S. Lee, Y.-S. Hwang, Human hair keratin and its-based biomaterials for biomedical applications, *Tissue Engineering and Regenerative Medicine*, 11 (2014) 255–265.

11. M. Shahbazi, A.S. Pannu, J. Alarco, P. Sonar, I. Mackinnon, Impact of hair-derived carbon substitution on structural and superconducting properties of MgB2, *AIP Advances*, 13 (2023).

12. D. Deng, M. Gopiraman, S.H. Kim, I.-M. Chung, I.S. Kim, Human hair: a suitable platform for catalytic nanoparticles, *ACS Sustainable Chemistry & Engineering*, 4 (2016) 5409–5414.

13. L. Eroğlu, E. Güneren, H. Akbaş, A. Demir, A. Uysal, Using human hair as suture material in microsurgical practice, *Journal of Reconstructive Microsurgery*, 19 (2003) 037–040.

14. B. Scharf, J. Hyvärinen, A. Poranen, M.M. Merzenich, Electrical stimulation of human hair follicles via microelectrodes, *Perception & Psychophysics*, 14 (1973) 273–276.

15. A. Zakeri, E. Bahmani, A.S.R. Aghdam, Plant extracts as sustainable and green corrosion inhibitors for protection of ferrous metals in corrosive media: a mini review, *Corrosion Communications*, 5 (2022) 25–38.

16. R. Höfer, J. Bigorra, Biomass-based green chemistry: sustainable solutions for modern economies, *Green Chemistry Letters and Reviews*, 1 (2008) 79–97.

17. M. Ismail, A. Abdulrahman, M.S. Hussain, Solid waste as environmental benign corrosion inhibitors in acid medium, *International Journal of Engineering Science and Technology*, 3 (2011) 1742–1748.

18. N. Odewunmi, S. Umoren, Z. Gasem, Watermelon waste products as green corrosion inhibitors for mild steel in HCl solution, *Journal of Environmental Chemical Engineering*, 3 (2015) 286–296.

19. A.N. Grassino, J. Halambek, S. Djaković, S.R. Brnčić, M. Dent, Z. Grabarić, Utilization of tomato peel waste from canning factory as a potential source for pectin production and application as tin corrosion inhibitor, *Food Hydrocolloids*, 52 (2016) 265–274.

20. A. Thakur, S. Sharma, R. Ganjoo, H. Assad, A. Kumar, Anti-corrosive potential of the sustainable corrosion inhibitors based on biomass waste: a review on preceding and perspective research, in: *Journal of Physics: Conference Series*, IOP Publishing, 2022, p. 012079.

21. D. Guedes, G.R. Martins, L.Y. Jaramillo, D. Simas Bernardes Dias, A.J.R. da Silva, M.T. Lutterbach, L.Y. Reznik, E.F. Sérvulo, C.S. Alviano, D.S. Alviano, Proanthocyanidins with corrosion inhibition activity for AISI 1020 carbon steel under neutral pH conditions of coconut (Cocos nucifera L.) husk fibers, *ACS Omega*, 6 (2021) 6893–6901.

22. T. Sathiyapriya, G. Rathika, Corrosion inhibition efficiency of bio waste on mild steel in acid media, *Oriental Journal of Chemistry*, 31 (2015) 1703–1710.

23. E. Ituen, A. James, O. Akaranta, Elephant grass biomass extract as corrosion inhibitor for mild steel in acidic medium, *Journal of Materials and Environmental Science*, 8 (2017) 1498–1507.

24. M.H. Hussin, A.A. Rahim, M.N.M. Ibrahim, N. Brosse, Improved corrosion inhibition of mild steel by chemically modified lignin polymers from Elaeis guineensis agricultural waste, *Materials Chemistry and Physics*, 163 (2015) 201–212.

25. A. Ait Aghzzaf, D. Veys-Renaux, E. Rocca, Pomegranate peels crude extract as a corrosion inhibitor of mild steel in HCl medium: passivation and hydrophobic effect, *Materials and Corrosion*, 71 (2020) 148–154.

26. M.F. Gromboni, A. Sales, M. de AM Rezende, J.P. Moretti, P.G. Corradini, L.H. Mascaro, Impact of agro-industrial waste on steel corrosion susceptibility in media simulating concrete pore solutions, *Journal of Cleaner Production*, 284 (2021) 124697.

27. A.C. Rodrigues, E.L.D. Silva, S.F. Quirino, A. Cuña, J.S. Marcuzzo, J.T. Matsushima, E.S. Gonçalves, M.R. Baldan, Ag@ activated carbon felt composite as electrode for supercapacitors and a study of three different aqueous electrolytes, *Materials Research*, 22 (2018).

28. C.N. Mbah, C.C. Onah, K.C. Nnakwo, Effectiveness of Irvingia wombolu extract on corrosion inhibition of mild steel in hydrochloric acid solution, *Engineering Research Express*, 2 (2020) 015039.

29. O. Olawale, C. Idefoh, B. Ogunsemi, J. Bello, Evaluation of groundnut leaves extract as corrosion inhibitor on mild steel in 1m sulphuric acid using response surface methodology (RSM), *International Journal of Mechanical Engineering and Technology (IJMET)*, 9 (2018) 829–841.

30. V. Aroulmoji, Mahendra Research & Development, https://mahendra.org/director-of-rd/

31. A. Shahmoradi, N. Talebibahmanbigloo, A. Javidparvar, G. Bahlakeh, B. Ramezanzadeh, Studying the adsorption/inhibition impact of the cellulose and lignin compounds extracted from agricultural waste on the mild steel corrosion in HCl solution, *Journal of Molecular Liquids*, 304 (2020) 112751.

32. M. Magni, E. Postiglione, S. Marzorati, L. Verotta, S.P. Trasatti, Green corrosion inhibitors from agri-food wastes: the case of Punica granatum extract and its constituent ellagic acid. A validation study, *Processes*, 8 (2020) 272.

33. S. Subhashini, R. Rajalakshmi, V. Kowshalya, Eco-friendly corrosion inhibitor from poultry waste for mild steel in acid medium, *Material Science Research of India*, 5 (2008) 423–428.

34. O. Olawale, J. Bello, B. Ogunsemi, U. Uchella, A. Oluyori, N. Oladejo, Optimization of chicken nail extracts as corrosion inhibitor on mild steel in 2M H2SO4, *Heliyon*, 5 (2019).

35. G.A. de Oliveira, V.M. Teixeira, J.N. da Cunha, M.R. dos Santos, V.M. Paiva, M.J.C. Rezende, A.F. do Valle, E. D'Elia, Biomass of Microalgae Chlorella sorokiniana as green corrosion inhibitor for mild steel in HCl solution, *International Journal of Electrochemical Science*, 16 (2021) 210249.

36. A.P. da Costa, G. Fontgalland, A.G. Neto, A.S. Sombra, R.R. do Valle, Dual-frequency magneto-dielectric resonator antenna based in a YIG matrix with control of HEM 11δ and TE 01δ modes, *Microwave and Optical Technology Letters*, 63 (2021) 310–321.

37. J. Halambek, I. Cindrić, A.N. Grassino, Evaluation of pectin isolated from tomato peel waste as natural tin corrosion inhibitor in sodium chloride/acetic acid solution, *Carbohydrate Polymers*, 234 (2020) 115940.

38. M. Shahini, M. Ramezanzadeh, B. Ramezanzadeh, Effective steel alloy surface protection from HCl attacks using Nepeta Pogonesperma plant stems extract, *Colloids and Surfaces A: Physicochemical and Engineering Aspects*, 634 (2022) 127990.

39. F. Gapsari, K.A. Madurani, F.M. Simanjuntak, A. Andoko, H. Wijaya, F. Kurniawan, Corrosion inhibition of honeycomb waste extracts for 304 stainless steel in sulfuric acid solution, *Materials*, 12 (2019) 2120.

40. O. Sanni, A. Popoola, Assessment of concentration, temperature and exposure time effect on waste product as a sustainable inhibitor for stainless steel corrosion: optimization using response surface method, *Journal of Bio-and Tribo-Corrosion*, 5 (2019) 1–14.

41. L.C. Go, W. Holmes, D. Depan, R. Hernandez, Evaluation of extracellular polymeric substances extracted from waste activated sludge as a renewable corrosion inhibitor, *PeerJ*, 7 (2019) e7193.

42. R. Mokkapati, V. Ratnakaram, J. Mokkapati, Utilization of agro-waste for removal of toxic hexavalent chromium: surface interaction and mass transfer

studies, *International Journal of Environmental Science and Technology*, 15 (2018) 875–886.

43. G.M. Al-Senani, M. Alshabanat, Study the corrosion inhibition of carbon steel in 1 M HCl using extracts of date palm waste, *International Journal of Electrochemical Science*, 13 (2018) 3777–3788.

44. S. Umoren, I. Obot, L. Akpabio, S. Etuk, Adsorption and corrosive inhibitive properties of Vigna unguiculata in alkaline and acidic media, *Pigment & Resin Technology*, 37 (2008) 98–105.

45. A. Pal, C. Das, A novel use of solid waste extract from tea factory as corrosion inhibitor in acidic media on boiler quality steel, *Industrial Crops and Products*, 151 (2020) 112468.

46. O. Sanni, O. Adeleke, K. Ukoba, J. Ren, T.-C. Jen, Application of machine learning models to investigate the performance of stainless steel type 904 with agricultural waste, *Journal of Materials Research and Technology*, 20 (2022) 4487–4499.

47. S.A. Umoren, M.M. Solomon, Polymeric corrosion inhibitors for oil and gas industry, *Corrosion Inhibitors in the Oil and Gas Industry*, (2020) 303–320.

48. S.A. Umoren, M.M. Solomon, A. Madhankumar, I.B. Obot, Exploration of natural polymers for use as green corrosion inhibitors for AZ31 magnesium alloy in saline environment, *Carbohydrate Polymers*, 230 (2020) 115466.

49. Shivani Singh, Rahul Singh, Neeta Raj Sharma, Ambrish Singh, Ethanolic extract of cockroach wing powder as corrosion inhibitor for N80 steel in an ASTM D1141-98(2013) standard artificial seawater solution, *International Journal of Electrochemical Science*, 16 (2021) Article ID: 210841.

50. Shivani Singh, Rahul Singh, Neeta Raj Sharma, Ambrish Singh, Extract from Clarias batrachus fins as environmental benign corrosion inhibitor for mild steel in acidic solution, *International Journal of Electrochemical Science*, 17 (2022) Article Number: 220341.

51. Bilan Lin, Junjie Shao, Chen Zhao, Xinxin Zhou, Fan He, Yuye Xu, *Passiflora edulis* Sims peel extract as a renewable corrosion inhibitor for mild steel in phosphoric acid solution, *Journal of Molecular Liquids*, 375 (2023) 121296.

52. Hassan Mohammed, Shafreeza bt. Sobri, Corrosion inhibition studies of cashew nut (*Anacardium occidentale*) on carbon steel in 1.0 M hydrochloric acid environment, *Materials Letters*, 229 (2018) 82–84.

53. A. Hossein Mostafatabar, Ali Dehghani, Pantea Ghahremani, Ghasem Bahlakeh, Bahram Ra mezanzadeh, Molecular-dynamic/DFT-electronic theoretical studies coupled with electrochemical investigations of the carrot pomace extract molecules inhibiting potency toward mild steel corrosion in 1 M HCl solution, *Journal of Molecular Liquids*, 346 (2022) 118344.

54. R.M. Saleh, A.A. Ismail, A.A. Ei Hosary, Corrosion inhibition by naturally occurring substances VII. The effect of aqueous extracts of some leaves and fruit peels on the corrosion of steel, Al, Zn and Cu in acids, *British Corrosion Journal*, 17 (1982) 131–135.

55. Namrata Chaubey, Vinod Kumar Singh, M.A. Quraishi, Papaya peel extract as potential corrosion inhibitor for Aluminium alloy in 1 M HCl:

electrochemical and quantum chemical study, *Ain Shams Engineering Journal*, 9 (2018) 1131–1140.

56. N.A. Odewunmi, S.A. Umoren, Z.M. Gasem, Watermelon waste products as green corrosion inhibitors for mild steel in HCl solution, *Journal of Environmental Chemical Engineering*, 3 (2015) 286–296.

57. Mohammed Tariq Saeed, Muhammad Saleem, Soofia Usmani, Izhar Ahmed Malik, Faisal Ahmad Al-Shammari, Kashif Mairaj Deen, Corrosion inhibition of mild steel in 1 M HCl by sweet melon peel extract, *Journal of King Saud University - Science*, 31 (2019) 1344–1351.

58. Anaele John Vitus, Ipeghan Jonathan Otaraku, Study of corrosion inhibition of pineapple peels extracts on mild steel in 1M HCl, *Journal of Scientific and Engineering Research*, 5 (2018) 311–317.

59. Bi-lan Lin, Jun-jie Shao, Yu-ye Xu, Yi-ming Lai, Zhong-ning Zhao, Adsorption and corrosion of renewable inhibitor of *Pomelo* peel extract for mild steel in phosphoric acid solution, *Arabian Journal of Chemistry*, 14 (2021) 103114.

60. Vinit K. Jha, Manisha Singh Chauhan, Shweta Pal, Shubhajit Jana, Gopal Ji, Rajiv Prakash, Experimental and DFT analysis of onion peels for its inhibition behavior against mild steel corrosion in chloride solutions, *Journal of the Indian Chemical Society*, 99 (2022) 100534.

61. S. Wang, B. Wu, Lanlan Qiu, Yuyao Chen, Jing Yuan, Songsong Chen, Mingyu Bao, Chunyan Fu, Xin Wang, Inhibition effect of tangerine peel extract on J55 steel in CO_2-saturated 3.5 wt. % NaCl solution, *International Journal of Electrochemical Science*, 12 (2017) 11195–11211.

62. Chandra Shekhar, Anirudha Jaiswal, Gopal Ji, Rajiv Prakash, Ethanol extract of waste potato peels for corrosion inhibition of low carbon steel in chloride medium, *Materials Today Proceedings*, 44 (2021) 2267–2272.

63. Monika Srivastava, Preeti Tiwari, S.K. Srivastava, Ashish Kumar, Gopal Ji, Rajiv Prakash, Low cost aqueous extract of *Pisum sativum* peels for inhibition of mild steel corrosion, *Journal of Molecular Liquids*, 254 (2018) 357–368.

64. Siau Ying Hong, Peck Loo Kiew, The inhibitive and adsorptive characteristics of orange peel extract on metal in acidic media, *Progress in Energy and Environment*, 11 (2019) 1–14.

65. Abhinay Thakur, Ashish Kumar, Savas Kaya, Fouad Benhiba, Shveta Sharma, Richika Ganjoo, Humira Assad, Electrochemical and computational investigations of the *Thysanolaena latifolia* leaves extract: an eco-benign solution for the corrosion mitigation of mild steel, *Results in Chemistry*, 6 (2023) 101147.

66. K.N. Kikanme, A.O. James, N.C. Ngobiri, Corrosion inhibition characteristics of *plukenetia conophora* shell extract on corrosion of pipeline steel in acidic solution, *Scientia Africana*, 19 (2020) 125–138.

67. Bo Huang, E.E. Ebenso, Ambrish Singh, Yuanhua Lin, *Saccharum sinense* bagasse extract as an effective corrosion inhibitor for J55 steel in 3.5% NaCl solution saturated with CO_2, *Anti-Corrosion Methods and Materials*, 62 (2015) 388–393.

68. Popoola Lekan Taofeek, Aderibigbe Tajudeen Adejare, Lala Mayowa, Mild steel corrosion inhibition in hydrochloric acid using cocoa pod husk-*Ficus exasperata*: extract preparation optimization and characterization, *Iranian Journal of Chemistry and Chemical Engineering*, 41 (2022) 482–492.

69. Richard Alexis Ukpe, Joint effect of halides and ethanol extract of Sorghum on the inhibition of the corrosion of aluminum in HCl, *Communication in Physical Sciences*, 4 (2019) 141–150.

70. Abhinay Thakur, Shveta Sharma, Richika Ganjoo, Humira Assad, Ashish Kumar, Anti-corrosive potential of the sustainable corrosion inhibitors based on biomass waste: a review on preceding and perspective research, *Journal of Physics: Conference Series*, 2267 (2022) 012079.

71. N.O. Eddy, A.O. Odiongenyi, E.E. Ebenso, R. Garg, R. Garg, Plant wastes as alternative sources of sustainable and green corrosion inhibitors in different environment, *Corrosion Engineering Science and Technology*, 58 (2023) 521–533.

72. P.C. Okafor, E.E. Ebenso, A.Y. El-Etre, M.A. Quraishi, Green approaches to corrosion mitigation, *International Journal of Corrossion*, 2012 (2012) 1–2.

73. J. Haruna, Method to improve Cu corrosion performance of Mo-DTC and active sulfur by adding sunflower oil, in, Google Patents, 2001.

74. M.A. Alkhaldi, A.-Q. Khadeeja, Method for inhibiting corrosion of steel with leaf extracts, in, Google Patents, 2019.

75. J.A.D.C.P. Gomes, J.C. Rocha, E. D'elia, Use of fruit skin extracts as corrosion inhibitors and process for producing same, in, Google Patents, 2015.

76. I.B. Obot, A. Meroufel, A.A. Sorour, I.B. Onyeachu, S.A. Umoren, A.F. Alenazi, M.M. Solomon, Corrosion inhibitor composition and methods of inhibiting corrosion, in, Google Patents, 2021.

77. M. Shahini, M. Keramatinia, M. Ramezanzadeh, B. Ramezanzadeh, G. Bahlakeh, Combined atomic-scale/DFT-theoretical simulations & electrochemical assessments of the chamomile flower extract as a green corrosion inhibitor for mild steel in HCl solution, *Journal of Molecular Liquids*, 342 (2021) 117570.

78. Q.-W. Zhang, L.-G. Lin, W.-C. Ye, Techniques for extraction and isolation of natural products: a comprehensive review, *Chinese Medicine*, 13 (2018) 1–26.

79. S. Tshepelevitsh, K. Hernits, J. Jenčo, J.M. Hawkins, K. Muteki, P. Solich, I. Leito, Systematic optimization of liquid–liquid extraction for isolation of unidentified components, *ACS Omega*, 2 (2017) 7772–7776.

80. M. Rawa-Adkonis, L. Wolska, A. Przyjazny, J. Namieśnik, Sources of errors associated with the determination of PAH and PCB analytes in water samples, *Analytical Letters*, 39 (2006) 2317–2331.

81. G.T. Galo, A.D.A. Morandim-Giannetti, F. Cotting, I.V. Aoki, I.P. Aquino, Evaluation of purple onion (*Allium cepa* L.) extract as a natural corrosion inhibitor for carbon steel in acidic media, *Metals and Materials International*, 27 (2021) 3238–3249.

82. L.B. Furtado, R. Nascimento, Plant extracts as green corrosion inhibitors, in: *Sustainable Corrosion Inhibitors II: Synthesis, Design, and Practical Applications*, ACS Publications, 2021, pp. 19–77.

83. J.E. McMurry, R.C. Fay, J.K. Robinson, *Chemistry*, Pearson, 2015.

84. P.B. Raja, A.K. Qureshi, A.A. Rahim, H. Osman, K. Awang, Neolamarckia cadamba alkaloids as eco-friendly corrosion inhibitors for mild steel in 1 M HCl media, *Corrosion Science*, 69 (2013) 292–301.

85. W. Routray, V. Orsat, Microwave-assisted extraction of flavonoids: a review, *Food and Bioprocess Technology*, 5 (2012) 409–424.

86. K.P. Ingle, A.G. Deshmukh, D.A. Padole, M.S. Dudhare, M.P. Moharil, V.C. Khelurkar, Phytochemicals: extraction methods, identification and detection of bioactive compounds from plant extracts, *Journal of Pharmacognosy and Phytochemistry*, 6 (2017) 32–36.

87. S.K. Sharma, A. Mudhoo, G. Jain, J. Sharma, Corrosion inhibition and adsorption properties of Azadirachta indica mature leaves extract as green inhibitor for mild steel in HNO_3, *Green Chemistry Letters and Reviews*, 3 (2010) 7–15.

88. S.O. Majekodunmi, Review of extraction of medicinal plants for pharmaceutical research, *Merit Research Journal of Medicine and Medical Sciences*, 3 (2015) 521–527.

89. C.N. Njoku, A.I. Ikeuba, C.C. Anorondu, I.C. Shammah, E. Yakubu, B.N. Elendu, C.S. Enechukwu, I.O. Uduma, P.C. Uzor, A review of the extraction and application of eco-friendly biomass for corrosion protection of metals, *Results in Chemistry*, (2023) 101286.

90. N. Azwanida, A review on the extraction methods use in medicinal plants, principle, strength and limitation, *Medicinal and Aromatic Plants*, 4 (2015) 2167–2412.

91. W. Daoudi, O. Dagdag, C. Verma, E. Berdimurodov, A. Oussaid, A. Berisha, A. Oussaid, M. Abboud, A. El Aatiaoui, Rosmarinus officinalis l. Oil as an eco-friendly corrosion inhibitor for mild steel in acidic solution: experimental and computational studies, *Inorganic Chemistry Communications*, 161 (2024) 112030.

92. L. Chaudhari, B.A. Jawale, S. Sharma, H. Sharma, C. Kumar, P.A. Kulkarni, Antimicrobial activity of commercially available essential oils against Streptococcus mutans, *The Journal of Contemporary Dental Practice*, 13 (2012) 71–74.

93. R. Yang, D. Li, A. Li, H. Yang, Adsorption properties and mechanisms of palygorskite for removal of various ionic dyes from water, *Applied Clay Science*, 151 (2018) 20–28.

94. J.R. Lim, L.S. Chua, A.A. Mustaffa, Ionic liquids as green solvent and their applications in bioactive compounds extraction from plants, *Process Biochemistry*, 122 (2022) 292–306.

95. P. Rahayu, C. Sundari, I. Farida, Corrosion inhibition using lignin of sugarcane bagasse, in: *IOP Conference Series: Materials Science and Engineering*, IOP Publishing, 2018, p. 012087.

96. A.M. Gizir, N. Turker, E. Artuvan, Pressurized acidified water extraction of black carrot [Daucus carota ssp. sativus var. atrorubens Alef.] anthocyanins, *European Food Research and Technology*, 226 (2008) 363–370.

97. A.-E. Segneanu, F. Cziple, P. Vlazan, P. Sfirloaga, I. Grozescu, V.D. Gherman, Biomass extraction methods, in: *Biomass Now-Sustainable Growth and Use*, 2013, IntechOpen, pp. 390–399. ISBN: 978-953-51-1105-4

98. J.R. Vergara-Salinas, J. Cuevas-Valenzuela, J.R. Pérez-Correa, Extraction of polyphenols by pressurized liquids, in: *Advances in Technologies for Producing Food-Relevant Polyphenols*, 2016. CRC Press. ISBN 9781498714976

99. Z. Chen, X. Zhang, W. Han, L. Gao, S. Li, A power generation system with integrated supercritical water gasification of coal and CO_2 capture, *Energy*, 142 (2018) 723–730.

100. O. Wrona, K. Rafińska, C. Możeński, B. Buszewski, Supercritical fluid extraction of bioactive compounds from plant materials, *Journal of AOAC International*, 100 (2017) 1624–1635.

101. L.A. Conde-Hernández, J.R. Espinosa-Victoria, A. Trejo, J.Á. Guerrero-Beltrán, CO_2-supercritical extraction, hydrodistillation and steam distillation of essential oil of rosemary (Rosmarinus officinalis), *Journal of Food Engineering*, 200 (2017) 81–86.

102. F.J. Barba, Z. Zhu, M. Koubaa, A.S. Sant'Ana, V. Orlien, Green alternative methods for the extraction of antioxidant bioactive compounds from winery wastes and by-products: a review, *Trends in Food Science & Technology*, 49 (2016) 96–109.

103. F. Chemat, G. Cravotto, *Microwave-Assisted Extraction for Bioactive Compounds: Theory and Practice*, Springer Science & Business Media, 2012.

104. M. Vinatoru, T. Mason, I. Calinescu, Ultrasonically assisted extraction (UAE) and microwave assisted extraction (MAE) of functional compounds from plant materials, *TrAC Trends in Analytical Chemistry*, 97 (2017) 159–178.

105. D. Oyekunle, O. Agboola, A. Ayeni, Corrosion inhibitors as building evidence for mild steel: a review, in: *Journal of Physics: Conference Series*, IOP Publishing, 2019, p. 032046.

106. S.K. Karn, A. Bhambri, I.R. Jenkinson, J. Duan, A. Kumar, The roles of biomolecules in corrosion induction and inhibition of corrosion: a possible insight, *Corrosion Reviews*, 38 (2020) 403–421.

107. V. Srivastava, S. Banerjee, M. Singh, Inhibitive effect of polyacrylamide grafted with fenugreek mucilage on corrosion of mild steel in 0.5 M H2SO4 at 35° C, *Journal of Applied Polymer Science*, 116 (2010) 810–816.

108. A. Pilbáth, L. Nyikos, I. Bertóti, E. Kálmán, Zinc corrosion protection with 1, 5-diphosphono-pentane, *Corrosion Science*, 50 (2008) 3314–3321.

CHAPTER 15

Biomass as a Sustainable Energy Source for Energy Storage

Luis Fernando Macias Gamboa, Raúl Segovia Pérez, Blanca M. Muñoz Flores and Víctor M. Jiménez Pérez

15.1 INTRODUCTION

Nowadays, human beings have the greatest challenges in their entire history in the face of imminent climate change generated by themselves due to the excessive use of fossil fuels, massive tree felling, polluting transport, unsustainable agriculture, livestock, and so on. Floods, droughts, and storms have always been part of human history; however, in the last decades, all these events and rising sea levels have been a cause of serious concern worldwide. It is well-known that the temperature of Earth increased by 1.1°C after the Industrial Revolution (approximately 1760–1840) and is still increasing, according to data from the World Meteorological Organization (WMO). We continue to emit more carbon dioxide than we should, and in 2018, Earth's atmospheric concentration reached a new all-time high: 407.8 parts per million. According to the WMO, this gas can remain there for centuries, and in the oceans even longer, thus "perpetuating" global warming. (https://wmo.int/news/media-centre/greenhouse-gas-concentrations-hit-record-high-again) Based on these facts, some

DOI: 10.1201/9781003466833-18

proposals to decrease carbon dioxide (CO_2) emissions have been carried out, such as planting trees, calculating the carbon footprint, reducing meat production, integrating circular economy habits, acquiring fresh and local products, implementing sustainable mobility, and using renewable energy sources. Renewable energies are energy, inexhaustible resources, such as wind, solar radiation, water, and biomass. Thus, biomass energy technologies are inescapable toward a more sustainable and resilient energy system, providing a renewable, locally available, and versatile energy source that can help us reduce the dependence on fossil fuels and mitigate the negative environmental impact. Biomass includes various materials, such as wood, agricultural residues, crop wastes, animal manure, algae, and municipal solid waste.

15.2 BIOMASS AS A RENEWABLE ENERGY SOURCE

15.2.1 Definition and Types of Biomasses

Biomass is a broad term encompassing organic matter derived from plants and animals. It holds immense potential as a renewable energy source, offering versatility and flexibility in its applications. Biomass refers to all organic matter existing in the biosphere, whether of animal or plant origin, as well as those materials obtained by natural or artificial transformation of these.[10]

Biomass primarily includes three types of organic materials:

- Plant-based materials: wood, crops, agricultural residues (straw, stalks), energy crops (fast-growing trees, grasses).

- Animal-based materials: manure, food waste, sewage sludge.

- Algae: a promising source due to its rapid growth and efficient CO_2 absorption.

15.2.2 Advantages and Challenges of Using Biomass as an Energy Source[10–12]

There are multiple advantages of using biomass as an energy source from the environmental point of view, such as its versatility and impact on economic development. Compared to fossil fuels, biomass is replenishable through sustainable practices such as forest management and dedicated energy crops. Good management of this replenishable source can ensure a stable energy supply. Also, it utilizes organic waste, reducing landfill occupancy and associated environmental problems—greening the forgotten. Biomass combustion releases the CO_2 captured during plant growth when used sustainably, potentially resulting in a carbon-neutral process. Maintaining this balance requires extensive planning of processes, communities, and resources. It can be converted into various forms, such as heat, electricity, solid/liquid fuels or biogas, to meet various energy needs. This maximizes the possible applications for which biomass has potential. Biomass production can create jobs and income opportunities in rural areas. It generates development opportunities for communities dedicated to maintaining the balance of biomass regeneration to keep this as a long-term sustainable path. Having seen the main advantages of biomass utilization, one might think that everything is already solved, so it is best to put all of humanity's efforts into its implementation. Although biomass is a very promising route, it is not free of challenges and pending management work.

Improper biomass harvesting can lead to deforestation, soil erosion, and biodiversity loss. Sustainable practices are crucial, as is a comprehensive monitoring system that considers the population's energy needs and the rate of sustainable regeneration. Similarly, large-scale biofuel production could compete with food crops for land, raising food security concerns. There are some challenges in terms of the necessary technologies and economic investment. First, burning biomass can release pollutants such as nitrogen oxides and particulate matter, requiring advanced filtration systems. Compared to traditional fossil fuels, biomass technologies may be less cost-competitive, requiring further development and subsidies. Depending on the source and conversion process, biomass collection, transportation, and storage can be complex and expensive. Biomass offers a promising avenue for renewables because of its versatility and carbon-neutral potential. However, addressing sustainability issues, managing land-use conflicts and logistics, and optimizing conversion technologies are crucial to its responsible and long-term deployment.

15.3 BIOMASS CONVERSION TECHNOLOGIES

As the world's population continues to grow, the demand for food and energy has skyrocketed, causing severe environmental pressure and jeopardizing the overall well-being of the human population and the environment. Traditional food and energy production methods have left a significant environmental footprint due to greenhouse gas emissions and increased waste production.[10,11] However, scientists have recently focused on developing new sustainable solutions for managing biomass waste and converting it into valuable products. Several biomass conversion technologies, such as combustion, gasification, and pyrolysis, have emerged to transform waste materials into valuable products such as biofuels, fertilizers, and high-value chemicals. These technologies present an alternative to conventional energy production methods and promote a sustainable vision by reducing dependence on non-renewable resources.[13,14]

15.3.1 Understanding the Differences between Biomass and Biofuel

While both terms are commonly used interchangeably, biomass and biofuels are distinct entities in renewable energy. Biomass encompasses all organic matter derived from living or recently dead organisms, including plants, animals, and microbial sources. Biofuel, on the other hand, refers to a fuel produced from biomass through various conversion processes, and its applications are more directed to its use as a heat or electricity generator.[15] Biofuels derived from biomass include firewood, wood chips, pellets, fruit pits, and nut shells. Of these, chopped firewood is the least processed and usually burned directly in household appliances such as stoves and boilers. Wood chips come from the shredding of agricultural and forestry biomass, and their size varies depending on the manufacturing process. Finally, pellets are the most manufactured biofuel and consist of small cylinders of 6–12 mm in diameter with a length of 10–30 mm obtained by pressing biofuels with binders.[16, 17]

15.3.2 Overview of Biomass Conversion Processes

Bioenergy is the term used to describe energy obtained from biomass feedstocks. Multiple steps are required to convert biomass into a valuable energy source, including harvesting, drying, storage, transportation, processing, etc. The conversion of biomass to energy is done by three leading technologies: thermochemical, biochemical, and physicochemical (Figure 15.1).

Thermo-chemical conversion encompasses four main process options: combustion, pyrolysis, gasification, and liquefaction. The two main process options within bio-chemical conversion are anaerobic digestion (production of biogas, a mixture of mainly methane and carbon dioxide) and fermentation (principally ethanol). Physio-chemical conversion mainly consists of extraction, where oilseeds are crushed to extract oil.[13, 15]

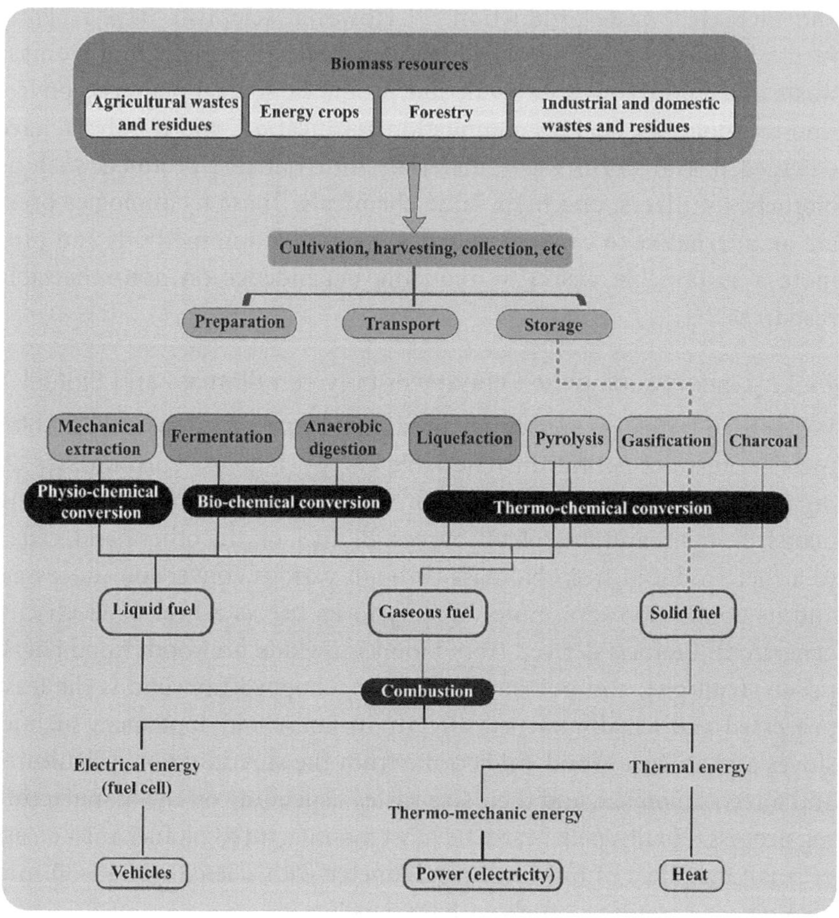

FIGURE 15.1 Schematic representation of biomass conversion from resources to end-use.[13–15]

15.3.3 Emerging Biomass Conversion Technologies

Biomass conversion constantly evolves, and new technologies are emerging as alternatives or complementary methods to improve efficiency, sustainability, and economic viability. The most outstanding emerging conversion technologies are:

15.3.4 Supercritical Water Gasification (SCWG)

One of the leading technical barriers to industrializing biomass-derived energy is the high energy input of conversion processes for wet bio-feedstock that require energy-costly drying/de-watering operation but relatively low energy density of the resulting biofuel products compared to inversion. In contrast, supercritical water gasification technology overcomes these problems as high water-containing feedstocks can be used directly, avoiding the costly drying process.[18] This method operates at high temperatures and pressure to efficiently convert biomass into syngas (a mixture of H_2 and CO), a versatile fuel precursor. SCWG is often performed at temperatures of 375–700°C and pressure of 22.1–25 MPa with different types of catalyst. Water can be by itself the catalyst of this process. Still, there are also routes for acid/alkaline catalysts, superficial catalysts (e.g. Ni, Ru, Rh, Pt, Pd, Al_2O_3, TiO_2), or even salt precipitation in SCW.[18, 19] SCWG is mainly used for lignocellulosic biomass (cellulose, hemicellulose, and lignin), where a hydrolysis process is described as a first step producing sugars, guaiacol, syringol, and phenolics derivates. These compounds enter the SCWG process, obtaining acids, alcohols, phenols, aromatics, and aldehydes, which are also catalyzed to transform into syngas.[19]

15.3.5 Fast Pyrolysis

Pyrolysis is a thermal decomposition process occurring in the absence of oxygen. Lower process temperature and more extended vapor residence favor the production of charcoal. High temperatures and longer residence times increase biomass conversion to gas, and a moderate term is optimum for producing liquids.[20] Unlike conventional pyrolysis, rapid heating rates are used to maximize the production of bio-oils, a liquid fuel that can be refined into transportation fuels. In fast pyrolysis, biomass decomposes quickly to generate primarily vapours and aerosols; after cooling and condensation, a dark brown homogenous mobile liquid is formed.[21, 22] Determinant factors for fast pyrolysis are very high heating rates and very high heat transfer rates at the biomass particle reaction interface, carefully controlled temperature of around 500°C to maximize liquid yield, short hot vapor residence times (typically less than 2s) minimizing secondary reactions, rapid removal of product char, and rapid cooling of the pyrolysis vapors to give the bio-oil product.[20, 22]

15.3.6 Plasma Gasification

Plasma is used at high temperatures to convert biomass into syngas, offering high efficiency and leading to cleaner outputs. Plasma gasification has gained attention as an environmentally friendly and efficient approach to convert carbon-based materials such as municipal solid waste, plastics, tires, and biomass in general into syngas, which can be further utilized for the synthesis of liquid fuels or power generation through combustion or fuel cell technologies.[23, 24] Plasma gasification has been considered an alternative to traditional biomass gasification technologies because it is performed at higher temperatures (up to 5,000°C) and heating rates. Another merit accompanied by plasma gasification is that the extent of the tar cracking reaction can be intensified to destroy tars and heavy components in high-temperature plasma and convert them into product gas.[24]

15.3.7 Microbial Electrolysis Cells (MECs)

MECs utilize microbes to convert waste organic matter directly into electricity. Compared with previous technologies for biohydrogen generation, the key competitive advantage of this technology is that the biomass can be utilized thoroughly for H_2 conversion; enhancing the hydrogen-production rate and lowering the energy input are the main challenges of MEC technology. In an MEC, electrochemically active microbes grow on the anode's surface, breaking organic matter into CO_2, electrons, and protons. The electrons and protons travel through the external circuit and solution, respectively, and combine at the cathode to generate hydrogen. Compared with fermentative hydrogen production, the MEC process can use various organic materials without being limited to fermentable substrates. This means that many different biomass derivatives, even those with low or negative economic value, such as wastewater, biowaste and agriculture residue, can be utilized.[25] These emerging technologies also incur an incursion in implementing advanced pretreatment techniques for biomass before a conversion process. These pretreatments are a way of conditioning or preparing organic matter to improve its compatibility, availability, and efficiency when subjected to one of the conversion processes. The first two types of pretreatments are acidic and alkaline. These involve catalysis mechanisms for biomass degradation into less complex chemical compounds. Another chemical-based pretreatment is ozonolysis, where the formation of highly reactive intermediates is promoted. Two pretreatments based on structural modification of the biomass are hydrothermal and steam explosion, which take advantage of the energy given to the medium to accelerate the conversion reaction. Something similar occurs when implementing ionic liquid, where the solubility of the main components of biomass in a solvent is promoted, resulting in total integration and a homogeneous reaction medium where conversion occurs. Other pretreatment methods are microwaves and sonication, where the high energy of the waves is concentrated and directed towards the material.[26,27]

15.4 ENERGY STORAGE AND ITS SIGNIFICANCE

15.4.1 The Need for Energy Storage Solutions

Energy storage is crucial in maximizing biomass's effectiveness and value as a renewable energy source. Since renewable energy sources such as solar and wind are inherently intermittent, their potential to provide stable and reliable power depends on efficient storage solutions. Multiple strategies can be applied to address this situation: electricity generation based on renewable sources such as solar panels and wind turbines depends on weather conditions. Because of this, there are periods of intermittency where it could be complemented with biomass-based energy and energy storage during peak generation, where surplus energy can be stored and released when demand requires it. This would imply a closer integration with the grid, representing a considerable challenge. However, providing facilities for gradual integration can prevent fluctuations and instabilities due to readjustment.[28]

Improving the integration of these biomass-based energy sources into the electricity system must be accompanied by optimizing the system's efficiency. It is important to integrate management systems to store surplus energy and release it when demand requires it. This reduces the need for additional power generation during specific periods, reducing not only resource consumption but also operational costs. Enabling and expanding energy storage capacity opens up opportunities for new applications such as microgrids and biomass-powered off-grid communities. This promotes energy independence and resilience in areas of limited or no access.[11]

15.4.2 Different Types of Energy Storage Devices

There are multiple energy storage strategies focused on biomass and mechanical or potential energy, electrochemical, chemical, thermal, and electrical potential. Dams or pumped storage of water, compressed air, and flywheels are within the mechanical or potential energy storage divisions. On a large scale, pumped water storage has a mature technology that takes advantage of gravity by storing energy from pumping water upstream and releasing it through turbines when necessary.[29] The electrochemical storage division includes secondary batteries such as Pb-A, Na-S, Li-ion, and Ni-Cd and flow batteries such as Redox Vanadium and Zn-Br. These devices take advantage of reversible oxidation-reduction or ion mobility reactions that release the energy initially transferred to them upon returning to their basal state. This type of storage has been extensively investigated in recent years for its potential to maintain high energy densities, resulting in exceptional amounts of energy in relatively small volumes compared to other storage technologies; as this technology advances, so does improvement in versatility, modularity, and fast responding times.[30–32] A storage technology similar to batteries is capacitors, which, unlike batteries, are not based on a chemical reaction but use electrostatics principles for direct electrical energy storage.[33]

On the other hand, chemical and thermal potential storage technologies surround biomass. This division can focus on a two-way strategy. Electricity can be converted to hydrogen or other biofuels for later use as flexible, long-term storage options, although these technologies are still under development. The opposite direction is obtaining biomass, the source of stored energy with the potential for combustion and use as electrical, thermal, and mechanical energy.[34–36]

15.5 APPLICATIONS OF BIOMASS IN ENERGY STORAGE

Numerous countries worldwide are focusing on using their scientific and technological resources and funding to develop available green energy sources; biomass has proven to be a highly effective energy storage solution that can significantly transform the energy landscape into a more sustainable and resilient one.[37] With advancements in research and technology, biomass-based energy storage options are expected to become even more efficient, low-cost, and eco-friendly. Biomass energy storage offers several advantages, such as reducing greenhouse gas emissions and minimizing reliance on fossil fuels. Effective energy storage technologies are crucial in maximizing the potential of biomass energy.[38] In this chapter, we will explore the various technologies for biomass energy storage and their key benefits. Electrochemical energy storage devices (EEDs) such as batteries and supercapacitors are considered a potential choice of energy storage devices that complement their high energy and power densities.[39]

15.5.1 Biomass-Based Batteries

The demand for energy storage technologies is rising due to their potential applications in electric devices that require batteries with high energy density, long cycle life, and satisfactory rate performance.[40] Lithium-ion batteries (LiBs), which have the highest energy density among commercially viable energy storage technologies, work on an intercalation mechanism.[41]

However, they are approaching the limits of theoretical energy density, which calls for developing a new energy storage system with a higher energy density. Lithium-sulfur (Li-S) batteries, which consist of a sulfur cathode and a lithium metal anode, are a promising solution offering a high energy density of approximately 2,600 W h K^{-1}.[32] Unfortunately, the real-world application of Li-S batteries is limited by the low conductivity of sulfur and discharge products (Li_2S/Li_2S_2), high volume expansion of sulfur during cycles, profound shuttle effect caused by the dissolved lithium polysulfides (LiPSs) in the electrolyte, and irreversible Li_2S/Li_2S_2 deposition and phase conversion. Lithium-sulfur (Li-S) batteries or sodium-ion batteries (SIBs), are a response to the premature depletion of non-renewable energy sources (Figure 15.2).[31, 42–45]

On the other hand, researchers have been exploring alternative sources such as biomass. The study of biomass is important for developing new technologies because it is easy to handle, environmentally friendly, cost-effective, and has the potential for large-scale production. Biomass, which has little or no economic value, can be used as a low-cost and efficient carbon source for creating advanced functional materials with significant technological and economic value.[46,47] The carbon precursor material's structure, morphology, and surface chemical composition highly affect its applications.

Carbon materials derived from biomass precursors can contain desired/ suitable properties e.g., SSA, different pore structures (micro and meso pores), the high number of functional groups on its surfaces, adjustable

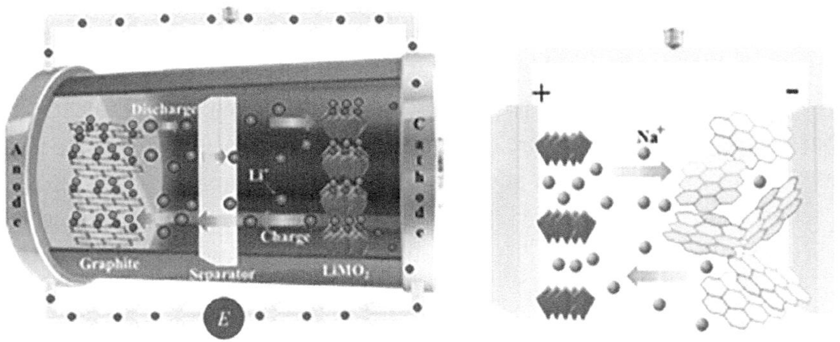

FIGURE 15.2   Schematic and working principles of LiBs and NiBs. Reproduced from [45] with permission from MDPI.

hydrophilicity, and conductivity. Thanks to all these properties, they can be successfully employed as anode materials for LIBs and NIBs.[48] Also, biomass carbon materials can be easily turned into hierarchically porous structures for battery technologies due to their excellent cycling stability and rate performance.[49] Zhao et al. synthesized hierarchically porous carbon (HPC) by multi-step carbonization of biomass reed flowers. Carbon materials are commonly classified into two main categories: graphitizable (soft carbons) and non-graphitizable (hard carbons). Soft carbons can be tuned into graphite while subjected to very high temperatures (> 2,000°C), while hard carbons are hardly graphitized even at 3,000°C.[50]

The literature shows that the properties of carbon materials are widely dependent on pyrolysis conditions, especially temperature. These features play a huge role in the electrochemical performances of carbon materials when employed as anodes for battery applications. It is well known that other pyrolysis and preparation parameters, such as type of biomass and its composition, pyrolysis time, used catalyst, heating rate, and reaction atmosphere, influence the carbon material's final properties, and all these parameters act concomitantly.[51-53]

15.5.2 Biomass-Based Supercapacitors

Supercapacitors are receiving significant attention, and researchers are exploring various electrode materials and polymer electrolytes to improve their performance. Supercapacitors are ultrafast electrochemical charge storage devices, composed of several key parts, such as a positive electrode, a negative electrode, an electrolyte, a non-conductive separator (to prevent a short circuit between the two electrodes), and two current collectors that connect the electrodes to an external circuit.[54] Based on their charge storage mechanisms, supercapacitors can be categorized into three fundamental types: electrical double-layer capacitors (EDLC), pseudo-capacitors (PC), and battery hybrid supercapacitors (BHS). Biomass-derived carbon-based electrode materials are being prepared using natural materials such as corn straws and peanut shells to reduce the expenditure of producing carbon materials and alleviate the environmental problems caused by excess biomass resources. The unique and precise structure of biomass materials allows them to be used as skeletons, binders, and, after carbonization, as support materials for supercapacitor electrodes.[33, 55]

These materials exhibit potential for use as supercapacitor electrodes, providing comparable performance at a lower cost than traditional materials like carbon nanotubes and graphene. Therefore, developing modern, durable, and compact supercapacitors with high specific energy and specific power will revolutionize the landscape of electric energy storage and supply for industrial and home uses.[56–58] The current study is primarily concerned with recent developments in research about biomass-derived carbon electrode materials for supercapacitor applications, including plant, fruit, vegetable, and microorganism-based carbon electrode materials. A summary of alternative synthesis methods for converting and activating biomass waste is also provided.

Supercapacitors have gained significant attention as a promising energy storage technology due to their exceptional power density, excellent charging and discharging rates, superior cycle life, and good reversibility. SCs are categorized into electrochemical double-layer capacitors and pseudo-capacitors based on their energy storage mechanism. The development of advanced electrode materials is crucial to creating robust supercapacitors. Supercapacitors have been identified as one of the most promising energy storage options for high-power applications, including hybrid electric vehicles, portable electronics, digital communications, and renewable energy systems. They offer many attractive properties, such as high power density, long cycle life, fast charge and discharge rates, and better safety. Innovative energy storage technologies that utilize biomass are also being researched.[59] The manufacturing of biomass-based supercapacitors is a developing field that promises to contribute to sustainability and the availability of materials for energy storage applications.

15.6 DEVELOPMENT OF BIOMASS MATERIALS FOR ENERGY STORAGE

Biomass material contains a large amount of lignin, hemicellulose, cellulose[37], etc. When subjected to high-temperature carbonization, it produces a unique pore structure that is difficult for humans to replicate. This particular pore structure gives them a variety of excellent performance. If widely used in various fields such as catalyst carrier, sewage treatment, gas storage and separation, and precious metal recovery, it can significantly reduce the preparation cost and difficulty of preparation of porous carbon materials in various fields, and waste. Turning pollutants into valuable waste can alleviate the energy crisis and environmental pressure.

15.6.1 Types of Biomass Materials

The most widely utilized material for energy production from biomass is wood. This can take various forms, such as pellets, wood chips, briquettes, sawdust, or firewood. We can obtain heat, fuels, and electricity through the combustion of these organic materials. However, some other sources of biomass have been explored, such as natural resources including rice husk[60], corncobs[61], tea waste[62], sugarcane bagasse[63], peanut shell[60], onion husk[64], olive pits[65], pinecones[66], watermelon rind[67], walnut shell[68], and coconut shell[69] as the primary precursor material for electronic and energy storage applications. Biomass is an organic material and an abundant source of raw material for the preparation of carbon and its composites; it is classified based on its origins, characteristics, and compositions, such as forestry, crops and residues, industrial residues, aquatic residues, animal residues, and municipal/urban waste. Efforts have been made to detail the various sources of biomasses (forestry, agriculture, industrial, and municipal); the biochar preparation techniques involving carbonization, pyrolysis, hydrothermal treatment, ionothermal, and salt-assisted methods; and their activation through physical and chemical processes along with necessary critical reaction parameters.

15.6.2 Relevant Material Properties for Energy Storage

Energy storage can be dated back to ancient times, and it is quite simple and natural. Mankind initially made fire using wood and charcoal, biomass energy storage carriers of solar energy. Fire brings warmth and brightness, cooked food, and bronze and iron wares; thus, charcoal energy acted as one of the ancient civilizations' most important driving powers. Therefore, various energy storage systems are being developed to utilize different energy sources properly.[37, 70, 71]

15.7 CASE STUDIES AND PRACTICAL APPLICATIONS

The following section explains how biomass is revolutionizing how we create sustainable sources and obtain materials with more excellent added value. These materials are used in advanced electrode material for energy storage devices, like supercapacitors, lithium (Li)-ion, sodium (Na)-ion, lithium sulfur (Li–S), and metal-air batteries.

15.7.1 Examples of Real-World Projects Using Biomass for Energy Storage

Biomass can be stored in various forms, such as wood logs, chips, and pellets. These forms can be burned to produce energy when needed. The characteristics of primary energy storage forms are that they have very high energy density and can provide long-term energy storage. However, since they only occur naturally, they cannot be used as a medium for storing secondary forms of energy.[71, 72] The use of biomass for energy production is prevalent in power plants such as Polaniec Biomass Power Plant, Ironbridge (United Kingdom), Alholmens Kraft (Finland), Kymijärvi II, (Finland), Vaasa Bio-gasification plant (Finland), Wisapower, (Finland), New Hope Power Partnership (US), Kaukaan Voima (Finland), and Seinäjoki (Finland). For example, the Polaniec Biomass Power Plant uses biomass for energy production, whereas GDF Suez owns and operates the Polaniec biomass power plant. The project is 80% fueled by wood chips and 20% by agricultural waste. The existing Polaniec power plant has eight 225MWe turbines fed by coal and biomass. During construction, the biomass unit was initially planned for a capacity of 190 MW, but it increased to 205 MW. It features a circulating fluidized bed (CFB) boiler, which can exclusively burn biomass fuel. These examples illustrate the successful use of biomass for energy generation in large-scale projects.

15.7.2 Challenges Faced and Lessons Learned

The utilization of biomass as an energy source faces various obstacles. First, the costs associated with the technology and infrastructure of biomass projects, such as bioenergy plants, can be significantly high, requiring considerable upfront investments. Moreover, biomass combustion can release pollutants, raising concerns regarding air quality. The bulky nature of biomass materials also presents logistical challenges. Several measures can be adopted to overcome these hurdles, such as implementing sustainable management practices, developing integrated systems to optimize resource utilization, and promoting innovation and research to improve the efficiency and profitability of technologies related to biomass conversion.[73, 74]

In summary, using biomass for energy storage has some challenges related to costs, conversion efficiency, the sustainability of biomass sources, and environmental management. However, biomass can play a vital role in meeting future energy requirements by implementing sustainable practices, developing advanced technologies, and receiving political support.

15.8 ENVIRONMENTAL AND SUSTAINABILITY ASPECTS

15.8.1 Environmental Impact Assessment of Biomass-Based Energy Storage

In recent years, there has been an increasing interest in using biomass as a source of renewable energy production due to concerns about energy independence and the environment. This is because there are concerns about energy independence and the environment. When we evaluate the sustainability of renewable energy technologies, it is crucial to consider their environmental impact. However, we must analyse the economic and environmental performance of different bioenergy solutions to encourage environmentally sustainable bioenergy strategies. To achieve a holistic sustainability evaluation, we must consider both aspects jointly to avoid any trade-offs between them. If we only consider one aspect, we may optimize the performance of one area at the expense of the other.[75]

15.8.2 Contributions to Sustainability and Climate Change Mitigation

Biomass has the potential to contribute to sustainability and climate change mitigation in several ways. First, biomass is a renewable energy source that can be used as a substitute for fossil fuels, thus reducing greenhouse gas (GHG) emissions and dependence on fossil fuels. Additionally, it can be cultivated and harvested sustainably without causing harm to the environment. However, it is crucial to note that if the production and use of biomass are not managed correctly, it can have negative environmental and social impacts. Therefore, it is necessary to produce and use biomass sustainably, minimizing adverse effects and maximizing environmental and social benefits.[36, 76]

ACKNOWLEDGMENTS

Authors thanks to FCQ-UANL, ProACTI (Grant: 53-BQ-2023). L. F. M. G. thanks to CONAHCYT for the fellowship "ayudante de investigador Nivel III."

AUTHORS' BIOGRAPHIES

Luis Fernando Macías Gamboa (Nuevo León, México, 2001) is currently a student of industrial chemistry at the School of Chemical Sciences of the Universidad Autónoma de Nuevo León (UANL) in Mexico. His passion for science began when he was 12 years old, and since then, he has represented his state and his country in mathematics, chemistry, physics, and programming competitions. Presently, he belongs to the Jimenez research group dedicated to materials chemistry, where he develops several research projects that include pollutant degradation, biomass utilization, and luminescent nanomaterials, consistently applying green chemistry.

Raúl Segovia Pérez (Hidalgo, México, 1991) received his BSc in biotechnology engineering at the Polytechnic University of Pachuca. He obtained his master's and PhD in chemistry from the Autonomous University of the State of Hidalgo. Following this, he worked as a postdoctoral researcher with Prof. Víctor Jiménez at the Chemistry Faculty of the Autonomous University of Nuevo Leon. His current research is focused on creating luminescent materials using eco-friendly methods.

Blanca M. Muñoz Flores (Tlaxcala, México, 1976) studied industrial chemistry at the Universidad Autónoma de Tlaxcala (1999). She received her PhD in organic chemistry (2008) at the CINVESTAV-IPN under the direction of Prof. Norberto Farfán. She joined the chemistry faculty at the Autonomous University of Nuevo León as a full professor. Her research group is currently working on luminescent materials as dyes in biomaterials. So far, she is the author of 62 research publications (over 900 citations, h-index 20) in leading scientific journals, 15 patents, and five book chapters.

Víctor M. Jiménez Pérez (México, 1974) received his BSc (1998) in chemistry from the Autonomous University of Tlaxcala. In 2005, he got his PhD in chemistry at CINVESTAV-CDMX. After the postdoctoral work with Prof. Herbert W. Roesky at Universitat Gottingen (2007 fellowship DAAD), he joined as a full professor of the chemistry faculty at the Autonomous University of Nuevo Leon. So far, he is the author of 82 publications (over 2,100 citations, h-index 24), 20 patents, and five book chapters. His research is focused on the design and green synthesis of new non-toxic luminescent materials from fine chemicals and biomass for biological, optoelectronic, chemical sensor, and tissue engineering applications.

REFERENCES

1. Akhade, S. A.; Singh, N.; Gutierrez, O. Y.; Lopez-Ruiz, J.; Wang, H.; Holladay, J. D.; Liu, Y.; Karkamkar, A.; Weber, R. S.; Padmaperuma, A. B.; Lee, M. S.; Whyatt, G. A.; Elliott, M.; Holladay, J. E.; Male, J. L.; Lercher, J. A.; Rousseau, R.; Glezakou, V. A. Electrocatalytic Hydrogenation of Biomass-Derived Organics: A Review. *Chemical Reviews*. American Chemical Society October 28, 2020, pp 11370–11419. https://doi.org/10.1021/acs.chemrev.0c00158.

2. Vassilev, S. V.; Vassileva, C. G. Water-Soluble Fractions of Biomass and Biomass Ash and Their Significance for Biofuel Application. *Energy and Fuels*. American Chemical Society April 18, 2019, pp 2763–2777. https://doi.org/10.1021/acs.energyfuels.9b00081.

3. Morgan, T. J.; Youkhana, A.; Turn, S. Q.; Ogoshi, R.; Garcia-Pérez, M. Review of Biomass Resources and Conversion Technologies for Alternative Jet Fuel Production in Hawai'i and Tropical Regions. *Energy and Fuels*. American

Chemical Society April 18, 2019, pp 2699–2762. https://doi.org/10.1021/acs. energyfuels.8b03001.

4. Ren, X. Y.; Feng, X. B.; Cao, J. P.; Tang, W.; Wang, Z. H.; Yang, Z.; Zhao, J. P.; Zhang, L. Y.; Wang, Y. J.; Zhao, X. Y. Catalytic Conversion of Coal and Biomass Volatiles: A Review. *Energy and Fuels*. American Chemical Society September 17, 2020, pp 10307–10363. https://doi.org/10.1021/acs. energyfuels.0c01432.

5. Bermúdez, J. M.; Francavilla, M.; Calvo, E. G.; Arenillas, A.; Franchi, M.; Menéndez, J. A.; Luque, R. Microwave-Induced Low Temperature Pyrolysis of Macroalgae for Unprecedented Hydrogen-Enriched Syngas Production. *RSC Advances* **2014**, *4* 72), pp 38144–38151. https://doi.org/10.1039/c4ra05372a.

6. Yu, Z.; Wu, H.; Li, Y.; Xu, Y.; Li, H.; Yang, S. Advances in Heterogeneously Catalytic Degradation of Biomass Saccharides with Ordered-Nanoporous Materials. *Industrial and Engineering Chemistry Research.* American Chemical Society September 30, 2020, pp 16970–16986. https://doi. org/10.1021/acs.iecr.0c01625.

7. Gérardy, R.; Debecker, D. P.; Estager, J.; Luis, P.; Monbaliu, J. C. M. Continuous Flow Upgrading of Selected C2-C6Platform Chemicals Derived from Biomass. *Chemical Reviews.* American Chemical Society August 12, 2020, pp 7219–7347. https://doi.org/10.1021/acs.chemrev.9b00846.

8. Pazhany, A. S.; Henry, R. J. Genetic Modification of Biomass to Alter Lignin Content and Structure. *Industrial and Engineering Chemistry Research.* American Chemical Society September 4, 2019, pp 16190–16203. https://doi. org/10.1021/acs.iecr.9b01163.

9. Chen, X.; Liu, L.; Zhang, L.; Zhao, Y.; Qiu, P.; Ruan, R. A Review on the Properties of Copyrolysis Char from Coal Blended with Biomass. *Energy and Fuels.* American Chemical Society April 16, 2020, pp 3996–4005. https://doi.org/10.1021/acs.energyfuels.0c00014.

10. Perea-Moreno, M. A.; Samerón-Manzano, E.; Perea-Moreno, A. J. Biomass as Renewable Energy: Worldwide Research Trends. *Sustainability (Switzerland)* **2019**, *11* (3). https://doi.org/10.3390/su11030863.

11. Mehedintu, A.; Sterpu, M.; Soava, G. Estimation and Forecasts for the Share of Renewable Energy Consumption in Final Energy Consumption by 2020 in the European Union. *Sustainability (Switzerland)* **2018**, *10* (5). https://doi. org/10.3390/su10051515.

12. Cherubini, F.; Strømman, A. H. Life Cycle Assessment of Bioenergy Systems: State of the Art and Future Challenges. *Bioresource Technology* January **2011**, pp 437–451. https://doi.org/10.1016/j.biortech.2010.08.010.

13. Adams, P.; Bridgwater, T.; Lea-Langton, A.; Ross, A.; Watson, I. Biomass Conversion Technologies. In *Greenhouse Gas Balances of Bioenergy Systems*; Elsevier, 2017; pp 107–139. https://doi.org/10.1016/B978-0-08-101 036-5.00008-2.

14. Mckendry, P. Energy Production from Biomass (Part 2): Conversion Technologies. *Bioresource Technology* **2002**, *83* (1), 47–54.

15. Kataya, G.; Cornu, D.; Bechelany, M.; Hijazi, A.; Issa, M. Biomass Waste Conversion Technologies and Its Application for Sustainable Environmental Development—A Review. *Agronomy*. Multidisciplinary Digital Publishing Institute (MDPI) November 1, 2023. https://doi.org/10.3390/agronomy 13112833.

16. Williams, O.; Taylor, S.; Lester, E.; Kingman, S.; Giddings, D.; Eastwick, C. Applicability of Mechanical Tests for Biomass Pellet Characterisation for Bioenergy Applications. *Materials* **2018**, *11* (8). https://doi.org/10.3390/ ma11081329.

17. Li, G.; Liu, C.; Yu, Z.; Rao, M.; Zhong, Q.; Zhang, Y.; Jiang, T. Energy Saving of Composite Agglomeration Process (CAP) by Optimized Distribution of Pelletized Feed. *Energies (Basel)* **2018**, *11* (9). https://doi.org/10.3390/ en11092382.

18. Hu, Y.; Gong, M.; Xing, X.; Wang, H.; Zeng, Y.; Xu, C. C. Supercritical Water Gasification of Biomass Model Compounds: A Review. *Renewable and Sustainable Energy Reviews*. Elsevier Ltd February 1, 2020. https://doi. org/10.1016/j.rser.2019.109529.

19. Reddy, S. N.; Nanda, S.; Dalai, A. K.; Kozinski, J. A. Supercritical Water Gasification of Biomass for Hydrogen Production. *International Journal of Hydrogen Energy*. Elsevier Ltd April 24, 2014, pp 6912–6926. https://doi. org/10.1016/j.ijhydene.2014.02.125.

20. Bridgwater, A. V. Review of Fast Pyrolysis of Biomass and Product Upgrading. *Biomass Bioenergy* **2012**, *38*, pp 68–94. https://doi.org/10.1016/j. biombioe.2011.01.048.

21. Contescu, C.; Adhikari, S.; Gallego, N.; Evans, N.; Biss, B. Activated Carbons Derived from High-Temperature Pyrolysis of Lignocellulosic Biomass. *C (Basel)* **2018**, *4* (3), p 51. https://doi.org/10.3390/c4030051.

22. Bridgwater, A. V; Meier, D.; Radlein, D. *An Overview of Fast Pyrolysis of Biomass*. www.elsevier.nl/locate/orggeochem.

23. Hlina, M.; Hrabovsky, M.; Kavka, T.; Konrad, M. Production of High Quality Syngas from Argon/Water Plasma Gasification of Biomass and Waste. *Waste Management* **2014**, *34* (1), pp 63–66. https://doi.org/10.1016/j. wasman.2013.09.018.

24. Kuo, P. C.; Illathukandy, B.; Wu, W.; Chang, J. S. Plasma Gasification Performances of Various Raw and Torrefied Biomass Materials Using Different Gasifying Agents. *Bioresource Technology* **2020**, *314*. https://doi. org/10.1016/j.biortech.2020.123740.

25. Liu, H.; Hu, H.; Chignell, J.; Fan, Y. Microbial Electrolysis: Novel Technology for Hydrogen Production from Biomass. *Biofuels*. January 2010, pp 129–142. https://doi.org/10.4155/bfs.09.9.

26. Haldar, D.; Purkait, M. K. A Review on the Environment-Friendly Emerging Techniques for Pretreatment of Lignocellulosic Biomass: Mechanistic Insight and Advancements. *Chemosphere*. Elsevier Ltd February 1, 2021. https://doi.org/10.1016/j.chemosphere.2020.128523.

27. Ashok Pandey; Sangeeta Negi; Parameswaran Binod; Christian Larroche. *Pretreatment of Biomass: Processes and Technologies*; Elsevier Science, 2014.

28. Wang, F.; Ouyang, D.; Zhou, Z.; Page, S. J.; Liu, D.; Zhao, X. Lignocellulosic Biomass as Sustainable Feedstock and Materials for Power Generation and Energy Storage. *Journal of Energy Chemistry* **2021**, *57*, pp 247–280. https://doi.org/10.1016/j.jechem.2020.08.060.

29. Rahman, M. M.; Oni, A. O.; Gemechu, E.; Kumar, A. Assessment of Energy Storage Technologies: A Review. *Energy Conversion and Management.* Elsevier Ltd November 1, 2020. https://doi.org/10.1016/j.enconman.2020.113295.

30. Olabi, A. G.; Onumaegbu, C.; Wilberforce, T.; Ramadan, M.; Abdelkareem, M. A.; Al—Alami, A. H. Critical Review of Energy Storage Systems. *Energy* **2021**, *214*. https://doi.org/10.1016/j.energy.2020.118987.

31. Peng, H. J.; Huang, J. Q.; Cheng, X. B.; Zhang, Q. Review on High-Loading and High-Energy Lithium-Sulfur Batteries. *Advanced Energy Materials.* Wiley-VCH Verlag December 20, 2017. https://doi.org/10.1002/aenm.201700260.

32. Yuan, H.; Liu, T.; Liu, Y.; Nai, J.; Wang, Y.; Zhang, W.; Tao, X. A Review of Biomass Materials for Advanced Lithium-Sulfur Batteries. *Chemical Science.* Royal Society of Chemistry 2019, pp 7484–7495. https://doi.org/10.1039/c9sc02743b.

33. Unknown, S.; Chand, P.; Joshi, A. Biomass Derived Carbon for Supercapacitor Applications: Review. *Journal of Energy Storage.* Elsevier Ltd July 1, 2021. https://doi.org/10.1016/j.est.2021.102646.

34. Wang, F.; Ouyang, D.; Zhou, Z.; Page, S. J.; Liu, D.; Zhao, X. Lignocellulosic Biomass as Sustainable Feedstock and Materials for Power Generation and Energy Storage. *Journal of Energy Chemistry* **2021**, *57*, pp 247–280. https://doi.org/10.1016/j.jechem.2020.08.060.

35. Zhu, Z.; Xu, Z. The Rational Design of Biomass-Derived Carbon Materials Towards Next-Generation Energy Storage: A Review. *Renewable and Sustainable Energy Reviews.* Elsevier Ltd December 1, 2020. https://doi.org/10.1016/j.rser.2020.110308.

36. Demirbas, M. F.; Balat, M.; Balat, H. Potential Contribution of Biomass to the Sustainable Energy Development. *Energy Conversion and Management* **2009**, *50* (7), pp 1746–1760. https://doi.org/10.1016/j.enconman.2009.03.013.

37. Liu, C.; Li, F.; Lai-Peng, M.; Cheng, H. M. Advanced Materials for Energy Storage. *Advanced Materials.* February 23, 2010. https://doi.org/10.1002/adma.200903328.

38. Thota, S. P.; Bag, P. P.; Vadlani, P. V.; Belliraj, S. K. Plant Biomass Derived Multidimensional Nanostructured Materials: A Green Alternative for EnergyStorage. *Engineered Science* **2022**, *18*, pp 31–58. https://doi.org/10.30919/es8d664.

39. Deng, J.; Li, M.; Wang, Y. Biomass-Derived Carbon: Synthesis and Applications in Energy Storage and Conversion. *Green Chemistry.* Royal Society of Chemistry 2016, pp 4824–4854. https://doi.org/10.1039/c6gc01172a.

40. Ferrero, G. A.; Fuertes, A. B.; Sevilla, M. From Soybean Residue to Advanced Supercapacitors. *Scientific Reports* **2015**, *5*. https://doi.org/10.1038/srep16618.

41. Choi, N. S.; Chen, Z.; Freunberger, S. A.; Ji, X.; Sun, Y. K.; Amine, K.; Yushin, G.; Nazar, L. F.; Cho, J.; Bruce, P. G. Challenges Facing Lithium Batteries and Electrical Double-Layer Capacitors. *Angewandte Chemie - International Edition*. October 1, 2012, pp 9994–10024. https://doi.org/10.1002/anie.201201429.

42. Fang, R.; Zhao, S.; Sun, Z.; Wang, D. W.; Cheng, H. M.; Li, F. More Reliable Lithium-Sulfur Batteries: Status, Solutions and Prospects. *Advanced Materials*. Wiley-VCH Verlag December 27, 2017. https://doi.org/10.1002/adma.201606823.

43. Tao, X.; Wan, J.; Liu, C.; Wang, H.; Yao, H.; Zheng, G.; Seh, Z. W.; Cai, Q.; Li, W.; Zhou, G.; Zu, C.; Cui, Y. Balancing Surface Adsorption and Diffusion of Lithium-Polysulfides on Nonconductive Oxides for Lithium-Sulfur Battery Design. *Nature Communications* **2016**, *7*. https://doi.org/10.1038/ncomms11203.

44. Seh, Z. W.; Yu, J. H.; Li, W.; Hsu, P. C.; Wang, H.; Sun, Y.; Yao, H.; Zhang, Q.; Cui, Y. Two-Dimensional Layered Transition Metal Disulphides for Effective Encapsulation of High-Capacity Lithium Sulphide Cathodes. *Nature Communications* **2014**, *5*. https://doi.org/10.1038/ncomms6017.

45. Molaiyan, P.; Dos Reis, G. S.; Karuppiah, D.; Subramaniyam, C. M.; García-Alvarado, F.; Lassi, U. Recent Progress in Biomass-Derived Carbon Materials for Li-Ion and Na-Ion Batteries—A Review. *Batteries*. MDPI February 1, 2023. https://doi.org/10.3390/batteries9020116.

46. Xia, Y.; Fang, R.; Xiao, Z.; Huang, H.; Gan, Y.; Yan, R.; Lu, X.; Liang, C.; Zhang, J.; Tao, X.; Zhang, W. Confining Sulfur in N-Doped Porous Carbon Microspheres Derived from Microalgaes for Advanced Lithium-Sulfur Batteries. *ACS Applied Materials & Interfaces* **2017**, *9* (28), pp 23782–23791. https://doi.org/10.1021/acsami.7b05798.

47. Xia, L.; Zhou, Y.; Ren, J.; Wu, H.; Lin, D.; Xie, F.; Jie, W.; Lam, K. H.; Xu, C.; Zheng, Q. An Eco-Friendly Microorganism Method to Activate Biomass for Cathode Materials for High-Performance Lithium-Sulfur Batteries. *Energy and Fuels* **2018**, *32* (9), pp 9997–10007. https://doi.org/10.1021/acs.energyfuels.8b01453.

48. Tang, W.; Zhang, Y.; Zhong, Y.; Shen, T.; Wang, X.; Xia, X.; Tu, J. Natural Biomass-Derived Carbons for Electrochemical Energy Storage. *Materials Research Bulletin*. Elsevier Ltd April 1, 2017, pp 234–241. https://doi.org/10.1016/j.materresbull.2016.12.025.

49. Dutta, S.; Bhaumik, A.; Wu, K. C. W. Hierarchically Porous Carbon Derived from Polymers and Biomass: Effect of Interconnected Pores on Energy Applications. *Energy and Environmental Science*. Royal Society of Chemistry November 1, 2014, pp 3574–3592. https://doi.org/10.1039/c4ee01075b.

50. Zhao, W.; Wen, J.; Zhao, Y.; Wang, Z.; Shi, Y.; Zhao, Y. Hierarchically Porous Carbon Derived from Biomass Reed Flowers as Highly Stable Li-Ion Battery Anode. *Nanomaterials* **2020**, *10* (2). https://doi.org/10.3390/nano10020346.

51. Zhang, L.; Liu, Z.; Cui, G.; Chen, L. Biomass-Derived Materials for Electrochemical Energy Storages. *Progress in Polymer Science.* Elsevier Ltd April 1, 2015, pp 136–164. https://doi.org/10.1016/j.progpolymsci.2014.09.003.

52. Wang, F.; Ouyang, D.; Zhou, Z.; Page, S. J.; Liu, D.; Zhao, X. Lignocellulosic Biomass as Sustainable Feedstock and Materials for Power Generation and Energy Storage. *Journal of Energy Chemistry* **2021**, *57*, pp 247–280. https://doi.org/10.1016/j.jechem.2020.08.060.

53. Senthil, C.; Lee, C. W. Biomass-Derived Biochar Materials as Sustainable Energy Sources for Electrochemical Energy Storage Devices. *Renewable and Sustainable Energy Reviews.* Elsevier Ltd March 1, 2021. https://doi.org/10.1016/j.rser.2020.110464.

54. Lin, S.; Wang, F.; Shao, Z. Biomass Applied in Supercapacitor Energy Storage Devices. *Journal of Materials Science.* Springer January 1, 2021, pp 1943–1979. https://doi.org/10.1007/s10853-020-05356-1.

55. Liu, Z.; Zhu, Z.; Dai, J.; Yan, Y. Waste Biomass Based-Activated Carbons Derived from Soybean Pods as Electrode Materials for High-Performance Supercapacitors. *Chemistry Select* **2018**, *3* (21), pp 5726–5732. https://doi.org/10.1002/slct.201800609.

56. Chernysheva, D. V.; Chus, Y. A.; Klushin, V. A.; Lastovina, T. A.; Pudova, L. S.; Smirnova, N. V.; Kravchenko, O. A.; Chernyshev, V. M.; Ananikov, V. P. Sustainable Utilization of Biomass Refinery Wastes for Accessing Activated Carbons and Supercapacitor Electrode Materials. *ChemSusChem* **2018**, *11* (20), pp 3599–3608. https://doi.org/10.1002/cssc.201801757.

57. Li, Z.; Guo, D.; Liu, Y.; Wang, H.; Wang, L. Recent Advances and Challenges in Biomass-Derived Porous Carbon Nanomaterials for Supercapacitors. *Chemical Engineering Journal.* Elsevier B.V. October 1, 2020. https://doi.org/10.1016/j.cej.2020.125418.

58. Senthil, R. A.; Yang, V.; Pan, J.; Sun, Y. A Green and Economical Approach to Derive Biomass Porous Carbon from Freely Available Feather Finger Grass Flower for Advanced Symmetric Supercapacitors. *Journal of Energy Storage* **2021**, *35.* https://doi.org/10.1016/j.est.2021.102287.

59. Shan, B.; Cui, Y.; Liu, W.; Zhang, Y.; Liu, S.; Wang, H.; Sun, L.; Wang, Z.; Wu, R. Fibrous Bio-Carbon Foams: A New Material for Lithium-Ion Hybrid Supercapacitors with Ultrahigh Integrated Energy/Power Density and Ultralong Cycle Life. *ACS Sustainable Chemistry & Engineering* **2018**, *6* (11), pp 14989–15000. https://doi.org/10.1021/acssuschemeng.8b03473.

60. Purkait, T.; Singh, G.; Singh, M.; Kumar, D.; Dey, R. S. Large Area Few-Layer Graphene with Scalable Preparation from Waste Biomass for High-Performance Supercapacitor. *Scientific Reports* **2017**, *7* (1). https://doi.org/10.1038/s41598-017-15463-w.

61. Wang, D.; Geng, Z.; Li, B.; Zhang, C. High Performance Electrode Materials for Electric Double-Layer Capacitors Based on Biomass-Derived Activated Carbons *Electrochimica Acta* **2015**, *173*, pp 377–384. https://doi.org/10.1016/j.electacta.2015.05.080.

62. Song, X.; Ma, X.; Li, Y.; Ding, L.; Jiang, R. Tea Waste Derived Microporous Active Carbon with Enhanced Double-Layer Supercapacitor Behaviors.

Applied Surface Science **2019**, *487*, pp 189–197. https://doi.org/10.1016/j. apsusc.2019.04.277.

63. Rufford, T. E.; Hulicova-Jurcakova, D.; Khosla, K.; Zhu, Z.; Lu, G. Q. Microstructure and Electrochemical Double-Layer Capacitance of Carbon Electrodes Prepared by Zinc Chloride Activation of Sugar Cane Bagasse. *Journal Power Sources* **2010**, *195* (3), pp 912–918. https://doi.org/10.1016/j. jpowsour.2009.08.048.

64. Wang, D.; Liu, S.; Fang, G.; Geng, G.; Ma, J. From Trash to Treasure: Direct Transformation of Onion Husks into Three-Dimensional Interconnected Porous Carbon Frameworks for High-Performance Supercapacitors in Organic Electrolyte. *Electrochimica Acta* **2016**, *216*, pp 405–411. https://doi. org/10.1016/j.electacta.2016.09.053.

65. Redondo, E.; Carretero-González, J.; Goikolea, E.; Ségalini, J.; Mysyk, R. Effect of Pore Texture on Performance of Activated Carbon Supercapacitor Electrodes Derived from Olive Pits. *Electrochimica Acta* **2015**, *160*, pp 178– 184. https://doi.org/10.1016/j.electacta.2015.02.006.

66. Karthikeyan, K.; Amaresh, S.; Lee, S. N.; Sun, X.; Aravindan, V.; Lee, Y. G.; Lee, Y. S. Construction of High-Energy-Density Supercapacitors from Pine-Cone-Derived High-Surface-Area Carbons. *ChemSusChem* **2014**, *7* (5), pp 1435–1442. https://doi.org/10.1002/cssc.201301262.

67. Lin, X. Q.; Yang, N.; Qiu-Feng, L.; Liu, R. Self-Nitrogen-Doped Porous Biocarbon from Watermelon Rind: A High-Performance Supercapacitor Electrode and Its Improved Electrochemical Performance Using Redox Additive Electrolyte. *Energy Technology* **2019**, *7* (3). https://doi.org/10.1002/ ente.201800628.

68. Xu, X.; Gao, J.; Tian, Q.; Zhai, X.; Liu, Y. Walnut Shell Derived Porous Carbon for a Symmetric All-Solid-State Supercapacitor. *Applied Surface Science* **2017**, *411*, pp 170–176. https://doi.org/10.1016/j.apsusc.2017.03.124.

69. Mi, J.; Wang, X. R.; Fan, R. J.; Qu, W. H.; Li, W. C. Coconut-Shell-Based Porous Carbons with a Tunable Micro/Mesopore Ratio for High-Performance Supercapacitors. In *Energy and Fuels*; 2012; Vol. 26, pp 5321–5329. https:// doi.org/10.1021/ef3009234.

70. Hittinger, E.; Whitacre, J. F.; Apt, J. What Properties of Grid Energy Storage Are Most Valuable? *Journal of Power Sources* **2012**, *206*, pp 436–449. https:// doi.org/10.1016/j.jpowsour.2011.12.003.

71. Mahlia, T. M. I.; Saktisahdan, T. J.; Jannifar, A.; Hasan, M. H.; Matseelar, H. S. C. A Review of Available Methods and Development on Energy Storage; Technology Update. *Renewable and Sustainable Energy Reviews*. Elsevier Ltd 2014, pp 532–545. https://doi.org/10.1016/j.rser.2014.01.068.

72. Aneke, M.; Wang, M. Energy Storage Technologies and Real Life Applications—A State of the Art Review. *Applied Energy*. Elsevier Ltd October 1, 2016, pp 350–377. https://doi.org/10.1016/j.apenergy.2016.06.097.

73. Torricelli, F.; Alessandri, I.; Macchia, E.; Vassalini, I.; Maddaloni, M.; Torsi, L. Green Materials and Technologies for Sustainable Organic Transistors. *Advanced Materials Technologies*. John Wiley and Sons Inc February 1, 2022. https://doi.org/10.1002/admt.202100445.

74. Zhu, J.; Yan, C.; Zhang, X.; Yang, C.; Jiang, M.; Zhang, X. A Sustainable Platform of Lignin: From Bioresources to Materials and Their Applications in Rechargeable Batteries and Supercapacitors. *Progress in Energy and Combustion Science*. Elsevier Ltd January 1, 2020. https://doi.org/10.1016/j.pecs.2019.100788.

75. Gao, M.; Shih, C. C.; Pan, S. Y.; Chueh, C. C.; Chen, W. C. Advances and Challenges of Green Materials for Electronics and Energy Storage Applications: From Design to End-of-Life Recovery. *Journal of Materials Chemistry A*. Royal Society of Chemistry 2018, pp 20546–20563. https://doi.org/10.1039/C8TA07246A.

76. Kaygusuz, K. Climate Change and Biomass Energy for Sustainability. *Energy Sources, Part B: Economics, Planning and Policy* **2010**, *5* (2), pp 133–146. https://doi.org/10.1080/15567240701764537.

CHAPTER 16

Hydrogen Production from Biomass

Anindita Bhuyan and Md. Ahmaruzzaman

16.1 INTRODUCTION

With rapid economic development and an ever-increasing global population, energy demand and consumption are skyrocketing. According to reports, the United Arab Emirates, a significant oil exporter, will not be able to meet its part of the demand for natural gas and oil by 2042 [1]. Egypt's fossil fuel reserves might run out in one to two decades [2]. Mao claimed that in 2000, 31% of China's energy needs were met by imports and that by 2010, that percentage had risen to 45–55% [3]. Fossil fuels continue to govern the energy sector, accounting for almost 87% of worldwide energy requirements [4]. The urgent need to minimize our reliance on fossil fuels has prompted numerous research studies focusing on growing and improving existing and novel technologies that utilize biomass-based wastes as substrates. The world is constantly impacted by anthropogenic climate change, brought on by the production and consumption of energy. Rising energy consumption has accelerated the depletion of the finite fossil fuel reserves at an alarming rate. In addition, the combustion of fossil fuels emits significant amounts of greenhouse gases, such as CO_2, NO_x, and SO_2, that contribute to acid rain and global warming. Transport alone is responsible for over 20% of the anthropogenic CO_2 emissions [5]. In response to the challenges mentioned, an ongoing effort has been undertaken to investigate clean, renewable options for long-term development. Sustainable energy production is a global issue because of the adverse effects and scarcity of fossil fuels, necessitating the development of appropriate and alternative energy sources. A sustainable approach combining

DOI: 10.1201/9781003466833-19

the advantages of renewability, energy efficiency, and sustainable development is needed to mitigate the effects of climate change on the environment. Depending on the current environmental situation, a shift towards a carbon-neutral economy appears to be a critical move [6].

Hydrogen is identified as a promising source of energy owing to its high energy density and clean-combustion characteristics. Hydrogen finds widespread applications in various fields, including stationary power, electronics, heat generation, vehicle fuel cells, chemical industry, and fuel synthesis. Although hydrogen is not naturally present, it can be easily manufactured using various methods, including either fossil fuels or renewable sources as feedstocks [7, 8]. Because hydrogen synthesis and its use in fuel cells is not associated with any harmful chemicals or contaminants, hydrogen is considered an ecologically sustainable energy source. The benefit of hydrogen as a fuel is attributed to its high calorific value compared to other fuels [9, 10]. Regarding net calorific value, 1 kg of hydrogen corresponds to 2.75 kg of petrol and 6 kg of methanol. Owing to its superior energy efficiency, hydrogen may outperform other renewable energy sources, including geothermal, tidal, solar, and wind power. The characteristics of hydrogen and the fact that it can be produced from both renewable and non-renewable resources make it an appealing energy source. An estimated 55 million tons of hydrogen are produced annually, and its use rises by around 6% annually. There are numerous methods for producing hydrogen from various raw materials. The primary method for producing hydrogen includes steam-reforming natural gas, which makes significant greenhouse gas emissions [11]. Natural gas steam reforming accounts for over 50% of the H_2 needed globally, followed by around 30% from naphtha and oil refinery and chemical industry off-gases, about 18% from coal gasification, and 4% from water splitting and other sources [12]. Although electrolytic and plasma techniques show excellent efficiency in H_2 production, they are regrettably considered energy-intensive methods. Similarly, methanol steam reforming, despite being a cost-efficient technology, utilizes non-renewable feedstock and generates CO_2 as a by-product.

Hydrogen production from renewable resources, such as biomass, is an alternative to conventional hydrogen generation methods. Hydrogen from biomass is a cost-effective, energy-efficient, and ecologically sustainable method for H_2 production. Considering the environmental concerns, sustainability, and process efficiency, biomass is currently the most abundant renewable energy source that contributes to almost 12% of the total energy requirement. In contrast, in some countries, it constitutes about 40–50% of the energy supply [13]. Biomass-based hydrogen generation technology consists primarily of thermochemical, biological, and electrolysis approaches [14, 15]. While biological and electrolysis techniques [16] are complex to expand and have poor energy output, thermochemical processes are gaining widespread attention [17].

This chapter presents the recent advances in hydrogen production from biomass using various thermochemical and biological processes. The advances in using hydrogen as an energy source and the advantages of using biomass as a feedstock are discussed. Furthermore, the common challenges and opportunities in the potential production, purification, and storage of hydrogen from biomass have been highlighted.

16.2 HYDROGEN AS AN ENERGY SOURCE

Energy is essential in our modern society and industrial life. Over many decades, fossil fuels such as crude oil, natural gas, and coal have met the world's energy needs. However, as the world's population grows and technology advances, fossil fuels will be unable to meet people's needs as the dominant energy source in the future due to a lack of energy supplies and a slew of environmental issues. The widespread concerns about the over-exploitation of fossil fuels urge using unconventional energy resources. Nonconventional energy sources include geothermal, solar, wind, hydro, nuclear, and biomass-derived energy. However, the potential for all renewable energy sources to meet the energy needs of the twenty-first century and beyond is considerable. Moreover, the intermittent nature of renewable energy sources such as solar and wind raises the need for an alternative energy storage medium.

Recently, hydrogen energy has emerged as a critical research field for environmentally friendly and sustainable energy production worldwide and can be utilized for numerous purposes such as fuel, transportation, energy storage media, methanol and ammonia synthesis, and energy carriers (Figure 16.1). Hydrogen constitutes the most abundant element of the universe, is found in water and all organic compounds, and has been known to mankind for more than 200 years. Hydrogen as a fuel has gained particular attention due to its colorless and odorless characteristics and high combustibility with a low ignition temperature. Moreover, hydrogen has a high specific heating value of 141.8 MJ/kg at 298 K, which is higher than that of many fuels, such as gasoline. A kilogram of hydrogen

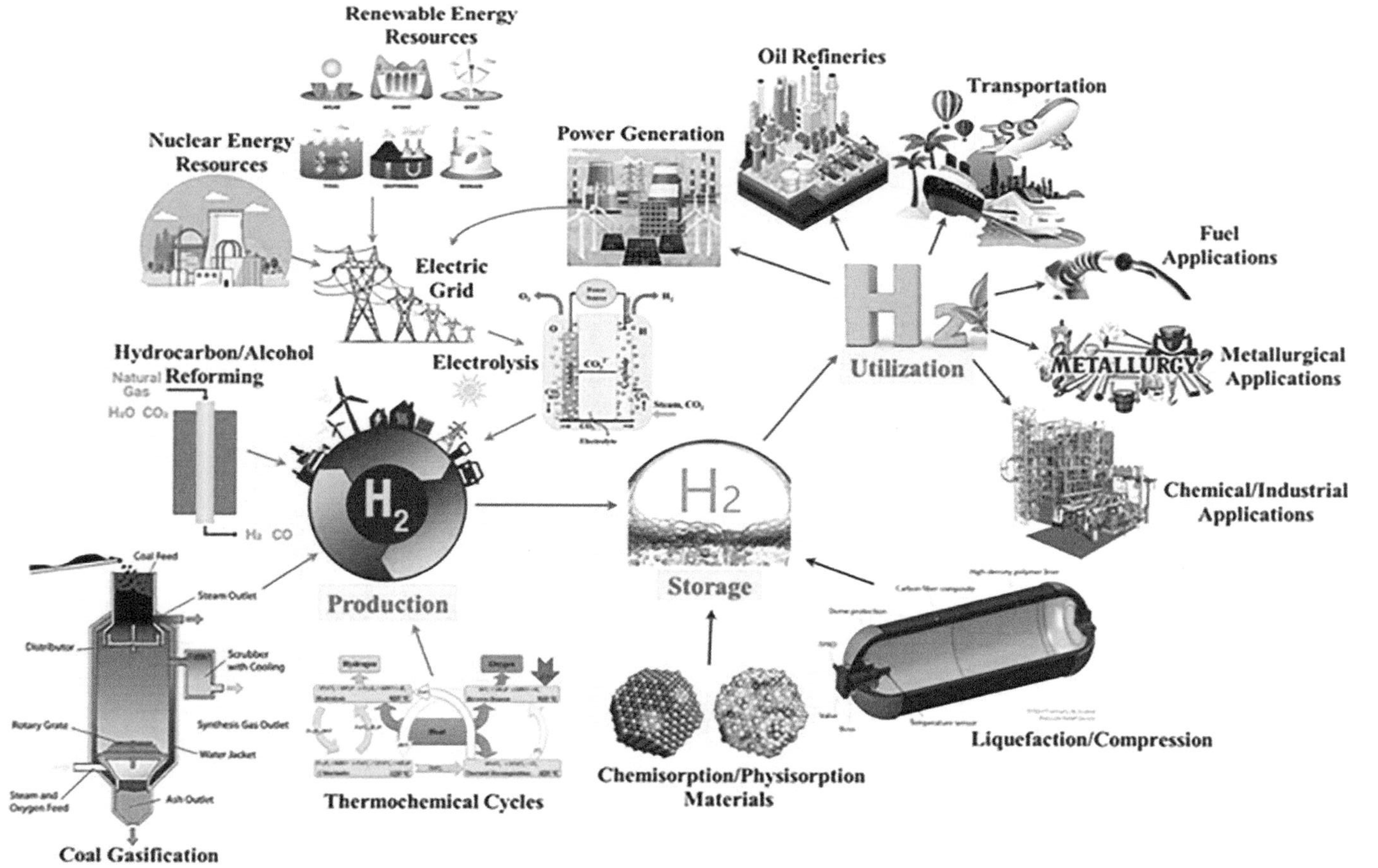

FIGURE 16.1 Diverse applications of hydrogen in an energy system. Reprinted from ref. [18]. Copyright 2022, with permission.

possesses 9.5 kg of energy, equal to 25 kg of petrol. Several nations are moving towards using hydrogen as fuel for fuel cell transport vehicles, including North America, Japan, Korea, India, China, and Europe. Furthermore, the "Hydrogen Council," which was initially composed of 13 prominent companies linked to energy and transportation, intends to augment investments in hydrogen-related items to 10 billion euros over the subsequent five years to promote hydrogen as a pivotal element of the forthcoming energy systems. The rapid technological developments based on hydrogen may mark the start of a hydrogen era.

The primary environmental and economic benefits of hydrogen as an alternate energy source are listed as follows:

i. Decreases dependence on fossil fuels since H_2 can be synthesized from various renewable resources like solar, wind, and biomass.

ii. Reduced greenhouse gas and improved air quality, since hydrogen, when burned, produces only heat and water vapor with no harmful emissions.

iii. Carbon-neutral energy production since hydrogen production from renewable resources produces minimal CO_2, mainly used by plants for photosynthesis, leaving behind lower carbon footprints than traditional fossil fuels.

iv. Increased energy efficiency leading to lower operating costs and reduced energy consumption.

v. New market generation for hydrogen-based products and technologies, generating new business prospects.

Despite the hype for hydrogen fuel emerging as an alternate form of renewable energy source, a significant challenge in employing hydrogen gas as a fuel is its limited availability in nature and the requirement for low-cost production technologies. Numerous techniques are known for producing hydrogen, including water electrolysis, pyrolysis, coal gasification, biomass gasification, partial oxidation of hydrocarbons, steam reforming of natural gas, autothermal reforming, thermolysis, and thermocatalysis. In all the processes, the production of H_2 based on the raw materials used can be divided into two major categories: conventional and renewable technologies. The traditional route utilizes fossil fuels as raw materials for H_2 production using various methods such as steam reforming, partial oxidation, and pyrolysis. Of all globally produced H_2, 98% is derived from non-renewable fossil fuels. The dependence on fossil fuels as the primary energy source has put immense pressure on their availability and environmental consequences. Consequently, the second and alternative category for hydrogen production that has drawn much attention recently is using renewable sources like biomass as an energy source.

16.3 BIOMASS FOR HYDROGEN PRODUCTION

Biomass is a renewable source of energy derived from plants and animals and can be broadly categorized into four classes:

i. Energy crops include agricultural, industrial, and aquatic crops and woody and herbaceous crops.

ii. Agricultural residue, crop waste, and animal waste.

iii. Industrial effluent, solid waste, municipal waste, and sewage sludge.

iv. Wood mill waste, forestry plant residues, and wastes.

Utilizing biomass feedstock for H_2 production is an excellent way to reduce wastes that would otherwise end up in landfills and cause environmental issues while producing high-value products. The annual production of lignocellulosic biomass obtained from agricultural food waste and residues, forestry, and aquatic by-products is estimated to be approximately 4.6 billion tons. In contrast, over 7 billion tons of biomass are generated from forests and grasslands, of which only 25% is used for energy production. Energy generation using biomass has gained popularity due to its abundance and low cost. Biomass can be converted into energy in various methods, including biogas generation, ethanol production, biodiesel production, and hydrogen production. Conversion of biomass to hydrogen increases net hydrogen production quantity, contributes to a sustainable economy, increases source flexibility, and reduces greenhouse gas. Literature suggests biomass is a carbon-neutral source due to its naturally maintained cycle. The two primary routes of hydrogen generation from biomass include thermochemical and biological routes. Although biological techniques are environmentally sustainable and require less energy since they function under moderate conditions, they generate low yields of hydrogen based on the raw materials employed. On the other hand, thermochemical processes are faster and provide a larger stoichiometric yield, with biomass gasification being a viable option based on economic and environmental factors. The thermochemical process comprises pyrolysis, combustion, liquefaction, and gasification, whereas biological water-gas shift reaction, direct bio-photolysis, indirect bio-photolysis, photo-fermentation, and dark-fermentation constitute the biological processes.

16.3.1 Thermochemical Processes

The thermochemical process is one of the most suitable and advanced methods of hydrogen production from biomass. Hydrogen production from syngas generated by such operations is a significant step towards reaching the environment of zero greenhouse gas emissions required for sustained development. The thermochemical process primarily includes gasification and pyrolysis. These methods generate gaseous mixtures such as CH_4 and CO, which are further processed to generate H_2 via steam reforming and the water-gas shift (WGS) reaction. Following pyrolysis and gasification, separating hydrogen gas from the produced gas mixture is crucial to ensure high-quality hydrogen production, recycling the unreacted gases and trapping greenhouse gases to avoid their release into the atmosphere (Figure 16.2).

16.3.1.1 Biomass Gasification

Gasification is a highly endothermic process at around 1,000°C in an oxygen-deficient medium. Gasification is based on the partial oxidation of the feedstock material to produce a mixture of hydrogen, methane, higher hydrocarbons, carbon monoxide, nitrogen, and carbon dioxide. Syngas with an H_2/CO ratio of 2 is obtained when steam and oxygen are added during gasification. The biomass gasification process can either be carried out in the presence or absence of a catalyst in a fluidized-bed reactor or fixed-bed reactor (FBR), with the former having superior performance [20]. The process varies depending on the oxidizing agent used and can be classified as air gasification, oxygen gasification, or steam gasification. The conversion process can be depicted as hydrocarbon.

$$\text{Biomass} + \text{stream} \xrightarrow{\text{Heat}} H_2 + CH_4 + CO + CO_2 + N_2 + \text{hydrocarbons} + \text{char} \quad (1)$$

Steam gasification of biomass has been widely accepted as the most suitable method for H_2 production. Steam gasification generally involves water-gas conversion, methane, and hydrocarbon reforming, and under appropriate

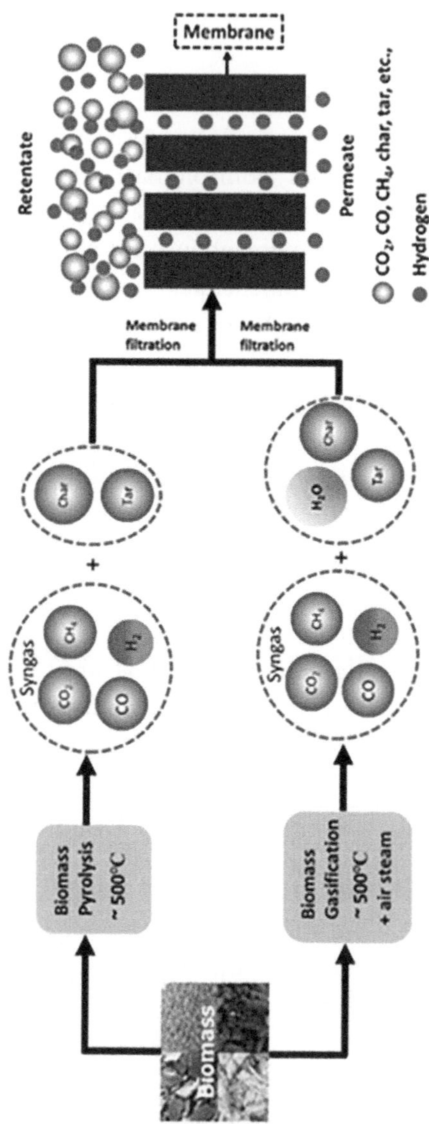

FIGURE 16.2 Hydrogen production from biomass feedstock by gasification and pyrolysis followed by membrane filtration [19].

conditions and the introduction of water vapor, the H_2 in the syngas can be further increased in the presence of catalysts and adsorbents [21];

$$C + H_2O \rightarrow H_2 + CO \ldots\ldots\ldots\ldots \Delta H_{298K} = 131 kJ/mol \tag{2}$$

$$CO + H_2O \rightarrow H_2 + CO_2 \ldots\ldots\ldots\ldots \Delta H_{298K} = -40.9 kJ/mol \tag{3}$$

$$CH_4 + H_2O \rightarrow 3H_2 + CO \ldots\ldots\ldots\ldots \Delta H_{298K} = 206.3 kJ/mol \tag{4}$$

$$C_x H_y + xH_2O \rightarrow xCO + (a + b/2)H_2 \tag{5}$$

The biomass feedstock and its properties are crucial factors that significantly impact hydrogen production. The chemical constituents of biomass (cellulose, hemicellulose, and lignin), elemental composition, mineral content, amount of volatile matter, moisture content, and other physical properties all impact the composition and yield of gas produced. According to Tian et al. [22], biomass with a more significant percentage of lignin produces more hydrogen, but cellulose and hemicellulose result in CO and CH_4 pollution.

The gasifying agent is another critical factor that significantly affects the calorific value, yield, and composition of H_2 in syngas. The usage of air, steam, or oxygen had a direct effect on the efficiency and yield of H_2 produced. The result findings of Wang et al. [23] show that when O_2 content in the air used as the gasifying agent was increased to 99.5% from 21%, the H_2 concentration in syngas improved as well. Moreover, using steam instead of air significantly increases H_2 production, which has a calorific value gas and no N_2 [24]. Singh et al. [25] examined the effects of varying temperature (700–900°C) and rate of steam flow (0.125–0.75 mL/min) on the yield, composition, and hydrogen concentration in the produced syngas. The study that utilized kitchen waste as fuel yielded a maximum H_2 yield of 1.23 m³/kg at a temperature of 800°C and a steam flow rate of 0.5 mL/min. Chen et al. [26] investigated the effect of different experimental conditions, such as temperature, equivalence ratio (ER), and residence time (RT), on the generation of H_2 and other volatile gases like CO, CH_4, and CO_2 via gasification of solid municipal wastes, using air as the gasifying agent. An optimal H_2 yield of 41.36 mol % (15.963 mol/kg/MSW) was obtained at 757.65°C, 0.241 ER, and 22.26 min R, and the reaction time strongly influenced the total yield as well as the composition of the produced gas, justifying Le Chatelier's principle. For instance, with the increase in temperature from 700–900°C, a significant decrease in the lower heating value (LHV) of the produced syngas (H_2/CO) was observed from 12.48 to 9.33 MJ/Nm³. This can be attributed to the decline in the production of CH_4 at higher temperatures, which generally contributes to the higher values of LHV of syngas.

Singh and Yadav [27] studied the effect of torrefaction, a thermochemical process used to remove water and any volatile contents from biomass, as a pretreatment method on the steam gasification of mixed food waste for syngas production. Torrefaction at 230°C resulted in a significant increase in the carbon content (12%), while the oxygen content decreased by 14%. Henceforth, with the increasing severity of the torrefaction from raw food waste, 230°C to 290°C, the syngas yield increases from 0.95 m³/kg, 2.39 m³/kg to 3.49 m³/kg. The increase in the syngas yield can be attributed to the increased char. However, the H_2 content increased from 61.96 to 65.62% at 260°C but dropped to 61.61%, further expanding the torrefaction temperature to 290°C. Additionally, the CO content decreases with increased torrefaction temperature.

The biggest challenge in biomass gasification is tar formation throughout the process. The undesired tar formation stimulates the generation of tar aerosols and conversion into complicated structures, both detrimental to steam reforming. Presently, three ways are used to reduce tar formation, viz.: suitable gasifier design, suitable operation and control system, and use of catalysts. In contrast to fluidized-bed or fixed-bed gasification methods, entrained flow gasification produces nearly tar-free syngas output by operating at higher temperatures and with smaller particles. According to Briesemeister et al., at 1,300°C, the tar production dropped to 0.2 g/Nm⁻³ or lower. Qin et al. discovered that raising the reaction temperature from 1,000 to 1,350°C in an entrained flow bed resulted in a 72% increase in producer gas production. Furthermore, the gasification process usually has a low thermal efficiency as the moisture in the biomass also needs to be evaporated.

16.3.1.2 Pyrolysis

Another potential technique for producing hydrogen is pyrolysis, also known as co-pyrolysis. The raw organic materials are heated to 500–900°C before gasification at 0.1–0.5 MPa pressure [28]. Since pyrolysis occurs without air or oxygen, the possibility of dioxin formation and the requirement for reactors for secondary reactions (water-gas shift reaction, preferential oxidation, etc.) can be virtually eliminated. Thus, there is a considerable reduction in emissions with this process. Fuel flexibility, a clean carbon by-product, relative simplicity and compactness, and a decrease in CO_2 emissions are the significant benefits of this method. The general reaction can be described as:

$$C_nH_m + heat \rightarrow nC + 0.5mH_2 \tag{6}$$

Based on the pyrolysis temperatures, the process can be divided into three categories, viz., low (< 500°C), medium (500–800°C), and high (> 800°C). The effect of pyrolysis temperature on the hydrogen yield was investigated by Demirbas and Arin [29] using different types of biomasses such as olive husk, tea factory waste, hazelnut shell, beech wood, cotton cocoon shell, and spruce wood. As the temperature increased from 377°C to 752°C, the hydrogen yield increased from 22–36 vol% to 41–55 vol% of the total gas volume. Fast pyrolysis is a relatively new method for converting high-energy organic matter value-added products such as H_2. Arregi et al. [30] studied the continuous fact pyrolysis of pine wood sawdust by a combination of biomass pyrolysis in a conical spouted bed reactor (CSBR), followed by in-line steam reforming of pyrolysis vapor in an FBR in the presence of commercially bought Ni catalyst. A maximum H_2 yield of 117 g/kg biomass was achieved at 600°C, a steam/biomass ratio of 4, and a space-time of 30 $g_{cat}min/g_{volatiles}$, accounting for 95.8% yield efficiency. Further, an increase in pyrolysis temperature to 700°C decreased H_2 production to 110 g/kg biomass. The space-time of the reaction processes directly influenced the H_2 production; hence, the highest conversion was obtained at the highest space-time studied. The possibility of obstruction by the generated carbon is one of the significant drawbacks of this strategy. However, experts claim that this may be reduced with proper design. In the current study, the char produced was continuously removed from the CSBR to maintain steady H_2 production.

Conventionally, few studies have reported the slow pyrolysis of woody biomass for syngas production with 42 vol% H_2. In a recent study, Jerzak et al. [31] reported that medium pyrolysis of wheat straw and pine bark biomass yielded a mixture of CO, CO_2, H_2, and C_xH_y, with an H_2 volume ranging from 9–12 vol%. In contrast, the H_2 concentration varied from 1.5–45 vol% in a fast pyrolysis process [32] and 9–44 vol% under a flash pyrolysis process [33].

Advanced pyrolysis processes such as catalytic hydropyrolysis and co-pyrolysis are extensively studied for efficient hydrogen yield in syngas. Catalytic pyrolysis generally uses a catalyst to speed up the chemical process, where it remains unaffected until the completion of the reaction. Generally speaking, the temperature range for catalytic pyrolysis is kept at 350–650°C. Yang et al. [34] evaluated the effect of Al_2O_3, Fe_2O_3, and SiO_2 in red mud on the catalytic H_2 generation using sewage sludge. Adding red mud enhances the catalytic cracking of C-C and C-H bonds during the breakdown of tar, boosting hydrogen yield by 268.5% at 700°C. Similarly, Liu et al. [35] studied the catalytic pyrolysis of water hyacinth using a Ni-based sepiolite catalyst. It was seen that the H_2 evolution increased from 41.45 to 77.2% in the presence of the catalyst when the reaction bed temperature was maintained at 800°C for 17 min residence time. The increase in the potential hydrogen yield from 238.84 g/kg biomass to 345.83 g/kg with an increase in the Ncatalyst's Ni content demonstrated the catalyst's role with the maximum conversion efficiency of 88.28% observed at 9% Ni content. In addition, Efika et al. [36] compared four Ni-based catalysts: NiO/SiO_2 (synthesized by two methods—incipient wetness method and sol-gel method), NiO/Al_2O_3, $NiO/CeO_2/Al_2O_3$, finding that the NiO/Al_2O_3 catalyst generated syngas having the highest H_2 concentration of 44.4 vol%, while $NiO/CeO_2/Al_2O_3$ had the lowest H_2 content of 43.1 vol%. Another Ni catalyst, $Ni/CaAlO_x$, with varying Ca/Al ratios of 1:1, 1:2, 1:3, 2:1, and 3:1, was investigated by Chen et al. [37]. It was found that the highest H_2 yield of 45 vol% was obtained at a Ca/Al ratio of 1:2; however, when the ratio was increased to 1:3, the coke formation increased to about 20 wt%. This was attributed to the lower alkaline metal (Ca) content in the catalyst corresponding to the lowest catalyst basicity. Upon increasing the Ca content and hence the basicity of the catalyst, a decrease in coke deposition on the catalyst surface was observed due to the promotion of steam-coke reaction during the steam reforming of the biomass. Wang et al. [38] studied the catalytic pyrolysis of plastic and paper waste for hydrogen production using Fe/CeO_2-CaO and Ru/ZSM-5-CaO catalysts. Under optimum conditions and 900°C, a maximum H_2 yield of 681.76, and 445.54 µmol/g was obtained for Fe/CeO_2-CaO and Ru/ZSM-5-CaO respectively, while the H_2 yield was only 429 µmol/g in the absence of a catalyst. Additionally, it was established that using CaO as a support material increases the adsorption capacity of the catalyst, thereby increasing the H_2 yield.

Overall, it was found that although transition metals are advantageous for catalytic pyrolysis, they possess drawbacks of rapid deactivation during the reforming process due to the agglomeration of metal clusters and coke formation. As a result, metals are supported on different base materials such as Al_2O_3 [39], SiO_2 [40], rice husk [41], and dolomite [41] to promote their catalytic activity. Notably, non-metallic catalysts such as activated carbon, composed of graphene and graphite, are found to foster reaction rates and are environmentally friendly materials. Suprianto et al. [42] studied the synergistic effect of activated carbon and curcumin for enhanced hydrogen production via the pyrolysis of coconut wood sawdust. The interaction of the curcumin molecules with the activated carbon surface increases the activity of the activated carbon. When combined in the ratio of 1:1, the hydrogen yield was increased by 25.6%.

Apart from gasification and pyrolysis, two less desirable thermochemical processes for producing hydrogen are combustion [43] and liquefaction [44, 45], which have low hydrogen production rates. The former produces gaseous pollutants as a by-product, necessitating challenging operating conditions of 5–20 MPa in an anaerobic environment.

16.3.2 Biological Processes

Biological hydrogen production by a photosynthetic process started receiving widespread attention in the 1970s after the oil crisis struck. However, most of the research is at a laboratory scale, and their practical application is still on the back foot. Hydrogen production by biological methods may be categorized into direct bio-photolysis, indirect photolysis, biological water-gas shift reaction, photo-fermentation, and dark fermentation (Figure 16.3). Hydrogen-producing enzymes, like nitrogenase and hydrogenase, primarily regulate biological techniques. The nitrogenase enzyme, composed of Fe and MoFe proteins, can trap electrons and reduce different substrates using magnesium adenosine triphosphate (MgATP). Hydrogen production via the nitrogenase-based system can be demonstrated as the following chemical reaction:

$$2H^+ + 2e^- + 4ATP \rightarrow H_2 + 4Pi + 4ADP \tag{7}$$

Where Pi is inorganic phosphate, and ADP refers to adenosine diphosphate.

Hydrogenase enzymes can further be divided into two categories: uptake hydrogenases (NiFe and NiFeSe hydrogenases) and reversible hydrogenases.

$$\text{Uptake hydrogenase: } H_2 \rightarrow 2e^- + 2H^+ \tag{8}$$

$$\text{Reversible hydrogenase: } H_2 \leftrightarrow 2e^- + 2H^+ \tag{9}$$

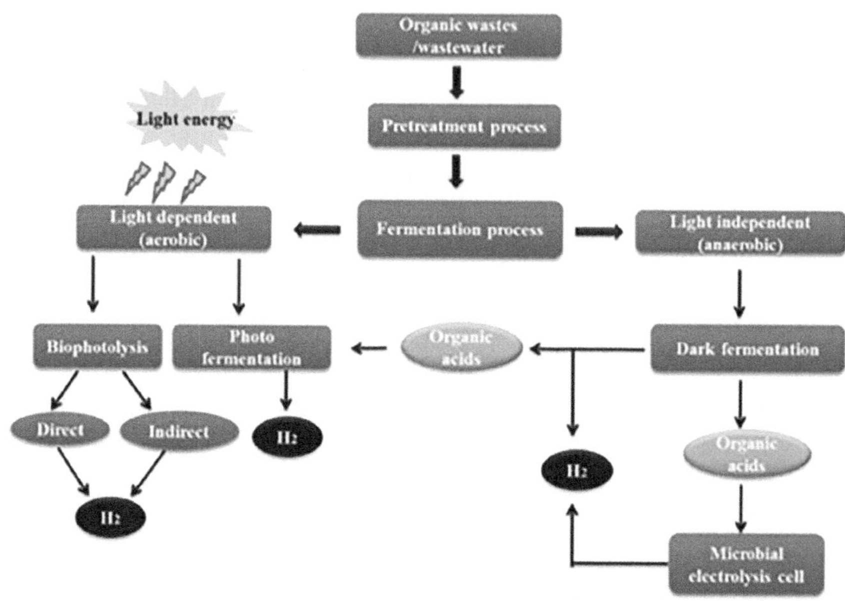

FIGURE 16.3   Different biological processes for hydrogen production using biomass. Reprinted from ref. [46]. Copyright 2022, with permission.

16.3.2.1 Direct Bio-Photolysis

The direct bio-photolysis process includes the conversion of water molecules into hydrogen using the catalytic activity of hydrogenase enzyme under aerobic conditions and solar energy [47]. The direct bio-photolysis reaction can be summarized as:

$$2H_2 \xrightarrow{\text{solarlight}} O_2 + 4H^+ + 4e^- \tag{10}$$

$$4H^+ + 4e^- \xrightarrow{\text{Hydrogenase}} 2H_2 \tag{11}$$

Cyanobacteria and green algae are the two primary organisms that assist the direct bio-photolysis process and hydrogen production [48]. Under dark conditions, green algae can either uptake hydrogen via CO_2 fixation or produce hydrogen upon light illumination under anaerobic conditions [49]. Since the hydrogenase enzyme is oxygen-sensitive, keeping the oxygen level below 0.1% is essential to sustain hydrogen production. Chlamydomonas reinhardtii, a type of unicellular green algae that can reduce oxygen during oxidative respiration, can be used effectively to create a low-oxygen environment [50]. Nevertheless, the efficiency of hydrogen production remains low because this process requires and consumes a large amount of substrate.

Recently, mutant strains derived from microalgae exhibited good O_2 tolerance capacity, consequently increasing H_2 production. A mutant, MTP4, was developed from a wild strain, Rhodobacter sphaeroides, by UV radiation for enhanced hydrogen production in a stable manner [51]. The photosynthetic bacterial mutant increases light penetration into the reactor, thereby strengthening the hydrogen production rate by 50% compared to the wild bacterial strain. The hydrogen production rate strongly depended on the reactor depth and cell concentration. Vasilyeva et al. [52] derived a stable mutant from the same bacterial strain of Rhodobacter sphaeroides with an altered light-harvesting system (P3 mutant) that exhibited H_2 evolution under light irradiation in the range of 800–850 nm. The enhanced H_2 evolution in the mutant was attributed to several possible reasons, including an impaired redox control system of the cell, altered properties of the nitrogenase, higher cell-reducing activity, and alteration in metabolic activities. Another mutant strain, ST410, derived from *Rhodobacter capsulatus* and lacking hydrogenase activity, was investigated by Ooshima et al. [53] from the viewpoint of hydrogen production. The hydrogen production rate strongly depended on the light intensity and yielded 66% H_2 compared to only 25% H_2 yield from wild strains.

In all cases, it was observed that the mutant variety enhanced the H_2 production by 1–2 times compared to the wild-type microorganisms. The pathway for hydrogen production using direct bio-photolysis in cyanobacteria and algae is depicted in Figure 16.4. The cost of direct bio-photolysis for hydrogen production was estimated by Benemann et al. [54] at about 20 USD/GJ, assuming a total solar conversion of 10% and a capital cost of about 60 USD/m². Recent studies indicate a production cost of 18.45 USD/Kg H_2. Notably, the potential cost is anticipated to significantly decrease shortly, with a price of 3.10 USD/Kg H_2, which could be an affordable alternative to photovoltaic-based electrolysis [55].

16.3.2.2 Indirect Bio-Photolysis

Hydrogen production via indirect bio-photolysis involves the accumulation of carbohydrates like glycogen and starch in the CO_2-fixation step to generate O_2 in the first step and H_2 in the second step under anaerobic conditions. Reports suggest that the concept of indirect bio-photolysis is based on four steps: i) photosynthesis biomass production, ii) concentration of

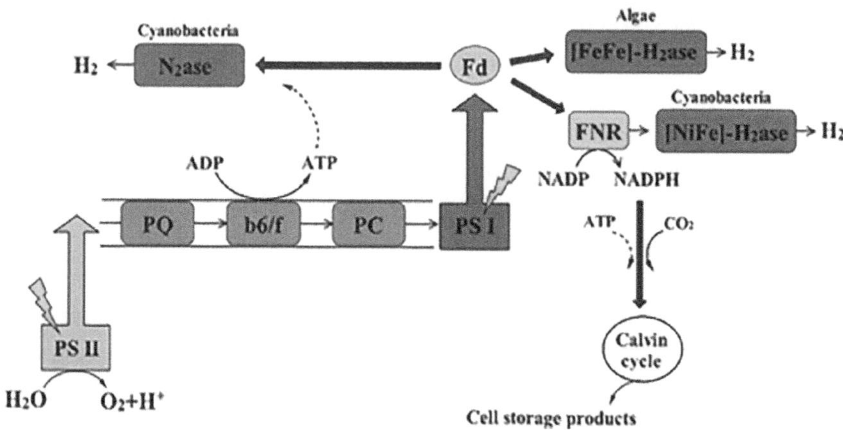

FIGURE 16.4 Pathway for direct bio-photolysis seen in cyanobacteria and green algae (PQ = plastoquinone, b_6 and f = cytochromes, and PC = plastocyanin and FNR = ferredoxin-NADP⁺ oxidoreductase). Reprinted from ref. [47]. Copyright 2020, with permission.

biomass, iii) aerobic dark fermentation to produce H_2 and acetates, and iv) hydrogen production from the acetates [48];

$$6CO_2 + 6H_2O \xrightarrow{\text{Light}} C_6H_{12}O_6 + 6O_2 \tag{12}$$

$$C_6H_{12}O_6 + 6H_2O \xrightarrow{\text{Light}} 12H_2 + 6CO_2 \tag{13}$$

Markov et al. studied the indirect bio-photolysis using Chanobacterium anabaena variabilis exposed to different light intensities of 45–55 µ/mol/m² and 170–180 µ/mol/m² for the two stages of the reaction. A hydrogen production rate of 12.5 ml H_2/g cdw h was obtained where cdw refers to cell dry weight. In another study, Troshina et al. established that keeping a pH of 6.8–8.3 produced an optimum hydrogen yield, and a temperature of about 30–40°C can enhance hydrogen generation by almost two times. A maximum hydrogen yield of 40.6 µmol H_2 mg/chl/h was recorded by Touloupakis et al. [56] utilizing Syechocystis sp. cells in an indirect light-driven photolysis process.

The primary issue with direct bio-photolysis, linked with hydrogenase's oxygen sensitivity, was solved during indirect bio-photolysis by isolating the O_2 photosynthesis step from the hydrogen generation step [57]. Moreover, during indirect bio-photolysis, the cyanobacteria and microalgae produce nitrogenase enzymes via nitrogen fixation, which catalyzes the H_2 production reaction using the stored starch and glycogen in the microorganisms. The nitrogenase enzymes in the heterocysts of the cyanobacteria are well protected from oxygen, resulting in an increase in hydrogen yield compared to microalgae with no heterocyst covering [58].

16.3.2.3 Biological Water-Gas Shift Reaction

The water-gas shift reaction involves the reaction of CO and water vapor to produce carbon dioxide and hydrogen. Some hetero phototrophic bacteria like Rhodospirillum rubrum can survive in the dark using CO as a carbon source and generate ATP by simultaneous reduction of H^+ to H_2 and oxidation of CO;

$$CO + H_2O \leftrightarrow CO_2 + H_2, \Delta G^0 = -20.1 kJ / mol \tag{14}$$

Under equilibrium conditions, CO_2 and H_2 form the dominating products. Microorganisms growing at the expense of the biological water-gas shift reaction process include gram-positive bacteria (Carboxydothermus hydrogenoformans) and gram-negative bacteria (R. rubrum and Rubrivax gelatinosus). The synthesis of CO dehydrogenase, CO-tolerant hydrogenase, and Fe-S protein under anaerobic conditions is stimulated by the CO gas, and the electrons generated from the oxidation of CO are transferred through the Fe-S protein into the hydrogenase enzyme for the production of hydrogen [59].

Hydrogen synthesis using the biological water-gas shift reaction is primarily done in laboratories, with only a few published studies. R. rubrum bacterium possesses the capacity to convert syngas into biohydrogen directly. Kerby et al. [60] observed that in the presence of Ni^{2+}, the R. rubrum strain doubled in less than 5 hours under anaerobic and dark conditions, producing H_2 by reducing protons coupled with CO oxidation to CO_2. However, R. rubrum requires sufficient light for optimal growth, and generation of H_2 is limited at CO partial pressures greater than 0.2 atm. Thereby, Jung et al. [61] explored an alternative bacterial strain, Citrobacter sp. Y19 for H_2 generation via the WGS process. They obtained a maximum H_2 yield of 27 mmol/g cell h, which was three-fold the amount received for R.rubrum. A considerably high yield of 25 mmol/g cell h observed at a high CO partial pressure of 1 atm indicates that the Y19 is sensitive to high CO partial pressure and holds significant prospects for H_2 production via the biological water-gas shift process. Moreover, the stability for

H_2 production could also be maintained for over 300 hours. Focusing on the studies on the growth of R. rubrum, mild light exposure is shown to exhibit a positive effect on the growth rate and the hydrogen production rate. Najafpour et al. [62] demonstrated the impact of different carbon sources, such as formate, malate, and acetate. It was observed that a high cell population was obtained when malate and fructose were used as the carbon source. However, a higher H_2 yield of 0.86 mmol H_2/mmol CO was found for acetate at adequate cell density, making it the best choice for H_2 production.

A detailed comparative study between the biological water-gas shift reaction and the conventional water-gas shift reaction was performed by Wolfrum et al. [63]. According to their research, the water-gas shift process was economically viable when the methane concentration was maintained under 3%. Younesi et al. [64] demonstrated the growth of R. rubrum under continuous gas flow using acetate in a stirred tank bioreactor. This study depicted a hydrogen yield of 16 mmol per gram substrate/cell/h and a conversion rate of 87 ± 2.4%, indicating the practical application of the biological water-gas shift reaction on a small scale.

Recently, Eckert et al. [65] studied the biological H_2 production wherein the photosynthetic PNSB, Rubrivivax gelatinosus CBS, acts as a support platform. The CO-mediated H_2 production was assisted by an energy-converting hydrogenase (Ech), whereas the H_2 oxidation was catalyzed by a membrane-bound hydrogenase (Mbh). The Ech hydrogenase supports H_2 generation over H_2 oxidation, indicating a potential application in an H_2 economy where CO is generated from gasified renewable biomass or other waste CO resources, such as the steel industry or municipal solid waste. Zhao et al. [66] drew a comparative water-gas shift activity of C. hydrogenoformans and R. gelatinosus at a CO concentration of 90–150 μM. Considering the stability, many research groups found a 92–95% H_2 yield using a gas-lift reactor with a continuous gas supply for three to four months. Despite a high hydrogen production rate, low cell density, low CO transfer rate, and membrane fouling remain significant limitations for long-term use. Moreover, optimizing the CO concentrations in the biomass feedstock to obtain maximum bacterial activity is an essential criterion.

16.3.2.4 Photo-Fermentation

Photosynthetic bacteria can generate hydrogen using solar energy and biomass through nitrogenase action. This process can be termed photo-fermentation. Recently, several initiatives have been taken to create hydrogen from industrial, municipal, and agricultural waste to improve waste management. The photo-fermentation processes are mediated by gram-negative purple non-sulfur (PNS) bacteria, such as Rhodobacter sphaeroides, R. capsulatus, R. sulfidophilus, and Rhodospirillum rubrum [67]. Under photoheterotrophic conditions, organic substances serve as an energy source for the growth of PNS for hydrogen production using light energy. The H_2 generation may occur through a photo-autotrophic, photo-heterotrophic, or chemo-heterotrophic route, depending on environmental variables such as carbon supplies, the intensity of light, and the degree of anaerobic activity. Photo-fermentation occurs via the catalytic action of two enzymes via the Krebs cycle, viz., nitrogenase and hydrogenase. The nitrogenase activity directly supports the effectiveness of the Krebs cycle in H_2 production. The nitrogenase enzyme catalyzes the nitrogen-fixation reaction in the presence of N_2, producing H_2, as shown in equation 15. However, the NH3 released as a by-product has an inhibitory effect on the nitrogenase activity, resulting in the generation of only one molecule of H_2 [67].

$$N_2 + 8H^+ + 8e^- + 16ATP \rightarrow H_2 + NH_3 + 16ADP + 16Pi \qquad (15)$$

The nitrogenase shifts its catalytic activity from nitrogen-fixing to Equation 16 without N_2. Four H_2 molecules are generated in this process.

$$8H^+ + 8e^- + 16ATP \rightarrow 4H_2 + 16ADP + 16Pi \qquad (16)$$

The hydrogen production pathway in PNSB is depicted in Figure 16.5.

Kars and Ceylan [68] utilized waste barley, containing carbohydrates in starch ($\approx 38\%$) along with 2.45% protein and 12% fat for hydrogen production using Rhodobacter sphaeroides O.U.001. The biomass was hydrolyzed into simple substrates using dilute H_2SO_4 to increase fermentability. The hydrolysate of the barley feedstock showed 48 grams equivalent/L glucose along with seven organic acids: lactate, acetate, formate, malate, citrate, fumarate, and propionate. Several multiple organic acids in the hydrolysate make it an excellent feedstock for photo-fermentative H_2 generation. The H_2 production rate was tested in four growth mediums containing 5–11 g/L sugar content. It was observed that an increase in sugar content increased cell density. Accordingly, hydrogen accumulation increased, and a 0.4 L H_2/L culture production rate was observed with

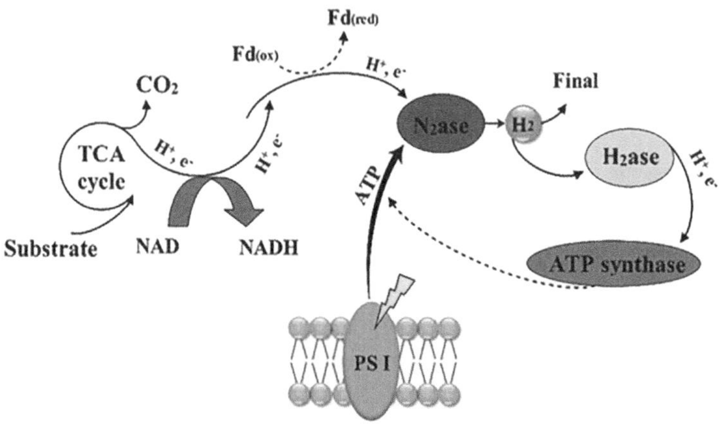

FIGURE 16.5 Pathway for hydrogen production in PNSB. Reprinted from ref. [47]. Copyright 2020, with permission.

the medium containing 11 g/L sugar content. Hydrogen generation from acid-hydrolyzed ground wheat starch by photo-fermentation was studied by Kapdan et al. [69] again using strains of Rhodobacter sphaeroides, namely RV, NRLL, and DSZM. The highest hydrogen yield of 1.23 mol H_2/ mol glucose (178 mL H_2) with a high specific hydrogen production rate of 46 mL H_2/g biomass/h was observed with the RV strain. The cumulative volume of hydrogen increases from 30 to 232 mL as the sugar content rises from 2.2 to 8.5 g/L. Low sugar contents result in substrate limitation, while a high sugar level of over 8.5 g/L results in reaction inhibition due to high concentrations of volatile fatty acid (VFA). Numerous other photofermentative production of H_2 using diverse biomass substrates such as sugarcane bagasse [70], vegetable waste [71], brewery wastewater [72], sugarbeet molasses [73], dairy wastewater [74], tofu wastewater [75], and olive mill wastewater [76] using Rhodobacetr sphaeroides has been reported in the literature.

Bianchi et al. [77] extracted a new strain of PNSB, Rhodopseudomonas palustris AV33, from the tropical lake of Averno located near Naples Italy, to test its activity towards the H_2 production from VFA obtained from stratified vegetable waste. The PNSB exhibited good H_2 production, although the efficiency rate was found to be less than that of pure VFA. Furthermore, Rp. palustris AV33's capacity to produce H_2 in the presence of lactate indicates its potential to be used in integrated H_2-producing systems that combine dark and photofermentation. Adessi et al. [78] utilized a mutant strain of the same bacteria, Rhodopseudomonas palustris CGA676, with a higher tolerating capacity for ammonia The mutant PNSB strain could efficiently generate hydrogen from a vegetable waste-derived (VWD) medium containing high concentrations of ammonia. However, diluting the substrate medium revealed a noticeable boost in the rate of H_2 production, which was maximized when using three-fold diluted VWD media. In conclusion, it was found that several factors such as biomass substrate, nitrogen source, presence of trace metals and minerals, and light source have a prominent effect on the rate of H_2 production, and the synergistic effect of all these factors cumulatively affects the yield of H_2 yield in the photo-fermentation process.

16.3.2.5 Dark Fermentation

Dark fermentation is believed to be an attractive approach for hydrogen production via biomass conversion, with a net energy ratio of 1.9. Dark fermentation produces biohydrogen primarily by strictly or facultative anaerobic bacteria and under anaerobic conditions. Hydrogen production through the dark-fermentation process is mediated through an acetate-assisted pathway, which can be depicted as;

$$C_6H_{12}O_6 + 2H_2O \rightarrow 4H_2 + 2CO_2 + 2CH_3COOH; .. \Delta G^0 = -20.1 \text{kJ/mol/mol} \quad (17)$$

As indicated in equation (17), although a maximum of four moles of H_2 is produced per mole of glucose, the experimental H_2 yield is always less than the theoretical value owing to the production of VFA and because microorganisms use a portion of glucose for growth and maintenance. Many anaerobic microbes rely on hydrogen as a primary source for metabolism. However, in the absence of any external electron acceptors, the excess electron generated during the metabolic processes reduces protons to yield hydrogen. The primary enzymes that catalyze the reversible reaction (Equation 18) include [FeFe] hydrogenase and [NiFe] hydrogenase.

$$2H^+ + 2e^- \leftrightarrow H_2 \quad (18)$$

The schematic illustration of the mechanistic pathway for converting biomass substrate into H_2 via the dark fermentation process is shown in Figure 16.6.

The Enterobacter sp., Bacillus sp., and Clostridium sp. are anaerobic fermentative bacteria that can create hydrogen via dark fermentation

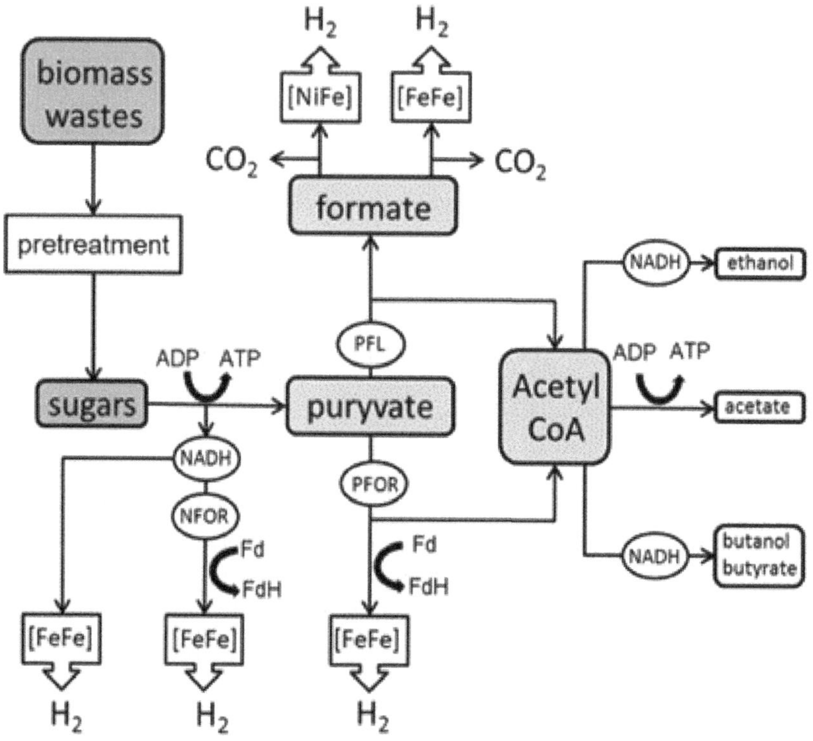

FIGURE 16.6 Schematic illustration of the metabolic pathways for H_2 production from biomass waste during dark fermentation. Reprinted from ref. [79]. Copyright 2018, with permission.

[80]. The most often utilized bacteria in the dark fermentation process are Clostridium sp., which includes C. buytricum, C. thermolacticum, C. bifermentants, and C. pasteurianum. Clostridium sp. belongs to a gram-positive bacterial strain with a high rate of hydrogen generation capacity; yet, due to its flexible metabolic activity, it produces a variety of by-products in addition to hydrogen production. The amount of VFAs and the yield of generated hydrogen vary according to the bacteria species.

Another factor that directly influences the rate of production of H_2 is the choice of raw materials and substrate. Alibardi et al. [81] demonstrated the influence of the composition of food and slaughterhouse waste on H_2 production. Different samples comprised different amounts of moisture, lipid, protein, cellulose, and starch. For samples rich in protein (48%), the lowest H_2 yield was obtained, equaling 25 cm^3 H_2/g volatile solid (VS), whereas, for samples rich in carbohydrates (67%), the H_2 production yield was three times higher with 85 cm³ H_2/g VS. Similar observations were reported by Karlsson et al. [82] who used a mixture of wastes (85% slaughterhouse waste, 5% manure, and 4% food waste) containing high protein content where only 16.5 cm³ H_2/g biomass was obtained, which constituted 2.5% of the total biogas produced during the fermentation process. On the contrary, Jayalakshmi et al. obtained an H_2 yield of 72 cm³/g VS using kitchen waste containing mostly carbohydrates and starch. Several other research studies have been conducted on the dark fermentation of food waste H_2 production [83–86].

Apart from food and kitchen waste, agricultural residue and waste such as rice straw [87], cornstalks [83], wheat straw [88], barley hull [89], corn leaves [90], sugarcane bagasse [91], and soybean straw [92] are rich in lignocellulose and starch and are also widely used in dark fermentation for hydrogen production. Although the cellulose and hemicellulose, comprising 70–80% of lignocellulose biomass, are desirable for dark fermentation, the lignin component hinders the fermentation process due to its resistance to biodegradation.

It has been observed that using a mixed bacterial culture containing Bacillus and Brevumdimonas sp. boosts the dark-fermentation H_2 production rate two-fold compared to pure cultures due to differences in their metabolic pathways. The maximum yield of 1.04 $molH_2/mol_{glucose}$ was observed in a single bioreactor using a mixed bacterial culture and starch as substrate [93]. Despite the advantage of a high production rate, the primary drawback of a mixed culture dark fermentation process is the presence of some unfavorable microorganisms such as methanogens, sulfur-reducing bacteria, and homoacetogenic bacteria. However, mixed cultures are preferable for H_2 production compared to pure cultures because of their applicability to a wide range of substrates and their economic benefits.

16.4 CHALLENGES IN HYDROGEN PRODUCTION AND STORAGE

The various hydrogen production technologies require varying reaction conditions and substrates and have different conversion efficiencies. Every technique discussed previously has its advantages and disadvantages. Moreover, hydrogen storage and its use as an energy source come with their challenges. Like all other fuels, hydrogen fuel poses some danger and requires engineered, controlled measures to ensure its safe use.

Hydrogen production by biomass gasification uses renewable feedstocks from agriculture and forestry but is associated with producing various toxic gaseous and liquid phase by-products. The major drawback of the gasification process is ash fouling, which can generate economic concerns due to a significant decrease in production efficiency. The financial disadvantage in the design and operation of this process thereby delays its commercialization on a larger scale. Furthermore, the investment in a biomass gasification reactor is fairly substantial. Another significant obstacle to gasification using air is eliminating N_2, an incombustible gas that lowers the calorific value of the gas produced. Alternatively, using steam or oxygen instead of air creates a purer gas with high concentrations of H_2, no N_2, and less char. However, the cost of oxygen makes it unfitting to use oxygen gasification at an industrial level [94]. On the contrary, the slow pyrolysis technique requires simple equipment and standard technologies. However, syngas obtained from the slow pyrolysis of biomass contains high CO concentrations, which require further treatment to extract the H_2 [95]. Similarly, although fast pyrolysis is a rapid process with economical scale-up possibility, it is more suited for bio-oil production than hydrogen production due to its operational factors. Moreover, biomass with low moisture content and fine size is preferred for fast pyrolysis. Catalytic biomass pyrolysis is generally preferred over conventional pyrolysis due to high selectivity, product efficiency, and low activation energy; however, this process experiences the major disadvantage of loss in catalytic activity due to coke formation. The catalytic activity may decrease with prolonged use and increase the production cost of the overall process, bearing in mind the catalyst's lifetime and regeneration. Also, an ineffective interaction between the catalyst and the biomass due to their carrying activation temperatures may lead to a poor deoxygenation process [96].

Biological hydrogen production via photo- and dark fermentation holds enormous promise, mainly when cellulosic and sugar-rich food waste is used as a feedstock. However, the main drawback of this technique is the low production rate and quantity. To become commercially competitive, biological hydrogen systems must be capable of producing H_2 at adequate rates for practical applications. To increase H_2 synthesis and yields, research involves optimizing reactor designs to quickly separate hydrogen from CO_2, maintaining low H_2 partial pressures, and genetically modifying microorganisms that synthesize H_2. The biomass feedstock must also be enzymatically fermentable due to the intricacy of lignocellulosic material and agricultural waste. This extra costly and time-consuming procedure hampers development at an industrial scale. The choice of substrates is critical because the feedstock's unpredictability significantly impacts both the process economy and the H_2 production.

Whether thermochemical or biological, every renewable hydrogen generation structure shares a fundamental difficulty in producing gas mixes containing H_2 and CO_2/CO. Anaerobic bacteria ferment carbohydrates, starch, or cellulose, producing equal amounts of H_2 and CO_2 under low pressure and density. The gas stream also contains considerable volumes of water vapor. Driven by the need to extract pure hydrogen from a mixture of gaseous products, developments in gas extraction technology, such as enhanced membrane separation techniques, can potentially increase recovery and efficiency while reducing the cost of producing hydrogen [97]. Furthermore, renewable hydrogen generation systems require integrated purification and storage systems. Although hydrogen has a good energy weight density, it has poor energy volume density and requires ample storage space. As a result, developing an effective and economical storage system remains a significant problem for H_2 storage. The high expenses of using hydrogen as an energy source are partly attributed to the storage and transportation of hydrogen. An interdisciplinary strategy that incorporates infrastructural investment, R&D, technology developments, and policy assistance is needed to tackle these challenges [98].

16.5 CONCLUSION

Hydrogen has been identified as a potential energy carrier for the coming generations. Numerous research on different hydrogen production technologies has been undertaken during the last few years in a quest for an alternate energy source. Biomass research has recently received immense attention due to its potential waste-to-energy application. Biomass feedstock holds promise as a reliable energy source for producing hydrogen owing to its renewability, abundance, and simplicity of use. Over the life cycle of biomass utilization for H_2 production, the net CO_2 emission is nearly zero by the photosynthesis of green plants.

Hydrogen generation from biomass may be carried out using various methods, such as gasification, pyrolysis, water gas shift reaction, biophotolysis, and biological fermentation. Thermochemical processes, such as pyrolysis and gasification, have already proven to be an economically viable solution for H_2 production from biomass. Gasification of biomass, when combined with an oxidizing agent, occurs at a temperature higher than pyrolysis and mostly yields gaseous compounds, making it a more efficient method of producing syngas. Biomass gasification exhibits excellent promise for field-scale applications. Nonetheless, the selectivity and efficiency of hydrogen synthesis must be enhanced to provide economical industrial applications with excellent atom economy. Compared to standard gasification, the concentrated solar thermochemical gasification process was anticipated to boost biomass feedstock conversion by 30% and raise overall productivity by 40% [99]. Remarkably, pyrolysis is much developed; specific pyrolysis techniques are already used commercially. Their contribution to the world's energy production is anticipated to help them rapidly realize their full potential. However, there are still a lot of technical difficulties in producing syngas with a high H_2 content by pyrolysis. Particular barriers to pyrolysis for the production of H_2 included volatility, gas separation and purification, production costs, catalyst selection, and the availability of renewable biomass [100]. Henceforth, it is evident that innovative membrane technologies are necessary to remove and purify H_2 from the generated gas stream. Recently, biological synthesis methods utilizing microorganisms have also emerged as a promising area of technological research for producing sustainable H_2 from biomass. Biological processes are typically classified as either light-dependent or light-independent. Light-dependent techniques involve using microalgae or cyanobacteria for photo-fermentation and bio-photolysis, whereas anaerobic purple non-sulfur bacteria are used in a light-independent process for dark fermentation. Considering the availability of energy worldwide, the synthesis of hydrogen using microorganisms shows considerable potential. Furthermore, compared to other manufacturing processes, it is less expensive and can provide higher yields at lower temperatures, often between 20 and 45°C.

The selection of biomass feedstock is also a crucial step in biohydrogen production. Reducing sugars like glucose, xylose, and lactose are easily degradable and are among the suitable substrates for hydrogen production, whereas protein-rich biomass results in low H_2 content. Lignocellulosic biomass, such as crops and food waste, is among the ideal substrates for hydrogen production and has immense potential. Renewable resources such as raw materials like cellulose, bagasse, molasses, and other farm wastes are becoming more popular due to the associated costs. Furthermore, biomass waste is a renewable and environmentally friendly energy source that has the potential to replace fossil fuels in H_2 generation processes.

Despite much research, hydrogen production from biomass still has some challenges related to low yield, gas separation, hydrogen purification, and storage. It is worth noting that the outcomes of this investigation can be efficiently applied to developing an approach for producing renewable, sustainable hydrogen from biomass feedstock. With the advancement of these technologies discussed in the chapter, biomass will play a significant part in developing a sustainable hydrogen economy shortly.

REFERENCES

1. A. Kazim, T.N. Veziroglu, Utilization of solar–hydrogen energy in the UAE to maintain its share in the world energy market for the 21st century, *Renew Energy* 24 (2001) 259–274. https://doi.org/10.1016/S0960-1481(00)00199-3.
2. M.A.H. Abdallah, S.S. Asfour, T.N. Veziroglu, Solar–hydrogen energy system for Egypt, *Int J Hydrogen Energy* 24 (1999) 505–517. https://doi.org/10.1016/S0360-3199(98)00108-6.
3. K. Yuan, W. Lin, Hydrogen in China: Policy, program and progress, *Int J Hydrogen Energy* 35 (2010) 3110–3113. https://doi.org/10.1016/J.IJHYDENE.2009.08.042.
4. L. Cao, I.K.M. Yu, X. Xiong, D.C.W. Tsang, S. Zhang, J.H. Clark, C. Hu, Y.H. Ng, J. Shang, Y.S. Ok, Biorenewable hydrogen production through biomass gasification: A review and future prospects, *Environ Res* 186 (2020) 109547. https://doi.org/10.1016/J.ENVRES.2020.109547.
5. T. Lepage, M. Kammoun, Q. Schmetz, A. Richel, Biomass-to-hydrogen: A review of main routes production, processes evaluation and techno-economical assessment, *Biomass Bioenerg* 144 (2021) 105920. https://doi.org/10.1016/J.BIOMBIOE.2020.105920.
6. N. Stern, Towards a carbon neutral economy: How government should respond to market failures and market absence, *J Gov Econ* 6 (2022) 100036. https://doi.org/10.1016/J.JGE.2022.100036.
7. F.M. Alptekin, M.S. Celiktas, Review on catalytic biomass gasification for hydrogen production as a sustainable energy form and social, technological, economic, environmental, and political analysis of catalysts, *ACS Omega* 7 (2022) 24918–24941. https://doi.org/10.1021/ACSOMEGA.2C01538/ASSET/IMAGES/LARGE/AO2C01538_0002.JPEG.
8. P. Nikolaidis, A. Poullikkas, A comparative overview of hydrogen production processes, *Renew Sustain Energy Rev* 67 (2017) 597–611. https://doi.org/10.1016/J.RSER.2016.09.044.
9. M.S. Graboski, R.L. McCormick, Combustion of fat and vegetable oil derived fuels in diesel engines, *Prog Energy Combust Sci* 24 (1998) 125–164. https://doi.org/10.1016/S0360-1285(97)00034-8.
10. J.E. Zajic, N. Kosaric, J.D. Brosseau, Microbial production of hydrogen, *Adv Biochem Eng* 9 (1978) 57–109. https://doi.org/10.1007/BFB0048091/COVER.
11. C.M. Kalamaras, A.M. Efstathiou, Hydrogen production technologies: Current state and future developments, *Conf Papers Energy* 2013 (2013) 1–9. https://doi.org/10.1155/2013/690627.

12. J.D. Holladay, J. Hu, D.L. King, Y. Wang, An overview of hydrogen production technologies, *Catal Today* 139 (2009) 244–260. https://doi.org/10.1016/J.CATTOD.2008.08.039.

13. N.Z. Muradov, T.N. Veziroğlu, From hydrocarbon to hydrogen–carbon to hydrogen economy, *Int J Hydrogen Energy* 30 (2005) 225–237. https://doi.org/10.1016/J.IJHYDENE.2004.03.033.

14. C.Y. Lin, T.M.L. Nguyen, C.Y. Chu, H.J. Leu, C.H. Lay, Fermentative biohydrogen production and its byproducts: A mini review of current technology developments, *Renew Sustain Energy Rev* 82 (2018) 4215–4220. https://doi.org/10.1016/J.RSER.2017.11.001.

15. H.K. Ju, S. Badwal, S. Giddey, A comprehensive review of carbon and hydrocarbon assisted water electrolysis for hydrogen production, *Appl Energy* 231 (2018) 502–533. https://doi.org/10.1016/J.APENERGY.2018.09.125.

16. R.X. Yang, K.H. Chuang, M.Y. Wey, Effects of temperature and equivalence ratio on carbon nanotubes and hydrogen production from waste plastic gasification in fluidized bed, *Energy Fuels* 32 (2018) 5462–5470. https://doi.org/10.1021/ACS.ENERGYFUELS.7B04109/ASSET/IMAGES/LARGE/EF-2017-04109S_0009.JPEG.

17. G. Ji, J.G. Yao, P.T. Clough, J.C.D. Da Costa, E.J. Anthony, P.S. Fennell, W. Wang, M. Zhao, Enhanced hydrogen production from thermochemical processes, *Energy Environ Sci* 11 (2018) 2647–2672. https://doi.org/10.1039/C8EE01393D.

18. V. Madadi Avargani, S. Zendehboudi, N.M. Cata Saady, M.B. Dusseault, A comprehensive review on hydrogen production and utilization in North America: Prospects and challenges, *Energy Convers Manag* 269 (2022) 115927. https://doi.org/10.1016/J.ENCONMAN.2022.115927.

19. A.T. Besha, M.T. Tsehaye, G.A. Tiruye, A.Y. Gebreyohannes, A. Awoke, R.A. Tufa, Deployable membrane-based energy technologies: The Ethiopian prospect, *Sustainability* 12 (2020) 8792.

20. F.M. Alptekin, M.S. Celiktas, Review on catalytic biomass gasification for hydrogen production as a sustainable energy form and social, technological, economic, environmental, and political analysis of catalysts, *ACS Omega* 7 (2022) 24918–24941. https://doi.org/10.1021/ACSOMEGA.2C01538/ASSET/IMAGES/LARGE/AO2C01538_0002.JPEG.

21. L. Cao, I.K.M. Yu, X. Xiong, D.C.W. Tsang, S. Zhang, J.H. Clark, C. Hu, Y.H. Ng, J. Shang, Y.S. Ok, Biorenewable hydrogen production through biomass gasification: A review and future prospects, *Environ Res* 186 (2020) 109547. https://doi.org/10.1016/J.ENVRES.2020.109547.

22. T. Tian, Q. Li, R. He, Z. Tan, Y. Zhang, Effects of biochemical composition on hydrogen production by biomass gasification, *Int J Hydrogen Energy* 42 (2017) 19723–19732. https://doi.org/10.1016/J.IJHYDENE.2017.06.174.

23. Z. Wang, T. He, J. Qin, J. Wu, J. Li, Z. Zi, G. Liu, J. Wu, L. Sun, Gasification of biomass with oxygen-enriched air in a pilot scale two-stage gasifier, *Fuel* 150 (2015) 386–393. https://doi.org/10.1016/J.FUEL.2015.02.056.

24. Y. Wu, W. Yang, W. Blasiak, Energy and exergy analysis of high temperature agent gasification of biomass, *Energies* 7 (2014) 2107–2122. https://doi.org/10.3390/EN7042107.

25. D. Singh, S. Yadav, N. Bharadwaj, R. Verma, Low temperature steam gasification to produce hydrogen rich gas from kitchen food waste: Influence of steam flow rate and temperature, *Int J Hydrogen Energy* 45 (2020) 20843–20850. https://doi.org/10.1016/J.IJHYDENE.2020.05.168.

26. G. Chen, I.A. Jamro, S.R. Samo, T. Wenga, H.A. Baloch, B. Yan, W. Ma, Hydrogen-rich syngas production from municipal solid waste gasification through the application of central composite design: An optimization study, *Int J Hydrogen Energy* 45 (2020) 33260–33273. https://doi.org/10.1016/J.IJHYDENE.2020.09.118.

27. D. Singh, S. Yadav, Steam gasification with torrefaction as pretreatment to enhance syngas production from mixed food waste, *J Environ Chem Eng* 9 (2021) 104722. https://doi.org/10.1016/J.JECE.2020.104722.

28. M. Ni, D.Y.C. Leung, M.K.H. Leung, K. Sumathy, An overview of hydrogen production from biomass, *Fuel Process Technol* 87 (2006) 461–472. https://doi.org/10.1016/J.FUPROC.2005.11.003.

29. A. Demirbas, G. Arin, Hydrogen from biomass via pyrolysis: Relationships between yield of hydrogen and temperature, *Energy Sources* 26 (2004) 1061–1069. https://doi.org/10.1080/00908310490494568.

30. A. Arregi, G. Lopez, M. Amutio, I. Barbarias, J. Bilbao, M. Olazar, Hydrogen production from biomass by continuous fast pyrolysis and in-line steam reforming, *RSC Adv* 6 (2016) 25975–25985. https://doi.org/10.1039/C6RA01657J.

31. W. Jerzak, A. Bieniek, A. Magdziarz, Multifaceted analysis of products from the intermediate co-pyrolysis of biomass with Tetra Pak waste, *Int J Hydrogen Energy* 48 (2023) 11680–11694. https://doi.org/10.1016/J.IJHYDENE.2021.06.202.

32. M. Tripathi, J.N. Sahu, P. Ganesan, Effect of process parameters on production of biochar from biomass waste through pyrolysis: A review, *Renew Sustain Energy Rev* 55 (2016) 467–481. https://doi.org/10.1016/J.RSER.2015.10.122.

33. K. Maliutina, A. Tahmasebi, J. Yu, S.N. Saltykov, Comparative study on flash pyrolysis characteristics of microalgal and lignocellulosic biomass in entrained-flow reactor, *Energy Convers Manag* 151 (2017) 426–438. https://doi.org/10.1016/J.ENCONMAN.2017.09.013.

34. J. Yang, X. Xu, S. Liang, R. Guan, H. Li, Y. Chen, B. Liu, J. Song, W. Yu, K. Xiao, H. Hou, J. Hu, H. Yao, B. Xiao, Enhanced hydrogen production in catalytic pyrolysis of sewage sludge by red mud: Thermogravimetric kinetic analysis and pyrolysis characteristics, *Int J Hydrogen Energy* 43 (2018) 7795–7807. https://doi.org/10.1016/J.IJHYDENE.2018.03.018.

35. S. Liu, J. Zhu, M. Chen, W. Xin, Z. Yang, L. Kong, Hydrogen production via catalytic pyrolysis of biomass in a two-stage fixed bed reactor system, *Int J Hydrogen Energy* 39 (2014) 13128–13135. https://doi.org/10.1016/J.IJHYDENE.2014.06.158.

36. C.E. Efika, C. Wu, P.T. Williams, Syngas production from pyrolysis–catalytic steam reforming of waste biomass in a continuous screw kiln reactor, *J Anal Appl Pyrolysis* 95 (2012) 87–94. https://doi.org/10.1016/J.JAAP.2012.01.010.

37. F. Chen, C. Wu, L. Dong, A. Vassallo, P.T. Williams, J. Huang, Characteristics and catalytic properties of Ni/CaAlOx catalyst for hydrogen-enriched syngas production from pyrolysis-steam reforming of biomass sawdust, *Appl Catal B* 183 (2016) 168–175. https://doi.org/10.1016/J.APCATB.2015.10.028.

38. C. Wang, Z. Jiang, Q. Song, M. Liao, J. Weng, R. Gao, M. Zhao, Y. Chen, G. Chen, Investigation on hydrogen-rich syngas production from catalytic co-pyrolysis of polyvinyl chloride (PVC) and waste paper blends, *Energy* 232 (2021) 121005. https://doi.org/10.1016/J.ENERGY.2021.121005.

39. M. Koike, C. Ishikawa, D. Li, L. Wang, Y. Nakagawa, K. Tomishige, Catalytic performance of manganese-promoted nickel catalysts for the steam reforming of tar from biomass pyrolysis to synthesis gas, *Fuel* 103 (2013) 122–129. https://doi.org/10.1016/J.FUEL.2011.04.009.

40. M. Ye, Y. Tao, F. Jin, H. Ling, C. Wu, P.T. Williams, J. Huang, Enhancing hydrogen production from the pyrolysis-gasification of biomass by size-confined Ni catalysts on acidic MCM-41 supports, *Catal Today* 307 (2018) 154–161. https://doi.org/10.1016/J.CATTOD.2017.05.077.

41. Q.M.K. Waheed, C. Wu, P.T. Williams, Pyrolysis/reforming of rice husks with a Ni–dolomite catalyst: Influence of process conditions on syngas and hydrogen yield, *J Energy Inst* 89 (2016) 657–667. https://doi.org/10.1016/J.JOEI.2015.05.006.

42. T. Suprianto, Winarto, W. Wijayanti, I.N.G. Wardana, Synergistic effect of curcumin and activated carbon catalyst enhancing hydrogen production from biomass pyrolysis, *Int J Hydrogen Energy* 46 (2021) 7147–7164. https://doi.org/10.1016/J.IJHYDENE.2020.11.211.

43. H. Dai, H. Dai, Green hydrogen production based on the co-combustion of wood biomass and porous media, *Appl Energy* 324 (2022) 119779. https://doi.org/10.1016/J.APENERGY.2022.119779.

44. A.F.M. Ibrahim, K.P.R. Dandamudi, S. Deng, Y.S. Lin, Pyrolysis of hydrothermal liquefaction algal biochar for hydrogen production in a membrane reactor, *Fuel* 265 (2020) 116935. https://doi.org/10.1016/J.FUEL.2019.116935.

45. J. Rajagopal, K.P. Gopinath, A. Krishnan, N. Vikas Madhav, J. Arun, Photocatalytic reforming of aqueous phase obtained from liquefaction of household mixed waste biomass for renewable bio-hydrogen production, *Bioresour Technol* 321 (2021) 124529. https://doi.org/10.1016/J.BIORTECH.2020.124529.

46. B. Ramprakash, P. Lindblad, J.J. Eaton-Rye, A. Incharoensakdi, Current strategies and future perspectives in biological hydrogen production: A review, *Renew Sustain Energy Rev* 168 (2022) 112773. https://doi.org/10.1016/J.RSER.2022.112773.

47. N. Akhlaghi, G. Najafpour-Darzi, A comprehensive review on biological hydrogen production, *Int J Hydrogen Energy* 45 (2020) 22492–22512. https://doi.org/10.1016/J.IJHYDENE.2020.06.182.

48. M. Ni, D.Y.C. Leung, M.K.H. Leung, K. Sumathy, An overview of hydrogen production from biomass, *Fuel Process Technol* 87 (2006) 461–472. https://doi.org/10.1016/J.FUPROC.2005.11.003.

49. D.H. Kim, M.S. Kim, Hydrogenases for biological hydrogen production, *Bioresour Technol* 102 (2011) 8423–8431. https://doi.org/10.1016/J. BIORTECH.2011.02.113.

50. A. Melis, L. Zhang, M. Forestier, M.L. Ghirardi, M. Seibert, Sustained photobiological hydrogen gas production upon reversible inactivation of oxygen evolution in the green AlgaChlamydomonas reinhardtii, *Plant Physiol* 122 (2000) 127–136. https://doi.org/10.1104/PP.122.1.127.

51. T. Kondo, M. Arakawa, T. Hirai, T. Wakayama, M. Hara, J. Miyake, Enhancement of hydrogen production by a photosynthetic bacterium mutant with reduced pigment, *J Biosci Bioeng* 93 (2002) 145–150. https:// doi.org/10.1016/S1389-1723(02)80006-8.

52. L. Vasilyeva, M. Miyake, E. Khatipov, T. Wakayama, M. Sekine, M. Hara, E. Nakada, Y. Asada, J. Miyake, Enhanced hydrogen production by a mutant of Rhodobacter sphaeroides having an altered light-harvesting system, *J Biosci Bioeng* 87 (1999) 619–624. https://doi.org/10.1016/S1389-1723(99)80124-8.

53. H. Ooshima, S. Takakuwa, T. Katsuda, M. Okuda, T. Shirasawa, M. Azuma, J. Kato, Production of hydrogen by a hydrogenase-deficient mutant of Rhodobacter capsulatus, *J Ferment Bioeng* 85 (1998) 470–475. https://doi. org/10.1016/S0922-338X(98)80064-0.

54. J.R. Benemann, *Process analysis and economics of biophotolysis of water.* IEA technical report from the IEA Agreement on the Production and Utilization of Hydrogen (1998). https://doi.org/10.2172/776255.

55. V.H.S. de Abreu, V.G.F. Pereira, L.F.C. Proença, F.S. Toniolo, A.S. Santos, A systematic study on techno-economic evaluation of hydrogen production, *Energies* 16 (2023) 6542. https://doi.org/10.3390/EN16186542.

56. E. Touloupakis, G. Rontogiannis, A.M. Silva Benavides, B. Cicchi, D.F. Ghanotakis, G. Torzillo, Hydrogen production by immobilized Synechocystis sp. PCC 6803, *Int J Hydrogen Energy* 41 (2016) 15181–15186. https:// doi.org/10.1016/J.IJHYDENE.2016.07.075.

57. E. Eroglu, A. Melis, Photobiological hydrogen production: Recent advances and state of the art, *Bioresour Technol* 102 (2011) 8403–8413. https://doi. org/10.1016/J.BIORTECH.2011.03.026.

58. J. Jiménez-Llanos, M. Ramírez-Carmona, L. Rendón-Castrillón, C. Ocampo-López, Sustainable biohydrogen production by Chlorella sp. microalgae: A review, *Int J Hydrogen Energy* 45 (2020) 8310–8328. https:// doi.org/10.1016/J.IJHYDENE.2020.01.059.

59. S.A. Ensign, P.W. Ludden, Characterization of the CO oxidation/H2 evolution system of Rhodospirillum rubrum: Role of a 22-kDa iron-sulfur protein in mediating electron transfer between carbon monoxide dehydrogenase and hydrogenase, *J Biol Chem* 266 (1991) 18395–18403. https://doi. org/10.1016/s0021-9258(18)55283-2.

60. R.L. Kerby, P.W. Ludden, G.P. Roberts, Carbon monoxide-dependent growth of Rhodospirillum rubrum, *J Bacteriol* 177 (1995) 2241–2244. https://doi. org/10.1128/JB.177.8.2241-2244.1995.

61. G.Y. Jung, J.R. Kim, J.Y. Park, S. Park, Hydrogen production by a new chemoheterotrophic bacterium Citrobacter sp. Y19, *Int J Hydrogen Energy* 27 (2002) 601–610. https://doi.org/10.1016/S0360-3199(01)00176-8.

62. G. Najafpour, H. Younesi, A.R. Mohamed, A Survey on various carbon sources for biological hydrogen production via the water-gas reaction using a photosynthetic bacterium (Rhodospirillum rubrum), *Energy Sources, Part A* 28 (2006) 1013–1026. https://doi.org/10.1080/009083190910541.

63. E. Wolfrum, G. Vanzin, A.S. Watt, *Biological water gas shift development* (2003). https://www.researchgate.net/publication/228502175 (accessed January 11, 2024).

64. H. Younesi, G. Najafpour, K.S. Ku Ismail, A.R. Mohamed, A.H. Kamaruddin, Biohydrogen production in a continuous stirred tank bioreactor from synthesis gas by anaerobic photosynthetic bacterium: Rhodopirillum rubrum, *Bioresour Technol* 99 (2008) 2612–2619. https://doi.org/10.1016/J.BIORTECH.2007.04.059.

65. C.A. Eckert, E. Freed, K. Wawrousek, S. Smolinski, J. Yu, P.C. Maness, Inactivation of the uptake hydrogenase in the purple non-sulfur photosynthetic bacterium Rubrivivax gelatinosus CBS enables a biological water–gas shift platform for H2 production, *J Ind Microbiol Biotechnol* 46 (2019) 993–1002. https://doi.org/10.1007/S10295-019-02173-7.

66. Y. Zhao, R. Cimpoia, Z. Liu, S.R. Guiot, Kinetics of CO conversion into H 2 by Carboxydothermus hydrogenoformans, *Appl Microbiol Biotechnol* 91 (2011) 1677–1684. https://doi.org/10.1007/S00253-011-3509-7/TABLES/1.

67. S. Ghosh, U.K. Dairkee, R. Chowdhury, P. Bhattacharya, Hydrogen from food processing wastes via photofermentation using purple non-sulfur bacteria (PNSB)—a review, *Energy Convers Manag* 141 (2017) 299–314. https://doi.org/10.1016/J.ENCONMAN.2016.09.001.

68. G. Kars, A. Ceylan, Biohydrogen and 5-aminolevulinic acid production from waste barley by Rhodobacter sphaeroides O.U.001 in a biorefinery concept, *Int J Hydrogen Energy* 38 (2013) 5573–5579. https://doi.org/10.1016/J.IJHYDENE.2013.03.013.

69. I.K. Kapdan, F. Kargi, R. Oztekin, H. Argun, Bio-hydrogen production from acid hydrolyzed wheat starch by photo-fermentation using different Rhodobacter sp, *Int J Hydrogen Energy* 34 (2009) 2201–2207. https://doi.org/10.1016/J.IJHYDENE.2009.01.017.

70. K. Anam, M.S. Habibi, T.U. Harwati, D. Susilaningsih, Photofermentative hydrogen production using Rhodobium marinum from bagasse and soy sauce wastewater, *Int J Hydrogen Energy* 37 (2012) 15436–15442. https://doi.org/10.1016/J.IJHYDENE.2012.06.076.

71. R. DE PHILIPPIS, L. Bianchi, G. Colica, C. Bianchini, M. Peruzzini, F. Vizza, From vegetable residues to hydrogen and electric power: Feasibility of a two step process operating with purple non sulfur bacteria, *J Biotechnol* 131S (2023) S122–S123. https://flore.unifi.it/handle/2158/251994 (accessed January 12, 2024).

72. K. Seifert, M. Waligorska, M. Laniecki, Brewery wastewaters in photobiological hydrogen generation in presence of Rhodobacter sphaeroides O.U. 001, *Int J Hydrogen Energy* 35 (2010) 4085–4091. https://doi.org/10.1016/J. IJHYDENE.2010.01.126.

73. G. Kars, Ü. Alparslan, Valorization of sugar beet molasses for the production of biohydrogen and 5-aminolevulinic acid by Rhodobacter sphaeroides O.U.001 in a biorefinery concept, *Int J Hydrogen Energy* 38 (2013) 14488–14494. https://doi.org/10.1016/J.IJHYDENE.2013.09.050.

74. K. Seifert, M. Waligorska, M. Laniecki, Hydrogen generation in photobiological process from dairy wastewater, *Int J Hydrogen Energy* 35 (2010) 9624–9629. https://doi.org/10.1016/J.IJHYDENE.2010.07.015.

75. H. Zhu, T. Suzuki, A.A. Tsygankov, Y. Asada, J. Miyake, Hydrogen production from tofu wastewater by Rhodobacter sphaeroides immobilized in agar gels, *Int J Hydrogen Energy* 24 (1999) 305–310. https://doi.org/10.1016/ S0360-3199(98)00081-0.

76. G. Padovani, C. Pintucci, P. Carlozzi, Dephenolization of stored olive-mill wastewater, using four different adsorbing matrices to attain a low-cost feedstock for hydrogen photo-production, *Bioresour Technol* 138 (2013) 172–179. https://doi.org/10.1016/J.BIORTECH.2013.03.155.

77. L. Bianchi, F. Mannelli, C. Viti, A. Adessi, R. De Philippis, Hydrogen-producing purple non-sulfur bacteria isolated from the trophic lake Averno (Naples, Italy), *Int J Hydrogen Energy* 35 (2010) 12216–12223. https://doi. org/10.1016/J.IJHYDENE.2010.08.038.

78. A. Adessi, J.B. McKinlay, C.S. Harwood, R. De Philippis, A Rhodopseudomonas palustris nifA* mutant produces H2 from NH4+-containing vegetable wastes, *Int J Hydrogen Energy* 37 (2012) 15893–15900. https://doi.org/10.1016/ J.IJHYDENE.2012.08.009.

79. R. Łukajtis, I. Hołowacz, K. Kucharska, M. Glinka, P. Rybarczyk, A. Przyjazny, M. Kamiński, Hydrogen production from biomass using dark fermentation, *Renew Sustain Energy Rev* 91 (2018) 665–694. https://doi. org/10.1016/J.RSER.2018.04.043.

80. D.H. Kim, M.S. Kim, Hydrogenases for biological hydrogen production, *Bioresour Technol* 102 (2011) 8423–8431. https://doi.org/10.1016/ J.BIORTECH.2011.02.113.

81. L. Alibardi, R. Cossu, Composition variability of the organic fraction of municipal solid waste and effects on hydrogen and methane production potentials, *Waste Manag* 36 (2015) 147–155. https://doi.org/10.1016/J. WASMAN.2014.11.019.

82. A. Karlsson, L. Vallin, J. Ejlertsson, Effects of temperature, hydraulic retention time and hydrogen extraction rate on hydrogen production from the fermentation of food industry residues and manure, *Int J Hydrogen Energy* 33 (2008) 953–962. https://doi.org/10.1016/J.IJHYDENE.2007.10.055.

83. J. Pan, R. Zhang, H.M. El-Mashad, H. Sun, Y. Ying, Effect of food to microorganism ratio on biohydrogen production from food waste via

anaerobic fermentation, *Int J Hydrogen Energy* 33 (2008) 6968–6975. https://doi.org/10.1016/J.IJHYDENE.2008.07.130.

84. N.H. Mohd Yasin, N.A. Rahman, H.C. Man, M.Z. Mohd Yusoff, M.A. Hassan, Microbial characterization of hydrogen-producing bacteria in fermented food waste at different pH values, *Int J Hydrogen Energy* 36 (2011) 9571–9580. https://doi.org/10.1016/J.IJHYDENE.2011.05.048.

85. A.E. Mars, T. Veuskens, M.A.W. Budde, P.F.N.M. Van Doeveren, S.J. Lips, R.R. Bakker, T. De Vrije, P.A.M. Claassen, Biohydrogen production from untreated and hydrolyzed potato steam peels by the extreme thermophiles Caldicellulosiruptor saccharolyticus and Thermotoga neapolitana, *Int J Hydrogen Energy* 35 (2010) 7730–7737. https://doi.org/10.1016/J.IJHYDENE.2010.05.063.

86. X. Gómez, A. Morán, M.J. Cuetos, M.E. Sánchez, The production of hydrogen by dark fermentation of municipal solid wastes and slaughterhouse waste: A two-phase process, *J Power Sources* 157 (2006) 727–732. https://doi.org/10.1016/J.JPOWSOUR.2006.01.006.

87. T.A.D. Nguyen, K.R. Kim, M.S. Kim, S.J. Sim, Thermophilic hydrogen fermentation from Korean rice straw by Thermotoga neapolitana, *Int J Hydrogen Energy* 35 (2010) 13392–13398. https://doi.org/10.1016/J.IJHYDENE.2009.11.112.

88. G.L. Cao, L. Zhao, A.J. Wang, Z.Y. Wang, N.Q. Ren, Single-step bioconversion of lignocellulose to hydrogen using novel moderately thermophilic bacteria, *Biotechnol Biofuels* 7 (2014) 1–13. https://doi.org/10.1186/1754-6834-7-82/TABLES/5.

89. L. Magnusson, R. Islam, R. Sparling, D. Levin, N. Cicek, Direct hydrogen production from cellulosic waste materials with a single-step dark fermentation process, *Int J Hydrogen Energy* 33 (2008) 5398–5403. https://doi.org/10.1016/J.IJHYDENE.2008.06.018.

90. G. Ivanova, G. Rákhely, K.L. Kovács, Thermophilic biohydrogen production from energy plants by Caldicellulosiruptor saccharolyticus and comparison with related studies, *Int J Hydrogen Energy* 34 (2009) 3659–3670. https://doi.org/10.1016/J.IJHYDENE.2009.02.082.

91. J. Bu, H.L. Wei, Y.T. Wang, J.R. Cheng, M.J. Zhu, Biochar boosts dark fermentative H2 production from sugarcane bagasse by selective enrichment/colonization of functional bacteria and enhancing extracellular electron transfer, *Water Res* 202 (2021) 117440. https://doi.org/10.1016/J.WATRES.2021.117440.

92. H. Han, L. Wei, B. Liu, H. Yang, J. Shen, Optimization of biohydrogen production from soybean straw using anaerobic mixed bacteria, *Int J Hydrogen Energy* 37 (2012) 13200–13208. https://doi.org/10.1016/J.IJHYDENE.2012.03.073.

93. M. Bao, H. Su, T. Tan, Biohydrogen production by dark fermentation of starch using mixed bacterial cultures of Bacillus sp and Brevumdimonas sp., *Energy Fuels* 26 (2012) 5872–5878. https://doi.org/10.1021/EF300666M/ASSET/IMAGES/EF-2012-00666M_M006.GIF.

94. S. Rapagná, H. Provendier, C. Petit, A. Kiennemann, P.U. Foscolo, Development of catalysts suitable for hydrogen or syn-gas production from biomass gasification, *Biomass Bioenerg* 22 (2002) 377–388. https://doi.org/10.1016/S0961-9534(02)00011-9.

95. A.K. Vuppaladadiyam, S.S.V. Vuppaladadiyam, A. Awasthi, A. Sahoo, S. Rehman, K.K. Pant, S. Murugavelh, Q. Huang, E. Anthony, P. Fennel, S. Bhattacharya, S.Y. Leu, Biomass pyrolysis: A review on recent advancements and green hydrogen production, *Bioresour Technol* 364 (2022) 128087. https://doi.org/10.1016/J.BIORTECH.2022.128087.

96. A.K. Vuppaladadiyam, M.Z. Memon, G. Ji, A. Raheem, T.Z. Jia, V. Dupont, M. Zhao, Thermal characteristics and kinetic analysis of woody biomass pyrolysis in the presence of bifunctional alkali metal ceramics, *ACS Sustain Chem Eng* 7 (2019) 238–248. https://doi.org/10.1021/ACSSUSCHEMENG.8B02967/ASSET/IMAGES/LARGE/SC-2018-029678_0005.JPEG.

97. D.B. Levin, R. Chahine, Challenges for renewable hydrogen production from biomass, *Int J Hydrogen Energy* 35 (2010) 4962–4969. https://doi.org/10.1016/J.IJHYDENE.2009.08.067.

98. A.M. Abdalla, S. Hossain, O.B. Nisfindy, A.T. Azad, M. Dawood, A.K. Azad, Hydrogen production, storage, transportation and key challenges with applications: A review, *Energy Convers Manag* 165 (2018) 602–627. https://doi.org/10.1016/J.ENCONMAN.2018.03.088.

99. Y. Fang, M.C. Paul, S. Varjani, X. Li, Y.K. Park, S. You, Concentrated solar thermochemical gasification of biomass: Principles, applications, and development, *Renew Sustain Energy Rev* 150 (2021) 111484. https://doi.org/10.1016/J.RSER.2021.111484.

100. M. Nasir Uddin, W.M.A.W. Daud, H.F. Abbas, Potential hydrogen and non-condensable gases production from biomass pyrolysis: Insights into the process variables, *Renew Sustain Energy Rev* 27 (2013) 204–224. https://doi.org/10.1016/J.RSER.2013.06.031.

CHAPTER 17

Modified Biopolymers and Their Wastes as Corrosion Inhibitors

Luana Barros Furtado, Rafaela C. Nascimento, Carolina Garín Correa and Camila Dias dos Reis Barros

17.1 INTRODUCTION

Corrosion is a problem that affects metallic structures in different areas, particularly the oil and gas and construction industries. Operations such as acid descaling, acid cleaning, pickling, and oil well acidizing exacerbate corrosion. Low concentrations of corrosion inhibitors are used to mitigate this phenomenon, responsible for many direct and indirect costs [1, 2]. Corrosion inhibitors, chemical compounds added to prevent or delay corrosion, are composed primarily of organic and inorganic compounds [3, 4].

The effects of conventional inhibitor toxicity prompted the search for more sustainable alternatives, thereby increasing the number of studies on green inhibitors. Green corrosion inhibitors should be eco-friendly, biodegradable, low-cost and/or obtained from renewable or natural sources and biomass waste [5–8]. Plants and their extracts contain phytochemical compounds such as phenolics, terpenoids, alkaloids, and flavonoids, which can inhibit corrosion. These extracts can be obtained from biomass waste or the extraction of seeds, peels, roots, pomace, and leaves.

Plants or animals produce natural biopolymers, one of the green corrosion inhibitor categories. The inhibitory activity of biopolymers is

DOI: 10.1201/9781003466833-20

attributed to active sites on their structure, which are favorable to surface adsorption and consist mainly of functional groups [7, 9]. Moreover, biopolymer efficiency and stability may be compromised over time [5, 10]. Biopolymer functionalization alters their physicochemical properties, which may improve their inhibitor activity and efficiency. One form of functionalization is graphitization, a chemical modification that enables polymeric chains to be inserted into the biopolymer structure. Chemical modification via a non-grafting process involves introducing strategic functional groups into the biopolymer structure. Both methods increase adsorption sites, enhancing their inhibition efficiency, which is desired regardless of the application [6, 11]. However, parameters such as inhibitor concentration, temperature, and structural characteristics need to be considered based on the desired application.

Some examples of biopolymers are chitosan, gum, cellulose, starch, pectin, dextrin, chitin, and alginates. Dextrin and β-cyclodextrin exhibit good solubility and adhesion properties [12]. In addition, cassava starch is an attractive renewable, biodegradable, low-cost polymer studied as a graft copolymer [10]. Chitosan has been the most widely reported biopolymer as a corrosion inhibitor. It is a linear polysaccharide composed of β-(1-4)-linked D-glucosamine and N-acetyl-D-glucosamine [7]. This macromolecular structure is obtained via chitin *N*-deacetylation [13], naturally produced by crustaceans such as crabs and shrimp (shells), insects (cuticles), and fungi (cell walls) [14]. Its high molecular weight contributes to efficient surface coverage. Chitosan exhibits two main characteristics that enhance its potential to produce high performance chemically modified corrosion inhibitors. First is its complexing ability, which allows it to form chelates of di and trivalent cations and complexes with metal ions. Second is the presence of amine and hydroxyl groups that provide a broad scope for functionalization [9, 15] and nanocomposite formation [7]. The presence of amine groups is also important due to their protonation in acidic media, showing polyelectrolyte-like behavior [7].

A possible chemical modification of chitosan is functionalization via nanocomposites (titanium dioxide (TiO_2) and magnetite (Fe_3O_4) hybrid nanocomposites) owing to the hydroxyl groups and amines in the biopolymer structure. These materials incorporate nano-sized inclusions into the biopolymer matrix, thereby improving its physicochemical properties, mechanical strength, and electrical and thermal conductivity. The number of active centers required for adsorption increases because of their large surface-to-volume ratio. Both hybrid nanocomposites showed better inhibition performance than that of the unmodified chitosan, reaching maximum efficiency of 97.19 and 95.49% (TiO_2 and Fe_3O_4, respectively) [7]. Although chitosan does not remain active for long periods and exhibits low water solubility [16, 17], it is widely studied in the literature, mainly its modifications and the formation of new compounds, which increase its solubility and long-term effectiveness [18].

Theoretical calculations have been carried out to confirm and corroborate the results obtained from experimental work. Computational studies can also predict interaction between biopolymers and metallic surfaces, saving time and resources in experimental tests. Calculations based on molecular dynamics can help understand inhibition mechanisms via adsorption geometry [8, 17]. Calculations based on the density functional theory (DFT) are also used to obtain chemical parameters that will help identify possible adsorption sites on the corrosion inhibitor and understand the adsorption mechanism [19].

In this respect, the present chapter includes a systematic review of the literature that describes improvements in the efficient application of functionalized biopolymers as corrosion inhibitors. The brief discussion of the chapter addresses parameters such as concentration, immersion time, temperature, and structural applications for different applications. Furthermore, a comparison between functionalized and non-functionalized biopolymers, as well as between grafted and modified biopolymers, is presented. Finally, gaps that should be considered in future studies are highlighted.

17.2 CURRENT APPLICATIONS

The application and anticorrosive efficacy of biopolymers depend on their molar mass, availability of electrons to contribute to and interact with metallic surfaces, and adhesion to the substrate if used as coating.

Due to their excellent film-forming capabilities, metal ion chelation, and non-toxic nature, film formation by biopolymers stabilizes the passive oxide layer and can protect the metallic surface from the diffusion of aggressive ions from corrosive media.

These inhibitors are used in specific media, which are strongly related to the intended industrial application. Most studies have reported the use of acid solutions, as shown in Figure 17.1. More than 50% of the articles

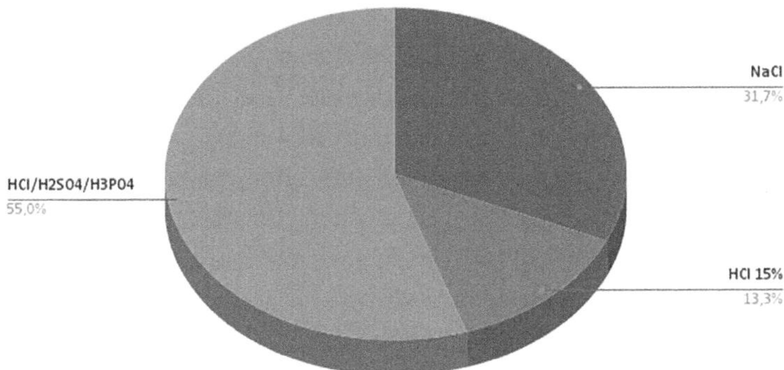

FIGURE 17.1 Distribution of the different aqueous media used to perform corrosion tests with modified biopolymers as corrosion inhibitors (percentage based on the reported media).

investigated mild acidic conditions, which represent applications such as acid pickling and descaling. There is less information in the literature on studies carried out under aggressive conditions, which focus on oil and gas acidizing operations. These processes usually have higher operating temperatures, which could decompose the inhibitor. Aggressive conditions still pose a challenge, underscoring the need to design chemically modified biopolymers to inhibit corrosion under higher acidity and temperature conditions.

17.2.1 Inhibitors in Acidic Media

One of the major problems with strong acids is corrosion in metallic equipment, requiring the use of inhibitors to protect them. Inhibitors can be used in different types of acidic media, such as 15% HCl, which simulates oil well acidizing conditions. This operation requires a higher acid concentration to increase rock matrix permeability. Depending on the rock matrix composition, other types of acidic media can be used in this operation, such as: 1.50% HF + 13.50% HCl (low permeability (< 5mD)), 12% HCl + 3% HF (Quartz (80%) and clay (< 5%)), 13.50% HCl + 1.50% HF (high feldspar content (> 20%)), 6.50% HCl + 1% HF (high clay content (> 10%)) [20].

Analysis of immersion time in harsh conditions shows that highly acidic conditions can not only degrade the biopolymer but also severely damage the metallic surface. This is likely the main reason that studies using 15% HCl immersed steel coupons for less than 24 hours. In those carried out under highly acidic conditions in 15% HCl, several authors synthesized biopolymer derivatives such as gum acacia-*graft*-poly (acrylamide), guar gum, xanthan gum-*graft*-poly (acrylamide), and dextrin-*graft*-polyvinyl acetate, among others. At concentrations between 0.3–0.5 g/L, maximum inhibition efficiency was between 92–98% after 24 hours of immersion [21–23]. However, Biswas et al. [21] studied the effect of immersion time (6, 12, 18, and 24 hours) of both gum acacia and the grafted biopolymer, demonstrating that the latter maintained high efficiencies even at longer exposure times (variation of approximately 5% in inhibition efficiency), while the non-functionalized biopolymer suffered a 20% loss of efficiency, due to the hydrolysis of glycosidic linkages or its biodegradable nature. In this case, biopolymer functionalization is clearly an advantageous process. However, Biswas, Pal, and Udayabhanu [23] observed very similar efficiencies for both the functionalized and non-functionalized biopolymers of 61.30 and 60.57%, respectively, at 333 K. Since this difference may not be statistically significant, grafting seems not to be a good approach due to the time and effort involved.

Another application of this topic is the acid cleaning and electropolishing of aluminum in pickling delicate, costly precision items [24–26]. Deng, Li, and Du [24] investigated starch as a corrosion inhibitor for aluminum in 1 M H_3PO_4. It is important to note that H_3PO_4 is more complex than HCl due to different acid anions such as $H_2PO_4^-$, HPO_4^{2-}, and PO_4^{3-}. Although some inhibitors are effective for HCl media, they may not be for H_3PO_4, justifying chemical modification of the biopolymer. The highest efficiency was 88.90% in 1.00 g/L by EIS.

In relation to the concentration effect of these biopolymers, studies conducted by Sangeetha, Meenakshi, and Sundaram [27] with fumaric acid derivatives in HCl media (O-fumaryl-chitosan (OFC)) revealed that the dissolution rate of mild steel (MS) and current density decline with increasing OFC concentration. This can be explained by the fact that adsorption increases as concentration rises, due to gradual replacement of water molecules present on the surface of the MS by OFC. Other literature studies also observed greater efficiency with increased concentration [10, 17, 25, 26, 28–30].

Although increased efficiency at higher concentrations is the most commonly observed phenomenon in biopolymer research, it is important to consider solubility. In this respect, Banerjee, Srivastava, and Singh [31] synthesized the polyacrylamide-grafted copolymer of Okra mucilage to act as a corrosion inhibitor for mild steel in 0.50 M H_2SO_4. Inhibition efficiency increased with concentration from 65.70 (1 ppm) to 91.80% (100 ppm). In addition, the grafted biopolymer was more efficient than its counterpart in 100 ppm (72.50%), indicating that grafting is a good strategy to improve inhibition performance. It is important to underscore that further inhibitor concentration increases would not be possible since it has a limited water solubility.

Analysis of the concentration effect from the EIS perspective shows that the size of Nyquist curves increases with rising inhibitor concentration, which is associated with a charge transfer mechanism and greater metal surface resistivity. Zhang et al. [32] found that as the concentration of two modified chitosan inhibitors (CS-1 and CS-2) increased, so did the diameter of the half circles in Nyquist plots, which was associated with the adsorption of more molecules on the surface of Q235 steel in acidic solution. In another study, Frecky and Mohamed [33] observed that the diameter of depressed Nyquist circles increases slowly with longer immersion, suggesting that the surface film formed with acetyl thiourea chitosan polymer (ATUCS) remains stable for 8 hours' immersion time. This indicates that the corrosion process is controlled by the corrosion of mild steel, the behavior characteristic of solid electrodes, and often refers to frequency dispersion, attributed to roughness and other inhomogeneities of the solid surface. This may be due to the slow transformation of the partially oxidized film to its reduced form, related to the stability of the oxidation states. Likewise, the increase in resistance charge transfer (R_{ct}) is evidence of efficient corrosion protection, as demonstrated in several studies with different types of natural origin inhibitors [34–37]

In regard to temperature, it has been established that it accelerates the corrosion of steel and other metallic materials in aqueous solutions. The pronounced temperature effect can hinder obtaining effective inhibitors to inhibit metallic corrosion. Accelerated corrosion at high temperatures leads to surface roughening, at the same time as hydrogen gas evolution rates increase significantly. These effects adversely affect the performance and adsorption of corrosion inhibitors. Unfortunately, even high-performance corrosion inhibitors tend to lose their anticorrosion activity at high temperatures [38]. Thus, it is important to compare the effect of temperature on modified and non-modified biopolymers. Chai et al. [28] chemically modified polyaspartic acid with cysteamine in order to investigate the final product as a corrosion inhibitor for mild steel in 0.50 M

H_2SO_4. The study compared unmodified polyaspartic acid with polyaspartic acid chemically modified using cysteamine. The results indicated that the best efficiency achieved was 93.90% for the chemically modified acid and 67.50% for its non-modified counterpart. This is due to the addition of the functional thiol group to the polyaspartic acid structure, which leads to an increase in anchoring sites on the metallic surface. In addition, sulfur heteroatom exhibits higher inhibition efficiency than that of nitrogen and oxygen. A rise in temperature increases hydrogen release and metal dissolution rate, reducing efficiency for both the modified and non-modified structures. An increase in temperature also leads to desorption of the inhibiting molecules, exposing the substrate to the aggressive medium and, consequently, reducing inhibition efficiency. However, the chemically modified structure decreases less (from 93.90 to 91.30%) from 298 to 318 K than its non-modified counterpart (67.50 to 48.30%). This is the consequence of more and stronger adsorption sites in the substrate of the chemically modified biopolymer.

17.2.2 Saturated Carbon Dioxide and Saline Systems

Another possible application for corrosion inhibitors is in carbon dioxide-saturated environments. This greenhouse gas is used in CO_2-enhanced oil recovery. When CO_2 dissolves in the media, the resulting acidic environment can cause severe corrosion in the pipelines. Wang et al. [5] reported that under the same pH, the electrochemical corrosion caused by carbonic acid is more severe than that caused by hydrochloric acid. Specifically, shear stress on the pipe wall can cause thinning and cracking of the corrosion product film, leading to local corrosion under the galvanic action of a large cathode and a small anode. This is a significant threat and causes economic losses to the safe operation of the pipeline system in the oil industry.

Another point to highlight is that NaCl is abundant in seawater and highly corrosive to metals and alloys. To address these issues, Singh et al. [39] applied guar gum-*graft*-2-acrylamido-2-methylpropanesulfonic acid coating on copper in NaCl solution. The media decreases inhibition efficiency as NaCl concentration increases, which, according to the authors, is associated with NaCl penetration into the coating on the copper surface.

Cui et al. [40] chemically modified two chitosan oligosaccharides to prevent this type of corrosion. The insertion of more nitrogen adsorption sites, as well as unsaturations and aromatic rings, increases inhibition efficiency. The highest efficiency was achieved in 100 ppm (88%). Mouaden et al. [41] also modified chitosan with thiocarbohydrazide to incorporate sulfur heteroatoms and increase adsorption sites on the metallic surface. The inhibitor was applied in 3.50% NaCl to protect stainless steel. The maximum efficiency was 98% in 1,000 ppm. A comparison between the two chitosan derivatives studied by Mouaden et al. [41] and Cui et al. [16] shows that incorporating sulfur heteroatoms leads to higher inhibition efficiencies than using nitrogen alone. However, Mouaden et al. [41] employed a higher concentration than Cui et al. [16], which may also explain the greater efficiency. It should be noted that the modifying agent should be low-cost, eco-friendly, and enhance biopolymer efficiency, resulting in high inhibition performance at low concentrations. It is important that after chemical modification, the improvement in corrosion inhibition reduces biopolymer concentration, thereby maintaining their performance. Thus, the performance of these biomolecules would be comparable to that of traditional synthetic inhibitors, providing a more sustainable solution for industrial corrosion issues.

The performance of inorganic inhibitors in combination with biopolymers in saline media has been reported to be satisfactory. Khosravi, Naderi, and Ramezanzadeh [42] studied polydopamine as a muscle-inspired, biocompatible adhesive polymer. The several functional groups present in natural biopolymers, such as catechol, quinone, indole, and amine, can adhere to the metallic substrate. However, the volume of these molecular structures can compromise adsorption performance due to the limited movement of organic compounds in aqueous solutions. The efficiency of these inhibitors is influenced by the pH and molecular weight. The strategy adopted was to introduce sodium molybdate as a doping agent to improve inhibition by forming complexes between polydopamine molecules, molybdate ions, and iron cations. The organic and inorganic inhibitors were studied alone and blended by varying the ratio and maintaining a total inhibitor concentration of 1,000 ppm. A comparison of the EIS results for the isolated inhibitors demonstrated that PDA is not efficient at short immersion times and that MoO_4^{2-} exhibited the highest efficiency. After 1 hour, 5 hours, and 24 hours, the efficiencies obtained in 1,000 ppm of PDA were 25, 42, and 51%, respectively. However, MoO_4^{2-} showed the opposite behavior, that is, efficiency declined by 91.00, 87.00, and 62.00%, respectively. Additionally, the high R_{ct} obtained for 1 hour and 5 hours of immersion time in the pure Mo-containing solution showed a declining trend in the longer immersion tests, revealing the instability of this inorganic inhibitor's protective layer. On the other hand, the increase in R_{ct} values for the hybrid organic/inorganic inhibitor (PDA/Mo) indicated better adsorption and chelation on the mild steel surface. An increasing trend was seen in the hybrid inhibitor (250 ppm of PDA:750 ppm of MoO_4^{2-}), which achieved the highest efficiency (86%) of all the ratios studied. The PDP results obtained for the same ratio (250:750 ppm PDA and MoO_4^{2-}) showed a decrease of i_{corr}, revealing the synergistic effect of these inhibitors, which provided 90% inhibition efficiency.

Eduok, Ohaeri, and Szpunar [40] used a chitosan-based organic/inorganic hybrid nanocomposite to inhibit X70 corrosion in CO_2-saturated NaCl electrolyte. First, chitosan-*graft*-allyl sulfonate was synthesized by a reaction catalyzed by acetic acid in alkaline pH using sodium persulfate initiator. The resulting polymer was subsequently inserted into graphene oxide nanosheets. X70 pipeline steel was chosen because of its extensive use in oilfield applications. CO_2-saturated acidic oilfield formation water was prepared by adding NaCl (82.40 g), $CaCl_2$ (1.30 g), $MgCl_2{\cdot}6H_2O$ (12.60 g), $NaHCO_3$ (1.10 g), and Na_2SO_4 (2.60 g) per liter of water containing 3 mM acetic acid. First, APM (*N*-(3-Aminopropyl)-imidazole) was synthesized by catalytic hydrogenation of imidazole and chitosan was then grafted with 2-methyl-1-vinylimidazole monomer. The resulting imidazole–carboxymethyl chitosan graft copolymer was combined with 500 ppm of APM to enhance inhibition. Both the imidazole-based compound and the grafted biopolymer inhibited steel corrosion in CO_2-saturated formation water. However, significantly superior protection was provided when both imidazoline-based products were combined [43].

Fayyad et al. [44] compared chitosan-derived inhibitors, chitosan (CS), and oleic acid-modified chitosan/graphene oxide film (CS/GO-OA) in NaCl medium and found that CS/GO-OA film provided an effective barrier against carbon steel corrosion by preventing ionic transportation from the NaCl solution. CS/GO-AO hydrophobic film had the lowest corrosion current and corrosion rate. In the Tafel test, the corrosion potential of CS/GO-OA film improved, which indicates superior protection when compared with other chitosan-derived studies. Research conducted with mild steel in NaCl medium has also shown that increasing eco-inhibitor concentration decreases the current density. This suggests that the current passing through the system by virtue of metal dissolution and charge transfer declines in the presence of inhibitors.

Singh, Ansari, and Quraishi [45] observed that an increase in concentration did not improve efficiency, which peaked in 40 mg/L (90%). However, a different strategy was adopted, namely, assessing the synergistic effect with KI. Inhibition efficiency of 96.80% was found in the presence of 0.50 mM of KI. The interaction with KI in a formulation achieved the highest efficiency.

The aforementioned literature review shows that a different approach is frequently used in saline media. Most of the studies that report good results modify the biopolymer with inorganic agents in order to facilitate adsorption to the metallic surface through the formation of complexes, for example. In addition, the biopolymer/inorganic blend maintained efficiencies even at longer exposure times due to the synergistic effect and chelation process.

17.2.3 Main Test Parameter Conclusions

Studies using modified biopolymers are summarized in Table 17.1, and inhibitor concentration was categorized according to the different media used in corrosion tests (Figure 17.2 (a-c)). Most studies applied the biopolymer inhibitor at concentrations below 300 ppm. However, for more aggressive acidic media, 91% used concentrations up to 500 ppm, while in less aggressive acidic media and saline media, only 76 and 73% of the investigations, respectively, employed this concentration. This may indicate that efficiency will not increase beyond a certain concentration. Thus, in more aggressive media, greater protection may be achieved depending on the type of functionalization. Another important point is that solubility did not significantly improve even after chemical modification, thereby limiting the concentration range of these inhibitors in aqueous solution. It is also noteworthy that increasing concentration can sometimes decrease efficiency due to steric or agglomeration effects.

17.3 CASE STUDIES

17.3.1 Biopolymer Functionalization Methods

Biopolymers used as corrosion inhibitors are mostly carbohydrate-based. These macromolecules are obtained by the polycondensation of one or more monosaccharides, which bind together via glycosidic bonds, forming long linear or branched chains. Their corrosion inhibition potential is associated with the macromolecular architecture combined with the chemical composition and electronic structure. The macrostructural factors consist of the nature of chains (branched or linear), the number of cyclic rings or a combination of both. Structural modification is desirable in most applications aimed at improving the inhibition performance of biopolymers [46]. As shown in Figure 17.3, polymer functionalization can be achieved by using grafting reactions (Route I) or inserting different chemical species

TABLE 17.1 Corrosion Test Details according to the Different Types of Modified Biopolymer Structure and Modifying Agents

Medium	Material	Main Structure/Source or Raw Material	Modifying Agent or Compound	Main Structural Characteristics	Inhibitor Concentration Range	Temperature (K)	Immersion Time Range	Maximum Efficiency (%)	Adsorption Type	References
1 M H_2SO_4	X80 steel	β-cyclodextrin	Polyacrylamide	O and N	50–200 mg/L	293–323	36 hours	93.16% in 200 mg/L at 293 K by EIS	Langmuir	[2]
15% HCl	MS	Dextrin	Poly (vinyl acetate)	O and Br	0.025–0.150 g/L	303–333	1 hour for OCP	98.39% in 0.15 g/L by PDP	Langmuir	[4]
CO_2 saturated groundwater	N80	Pectin	Glycine (Gly), and lysine (Lys)	O and N	100–800 mg/L	333	12–120 hours (EIS)	800 mg/L: P-Gly (99.17%) and P-Lys (99.43%) by PDP	Langmuir	[5]
15% HCl	MS	Chitosan	Hybrid acrylamide-based titanium dioxide (CS-g-PAM/TiO_2) and magnetite (CS-g-PAM/Fe_3O_4) nanocomposites	O, N, Fe_3O_4, and TiO_2	0.025–0.500 g/L	298–333 (WL)	6 hours (WL)	97.19% (CS-g-PAM/TiO_2) and 95.49% (CS-g-PAM/Fe_3O_4) at 298 K by EIS	Langmuir	[7]
0.20 M H_2SO_4	MS	Chitosan	2-hydroxyethylmethacrylate	O and N	100–500 ppm	298	Not specified	90.67% in 500 ppm by PDP	No	[9]
1 M H_2SO_4	Cold rolled steel	Cassava starch	Poly (acrylamide)	O and N	10–100 mg/L	293–323	6 hours	92.40% in 100 mg/L at 293 K by WL	Langmuir	[10]
1 M HCl	MS	Chitosan	D-glucose	O and N	10^{-6}–10^{-3} M	298–328	Not specified	97.60% in 10^{-3} M at 298 K by PDP	Langmuir	[11]
0.50 M H_2SO_4	X70	β-cyclodextrin	Poly (acrylamide)	O and N	80–150 mg/L	293–323	Not informed	79.30% in 150 mg/L at 303 K	Langmuir	[12]
15% HCl	MS	Chitosan	Vanillin	O, N and Ar	100–500 mg/L	298	6 hours	98.30% in 500 mg/L at 298 K by WL	Langmuir	[13]
1 M HCl	MS	Chitosan	Monochloroacetic acid	O and N	50–250 mg/L	298	24 hours	93.00% in 200 mg/L by EIS	Langmuir	[14]
0.50 M HCl	MS	Chitosan	L-arginine	O and N	100–500 ppm	303	Not specified	91.47% in 500 ppm by PDP	No	[15]
CO_2 saturated brine (NaCl) with 3 M acetic acid	5L X70	Chitosan	Allyl sulfonate (later grafted onto graphene oxide nanosheets)	O, N and $NaSO_3$	5–25 ppm	333	Not specified	89.70% in 25 ppm by PDP	No	[40]

3.50% NaCl-CO$_2$ saturated	P110	Chitosan derivatives	Glycidyl trimethyl ammonium chloride, benzaldehyde, and propionaldehyde	O and N	30–100 mg/L	353	3 days	88.59% in 100 mg/L at 353 K, for chitosan modified with benzaldehyde, under static conditions	–	[16]
1 M HCl	CS	Chitosan	4-amino-5-methyl-4H-1,2,4-triazole-3-thiol	O, N, and S	50–250 mg/L	303–343	Not informed	91.42% in 200 mg/L at 303 K	Langmuir	[17]
1 M HCl	MS	Alginate	8-hydroxyquinoline	O, N, and Ar	$1*10^{-3}$–$10*10^{-3}$ g/L	298–328	–	92.60% in $10*10^{-3}$ g/L by PDP	Langmuir	[19]
15% HCl	MS	Gum acacia	Polyacrylamide	O and N	0.05–0.40 g/L	298	6, 12, 18, 24 hours	90.49% in 0.40 g/L using a gum acacia: acrylamide ratio of 1:3 after 12 hours by WL	Langmuir	[21]
15% HCl	P110	Guar gum	Ethyl acrylate	O	100–500 mg/L	308–348	Not informed	92.30% in 500 mg/L by WL	Langmuir	[22]
15% HCl	MS	Xanthan gum	Poly (acrylamide)	O and N	0.05–0.50 g/L	303–333	Not informed	92.80% in 0.40 g/L at 303 K	Langmuir	[23]
1 M H$_3$PO$_4$	Al	Cassava starch	Acrylamide	O and N	0.10–1.00 g/L	293	6 hours	88.90% in 1.00 g/L by EIS	–	[24]
1 M HCl	C-Mn steel	Starch	Glycerin	O	5–300 mg/L	298–323	Not informed	94.00% in 300 mg/L at 298 K by WL	Langmuir	[25]
0.50 M HCl	MS	Glucomannan	Amino acids (L-histidine and L-arginine)	O and N	100–2000 ppm	298	8 hours	92.40% in 2,000 ppm by WL	Langmuir	[26]
1 M HCl	MS	Chitosan	O-fumaryl	O and N	100–500 ppm	328–348	–	93.20% in 500 ppm by EIS	Langmuir	[27]
0.50 M H$_2$SO$_4$	MS	Polyaspartic acid	Cysteamine	O, N and S	10–100 mg/L	298–318	12 hours	93.90% in 100 mg/L at 318 K by WL	Langmuir	[28]
1 M HCl	CS	Polyaspartic acid	Schiff base	O, N, Ar and S	10–80 ppm	293–323	6 hours	90.39% in 80 ppm	Langmuir	[29]
0.50 M HCl	Q235 CS	Chitosan	B-Cyclodextrin	O and N	80–230 ppm	298	24 hours	94.23% in 230 ppm by WL	Langmuir	[30]
1 M HCl	Q235 MS	Chitosan derivatives	Benzyl bromide and isonicotinaldehyde	O and N	0–200 ppm	298	–	98.00% in 150 mg/L by EIS	Langmuir	[32]
0.50 M H$_2$SO$_4$	MS	Hibiscus esculentus fruit	Polyacrylamide	O and N	1–100 ppm	298–338	3–72 hours (WL)	91.80% in 100 ppm for 24 hours at 298 K by WL	Langmuir	[31]

(Continued)

TABLE 17.1 *(Continued)* Corrosion Test Details according to the Different Types of Modified Biopolymer Structure and Modifying Agents

Medium	Material	Main Structure/Source or Raw Material	Modifying Agent or Compound	Main Structural Characteristics	Inhibitor Concentration Range	Temperature (K)	Immersion Time Range	Maximum Efficiency (%)	Adsorption Type	References
0.50 M H_2SO_4	MS	Chitosan	Acetyl thiourea polymer	O and N	0.19–0.76 mM	298	–	94.50% in 76 mM by EIS	Langmuir	[33]
1 M HCl	MS	Polydopamine (PDA) nanoparticles	–	O and N	5 mg/L	298	24 hours	99.00% by EIS	–	[34]
1 N HCl	MS	Chitosan	Stearic acid grafted chitosan, 1-ethyl-3-(3-dimethylaminopropyl) carbodiimide	O and N	–	298	5–20 days	89.78% in 5 days by WL	–	[35]
3.50% NaCl	Cu	Graphite powder	Grapheneoxideandpolymethyl-methacrylate	O, N and Br	2–10 μM	295	1–100 hours	81.00% PMMA-g-GO film on Cu 99.99% by PDP	–	[36]
15% HCl	MS	Hydroxyethyl cellulose	Polyurethane	O, N and Ar	6–50 μM	293–353	6 hours	94.00% in 50 μM at 298 K by WL	Langmuir	[38]
3.50% NaCl	Copper	Guar gum	2-Acrylamido-2-Methylpropanesulfonic Acid	O and N	100–600 mg/L	308–338	24 hours	95.00% in 600 mg/L by WL	Langmuir	[39]
3.50% NaCl	Stainless steel	Chitosan	Thiocarbohydrazide	O, N and S	10–1,000 ppm	298	Not informed	98.00% in 1,000 ppm by PDP	Langmuir	[41]
3.50% NaCl	MS	Polydopamine (PDA)	Sodium molybdate	O, N and Ar	250–1000 ppm PDA 250–1000 ppm Mo	Not specified	1–48 hours	90.00% (250PDA:750 Mo, 24 hours) by PDP	No	[42]
CO_2 saturated formation water**	5L X70	Chitosan	*N*-(3-Aminopropyl) imidazole (APM)	O and N	100–500 ppm APM, 50–200 ppm copolymer	333	Not specified	93.5% in 500 ppm APM + 200 ppm copolymer by PDP	No	[43]
NaCl 3.50%	CS	Chitosan	Oleic acid and graphene oxide	O and N	–	298	48 hours	Not informed	–	[44]
3.50% CO_2 saturated	P110	Guar gum	Methyl methacrylate	O	100–400 mg/L	323–383	Not informed	90.00% in 400 mg/L by WL and 96.80% in 400 mg/L + 0.50 mM KI	Langmuir	[45]
1 M HCl	J55 Steel	Dextrin	Caprolactam	O and N	100–400 mg/L	298–338	24 hours (WL)	82.38% in 300mg/L at 298 K by EIS	Langmuir	[47]
1 M HCl	X70	Chitosan	Poly (N-vinyl imidazole)	O and N	200–1,000 ppm	RT	–	Not informed	–	[48]

2 M HCl	CS	Chitosan	1,3-propanesultone	O, N, and S	50–5,000 ppm	298	24 hours	95.60% in 5,000 ppm by WL	–	[49]
0.50 M HCl	MS	Glucomannan	4-aminoazobenzene	O, N, and Ar	0.015–0.038 mmol/L	298–328	0.5–10 hours by EIS	95.00% in 0.038 mmol/L by PDP	Langmuir	[52]
1 M HCl	Al	Xanthan gum	Polyaniline	Ar, O, and N	0.10% wt.	RT	–	91.33% at 303 K by WL	–	[51]
1 M HCl	MS	Chitosan	Glycidyl trimethyl ammonium chloride	O and N	100–500 ppm	303–323	2 hours	91.42% in 500 ppm by WL	Langmuir	[53]
3.00% NaCl solution	Galvanized steel plates	Chitosan	Beidellite	O and N	Coating	Not specified	16 days	Not applicable	Not applicable	[54]
$CaCl_2$, $MgSO_4.7H_2O$, NaCl, $NaHCO_3$	CS	Polyaspartic acid	3-amino-1H-1,2,4-Triazole-5-carboxylic acid hydrate	O, N, and Na	4–28 mg/L	318	72 hours	75.00% in 16 mg/L	–	[55]
3.50% NaCl	MS	Pectin	Acrylamide and acrylic acid	O and N	300–900 ppm	Not informed	–	85.53% in 800 ppm with acrylamide and 75.48% in 300 ppm with acrylic acid by PDP	Langmuir	[56]
1 M HCl	Low carbon steel	Alginate	Cu, Co, Zn	O and N	5×10^{-5}–5×10^{-3} M	298–333	24–300 hours	96.60% in 5×10^{-3} M at 333 K by WL (alginate-Cu)	Langmuir	[57]
15% H_2SO_4 and 15% HC	St37	Gum arabic (GA)	Ag nanoparticles	O	0–1,000 ppm	298 and 333	–	91.70% in HCl and 88.86% in H_2SO_4 in 1,000 ppm at 298 K by WL	Langmuir	[58]
MWW***	CS	N-carboxylated chitosan	2-hydroxypho-sphonocarboxylic Acid, gluconic acid sodium, $ZnSO_4$	O, N and P	–	313	72 hours	97.00% in 100 mg/L by WL	Langmuir	[50]
1 M HCl	MS	Guar gum	Acrylamide (AM),	O and N	500 ppm	303	1–96 hours	(> 90%) in 500 ppm after 50 hours' immersion by WL	Langmuir	[59]
200 ppm NaCl	MS	Chitin or poly(N-acetyl-1,4-b-D-glucopyranosamine)	Phosphorous Pentoxide and Zn^{2+}	O, N, and P	0–400 ppm	298	7 days	93.16% in 200 ppm PCT and 100 ppm Zn^{2+}	Langmuir	[60]

(Continued)

TABLE 17.1 *(Continued)*   Corrosion Test Details according to the Different Types of Modified Biopolymer Structure and Modifying Agents

Medium	Material	Main Structure/Source or Raw Material	Modifying Agent or Compound	Main Structural Characteristics	Inhibitor Concentration Range	Temperature (K)	Immersion Time Range	Maximum Efficiency (%)	Adsorption Type	References
50.0 g/L of NaCl, 2.0 g/L of MgCl$_2$*6H$_2$O, 6.0 g/L of Na$_2$SO$_4$, 4.0 g/L of CaCl$_2$	N 80 Carbon steel	Polyaspartic acid	B-cyclodextrin and polysuccinimide	O and N	100–800 mg/L	333	72 hours	68.40% in 800 mg/L	–	[61]
3.50% NaCl	A3 CS	Chitosan	Polyaspartic acid	O and N	5–35 mg/L	318	72 hours	82% in 35 mg/L by WL	–	[62]
0.50–3.00 M HCl	Al	Cassava starch	Sodium sulfonate and acryl amide	O, N, and S	1–100 mg/L	293–353	0.5–6 hours	98.20% in 100 mg/L at 293 K by PDP	Langmuir	[63]
NaCl 50 mM	AA2024	Chitosan	2-mercaptobenzothiazole poly(ethylene-alt-maleic anhydride) (PEMA) and poly(maleic anhydride-alt-1-octadecene) (POMA)	O and N	1 and 10 wt%	298	25 min 1 day	Not informed	–	[64]
3.50% NaCl	CS	Chitosan-oligosaccharide	Sodium silicate	O, N, and Si	MCO (chitosan-oligosaccharide) 100–700 ppm MCO 350 ppm + AS (sodium silicate) 350 ppm	300	48 hours	350 ppm MCO + 350 ppm SA 96.71% by WL	–	[65]
3.50% NaCl	MS	Oil-based microcapsules using phosphorylated ethyl cellulose as shell material	Phosphorylated ethyl cellulose (P-EC) Microcapsules of ethyl cellulose (EC)	O and P	–	298	28 days	–	–	[66]
0.20 M HCl	Al	i-Carrageenan a natural polymer	Pefloxacin mediator	O, N, and S	0–1,600 ppm	283–313	2 hours	68.90% in 1,200 ppm of i-Carrageenan and 400 ppm of pefloxacin at 283 K by WL	Langmuir	[67]

200 ppm NaCl	MS	Xanthan gum (XG)	Phosphate groups	O and P	0–200 ppm	298–303	7 days	87.89% in 150 ppm PXG at 303 K by WL	–	[68]
1 M HCl	MS	Biopolymer blend of starch and pectin	–	O	–	303–333	–	74.00% at 303 K by PDP	–	[69]
1 M HNO_3	Al	Cassava starch polymer (CS)	Acryl amide (AA)	O and N	0.10–1.00 g/L	293–323	3–48 hours	93.30% in 1.00 g/L CSGC at 293 K by WL	Langmuir	[70]
$CaCl_2 \cdot 5H_2O$, $MgSO_4 \cdot 7H_2O$, NaCl NaOH	CS	Polyaspartic acid	5-aminoorotic acid	O and N	1.00–20.00 mg/L	318	72 hours	60.90% in 12 mg/L by WL	–	[71]

* MS—mild steel; RT—room temperature
Ar—Aromatic, O—Oxygen, N—nitrogen, S—sulfur
WL—Weight-loss measurements

** Formation water: NaCl (82.4 g), $CaCl_2$ (1.3 g), $MgCl_2 \cdot 6H_2O$ (12.6 g), $NaHCO_3$ (1.1 g), and Na_2SO_4 (2.6 g) per liter of water containing 3 mM acetic acid.

*** Municipal wastewater

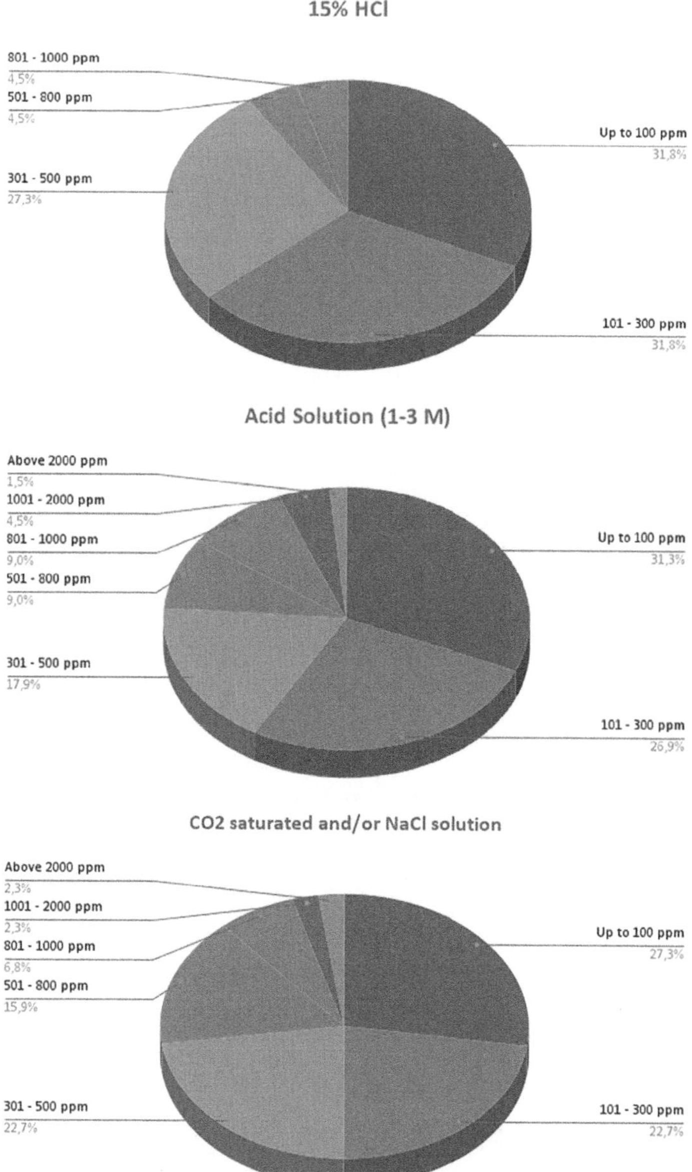

FIGURE 17.2 Percentage of articles based on the effect of inhibitor concentration: (a) 15% HCl media; (b) 1–3 M acidic media; (c) CO_2- saturated and/or NaCl solution.

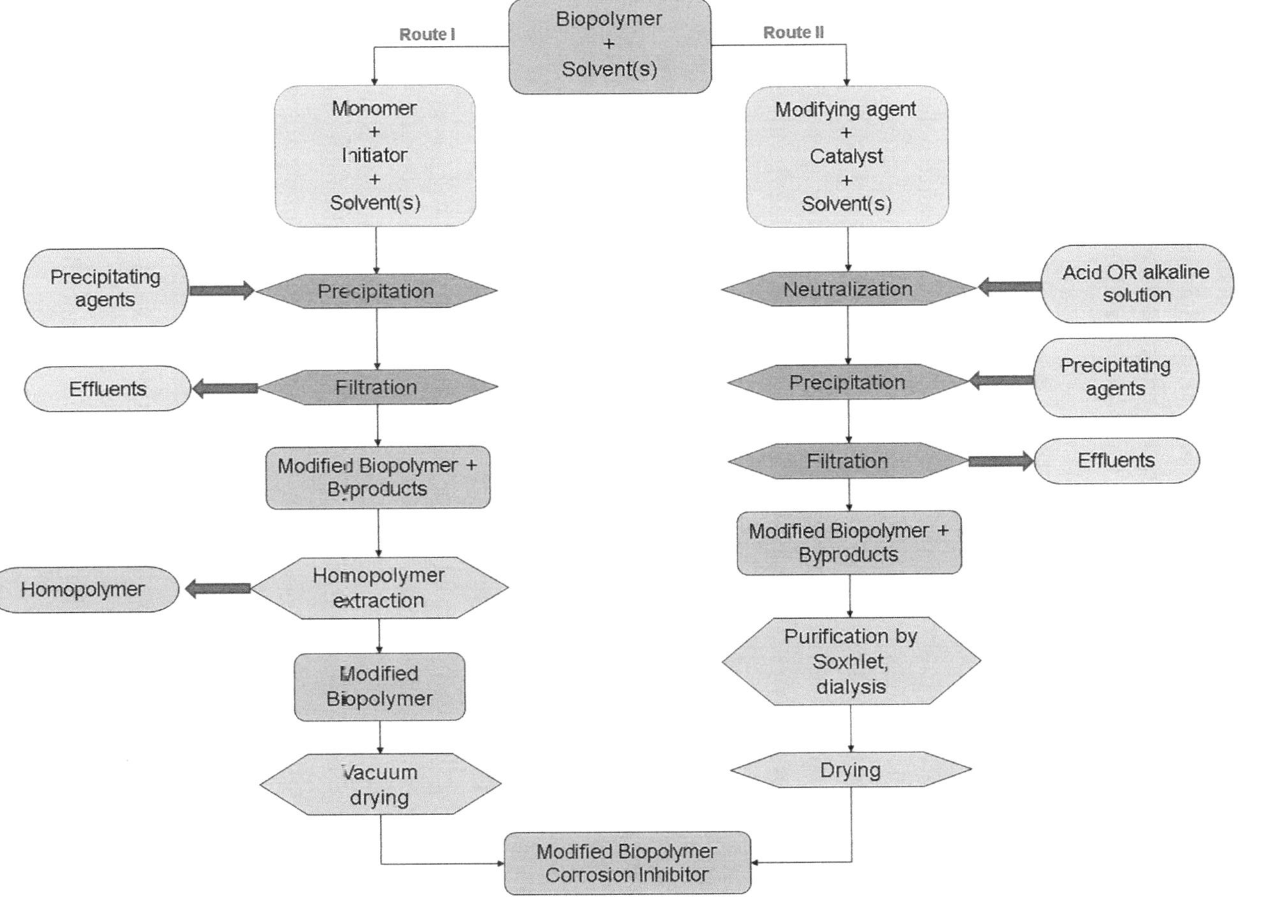

FIGURE 17.3 Representation of the main steps to enhance the functional groups in biopolymers.

(Route II). Graft copolymerization is usually carried out using the solution technique, where long-chain branches are inserted into the main biopolymer (backbone) chain. Chemical modification follows the traditional methods used to synthesize organic compounds. However, a previous step obtains a homogeneous solution with the biopolymer, which involves the use of acid or alkaline solutions. A further step is required to neutralize the reaction products.

Grafting enhances the presence of functional groups providing the copolymer with stronger adsorption capacity than that of simple polysaccharides. [47]. On the other hand, inorganic modifying agents or inorganic salts are another interesting possibility to improve absorbance potential on metallic surfaces by complexes and chelates. The several functional groups present in the natural biopolymer, such as hydroxyl groups, can be modified by reacting with organic compounds [7]. Another approach is to combine both strategies, blending biopolymers (or modified) with salts that can form complexes between the functional groups, inorganic salts, and iron cations from the metallic substrate [42]. Regardless of the strategy, the organic inhibitor obtained should have a higher number of active centers required for adsorption.

Most steps used to obtain grafted biopolymers follow traditional and simple methods such as reactant dissolution (usually in water or common organic solvents), mixing reactants with an initiator, copolymer precipitation, filtration, and homopolymer extraction. The steps involved in obtaining the functionalized biopolymers are summarized in Figure 17.3. The use of complex or expensive methods, such as dialysis, to purify the products obtained is not common because it requires higher investment and more time, even though Wang et al. [5] invested in the high purity of the modified biopolymer. Amino acid-modified pectin conjugates were prepared by adding pectin and amino acids with a mass ratio of 1:1 to deionized water without a catalyst. Graft copolymerization products were dialyzed using a 3,500 Da cut-off membrane to remove the non-reacted amino acids. Careful preparation and purification produced inhibitors with excellent adsorption stability that lasted for 120 hours [5].

Grafting was also used by Singh et al. [22] to modify guar gum with ethyl acrylate in order to prevent corrosion on P110 carbon steel in 15% HCl. Grafting was performed by adding ethyl acrylate to previously dissolved guar gum and potassium persulfate at 308 K for 1 hour. At the end of the reaction, the product was precipitated in acetone, filtered, and vacuum dried at 323 K for 24 hours. This process was responsible for incorporating more oxygen heteroatoms into the chemical structure. The grafted biopolymer achieved high efficiency (92.61%) at 308 K for 6 hours' immersion by EIS. Singh, Ansari, and Quraishi [45] also used guar gum modified with methyl methacrylate to protect P110 carbon steel in CO_2-saturated 3.50 wt.% NaCl. The same steps and synthesis procedure were performed. Potassium persulfate was also used as an initiator by Liu et al. [47] to graft caprolactam onto dextrin (Dxt-g-CPL). Greater inhibition efficiency (83.27%) was observed for the copolymer (Dxt-g-CPL) against the 67.30% non-grafted dextrin at the same concentration as the inhibitor (0.30 g/L). In this sense, introducing a polymeric chain into the biopolymer structure involves four main parameters: the initiator choice, reaction temperature, reaction time, and the modifying agent:biopolymer ratio. Temperature should be chosen as a function not only of initiator activation but also the thermal stability of the biopolymer. Reaction time should be considered in terms of the extension of the desired modification in the biopolymer structure. With respect to the modifying agent, the biopolymer ratio should be selected considering solubility and steric effects because of the modified biopolymer's robust structure.

Another interesting approach is to combine chemical modification with grafting using potassium persulfate as an initiator. Eduok, Ohaeri, and Szpunar [48] chemically modified chitosan (CMCh), which was subsequently grafted with poly(N-vinyl imidazole). Chemical modification was performed by reacting chitosan with chloroacetic acid in alkaline medium (pH 8), stirring the reacting system for 6 hours at 313 K. The product was washed with methanol (70.00%), filtered, and vacuum dried at 328 K for 8 hours. It was also acidified in a nitric acid/methanol mixture, and the excess acid was removed by washing with a methanol/water mixture. Next, CMCh was grafted with poly (N-vinyl imidazole) at 328 K for 30 min using potassium persulfate as an initiator. The grafted biopolymer was precipitated in acetone and dried at 333 K. Thermal analyses indicated that the 60% weight loss observed for chitosan declined after chemical modification, and the thermal degradation rate also decreased after grafting. Electrochemical tests showed that using 500 ppm of CMCh and 200 ppm of the grafted biopolymer did not reduce current density, demonstrating that grafting is more efficient for this biopolymer than chemical modification itself. This is consistent with the fact that the chemical modification provided in this case only introduced oxygen heteroatoms, while grafting provided both oxygen and nitrogen as adsorption sites. Based on this example, it is important to note that different inhibition results may be achieved depending on the functionalization agent, either a polymeric structure or a simple modifying group. In some cases, smaller modifying groups that contain strategic adsorption groups may be more effective than longer polymeric chains.

Another point to highlight is the purification step. Dalhatu et al.'s work [9] used Tert-Butyl hydroperoxide (THB) as an initiator to obtain grafted chitosan. 2-hydroxyethylmethacrylate (PHEMA) was grafted onto chitosan (CS) under indirect ultrasound irradiation. The ungrafted PHEMA homopolymer (homo PHEMA) was removed by Soxhlet extraction to ensure that only the copolymer was applied as corrosion inhibitor. In another study, Dalhatu et al. [15] grafted L-arginine onto chitosan using the traditional thermal method, also removing the homopolymer by Soxhlet extraction. Homopolymer removal after functionalization is rarely reported in the literature. However, it is an important issue because the substance responsible for the inhibitory process has been confirmed.

In relation to chemical modification via a non-grafting process, it is also important to consider removing the excess modifying agent for the reasons previously explained. Farhadian et al. [49] synthesized a sulfonated chitosan and evaluated the compound as a corrosion inhibitor for carbon steel and methane hydrate inhibitors. Sulfonated chitosan was obtained by mixing chitosan in acetic acid for 2 hours at room temperature. Next, 1,3-propanesultone was added to the mixture for 7 hours at 343 K. The mixture was then cooled to room temperature and the product precipitated in acetone. The excess modifying agent was removed by washing with methanol and the product vacuum dried.

In terms of functional groups incorporated into the biopolymer structure, it is important to highlight that there is an optimal limit, after which efficiency will not increase. Luo et al. [50] chemically modified natural glucomannan with 4-aminoazobenzene to add nitrogen and aromatic ring adsorption sites to the only oxygenated biopolymer structure. The modification procedure consisted of adding the biopolymer, succinic acid, and sulfuric acid with thionyl chloride solvent, followed by heating at 363 K for 12 hours. The solvent was then removed and the product purified via silica gel column chromatography. The product yield was 80%. The chemically modified biopolymer containing only oxygenated compounds exhibited the highest efficiency (87%), which is greater than that of the non-modified biopolymer (72.80%), indicating that adding oxygen heteroatoms to the chemical structure increases the number of adsorption sites on the metallic surface. However, the further addition of nitrogen and aromatic rings increases efficiency to 95%. This is due to the already large number of adsorption sites and maximum surface coverage. In addition, steric effects may result from the incorporation of these functional groups and side polymeric chains, hindering adsorption to the metallic surface.

Thus, grafting and other types of modification to the biopolymer structure are responsible for inserting adsorption sites, thereby improving inhibition efficiency. Table 17.1 shows that the most widely used biopolymer backbone is chitosan, followed by polyaspartic acid and β-cyclodextrin, while the most used modifying agent is polyacrylamide, due to the presence of both nitrogen and oxygen heteroatoms and the solubility of the modified biopolymer.

17.3.2 Grafted/Chemically Modified versus Non-Modified Biopolymers

When comparing grafted and non-grafted biopolymers it is also important to confirm the importance of this type of structural modification. Deng, Li, and Du. [24] investigated starch, a natural biopolymer, as a corrosion inhibitor for aluminum in 1 M H_3PO_4. The weight loss of the grafted biopolymer, cassava starch/acrylamide mixture, and isolated compounds (cassava starch, acrylamide) was compared. Cassava starch achieved only 32.50% efficiency, while acrylamide and the grafted polymer reached 45.60 and 90.60%, respectively. In addition, the physical mixture was 57.60% efficient, indicating that inhibition cannot be significant solely by mixing the compounds. As such, chemical modification is necessary to provide the desired increase in efficiency in relation to the isolated biopolymer. The EIS technique showed a first capacitive loop at high frequencies, which is related to the Al_2O_3 film on an aluminum surface. This film was formed by electrode oxidation. The inductive loop at medium frequency is related to the adsorption-desorption of intermediates on the aluminum surface (H_{ads}^+, $H_2PO_4^-$, HPO_4^{2-}, and PO_4^{3-} complex between aluminum and the biopolymer).

Biswas, Pal, and Udayabhanu [23] synthesized xanthan gum-*graft*-poly (acrylamide) as a corrosion inhibitor for mild steel in 15% HCl. The non-grafted polymer displayed inhibition efficiency of 90.80% in 0.50 g/L at 298 K by weight-loss measurements, while its grafted counterpart exhibited 93.18% under the same conditions. This rise in efficiency was not as high as expected given the number of steps in the grafting process. Moreover, both the grafted and non-grafted polymers showed a 60% decrease in efficiency as temperature increased. Thus, it should be questioned if this type of modifying agent is adequate for this biopolymer. The protonated molecules adsorb to the metallic surface with the aid of chloride ions. The grafted biopolymer is more protonated due to the presence of the acrylamide group, which facilitates adsorption to the metallic surface. However, even with the increased protonation of the grafted biopolymer, surface coverage did not improve significantly.

Biswas et al. [4] synthesized dextrin-*graft*-poly(vinyl acetate) as a corrosion inhibitor for mild steel in 15% HCl. The grated biopolymer exhibited higher molecular weight (5.70×10^3 g/mol for dextrin and 8.30×10^4 g/mol for the grafted biopolymer) and greater polydispersion. This higher molecular weight may lead to better surface coverage, which was confirmed by the greater efficiency (98.39% for the grafted biopolymer and 84.56% for non-modified dextrin). In addition, the PDP technique showed a concave-like inflection at higher concentrations (0.10–0.15 g/L), which is related to greater adsorption to the metallic substrate and/or inhibitor/Fe complexes. The grafted biopolymer has extra moiety, bromine atom, and polar vinyl acetate groups, which contribute to inhibition, when compared to dextrin. As such, incorporating unsaturations, more oxygenated adsorption sites, and bromo significantly increase inhibition efficiency, confirming that it is important to select the modifying agent for a specific biopolymer.

Cui et al. [16] chemically modified two chitosan oligosaccharides. The modified biopolymer was subjected to two different transformations, consisting of reactions with benzaldehyde for 10 hours at 353 K or propionaldehyde for 24 hour at 303 K. Comparison between the non-modified and modified chitosan showed an increase in efficiency of approximately 30% at the highest concentration. In addition, the biopolymer modified with an aromatic ring showed greater efficiency, likely due to the higher electron density, resulting in a better surface coverage of the metallic surface (88.59 compared to 85.70%). It is important to note that the corrosion rate (CR) is higher under a dynamic than a static condition. The dynamic environment increases the mass transfer rate of corrosive species, causing accelerated cathodic reactions. In addition, the increased mass transfer of the corrosion product (Fe^{2+}) far from the steel surface hinders $FeCO_3$ protective film formation. An important characterization technique of the inhibiting film is contact angle analysis. The contact angle of the first modified compound increased from 14.70 to 51.40°. However, the presence of hydrophilic groups, such as -OH, $-NH_2$, and $\left((CH_3)_4 N^+ \right)$, resulted in a non-ideal inhibition process. The second modification increased the contact angle to 80.70 and 95.50° after alkane and benzene structures were incorporated, respectively. The higher hydrophobicity of the benzene ring resulted in more efficient inhibition.

The previous paragraphs in this section reveal important aspects when the functionalized biopolymer is compared with its non-modified counterpart. First, chemical modification via synthesis steps is required and a simple physical mixture will not be enough to improve efficiency in relation to the original biopolymer. The modifying agent may not always be suitable for a specific application, especially in aggressive acidic media such as 15% HCl, as observed by Biswas, Pal, and Udayabhanu [23]. Thus, both the biopolymer and the modifying agent should be strategically selected for the desired application, and the improvement in efficiency should be statistically significant to compensate for all the functionalization steps.

17.3.3 Metallic Surface Characterization

Organic corrosion inhibitors can prevent interaction between ionic species from the electrolyte solution and metallic surfaces by adsorbing and depositing a protective layer over the metallic surface [42]. Unfortunately, the limited movement of the macromolecular organic compounds in solution alters the efficiency of these inhibitors to reach the metallic surface [13]. In addition, biopolymers, such as gums, can adsorb water in the outer layer, forming micelles without sufficient stirring, which can affect the corrosion inhibition. Although these issues can be overcome by enhancing the presence of functional groups through chemical modification, it is still significantly influenced by the molecular weight and the solution pH [42].

Singh et al. [39] studied copper surface protection with 600 mg/L of guar gum grafted 2-acrylamido-2-methylpropanesulfonic acid (GG-AMPS) using scanning electrochemical microscopy (SECM), scanning electron microscopy (SEM) and atomic force microscopy (AFM). SECM detected very high currents on 2D maps and 3D structures for the non-inhibited copper surface and lower currents for the inhibited surface. SEM and AFM images showed that copper surface roughness decreased with the addition of GG-AMPS. Analysis of the inhibited and smoother copper surfaces revealed the presence of a hydrophobic film that repelled the saline solution.

Polydopamine was synthesized and combined with sodium molybdate due to its capacity to adhere to the substrate through its functional groups such as catechol, quinone, indole, and amine. Scanning electron microscopy (SEM) and energy dispersive x-ray spectroscopy (EDS) analysis confirmed the morphological and chemical effects on mild steel after inhibitor adsorption. The surface film formed in the inhibitor-free 3.50% NaCl solution, inorganic inhibitor sodium molybdate (1,000 ppm solution Na_2MoO_4), and hybrid polydopamine-molybdate inhibitor (250:750 ppm PDA/MoO_4^{2-}) were observed. The inhibitor-free solution formed a porous film over the surface containing snowflake-shaped particles, which are related to iron oxides and hydroxides, confirmed by the high Fe and O elements detected. The inorganic Mo solution sample revealed considerably high Mo and O percentages, indicating that molybdate ions successfully adsorbed on the surface. However, the high film porosity prevented it from efficiently protecting the surface, evident in its weak inhibitive performance in electrochemical tests. The hybrid organic/inorganic inhibitor (250:750 ppm PDA/MoO_4^{2-}) deposited a dense uniform layer containing spherical particles, a completely different shape associated with PDA/Mo/Fe complexes. The high Fe percentage and presence of Mo, C, N, and O confirmed the formation of a compact protective layer of both inorganic and organic inhibitors on the steel specimen [42].

Liu et al. [47] analyzed dextrin (Dxt) and its graft copolymer with caprolactam (Dxt-g-CPL) films on J55 steel surfaces after immersion in 1 M HCl solution. The contact angle of the steel surface increased from 23.40°, when immersed only in 1 M HCl solution, to 71.00 and 98.40° with 300 mg/L Dxt and 300 mg/L Dxt-g-CPL, respectively. This was attributed to the hydrophilic group present in the corrosion inhibitor that formed an adsorption film, while the hydrophobic group formed a hydrophobic barrier, changing the hydrophilicity of the metallic substrate. This hydrophobic barrier prevents contact between the surface and the corrosive media. Thus, Dxt-g-CPL showed the greatest change in contact angle, with a better anti-corrosion effect.

Techniques such as SEM and AFM reveal corrosion morphology, indicating whether it is uniform or localized, while SEM-EDS can be applied to confirm the presence of an inhibiting film on the metallic surface. The hydrophobicity of the inhibiting film can be evaluated by contact angle analysis. Additionally, FTIR can be applied to understand the interaction between functional groups of the biopolymer and the metallic substrate by comparing the shift bands.

17.3.4 Inhibition Mechanism

Macromolecules repel corrosive substances through long hydrophobic carbon chains that can form a hydrophobic barrier, which in turn prevents water molecules from entering the metallic substrate matrix. It can also slow the diffusion of iron ions on the steel surface to the corrosive solution. Mixed corrosion inhibitors are usually an organic compound that contains O, N, P, S, and other polar atoms, as well as nonpolar alkyl-based groups. Adsorption generally occurs between the polar groups and the metal surface, while the orientation of the nonpolar groups is away from the surface, forming a hydrophobic layer [5].

Babaladimath, Badalamoole, and Nandibewoor [51] grafted xanthan gum (XG) with polyaniline (PANI) to act as a corrosion inhibitor for Al in 1 M HCl. Al undergoes oxidation, forming an oxide layer. The stability of this layer declines in the presence of aggressive chloride ions ($Al(OH)_2 Cl_2^-$) , which penetrate the oxide film and lead to pitting. The inhibitor containing heteroatoms adsorbs onto the metallic surface, preventing oxychloride complex formation and, consequently, metal dissolution. This adsorption occurs in four ways: i) electrostatic interaction between the inhibitor molecules and the metallic surface; ii) interaction through the unshared electron pair of inhibitor molecules; iii) interaction of the conjugated π-electron with the substrate; iv) mixed interaction, whereby the aforementioned types occur simultaneously [4]. In this sense, the lone electron pair in amines may become protonated, adsorbing onto the cathodic sites of Al, reducing hydrogen evolution. Additionally, the grafted biopolymer may adsorb onto anodic sites through π electrons in the PANI chains and lone electron pairs of electrons in the free hydroxyl groups of XG, preventing anodic dissolution of Al.

Chai et al. [28] chemically modified polyaspartic acid with cysteamine to investigate the final product as a corrosion inhibitor for mild steel in 0.50 M H_2SO_4. The study compared unmodified polyaspartic acid with polyaspartic acid chemically modified using cysteamine. Mild steel surfaces adsorb SO_4^{2-} layers, which will electrostatically interact with protonated N atoms. Additionally, chemical adsorption will occur between N and S heteroatoms of the modified biopolymer and the vacant d-orbitals of the metallic surface.

Deng, Li, and Du [24] investigated starch, a natural biopolymer, as a corrosion inhibitor for aluminum in 1 M H_3PO_4. The grafted biopolymer contains N and O atoms, and the biopolymer can be protonated in the acid solution. Thus, the modified biopolymer exists in both neutral and protonated forms. Acid anions ($H_2PO_4^-$, HPO_4^{2-}, and PO_4^{3-}) from the corrosive process adsorb onto the metallic surface and interact electrostatically with the protonated biopolymer. Moreover, the neutral and protonated forms of the biopolymer can donate electrons to the vacant p-orbitals of Al through

O and N. Both inhibitor forms also have hydroxyl and carboxyl groups that can coordinate with Al^{3+}.

Although studies aimed at modifying biopolymers have shown considerable improvement in anticorrosive properties after chemical functionalization, there is still work to be done. Gaps were identified in the biopolymer and modified biopolymer scenarios. The experimental studies were mostly carried out under mild conditions with low temperatures (below 333 K) and diluted acids. Future research should address how to improve inhibitor efficiency in more aggressive, highly acidic and high-temperature environments. Designing a suitable chemical modification strategy is important in achieving optimal efficiency in more aggressive conditions. The advantage is that the mobility of these macromolecules may improve as temperature increases. However, biopolymer degradation must be avoided, making proper chemical modification essential. This hypothesis should also be investigated in future studies.

17.3.5 Computational Studies

Research aimed at preventing corrosion by applying inhibitors has evolved over the years. Traditionally, experimental investigation of potential molecules was carried out using weight loss and electrochemical tests to determine inhibition efficiency for a specific medium and material. Later, experimental studies were combined with computational modeling to understand the chemical reactivity and inhibition potential of certain compounds [72]. Currently, efforts are largely aimed at combining well-established experimental methods with theoretical approaches to save time and resources and improve knowledge of chemical mechanisms.

Modeling biopolymers is a challenge due to the large size and volume of the bioactive compounds. The computational cost of modeling large molecules and acidic or saline media is a limitation of in silico methods. In order to overcome these issues, certain assumptions are made to enable the construction of a reliable model. It is well known that the active sites of organic molecules are responsible for chemical or physical interactions with metallic surfaces. These active sites are composed of functional groups and electron-dense regions. Other effects that can enhance corrosion inhibition have yet to be evaluated by computer modeling include hydrophobic effect, steric hindrance, and polarizability, among others. Nevertheless, analyzing monomer units and modifying agents seems to be a reasonable approach to minimally represent these large biomolecules.

The inhibition potential of modified biopolymers has been investigated by comparing the unmodified biopolymer, modifying agent, and one or more units of the two combined. A number of studies have used only theoretical calculations based on density functional theory (DFT) to establish a correlation between the structure and efficiency of corrosion inhibitors [7, 13, 16, 23, 49, 15, 29]. Several investigations combine results obtained by quantum parameters with molecular dynamics (MD) simulations to broaden the studies, using aqueous media and a metallic surface that interacts with the biopolymers [5, 11, 17, 19, 22, 32, 63, 70]. Although molecular orbitals obtained via DFT calculations provide useful information, few studies have used only MD simulations to orient the target compound towards the Fe surface (110) [4]. Indeed, the number of published papers on corrosion inhibitors that apply MD simulations has only increased considerably in the last 5 years, from fewer than 30 a year to over 100 since 2019 [72].

Some important quantum chemical descriptors have been explored via DFT modeling of modified biopolymers. Most are based on frontier molecular orbital energies, such as the highest occupied (E_{HOMO}) and lowest unoccupied molecular orbitals (E_{LUMO}) and the energy gap (ΔE) between them. Other important reactivity descriptors derived from frontier orbital energies include ionization energy (I), electron affinity (A) [22], absolute electronegativity (χ), and absolute hardness (η) [17]. In addition to the relevance of hardness and electronegativity, quantum mechanical descriptors also provide parameters such as the number of electrons transferred (ΔN), which suggests the intensity and direction of electron transfer between the biopolymer and metallic surface [23].

The inhibition properties of corrosion inhibitors are dependent on frontier molecular orbitals. The chemical prediction of corrosion inhibition has been attributed mainly to the concept of acid-base and donor-acceptor characteristics, with the electron-donating species as the inhibitor molecule and the electron acceptor the metallic surface. Thus, the vacant d orbitals on the metal preferentially interact with the electron-rich reactive centers in the biopolymers. This prevents corrosion because the inhibitor molecules are deposited on the metallic surface, blocking the active sites of the substrate [72]. The electron donating tendency is typically associated with HOMO energy, whereby the higher the E_{HOMO} (less negative), the greater the electron donating ability of the inhibitor molecule. This indicates that suitable acceptor atoms (such as those present in metallic substrates) with low energy and empty electron orbitals would be able to receive these electrons. On the other hand, the electron accepting tendency of an inhibitor molecule can be predicted by LUMO energy, whereby the probability of a corrosion inhibitor accepting electrons increases with a decline in E_{LUMO} [23]. In other words, high E_{HOMO} or low E_{LUMO} values indicate that inhibitor molecules can more easily share electrons with the metallic surface and are therefore better able to inhibit corrosion [26].

Another important quantum descriptor is the energy gap between the frontier orbitals. According to the hard-soft acid-base principle, hard acids prefer to coordinate with hard bases, while soft acids coordinate better with soft bases. Iron, the main metal atom of steel, is considered a soft acid that binds with soft molecules, producing a smaller HOMO/LUMO energy gap (ΔE) [7]. A small ΔE indicates better inhibition efficiency due to the low energy required to remove an electron from the last occupied orbital [23]. The energy gap represents the charge transition process, whereby inhibitor-metal system interaction is stronger for smaller ΔE values [26]. Ionization energy (I) and electron affinity (A) can be calculated using HOMO and LUMO orbital energies, as shown in Equations 1 and 2 [22].

$$I = -E_{HOMO} \tag{1}$$

$$A = -E_{LUMO} \tag{2}$$

Although conclusions can be drawn based on these parameters, they are typically used to obtain parameters such as electronegativity (χ) and hardness (η), with the latter also used to obtain softness (σ). Equations 3–5 are the most widely used for these calculations [22].

$$\chi = \frac{I + A}{2} \tag{3}$$

$$\eta = \frac{I - A}{2} \tag{4}$$

$$\sigma = 1 / \eta \tag{5}$$

The electron transfer number (ΔN) can be calculated using the values of electronegativity and hardness of the inhibitor and the metallic surface, largely represented as iron (Fe) [23], as shown in Equation 6.

$$\Delta N = \frac{\chi_{Fe} - \chi_{inh}}{2\left(\eta_{Fe} + \eta_{inh}\right)} \tag{6}$$

where χFe and χinh are the absolute electronegativity of iron and the corrosion inhibitor, respectively; and ηFe and ηinh the calculated absolute hardness of iron and the corrosion inhibitor molecule, respectively. The theoretical iron (Fe) values used for χ and η are 7 eV and 0 eV, respectively [23].

An interpretation of ΔN previously proposed by Lukovits, Kálmán, and Zucchi [73] is widely accepted by many authors. The calculated ΔN value indicates the intensity and direction of electron transfer from the organic molecule to metal when $\Delta N > 0$ and the opposite action in the case of $\Delta N < 0$ [26]. Another approach that has been widely used to interpret electron transfer is comparing the monomers and modified biopolymer, whereby $\Delta N < 3.6$ is associated with greater inhibition efficiency and better electron donating ability of the inhibitor to the metallic surface [23].

The relationship between the modeled inhibitor molecule and the metallic surface can also be described by estimating electron back donation. When both donation and back donation occur simultaneously, the energy change is directly proportional to the hardness (η) of the corrosion inhibitor molecule. This relationship is given by Equation 7 [7].

$$\Delta E_{Back-donation} = -\frac{\eta}{4} \tag{7}$$

Kesari et al. [7] considered this an important parameter due to the negative values observed for chitosan (–0.03 eV) and chitosan-*graft*-polyacrylamide (–0.02 eV), indicating back donation for both polymers.

Another important parameter obtained by DFT is the dipole moment, widely used to describe the polarity or polarizability of a molecule. The higher the dipole moment (α), the greater the inhibition efficiency of a molecule. This can be attributed to the increased bonding strength between the metallic surface and inhibitor molecule. It has also been suggested that inhibition efficiency increases when the value of the dipole moment declines [72]. However, opinions on the influence of the dipole moment differ. It has been suggested that a smaller dipole moment will increase the accumulation of corrosion inhibitor molecules on the metallic surface [11]. By contrast, other studies [26, 63] have shown that a larger dipole moment improves inhibition efficiency by enhancing dipole-dipole interaction between organic molecules and the metallic substrate.

The main findings for DFT calculations of biopolymer monomers and their modifying agents are presented in Table 17.2. Analysis of this data demonstrates that corrosion inhibition efficiency can be estimated as a function of the hardness, electronegativity, and energy gap values of the molecules studied. These results were compared to the experimental findings, indicating that computational results were in line with laboratory tests. Chitosan is among the most reported biopolymer backbones and acryl amide the major modifying agent. Despite the methodology differences, some conclusions can be drawn. The energy gap between the frontier orbitals HOMO-LUMO is usually higher for chitosan when compared to its analogs. The lower values obtained for modified chitosan demonstrated stronger inhibitor-metal system interaction, indicating that grafting or chemical modification improved inhibitory performance. The same trend is observed for hardness. Since metal is considered soft and would not interact well with hard molecules, modification enhanced this molecular property. Several studies [11, 13, 17] have calculated the quantum parameters for protonated species that further improved these properties.

Xanthan gum and xanthan gum-*graft*-poly(acrylamide) (XG-*g*-PAM) were described using their respective optimized monomers to indicate how grafting affects inhibition performance. The grafted gum, XG-*g*-PAM, exhibited lower ΔE and ΔN values but higher α when compared to its non-grafted counterpart. This demonstrates the greater ability of XG-*g*-PAM to donate electrons to the metallic surface (ΔN), confirming the better inhibition efficiency obtained in corrosion tests [23].

The adsorption mechanism of a cassava starch-sodium allylsulfonate-acryl amide graft copolymer (CS-SAS-AAGC) was theoretically analyzed by quantum chemical calculation and MD simulation. Starch grafted with a binary or ternary graft copolymer may have more adsorption centers on metallic surfaces than starch grafted with one vinyl monomer. The corrosion inhibitor was obtained by grafting sodium allylsulfonate (SAS) and acryl amide (AA) onto cassava starch (CS). Since starch contains a large number of glucose units, a single glucose (Glc) molecule was used to simplify the model. Similarly, only one SAS and AA molecule were studied, although CS-SAS-AAGC contains many SAS and AA units. The optimized molecular structures of Glc, Glc-SAS-AA, and Glc-AA-SAS at the level of GGA/BLYP/DND/COSMO were obtained. The chair conformation of Glc was maintained and remained unchanged in Glc-SAS-AA and Glc-AA-SAS [63]. A similar approach was used to study a cassava starch copolymer

TABLE 17.2 Chemical Parameters Obtained by DFT Calculations of Biopolymer Monomers and Their Modifying Agents

Corrosion Inhibitor Model	E_{HOMO} (eV)	E_{LUMO} (eV)	ΔE (eV)	Hardness (η) (eV)	Electronegativity (χ) (eV)	ΔN Electrons transferred (eV)	Dipole moment μ (Debye)	Reference
Pectin-amino acid derivatives with glycine (P-Gly) and Lysine (P-Lys)	−7.116 (P)	−0.381 (P)	6.735 (P)	3.368 (P)	3.749 (P)	0.483 (P)	1.817 (P)	[5]
	−7.153 (P-Gly)	−0.435 (P-Gly)	6.718 (P-Gly)	3.359 (P-Gly)	3.794 (P-Gly)	0.477 (P-Gly)	4.612 (P-Gly)	
	−6.583 (P-Lys)	−0.697 (P-Lys)	5.886 (P-Lys)	2.943 (P-Lys)	3.641 (P-Lys)	0.571 (P-Lys)	2.868 (P-Lys)	
Chitosan (CS), chitosan-*graft*-acrylamide (CS-g-PAM)	−0.230 (CS)	−0.060 (CS)	0.290 (CS)	0.140 (CS)	0.080 (CS)	16.110 (CS)	1.100 (CS)	[7]
	−0.210 (CS-g-PAM)	−0.020 (CS-g-PAM)	0.230 (CS-g-PAM)	0.110 (CS-g-PAM)	0.090 (CS-g-PAM)	19.810 (CS-g-PAM)	4.510 (CS-g-PAM)	
Chitosan-*graft*-D-glucose (COS-g-Glu) neutral and protonated forms	−6.383 (neutral)	0.182 (neutral)	6.565 (neutral)			0.262 (neutral)	8.110 (neutral)	[11]
	−12.070 (protonated)	−7.956 (protonated)	4.114 (protonated)			−1.262 (protonated)	28.420 (protonated)	
Vanillin (Van) modified chitosan, neutral (Van-Cht) and protonated (Van-Cht⁺) forms	−6.095 (Van)	−1.462 (Van)	4.630 (Van)	2.320 (Van)	3.780 (Van)			[13]
	−4.963 (Van-Cht)	−0.914 (Van-Cht)	4.040 (Van-Cht)	2.020 (Van-Cht)	2.940 (Van-Cht)			
	−6.133 (Van-Cht+)	−2.841 (Van-Cht+)	3.290 (Van-Cht+)	1.650 (Van-Cht+)	4.490 (Van-Cht+)			
Chitosan (CS) and chitosan-*graft*- L-arginine (CS-g-L-arg)	−9.850 (CS)	2.130 (CS)	11.980 (CS)	5.990 (CS)	3.860 (CS)	9.404 (CS)		[15]
	−9.120 (CS-g-L-arg)	0.290 (CS-g-L-arg)	9.410 (CS-g-L-arg)	4.705 (CS-g-L-arg)	4.415 (CS-g-L-arg)	6.081 (CS-g-L-arg)		
N-propyl chitosan (PHC) and N-benzyl chitosan (BHC) quaternary ammonium salts	−4.594 (PHC)	−0.627 (PHC)	3.967 (PHC)	1.984 (PHC)	2.611 (PHC)	1.106 (PHC)		[16]
	−4.271 (BHC)	−1.480 (BHC)	2.791 (BHC)	1.396 (BHC)	2.876 (BHC)	1.477 (BHC)		
Chitosan (CS), triazole (AMT) and triazole-modified chitosan (CS-AMT)	−6.067 (AMT)	−0.465 (AMT)	5.603 (AMT)	2.801 (AMT)	3.266 (AMT)			[17]
	−6.593 (CS)	1.724 (CS)	8.317 (CS)	4.159 (CS)	2.435 (CS)			
	−5.582 (CS-AMT)	−0.435 (CS-AMT)	5.147 (CS-AMT)	2.574 (CS-AMT)	3.009 (CS-AMT)			
	−6.815 (CS-AMT) protonated	−1.899 (CS-AMT) protonated	4.916 (CS-AMT) protonated	2.458 (CS-AMT) protonated	4.357 (CS-AMT) protonated			
Alginate (Alg) and Alginate-*graft*-8-hydroxyquinoline (HQ-g-Alg)	−6.334 (Alg)	−0.317 (Alg)	6.017 (Alg)			0.248 (Alg)	2.127 (Alg)	[19]
	−2.218 (HQ-g-Alg)	−0.274 (HQ-g-Alg)	1.944 (HQ-g-Alg)			1.838 (HQ-g-Alg)	15.523 (HQ-g-Alg)	

(*Continued*)

TABLE 17.2 *(Continued)*   Chemical Parameters Obtained by DFT Calculations of Biopolymer Monomers and Their Modifying Agents

Corrosion Inhibitor Model	E_{HOMO} (eV)	E_{LUMO} (eV)	ΔE (eV)	Hardness (η) (eV)	Electronegativity (χ) (eV)	ΔN Electrons transferred (eV)	Dipole moment μ (Debye)	Reference
Guar gum (GG) and modified guar gum-ethyl acrylate (GG-EEA)	−7.351 (GG) −6.947 (GG-EEA)	−0.446 (GG) −0.126 (GG-EEA)	6.904 (GG) 6.821 (GG-EEA)	3.452 (GG) 3.410 (GG-EEA)	3.899 (GG) 3.536 (GG-EEA)			[22]
Xanthan gum (XG) and copolymer grafted with acryl amide (XG-g-PAM)	−6.589 (XG) −6.534 (XG-g-PAM)	1.657 (XG) 0.746 (XG-g-PAM)	8.246 (XG) 7.280 (XG-g-PAM)			0.550 (XG) 0.564 (XG-g-PAM)	4.349 (XG) 6.072 (XG-g-PAM)	[23]
Polyaspartic acid (PASP) and modified polyaspartic acid-Schiff base (PASP/SB)	−5.767 (PASP) −5.508 (PASP/SB)	−1.916 (PASP) −2.889 (PASP/SB)	3.851 (PASP) 2.619 (PASP/SB)	1.926 (PASP) 1.309 (PASP/SB)	3.842 (PASP) 4.199 (PASP/SB)	0.820 (PASP) 1.070 (PASP/SB)		[29]
Chitosan (CS) and its derivatives (CS-1 and CS-2)	−6.564 (CS) −6.611 (CS-1) −6.609 (CS-2)	1.038 (CS) −1.927 (CS-1) −2.034 (CS-2)	7.603 (CS) 4.684 (CS-1) 4.575 (CS-2)	3.801 (CS) 2.342 (CS-1) 2.287 (CS-2)	2.763 (CS) 4.269 (CS-1) 4.322 (CS-2)	0.557 (CS) 0.583 (CS-1) 0.585 (CS-2)		[32]
Modified hydroxyethylcellulose (CHEC)	−4.469	−3.906	0.563					[38]
Sulfonated chitosan (SCS) with and without acetyl group	−6.648 (SCS without acetyl) −6.576 (SCS with acetyl)	−1.706 (SCS without acetyl) −0.043 (SCS with acetyl)	4.941 (SCS without acetyl) 6.533 (SCS with acetyl)	2.471 (SCS without acetyl) 3.267 (SCS with acetyl)	−4.177 (SCS without acetyl) −3.309 (SCS with acetyl)			[49]
Glucose (Glc), sodium allylsulfonate (SAS) and acryl amide (AA)	−6.013 (Glc) −5.549 (Glc-SAS-AA) −5.045 (Glc-AA-SAS)	0.466 (Glc) −0.600 (Glc-SAS-AA) −0.710 (Glc-AA-SAS)	6.479 (Glc) 4.947 (Glc-SAS-AA) 4.335 (Glc-AA-SAS)	3.240 (Glc) 2.474 (Glc-SAS-AA) 2.168 (Glc-AA-SAS)	2.774 (Glc) 3.076 (Glc-SAS-AA) 2.878 (Glc-AA-SAS)	0.652 (Glc) 0.793 (Glc-SAS-AA) 0.951 (Glc-AA-SAS)	5.450 (Glc) 34.920 (Glc-SAS-AA) 51.922 (Glc-AA-SAS)	[63]
Glucose (Glc) as starch and 1, 2, 3 units of acryl amide (AA)	−6.013 (Glc) −6.010 (Glc-AA) −5.910 (Glc-AA2) −5.885 (Glc-AA3)	0.466 (Glc) −0.706 (Glc-AA) −0.788 (Glc-AA2) −0.809 (Glc-AA3)	6.479 (Glc) 5.304 (Glc-AA) 5.122 (Glc-AA2) 5.076 (Glc-AA3)				5.450 (Glc) 6.156 (Glc-AA) 6.910 (Glc-AA2) 11.974 (Glc-AA3)	[70]

Cassava starch (CS)-sodium allylsulfonate (SAS)-acryl amide (AA) ternary copolymer (CS-SAS-AA) starch as (amylose disaccharide (ALDS) and amylopectin disaccharide (APDS)	−5.947 (ALDS)	0.297 (ALDS)	6.244 (ALDS)	3.122 (ALDS)	2.825 (ALDS)	0.669 (ALDS)	[74]
	−6.008 (APDS)	0.034 (APDS)	6.042 (APDS)	3.021 (APDS)	2.987 (APDS)	0.664 (APDS)	
	−5.280 (ALDS-SAS-AA)	−0.571 (ALDS-SAS-AA)	4.709 (ALDS-SAS-AA)	2.040 (ALDS-SAS-AA)	2.926 (ALDS-SAS-AA)	0.865 (ALDS-SAS-AA)	
	−5.440 (ALDS-AA-SAS)	−0.663 (ALDS-AA-SAS)	4.777 (ALDS-AA-SAS)	2.388 (ALDS-AA-SAS)	3.052 (ALDS-AA-SAS)	0.826 (ALDS-AA-SAS)	
	−5.689 (APDS-SAS-AA)	−0.595 (APDS-SAS-AA)	5.094 (APDS-SAS-AA)	2.547 (APDS-SAS-AA)	3.142 (APDS-SAS-AA)	0.757 (APDS-SAS-AA)	
	−5.367 (APDS-AA-SAS)	−2.417 (APDS-AA-SAS)	2.950 (APDS-AA-SAS)	1.475 (APDS-AA-SAS)	3.892 (APDS-AA-SAS)	1.054 (APDS-AA-SAS)	

grafted with acryl amide [70]. Glucose was used to represent starch and one (Glc-AA), two (Glc-AA2), and three units of acryl amide monomer (Glc-AA3) were combined to form the biopolymer model. Both studies found smaller energy gaps between HOMO-LUMO for modified glucose, showing enhanced glucose-metal system interaction. Moreover, adding an acryl amide monomer to glucose increased the dipole moment (α), which is widely associated with greater polarity or polarizability of modified glucose. A larger dipole moment indicates greater inhibition efficiency for the modified biopolymer (Glc-AA3).

Analysis of the use of DFT methods to evaluate inhibition potential demonstrates that the same chemical descriptors and findings obtained for traditional corrosion inhibitors have been applied for biopolymers. No coarse grain models or hybrid methods have been used to describe these large molecules and combining the two would likely be a better approach. The chemical reactivity of functionalized biopolymers could be better described if computational studies combined methods established in the literature (DFT models) with coarse grain calculations. Whereas the former only allows for small molecules (containing few atoms), the latter enables large structures to be studied by considering molecule fragments as opposed to individual electrons. A hybrid model that combines these two methods would provide a new set of descriptors more representative of real systems.

The chemical reactivity and inhibition potential of modified biopolymers has yet to be properly described due to the size limitation of monomer molecules. Another important aspect is the methodology applied. DFT calculations with appropriate functionals that consider dispersion correction should be used to calculate dimers or even two monomers interacting. As model size and the presence of side alkyl chains increases, the formation of instantaneous dipole moments in the electron distribution of molecules becomes more important. Since grafting aims to insert long alkyl chains onto the biopolymer backbone, incorporating dispersion correction into traditional functionals plays a more important role.

17.3.5.1 Modeling Surfaces

Molecular dynamic (MD) simulations are powerful tools to theoretically study:

i. The adsorption orientation of the inhibitor on the metallic surface, mimicking the real environment (aqueous or acid media).

ii. The calculated adsorption energy or interaction/binding energy using a more realistic approach.

Biswas et al. [4] used MD simulations to study dextrin-*graft*-poly(vinyl acetate). The Fe (110) surface was relaxed by minimizing its energy and enlarged to a (10 × 10) super cell, providing a larger surface to interact with the inhibitors analyzed. A super cell containing one inhibitor molecule and 500 H_2O, 5 H_3O^+, 5 Cl^- was built to simulate the acidic media. The simulation was carried out at 303 K, using the NVT ensemble (constant number of atoms, constant-volume, constant-temperature) and condensed-phase optimized molecular potentials for atomistic simulation studies (COMPASS) force field. Interactions between modified biopolymers and steel (represented as Fe (110) surface) can be simulated and rationalized based on the interaction and binding energies calculated by Equations 8 and 9.

$$E_{interaction} = E_{total} - \left(E_{surface+solution} + E_{inhibitor}\right) \tag{8}$$

$$E_{Binding} = -E_{interaction} \tag{9}$$

where E_{total} is the energy for the entire system, $E_{surface + solution}$ the total energy of the surface (Fe (110)) and the solution with no inhibitor molecule, and $E_{inhibitor}$ the total energy of the inhibitor studied.

Interaction between the surface and the inhibitor can also be obtained via adsorption energy. Li et al. [70] used this parameter to evaluate an adsorption system, placing the inhibitor layer (cassava starch copolymer with acryl amide) onto an Al (111) supercell. Another study [63] used the same strategy to adsorb glucose, sodium allylsulfonate and acryl amide onto an Al (111) supercell containing 900 Al atoms. The adsorption energy was calculated by Equation 10.

$$\Delta E_{ads} = E_{total} - \left(E_{inh} + E_{surface}\right) \tag{10}$$

where E_{total} is the total energy of the system containing the surface with the adsorbed corrosion inhibitor and E_{inh} and $E_{surface}$ the energies of the corrosion inhibitor and the Al(111) surface, respectively.

Li et al. [74] also built a supercell using iron to represent steel, containing 1331 Fe atoms. MD simulations were performed to study the adsorption behavior of a ternary copolymer (CS-SAS-AAGC) obtained from cassava starch (CS)-sodium allylsulfonate (SAS)-acryl amide (AA). The biopolymer inhibitor models were based on starch represented by amylose disaccharide (ALDS) and amylopectin disaccharide (APDS) combined with a monomer of sodium allylsulfonate and acryl amide. The adsorption modes were calculated for four grafted biopolymer forms of ALDS-SAS-AA, ALDS-AA-SAS, APDS-SAS-AA, or APDS-AA-SAS on an Fe (1) surface.

Despite efforts to describe corrosion inhibitors using models suited to DFT and MD calculations, designing accurate models at affordable computational costs remains a challenge. MD simulations are not widely used and only aqueous or acid media have been modeled. Saline media or other environments have yet to be simulated to determine whether the behavior of modified biopolymers changes in these environments.

17.4 CONCLUSIONS AND FINAL CONSIDERATIONS

- The main factors that influence inhibitor performance are molecular size, weight, composition, and the nature of the anchoring groups. Environmental factors that influence performance are pH, concentration, exposure time, and temperature.

- Copolymers can improve chemical and physical properties of substances used as corrosion inhibitors, resulting in larger size, greater density, and more anchoring functional groups. The higher number of anchoring sites is not the only important factor, since corrosion inhibitor concentration and selecting an adequate modifying agent also contribute to the desired result. The most widely used biopolymers in the literature reviewed were chitosan, polyaspartic, and ß-cyclodextrin, while the modifying agent was polyacrylamide.

- Chitosan is the most extensively studied copolymer in the literature, with promising combinations in terms of water solubility, surface adsorption, and efficiency. Graphitization is the most widely researched chitosan modification method. Among the factors that enhance chitosan performance are the ability to form chelates and complex interactions with metal ions, the presence of amine and hydroxyl groups supporting the functionalization, and nanocomposite formation.

- Nanocomposites are interspersed in the biopolymer structure via hybridization, improving corrosion inhibition activity when compared to that obtained individually. Additionally, greater dispersion and reduced film porosity make modified biopolymers a better option than their unmodified counterparts.

- Factors such as concentration, immersion time, and temperature directly influence inhibitor performance. In general, an increase in concentration improves efficiency; however, this process is limited by surface coverage. As such, efficiency does not improve after a certain concentration level once surface coverage has been achieved.

- Ideally, efficiency should be maintained over long periods of immersion, which depends on the application and aggressiveness of the media. Most studies cover an 18-hour period. A decline in efficiency was also observed, likely due to hydrolysis and biodegradation.

- The selection of suitable inhibitors is also influenced by the operating temperature range of the system. High temperatures may favor desorption of the inhibiting molecules from the surface or inhibitor degradation, thereby reducing inhibition efficiency. This effect is less pronounced for chemically modified and grafted inhibitors when compared to their unmodified counterparts, due to the presence of more and stronger adsorption sites and electrostatic interactions on the metal surface.

- Computational studies contribute significantly to understanding the chemical descriptors that favor interaction between the metallic surface and corrosion inhibitor, supporting the proposed inhibitory mechanism. The most widely used computational tools were density functional theory and molecular dynamic simulations, which make it possible to identify the most reactive active sites and demonstrate how chemical modification improves inhibition performance.

- Surface characterization techniques are used to assess surface coverage by the inhibitors. SEM/EDS analysis identifies the presence of functional groups on the surface, as well as surface morphological and chemical modifications resulting from interaction with and without the inhibitor. Techniques such as AFM can also be used to assess the roughness effect and contact angle measurements provide information on inhibitor film hydrophobicity.

- Futures perspectives: homopolymer removal and the development of modified macromolecules that are more resistant to aggressive media should be the focus of future research. The scale-up of functionalization techniques should be addressed in conjunction with the production costs of these compounds.

REFERENCES

1. M. Ramezanzadeh, G. Bahlakeh, Z. Sanaei, B. Ramezanzadeh. Corrosion inhibition of mild steel in 1 M HCl solution by ethanolic extract of eco-friendly Mangifera indica (mango) leaves: Electrochemical, molecular dynamics, Monte Carlo and ab initio study. *Applied Surface Science*, 2019, **463**, 1058–1077.
2. C. Wang, C. Zou, Y. Cao. Electrochemical and isothermal adsorption studies on corrosion inhibition performance of β-cyclodextrin grafted polyacrylamide for X80 steel in oil and gas production. *Journal of Molecular Structure*, 2021, **1228**, 129737.
3. Z. Ma, M. Sun, A. Li, G. Zhu, Y. Zhang. Anticorrosion behavior of polyvinyl butyral (PVB)/polymethylhydrosiloxane (PMHS)/chitosan (Ch) environment-friendly assembled coatings. *Progress in Organic Coatings*, 2020, **144**, 105662.
4. A. Biswas, D. Das, H. Lgaz, S. Pal, U. G. Nair. Biopolymer dextrin and poly (vinyl acetate) based graft copolymer as an efficient corrosion inhibitor for mild steel in hydrochloric acid: Electrochemical, surface morphological and theoretical studies. *Journal of Molecular Liquids*, 2019, **275**, 867–878.
5. D.-Y. Wang, J.-H. Wang, H.-J. Li, Y.-C. Wu. Pectin-amino acid derivatives as highly efficient green inhibitors for the corrosion of N80 steel in CO_2-saturated formation water. *Industrial Crops & Products*, 2022, **189**, 115866.
6. M. T. Ramesan, K. Surya. Studies on electrical, thermal and corrosion behaviour of cashew tree gum grafted poly(acrylamide). *Polymers from Renewable Resources*, 2016, **7**, 81–99.
7. P. Kesari, G. Udayabhanu, A. Roy, S. Pal. Chitosan based titanium and iron oxide hybrid bio-polymeric nanocomposites as potential corrosion inhibitor for mild steel in acidic medium. *International Journal of Biological Macromolecules*, 2023, **225**, 1323–1349.
8. Q. H. Zhang, N. Xu, Z. N. Jiang, H. F. Liu, G. A. Zhang. Chitosan derivatives as promising green corrosion inhibitors for carbon steel in acidic environment: Inhibition performance and interfacial adsorption mechanism. *Journal of Colloid and Interface Science*, 2023, https://doi.org/10.1016/j.jcis.2023.02.141.
9. S. N. Dalhatu, S. Khoonsap, S. Suwannarat, S. Amnuaypanich, S. Amnuaypanich. Biopolymer-based corrosion inhibitor from chitosan grafted with poly(2-hydroxyethylmethacrylate) (CS-g-PHEMA) synthesized by ultrasound-assisted method. *Journal of Science and Technology*, 2022, **44**, 1225–1231.
10. X. Li, S. Deng. Cassava starch graft copolymer as an eco-friendly corrosion inhibitor for steel in H_2SO_4 solution. *Korean Journal of Chemical Engineering*, 2015, **32**, 2347–2354.

11. M. Rbaa, F. Benhiba, R. Hssisou, Y. Lakhrissi, B. Lakhrissi, M. E. Touhami, I. Warad, A. Zarrouk. Green synthesis of novel carbohydrate polymer chitosan oligosaccharide grafted on D-glucose derivative as bio-based corrosion inhibitor. *Journal of Molecular Liquids*, 2021, **322**, 114549.

12. C. Zou, X. Yan, Y. Qin, M. Wang, Y. Liu. Inhibiting evaluation of β-Cyclodextrin-modified acrylamide polymer on alloy steel in sulfuric solution. *Corrosion Science*, 2014, **85**, 445–454.

13. M. A. Quraishi, K. R. Ansari, D. S. Chauhan, S. A. Umoren, M. A. J. Mazumder. Vanillin modified chitosan as a new bio-inspired corrosion inhibitor for carbon steel in oil-well acidizing relevant to petroleum industry. *Cellulose*, 2020, **27**, 6425–6443.

14. S. Cheng, S. Chen, T. Liu, X. Chang, Y. Yin. Carboxymenthylchitosan as an ecofriendly inhibitor for mild steel in 1 M HCl. *Materials Letters*, 2007, **61**, 3276–3280.

15. S. N. Dalhatu, K. A. Modu, A. A. Mahmoud, Z. U. Zango, A. B. Umar, F. Usman, J. O. Dennis, A. Alsadig, K. H. Ibnaouf, O. A. Aldaghri. L-arginine grafted chitosan as corrosion inhibitor for mild steel protection. *Polymers*, 2023, **15**, 398.

16. G. Cui, J. Guo, Y. Zhang, Q. Zhao, S. Fu, T. Han, S. Zhang, Y. Wu. Chitosan oligosaccharide derivatives as green corrosion inhibitors for P110 steel in a carbon-dioxide-saturated chloride solution. *Carbohydrate Polymers*, 2019, **203**, 386–395.

17. D. S. Chauhan, M. A. Quraishi, A. A. Sorour, S. K. Saha, P. Banerjee. Triazole-modified chitosan: A biomacromolecule as a new environmentally benign corrosion inhibitor for carbon steel in a hydrochloric acid solution. *RSC Advances*, 2019, **9**, 14990–15003.

18. H. Ashassi-Sorkhabi, A. Kazempour. Chitosan, its derivatives and composites with superior potentials for the corrosion protection of steel alloys: A comprehensive review. *Carbohydrate Polymers*, 2020, **237**,116110.

19. M. Fardioui, M. Rbaa, F. Benhiba, M. Galai, T. Guedira, B. Lakhrissi, I. Warad, A. Zarrouk. Bio-active corrosion inhibitor based on 8-hydroxyqinoline-grafted-Alginate: Experimental and computational approaches. *Journal of Molecular Liquids*, 2021, **323**, 114615.

20. L. V. Hong, H. B. Mahmud. A preliminary screening and characterization of suitable acids for sandstome matrix acidizing technique: A comprehensive review. *Journal of Petroleum Exploration and Production Technology*, 2019, **9**, 753–778.

21. A. Biswas, P. Mourya, D. Moldal, S. Pal, G. Udayabhanu. Grafting effect of gum acacia on mild steel corrosion in acidic medium: Gravimetric and electrochemical study. *Journal of Molecular Liquids*, 2018, **251**, 470–479.

22. A. Singh, K. R. Ansari, M. A. Quraishi, S. Kaya, S. Erkan. Chemically modified guar gum and ethyl acrylate composite as a new corrosion inhibitor for reduction in hydrogen evolution and tubular steel corrosion protection in acidic environment. *International Journal of Hydrogen Energy*, 2021, **46**, 9452–9465.

23. A. Biswas, S. Pal, G. Udayabhanu. Experimental and theoretical studies of xanthan gum and its graft co-polymer as corrosion inhibitor for mild steel in 15% HCl. *Applied Surface Science*, 2015, **353**, 173–183.

24. S. Deng, X. Li, G. Du. An efficient corrosion inhibitor of cassava starch graft copolymer for aluminum in phosphoric acid. *Chinese Journal of Chemical Engineering*, 2021, **37**, 222–231.

25. S. Lahrour, A. Benmoussat, B. Bouras, A. Mansri, L. Tannouga, S. Marzorati. Glycerin-grafted starch as corrosion inhibitor of C-Mn steel in 1 M HCl solution. *Applied Sciences*, 2019, **9**, 4684.

26. K. Zhang, W. Yang, X. Yin, Y. Chen, Y. Liu, J. Le, B. Xu. Amino acids modified konjac glucomannan as green corrosion inhibitors for mild steel in HCl solution. *Carbohydrate Polymers*, 2018, **181**, 191–199.

27. Y. Sangeetha, S. Meenakshi, C. Sairam Sundaram. Interactions at the mild steel acid solution interface in the presence of O-fumaryl-chitosan: Electrochemical and surface studies. *Carbohydrate Polymers*, 2016, **136**, 38–45.

28. C. Chai, Y. Xu, D. Li, X. Zhao, Y. Xu, L. Zhang, Y. Wu. Cysteamine modified polyaspartic acid as a new class of green corrosion inhibitor for mild steel in sulfuric acid medium: Synthesis, electrochemical, surface study and theoretical calculation. *Progress in Organic Coatings*, 2019, **129**, 159–170.

29. C. Wang, J. Chen, J. Han, C. Wag, B. Hu. Enhanced corrosion inhibition performance of novel modified polyaspartic acid on carbon steel in HCl solution. *Journal of Alloys and Compounds*, 2019, **771**, 736–746.

30. Y. Liu, C. Zou, X. Yan, R. Xiao, T. Wang, M. Li. B-Cyclodextrin modified natural chitosan as a green inhibitor for carbon steel in acid solutions. *Industrial and Engineering Chemistry Research*, 2015, **54**, 5664–5672.

31. S. Banerjee, V. Srivastava, M. M. Singh. Chemically modified natural polysaccharide as green corrosion inhibitor for mild steel in acidic medium. *Corrosion Science*, 2012, **59**, 35–41.

32. Q. H. Zhang, B. S. Hou, Y. Y. Li, G. Y. Zhu, H. F. Liu, G. A. Zhang. Two novel chitosan derivatives as high efficient eco-friendly inhibitors for the corrosion of mild steel in acidic solution. *Corrosion Science*, 2020, **164**, 108346.

33. A. M. Fekry, R. R. Mohamed. Acetyl thiourea chitosan as an eco-friendly inhibitor for mild steel in sulphuric acid médium. *Electrochimica Acta*, 2010, 55, 1933–1939.

34. A. Habibiyan, B. Ramezanzadeh, M. Mahdavian, M. Kasaeian. Facile size and chemistry-controlled synthesis of mussel-inspired biopolymers based on polydopamine nanospheres: Application as ecofriendly corrosion inhibitors for mild steel against aqueous acidic solution. *Journal of Molecular Liquids*, 2020, **298**, 111974.

35. KO Shamsheera, R. Prasad Anupama, PK Jaseela, J. Abraham. Development of self-assembled monolayer of stearic acid grafted chitosan on mild steel and inhibition of corrosion in hydrochloric acid. *Chemical Data Collections*, 2020, **28**, 100402.

36. K. Qi, Y. Sun, H. Duan, X. Guo. A corrosion-protective coating based on a solution-processable polymer-grafted graphene oxide nanocomposite. *Corrosion Science*, 2015, **98**, 500–506.

37. M. S. Hasanin, S. A. Al Kiey. Environmentally benign corrosion inhibitors based on cellulose niacin nano-composite for corrosion of copper in sodium chloride solutions international. *Journal of Biological Macromolecules* 2020, **161**, 345–354.

38. A. Farhadian, S. A. Kashani, A. Rahimi, E. E. Oguzie, A. A. Javidparvar, S. C. Nwanonenyi, S. Yousefzadeh, M. R. Nabid. Modified hydroxyethyl cellulose as a highly efficient eco-friendly inhibitor for suppression of mild steel corrosion in a 15% HCl solution at elevated temperatures. *Journal of Molecular Liquids*, 2021, **338**, 116607.

39. A. Singh, M. Liu, E. Ituen, Y. Lin. Anti-corrosive properties of an effective guar gum grafted 2-acrylamido-2-methylpropanesulfonic acid (GG-AMPS) coating on copper in a 3.5% NaCl solution. *Coatings*, 2020, **10**, 241.

40. U. Eduok, E. Ohaeri, J. Szpunar. Anticorrosion allyl sulfonate graft chitosan/graphene oxide nanocomposite material. *Materials Advances*, 2021, **2**, 1621–1634.

41. K. E. Mouaden, D. S. Chauhan, M. A. Quraishi, L. Bazzi. Thiocarbohydrazide-crosslinked chitosan as a bioinspired corrosion inhibitor for protection of stainless steel in 3.5% NaCl. *Sustainable Chemistry and Pharmacy*, 2020, **15**, 100213.

42. H. Khosravi, R. Naderi, B. Ramezanzadeh. Synthesis and application of molybdate-doped mussel-inspired polydopamine (MI-PDA) biopolymer as an effective sustainable anti-corrosion substance for mild steel in NaCl solution. *Biomass Convert & Biorefinery*, 2022, https://doi.org/10.1007/s13399-022-03514-w.

43. U. Eduok, E. Ohaeri, J. Szpunar. Conversion of imidazole to n-(3-aminopropyl) imidazole toward enhanced corrosion protection of steel in combination with carboxymethyl chitosan grafted poly(2-methyl-1-vinylimidazole). *Industrial & Engineering Chemistry Research*, 2019, **58**, 7179–7192.

44. E. M. Fayyad, K. Kumar Sadasivuni, D. Ponnamma, M. Al Ali Al-Maadeed. Oleic acid-grafted chitosan/graphene oxide composite coating for corrosion protection of carbon steel. *Carbohydrate Polymers*, 2016, **151**, 871–878.

45. A. Singh, K. R. Ansari, M. A. Quraishi. Inhibition effect of natural polysaccharide composite on hydrogen evolution and P110 steel corrosion in 3.5 wt.% NaCl solution saturated with CO_2: Combination of experimental and surface analysis. *International Journal of Hydrogen Energy*, 2020, **45**, 25398–25408.

46. S.A Umoren, U. M. Eduok. Application of carbohydrate polymers as corrosion inhibitors for metal substrates in different media: A review. *Carbohydr Polym.* 2016, **140**, 314–341. https://doi.org/10.1016/j.carbpol.2015.12.038.

47. M. Liu, D. Xia, A. Singh, Y. Lin. Analysis of the anti-corrosion performance of dextrin and its graft copolymer on J55 steel in acid solution. *Processes*, 2021, **9**, 1642.

48. U. Eduok, E. Ohaeri, J. Szpunar. Electrochemical and surface analyses of X70 steel corrosion in simulated acid pickling medium: Effect of poly (N-vinyl imidazole) grafted carboxymethyl chitosan additive. *Electrochimica Acta*, **278**, 2018, 302–312.

49. A. Farhadian, M. A. Varfolomeev, A. Shaabani, S. Nasiri, I. Vakhitov, Y. F. Zaripova, V. V. Yarkovoi, A. V. Sukhov. Sulfonated chitosan as green and high cloud point kinetic methane hydrate and corrosion inhibitor: Experimental and theoretical studies. *Carbohydrate polymers*, 2020, **236**, 116035.

50. Y. Zheng, Y. Gao, H. Li, M. Yan, R. Guo, Z. Liu. Corrosion inhibition performance of composite based on chitosan derivative. *Journal of Molecular Liquids,* 2021, **324**, 114679.

51. G. Babaladimath, V. Badalamoole, S. T. Nandibewoor. Electrical conducting xanthan gum-graft-polyaniline as corrosion inhibitor for aluminum in hydrochloric acid environment. *Materials Chemistry and Physics*, 2018, **205**, 171–179.

52. X. Luo, C. Ci, J. Li, K. Lin, S. Du, H. Zhang, X. Li, Y. F. Cheng, J. Zang, Y. Liu. 4-aminoazobenzene modified natural glucomannan as a green eco-friendly inhibitor for the mild steel in 0.5 M HCl solution. *Corrosion Science*, 2019, **151**, 132–142.

53. Y. Sangeetha, S. Meenakshi, C. SairamSundaram. Corrosion mitigation of N-(2-hydroxy-3-trimethyl ammonium) propyl chitosan chloride as inhibitor on mild steel. *International Journal of Biological Macromolecules*, 2015, **72**, 1244–1249.

54. A. A. Aghzzaf, B. Rhouta, J. Steinmetz, E. Rocca, L. Aranda, A. Khalil, J. Yvon, L. Daoudi. Corrosion inhibitors based on chitosan-heptanoate modified beidellite. *Applied Clay Science*, 2012, **65–66**, 173–178.

55. B. Zhang, D. Zhou, X. Lv, Y. Xu, Y. Cui. Synthesis of polyaspartic acid/ 3-amino-1H-1,2,4-triazole-5-carboxylic acid hydrate graft copolymer and evaluation of its corrosion inhibition and scale inhibition performance. *Desalination*, 2013, **327**, 32–38.

56. R. Geethanjali, A. A. F. Sabirneez, S. Subhashini. Water soluble and biodegradable pectin-grafted polyacrylamide and pectin-grafted polyacrylic acid: Electrochemical investigation of corrosion-inhibition behavior on mild steel in 3.5% NaCl media. *Indian Journal of Material Science*, 2014, **9**, 356075.

57. S. M. Tawfik. Alginate surfactant derivatives as an ecofriendly corrosion inhibitor for carbon steel in acidic environments. *RSC Advances*, 2015, **5**, 104535–104550.

58. M. M. Solomon, H. Gerengi, S. A. Umoren, N. B. Essien, U. B. Essien, E. Kaya. Gum Arabic-silver nanoparticles composite as a green anticorrosive formulation for steel corrosion in strong acid media. *Carbohydrate Polymers*, 2018, **181**, 43–55.

59. P. Roy, P. Karfa, U. Adhikari, D. Sukul. Corrosion inhibition of mild steel in acidic medium by polyacrylamide grafted Guar gum with various grafting percentage: Effect of intramolecular synergism. *Corrosion Science,* 2014, **88**, 246–253.

60. V. Kumar K., B. V. Appa Rao. Chemically modified biopolymer as an eco-friendly corrosion inhibitor for mild steel in a neutral chloride environment. *New Journal of Chemistry*, 2017, **41**, 6278–6289.

61. L. Fu, J. Lv, L. Zhou, Z. Li, M. Tang, J. Li. Study on corrosion and scale inhibition mechanism of polyaspartic acid grafted β-cyclodextrin. *Materials Letters*, 2020, **264**, 127276.

62. D. Zeng, T. Chen, S. Zhu. Synthesis of polyaspartic acid/chitosan graft copolymer and evaluation of its scale inhibition and corrosion inhibition performance. *International Journal of Electrochemical Science*, 2015, **10**, 9513–9527.

63. X. Li, S. Deng, T. Lin, X. Xie, G. Du. Cassava starch-sodium allylsulfonate-acryl amide graft copolymer as na effective inhibitor of aluminum corrosion in HCl solution. *Journal of the Taiwan Institute of Chemical Engineers*, 2018, **86**, 252–269.

64. J. Carneiro, J. Tedim, S. C. M. Fernandes, C. S. R. Freire, A. Gandini, M. G. S. Ferreira, M. L. Zheludkevich. Functionalized chitosan-based coatings for active corrosion protection. *Surface & Coatings Technology*, 2013, **226**, 51–59.

65. C. Wang, J. Chen, B. Hu, Z. Liu, C. Wang, J. Han, M. Su, Y. Li, C. Li. Modified chitosan-oligosaccharide and sodium silicate as efficient sustainable inhibitor for carbon steel against chloride-induced corrosion. *Journal of Cleaner Production*, 2019, **238**, 117823.

66. A. Ouarga, H. Noukrati, I. Iraola-Arregui, A. Elaissaric, A. Barroug, H. B. Youcef. Development of anti-corrosion coating based on phosphorylated ethyl cellulose microcapsules. *Progress in Organic Coatings*, 2020, **148**, 105885.

67. M. M. Fares, A. K. Maayta, J. A. Al-Mustafa. Corrosion inhibition of iota-carrageenan natural polymer on aluminum in presence of zwitterion mediator in HCl media. *Corrosion Science*, 2012, **65**, 223–230.

68. V. Kumar, and B. V. Appa Rao. Phosphorylated xanthan gum, an environment-friendly, efficient inhibitor for mild steel corrosion in aqueous 200 ppm NaCl. *Materials Today: Proceedings*, 2019, **15**, 155–165.

69. Y. Sushmitha, R. Padmalatha. Material conservation and surface coating enhancement with starch-pectin biopolymer blend: A way towards green. *Surfaces and Interfaces*, 2019, **16**, 67–75.

70. X. Li, S. Deng, T. Lina, X. Xie. Cassava starch graft copolymer as a novel inhibitor for the corrosion of aluminium in HNO_3 solution. *Journal of Molecular Liquids*, 2019, **282**, 499–514.

71. Y. Xu, B. Zhang, L. Zhao, Y. Cui. Synthesis of polyaspartic acid/5-aminoorotic acid graft copolymer and evaluation of its scale inhibition and corrosion inhibition performance. *Desalination*, 2013, **311**, 156–161.

72. E. E. Ebenso, C. Verma, L. O. Olasunkanmi, E. D. Akpan, D. K. Verma, H. Lgaz, L. Guo, S. Kayah, M. A. Quraishi. Molecular modelling of compounds used for corrosion inhibition studies: A review. *Physical Chemistry Chemical Physics*, **23**, 2021, 19987.

73. I. Lukovits, E. Kálmán, F. Zucchi. Corrosion inhibitors-correlation between electronic structure and efficiency. *Corrosion*, **57**, 2001, 3–8.

74. X. Li, S. Deng, T Lin, X. Xie. Cassava starch ternary graft copolymer as a corrosion inhibitor for steel in HCl solution. *Journal of Materials Research and Technology*, 2020, **9**, 2196–2207.

Index

For Product Safety Concerns and Information please contact our
EU representative GPSR@taylorandfrancis.com Taylor & Francis
Verlag GmbH, Kaufingerstraße 24, 80331 München, Germany